Upconverting Nanomaterials

Perspectives, Synthesis, and Applications

Nanomaterials and Their Applications

Series Editor: M. Meyyappan

Upconverting Nanomaterials: ***Perspectives, Synthesis, and Applications***
Claudia Altavilla, Editor

Biosensors Based on Nanomaterials and Nanodevices
Jun Li and Nianqiang Wu, Editors

Plasma Processing of Nanomaterials
R. Mohan Sankaran, Editor

Graphene: ***Synthesis and Applications***
Wonbong Choi and Jo-won Lee, Editors

Inorganic Nanoparticles: ***Synthesis, Applications, and Perspectives***
Claudia Altavilla, Enrico Ciliberto, Editors

Carbon Nanotubes: ***Reinforced Metal Matrix Composites***
Arvind Agarwal, Srinivasa Rao Bakshi, and Debrupa Lahiri

Inorganic Nanowires: ***Applications, Properties, and Characterization***
M. Meyyappan and Mahendra Sunkara

Upconverting Nanomaterials

Perspectives, Synthesis, and Applications

Edited by
Claudia Altavilla
IPCB CNR Institute for Polymers, Composites and Biomaterials - Italy

CRC Press
Taylor & Francis Group
Boca Raton London New York

CRC Press is an imprint of the
Taylor & Francis Group, an **informa** business

CRC Press
Taylor & Francis Group
6000 Broken Sound Parkway NW, Suite 300
Boca Raton, FL 33487-2742

First issued in paperback 2020

ISBN-13: 978-1-4987-0774-9 (hbk)
ISBN-13: 978-0-367-65587-7 (pbk)

Library of Congress Cataloging-in-Publication Data

Names: Altavilla, Claudia, editor.
Title: Upconverting nanomaterials : perspectives, synthesis, and applications / Claudia Altavilla, editor.
Description: Boca Raton : Taylor & Francis, 2016. | Series: Nanomaterials and their applications ; 6 | "A CRC title." | Includes bibliographical references and index.
Identifiers: LCCN 2016014112 | ISBN 9781498707749 (alk. paper)
Subjects: LCSH: Phosphors. | Luminescence. | Nanostructured materials. | Rare earths.
Classification: LCC QC476.7 .U63 2016 | DDC 620.1/15--dc23
LC record available at https://lccn.loc.gov/2016014112

Visit the Taylor & Francis Web site at
http://www.taylorandfrancis.com

and the CRC Press Web site at
http://www.crcpress.com

To my father Nicolò

and to my daughters

Nicla and Marida,

I love you.

Contents

Series Preface ... ix
Foreword ... xi
Preface ... xv
Acknowledgments ... xvii
Editor ... xix
Contributors ... xxi

Section I Syntheses, Mechanism, and Functionalization

1. **A Perspective on Lanthanide Chemistry** ... 3
Simon A. Cotton

2. **Principle of Luminescence Upconversion and Its Enhancement in Nanosystems** ... 19
Hong Zhang, Langping Tu, and Xiao Min Liu

3. **Synthesis of Upconverting Nanomaterials: Designing the Composition and Nanostructure** ... 37
Adolfo Speghini, Marco Pedroni, Nelsi Zaccheroni, and Enrico Rampazzo

4. **Functionalization Aspects of Water Dispersible Upconversion Nanoparticles** ... 69
Markus Buchner, Verena Muhr, Sandy-Franziska Himmelstoß, and Thomas Hirsch

5. **Synergistic Effects in Organic-Coated Upconversion Nanoparticles** ... 101
Laura Francés-Soriano, María González-Béjar, and Julia Pérez-Prieto

6. **Tuning Optical Properties of Lanthanide Upconversion Nanoparticles** ... 139
Yuanwei Zhang, Zhanjun Li, Xiang Wu, and Gang Han

7. **Upconversion Enhancement Using Epitaxial Core–Shell Nanostructures** ... 163
Shuwei Hao, Jing Liu, Meiling Tan, and Guanying Chen

Section II Applications

8. **Active-Core–Active-Shell Upconverting Nanoparticles: Novel Mechanisms, Features, and Perspectives for Biolabeling** 195
K. Prorok, D. Wawrzyńczyk, M. Misiak, and A. Bednarkiewicz

9. **Upconversion Nanoparticles for Phototherapy** 255
Akshaya Bansal and Zhang Yong

10. **Upconverting Nanoparticles for Security Applications** 291
A. Baride and J. Meruga

11. **Nanothermometry Using Upconverting Nanoparticles** 319
Eva Hemmer and Fiorenzo Vetrone

Index 359

Series Preface

Over the last decade, a variety of nanomaterials such as nanoparticles, nanowires, and nanotubes have been created or modified to obtain superior properties with greater functional versatility. Thus, all these nanomaterials occupy an active area of multidisciplinary research that covers different fields of applied sciences such as physics, chemistry, engineering, and medicine. The volume of studies dealing with these topics represents one of the most impressive phenomena in all scientific history.

Nanomaterials and Their Applications Series, consisting of edited books, presents the insights of experts on theoretical concepts, emerging experimental technologies, and potential applications at the vanguard on nanotechnology and material science.

Since 2009 with the publication of the first book *Inorganic Nanowires: Applications, Properties, and Characterization* six additional books were added. They cover new and exciting multidisciplinary areas of nanomaterials research such as carbon nanotubes, inorganic nanoparticles, graphene, nano biosensors and plasma treatment for nanomaterials.

The books are edited by experts and consist of chapters written by leaders in a particular field of nanomaterials research from all over the world. Each volume is richly illustrated and contains comprehensive and up-to-date references.

This new volume *Upconverting Nanomaterials: Perspectives, Synthesis, and Applications,* explores the cutting-edge research knowledge on the topic of upconverting nanosystems and simultaneously provides the necessary fundamental background for non-specialist readers.

As with the other volumes of the series, this book will be of immediate value to scientists, engineers, and researchers within the field, as well as to students on advanced courses.

My task as series editor is to produce outstanding products that contribute to the broad field of nanotechnology and I am pleased to present this volume to our readers.

Meyya Meyyappan
Series Editor

Foreword

Upconversion: *Quo Vadis*

> Lanthanons—These elements perplex us in our researches, baffle us in our speculations, and haunt us in our very dreams. They stretch like an unknown sea before us; mocking, mystifying and murmuring strange revelations and possibilities.
>
> **Sir William Crookes**
> *(in an address to the Royal Society, 1887)*

> Lanthanum has only one important oxidation state in aqueous solution, the +3 state. With few exceptions, that statement tells the whole boring story about the other fourteen elements
>
> **Pimental and Spratley**
> *(Understanding Chemistry, 1971)*

Remarkably, it is Sir William Crookes' words, spoken nearly 100 years before the infamous quote by Pimental and Spratley, that have held true and continue to globally resonate in the minds of scientists. Indeed, the chemistry of lanthanides has evolved and developed well beyond the +3 state. Their chemistry and optical properties have transcended single crystals and micron-sized powders and have even allowed scientists to foray into the nanoscale. It is here that lanthanide-doped nanomaterials have garnered significant attention, particularly in the past two decades, principally due to their versatile optical properties. Unlike other materials, lanthanide luminescence does not exhibit size-dependent optical properties, which typically yields a varying spectrum of emission colors in quantum dots, for example. Of greater interest is the fact that the lanthanides can emit light under direct excitation conditions as in single photon excitation processes, or under a multiphoton regime, where they are able to convert low energy light such as near-infrared (NIR), to ultraviolet (UV) and visible emissions in a process known as upconversion. At the nanoscale, lanthanide-doped upconverting materials have ushered the way for development of novel applications ranging from display devices and optical sensors, imaging and diagnostic optical probes, to drug delivery and therapeutics.

A significant driver has been the ability to prepare these luminescent nanomaterials in a colloidal form allowing for their dispersion in liquids. Indeed, a large body of work, in the past decade, has focused on the synthesis of these colloidal nanomaterials with the aim of generating high quality, uniformly sized, single phase nanoparticles with high luminescence efficiency. The scientific community has certainly witnessed an evolution in the synthesis procedures. Researchers have sought out ways to understand how these nanoparticles are formed and the influence of the reaction parameters that can allow for the tailoring of the physical, chemical, and surface properties. These efforts have spanned a plethora of synthetic routes, which include thermal decomposition, hydrothermal, co-precipitation, microwave-assisted, ionic liquids, and microemulsion syntheses. As a result, the understanding of the optical properties of these nanomaterials has matured significantly and has spanned the study of the optical signatures of lanthanide ions in the hosts, energy transfer phenomena between ions, and the tailoring of the specific optical signatures for multiplexed imaging, as well as white light generation among others.

Despite the progress achieved to-date, many challenges remain afoot hindering lanthanide-doped upconverting nanoparticles (UCNPs) (and other nanomaterials) from achieving the promise and potential they hold. For example, although the synthesis of nanomaterials has greatly evolved with time and has benefited from the tireless efforts of nanoscientists, our understanding of the surface chemistry has not. Many questions remain unanswered concerning the nature of the surface capping ligand, the nature of the interaction between the surface-lying ions and the ligand, the impact of the surface charge, as well as the surface energetics. The surface chemistry represents a critical parameter upon consideration of applications of these nanoparticles. The surface does not only represent a barrier between the lanthanide ions and the environment but can also act as the interface between them. The surface of the particle can be modified to achieve dispersibility in lipophilic or hydrophilic media allowing scientists to appropriately design these materials for a given application. Of particular importance is the surface functionalization for integration into bio and nanomedicine. It is there that the surface chemistry can be exploited to actively target specific cells in the body, as well as deliver a therapeutic agent (e.g., a drug molecule or delivery of singlet oxygen via photodynamic therapy for example).

With respect to the optical properties of lanthanide-doped nanomaterials, one of the major challenges is the quantum yield of the upconverted emission. While direct excitation has shown upwards of 25% quantum yield, the multiphoton process is appreciably weaker peaking at 1%–2%. Indeed, the community has not remained idle and has undertaken the task of improving the efficiency of this process through modification of the host composition, optimization of the dopant level, the co-doping of ions, investigation of core–shell architectures and metal-enhanced luminescence, among other approaches. Integration into biological applications dictates the necessity for higher emission quantum yields particularly as we look to probe deep tissue systems. As an alternative to upconversion, scientists have also explored direct excitation of lanthanide-doped nanomaterials, which can be achieved using NIR light in the first biological window (650–950 nm) with the resulting infrared (IR) emission appearing at lower energies in the second biological window (1100–1400 nm). Certainly, this could represent a compromise to the current quantum efficiency challenge; however, it is not an ideal solution since it cannot be used for the generation of *in situ* UV or visible light that can be solicited to generate a specific stimulus in biological applications. Currently, it remains unclear if synthesis alone can remediate the emission quantum yields of the upconversion process, or whether new hosts extending beyond the fluorides, currently considered the front-runners, must be investigated. Thus, beyond the investigation of new host materials, a fundamental understanding of the role and importance of (i) the site symmetry of the dopant ions in effecting the intensity ratios of the different transitions which ultimately effects the quantum yield for a specific transition, (ii) the electron–phonon coupling, (iii) the crystalline phase, morphology, and size, (iv) the interionic distance between dopants which plays a major role in the energy transfer process, and (v) not only the role of the specific capping ligand but also the extent of the surface coverage.

Another area of concern is the interplay between the IR heating and photon generation in the upconversion process. This is of utmost importance in some of the proposed photoisomerization reactions for applications in drug delivery. Therefore, it is essential to evaluate and quantify the amount of heat and light produced in the upconversion process and to carry out appropriate control experiments to ascertain the photon to heat ratio. Over the past years, reports using targeted drug delivery by functional UCNPs have appeared that make use of the *o*-nitrobenzyl photolabile protecting group that undergoes photolytic cleavage upon irradiation at 365 nm. UCNPs doped with Yb^{3+} and Tm^{3+} show emission in the region 345–380 nm, which overlaps the absorption of the *o*-nitrobenzyl making it an ideal photosensitive release system. Overall this would improve therapeutic efficacy, and minimize toxicity of chemotherapy to healthy organs and cells. Controlled rates of drug release may be achieved via the selection of the NIR excitation power. Such a vector making use of photomediated delivery can be envisaged for applications over a wide range of therapeutics.

One of the major challenges, yet often overlooked, is the effects of nanomaterials (lanthanide-doped UCNPs or others) on health and the environment. Indeed, there is a plethora of other parameters, which can act alone or in synergy. Such parameters include the chemical composition, the size (intrinsic or hydrodynamic), redox properties, surface capping agents, chemical and physical stability, (bio)degradability or simply the interactions with other engineered nanoparticles, and the system under study.

Broadly speaking, a thorough understanding of the interactions of nanoparticles with biological molecules and systems is needed in order to achieve an in-depth understanding of the fate of nanoparticles in the body. For example, nanoparticles, which are typically modified with specific capping agents and bioactive molecules may, upon entering any biological fluid, interact with proteins and other biomolecules, thus forming a specific nanoparticle–protein corona whose composition is determined by both the nanoparticle and protein physicochemical properties. Such nanoparticles have a very dynamic nature with a composition that varies over time, which in turn may constitute a novel and yet largely unexplored chemical entity. Thus, it is imperative to understand the binding of proteins to nanoparticles taking into consideration affinities, stoichiometries, and the driving forces.

In addition to the properties of the nanomaterials, a strong emphasis must be placed on the development of novel characterization and detection tools that can facilitate the path of integration of nanoparticles in applications. Thus, the challenge also lies with instrument and device manufacturers to develop novel and sensitive detection systems. Scientists and developers must work collaboratively to realistically achieve this goal. This will directly impact application development where improved, and sensitive, detection methods can allow scientists working on the development of bioimaging probes to diagnose diseases and ailments at a much earlier stage than the current standards allowing for earlier therapeutic intervention and a greater chance of success. Ultimately, this will improve the quality of care for patients. The impact can also be far reaching in application areas outside of biology. For example, we can look forward to the development of more efficient sensors, displays, or lighting materials.

The chapters in this volume are a manifestation of the potential and broad applications of UCNPs in a variety of fields. Nevertheless, as we have outlined, the community faces a number of challenges that must be resolved in order to bring to fruition the many potential applications for the benefit of society. As the research efforts continue, our understanding of these lanthanide-doped UCNPs with novel optical properties will also continue to grow. We are very hopeful that with the collaboration between physicists, chemists, biologists, materials scientists, as well as the medical community, major advancements will be made that will shed new light on the future of these nanoparticles.

We are absolutely delighted to have had the opportunity to write these opening remarks. The authors of the many chapters that follow have captured the current state of the field of lanthanide-doped upconverting nanomaterials. They have critically covered the progress achieved to-date and have offered their perspectives as to how the field must and will continue to evolve in the coming years. We hope that the readers share our view that the future of these luminescent nanomaterials is certainly bright!

R. Naccache and J. A. Capobianco
Department of Chemistry and Biochemistry and
Centre for Nanoscience Research
Concordia University
Montreal, Quebec, Canada

Preface

Rare earth (RE)-doped UCNPs can efficiently convert near-infrared light into visible or ultraviolet luminescence using a stepwise multiphoton process in a system of real energy levels of Ln(III) ions which are embedded in an appropriate host lattice.

The Ln^{3+} ions can exhibit sharp luminescence emissions via intra-4f or 4f–5d transitions. Their remarkable luminescence properties, such as narrow bandwidth, long-time emission, and anti-Stokes emission, make these materials ideal candidates for different applications such as lasers, solar cells, analytical sensors, and security inks. Moreover, UCNPs are promising alternatives to organic dyes and quantum dots (both down-converting materials) in the field of biomedical applications such as optical imaging, photodynamic therapy, and bioanalytical assays. Along with the low cytotoxicity provided by the inorganic host lattice, this unique frequency converting capability imparts a number of advantages to UCNPs, such as absence of autofluorescence, deep light penetration in tissues, and minimum photodamage to living organisms. These merits make them attractive in bioanalytical and biophysical studies.

Over the past decade, high-quality RE-doped UCNPs have been successfully synthesized with the rapid development of nanotechnology. The synthesis methods are usually phase-based processes, such as thermal decomposition, hydrothermal reaction, and ionic liquids-based synthesis. At present, low luminescence efficiency is one of the main limiting factors. Many efforts are focused on improving absorption efficiency of these systems by different strategies, by using "antenna" molecules, changing host matrix or doping materials, or synthesizing new core–shell systems.

The aim of this book, that required more than 3 years to produce, is to provide the necessary background in the most important aspects of UCNPs in a self-contained textbook, useful not only for PhD students but also for senior scientists.

The book contains 11 chapters, and can be ideally divided into two sections. Section I: Syntheses, Mechanism, and Functionalization from Chapters 1 through 7, covers syntheses, mechanism, and functionalization aspects. Section II: Applications from Chapters 8 through 11, is entirely dedicated to applications in different fields of research. Each chapter has been written by scientists of great renown in their field. Chapter 1 traces the history of lanthanide chemistry and shows how one generation's discoveries often have to wait until the next generation to be applied. Chapters 2 and 3 are, respectively, focused on the mechanisms of luminescence upconversion and on synthesis, designing, and composition of these nanostructures. Chapters 4 and 5, respectively, cover the important role of surface functionalization, in order to control dispersability in water and to introduce additional functionalities. Chapters 6 and 7 are focused on the possibilities of controlling and enhancing the optical properties of UCNPs.

Chapter 8 explores the use of biolabels based on UCNPs as an elegant solution for the limitations of the corresponding downconverting systems such as organic dyes and quantum dots.

Chapter 9 describes the various phototherapy techniques, how UCNPs have been used in such therapies, the limitations that persist and potential solutions. Chapter 10 discusses the multifaceted UCNPs for security applications that include micro and macro multicolored covert QR codes and bar codes, security inks, etc. Last but not least, Chapter 11 provides some theoretical background for optical nanothermometers and subsequently highlights the latest developments in the field of optical nanothermometry, with special emphasis on those probes that are based on lanthanide-doped UCNPs.

This book was a challenging project, and I am glad to be the editor. I hope this book fulfills its purpose of inspiring new generations of researchers and scientists and guiding them as Ariadne's thread into the world of upconverting nanomaterials that is still immature but full of potential.

Acknowledgments

This project was born 3 years ago, as a result of a studying period on luminescence materials.

The necessity to rationalize my own thinking on this fascinating field of research became a proposal for a book after a discussion with Dr. Meyya Meyyappan, CRC series editor and chief scientist for exploration technology at NASA's Ames Research Center in California. I had the honor of meeting him a few years ago during one of his visits to Italy. He evaluated the proposal for my first book (*Inorganic Nanoparticles: Synthesis Applications and Perspectives*, 2010, CRC Press) without any prejudice with regard to my position (I'm a postdoc).

A special thank you to Nora Konopka and all her staff who supported me through this book and the previous one.

I'm grateful to Professor Capobianco and Professor Naccache who have written the Foreword of this book, adding value to this project.

Thanks to Dr. Simon Cotton for the interesting and pleasant conversations on chemistry and on lanthanides elements. Thanks to Professor Vetrone, one of the most important experts in this field, who promptly agreed to write a chapter during his visit to the IPCB-CNR, in Portici, Italy where I currently work. Finally, it is my pleasant duty to thank all the contributors who spent their valuable time writing exciting and instructive chapters for this book. They are leading experts in the field and have here covered cutting-edge research as well as providing the necessary fundamental background for nonspecialist readers.

Claudia Altavilla
Portici, Napoli

Editor

Claudia Altavilla graduated in chemistry (cum laude) in 2001 and she earned her PhD in chemistry in 2006 from the University of Catania, Italy.

She worked as a visiting scientist at Ludwig Maximillians Universitat of Munich, Germany with Professor Wolfgang Parak and at the University of Florence, Italy with Professor Dante Gatteschi. From 2006 to 2013, she worked as a research fellow at the NANOMATES Centre of the University of Salerno, Italy. Currently, she is a research fellow at the IPCB-CNR (**I**nstitute for **P**olymers, **C**omposites and **B**iomaterials), in Portici, Naples, Italy. Her research interests include biopolymers, upconverting, magnetic, and 2D graphene-like materials for applications in materials science, civil engineering, lubricants, sensors, and biomedicine. She is a referee for international journals on materials science and nanotechnology. She is the author of several papers and monographs and holds two international patents. She is the editor of *Inorganic Nanoparticles: Synthesis Applications and Perspectives*, CRC, November 2012.

Contributors

Akshaya Bansal
NUS Graduate School for Integrative Sciences & Engineering (NGS)
National University of Singapore
Singapore

A. Baride
Department of Chemistry
University of South Dakota
Vermillion, South Dakota

A. Bednarkiewicz
Wrocław Research Center EIT+, ulica Stabłowicka
Wrocław, Poland

and

Institute of Low Temperature and Structure Research
Polish Academy of Sciences, ulica Okólna
Wrocław, Poland

Markus Buchner
Institute of Analytical Chemistry, Chemo- and Biosensors
University of Regensburg
Regensburg, Germany

Guanying Chen
School of Chemical Engineering and Technology
Harbin Institute of Technology
Harbin, Heilongjiang province, China

and

Institute for Lasers, Photonics, and Biophotonics
State University of New York at Buffalo
Buffalo, New York

Simon A. Cotton
School of Chemistry
University of Birmingham
Edgbaston, Birmingham, UK

Laura Francés-Soriano
Instituto de Ciencia Molecular (ICMOL)
Departamento de Química Orgánica, Universidad de Valencia
Paterna, Valencia, Spain

María González-Béjar
Instituto de Ciencia Molecular (ICMOL)
Departamento de Química Orgánica, Universidad de Valencia
Paterna, Valencia, Spain

Gang Han
Department of Biochemistry and Molecular Pharmacology
University of Massachusetts Medical School
Worcester, Massachusetts

Shuwei Hao
School of Chemical Engineering and Technology
Harbin Institute of Technology
Harbin, Heilongjiang province, China

Eva Hemmer
Institut National de la Recherche Scientifique—Énergie, Matériaux et Télécommunications
Université du Québec
Varennes, Québec, Canada

Sandy-Franziska Himmelstoß
Institute of Analytical Chemistry, Chemo- and Biosensors
University of Regensburg
Regensburg, Germany

Thomas Hirsch
Institute of Analytical Chemistry, Chemo- and Biosensors
University of Regensburg
Regensburg, Germany

Zhanjun Li
Department of Biochemistry and Molecular Pharmacology
University of Massachusetts Medical School
Worcester, Massachusetts

Jing Liu
School of Chemical Engineering and Technology
Harbin Institute of Technology
Harbin, Heilongjiang province, China

Xiao Min Liu
Changchun Institute of Optics, Fine Mechanics and Physics
Chinese Academy of Sciences
Changchun, China

J. Meruga
South Dakota School of Mines & Technology
Rapid City, South Dakota

M. Misiak
Wrocław Research Center EIT+, ulica Stabłowicka
Wrocław, Poland

Verena Muhr
Institute of Analytical Chemistry, Chemo- and Biosensors
University of Regensburg
Regensburg, Germany

Marco Pedroni
Dipartimento di Biotecnologie and INSTM
UdR Verona, Università degli Studi di Verona
Verona, Italy

Julia Pérez-Prieto
Instituto de Ciencia Molecular (ICMOL)
Departamento de Química Orgánica, Universidad de Valencia
Paterna, Valencia, Spain

K. Prorok
Wrocław Research Center EIT+, ulica Stabłowicka
Wrocław, Poland

Enrico Rampazzo
Dipartimento di Chimica "G. Ciamician"
Università di Bologna
Bologna, Italy

Adolfo Speghini
Dipartimento di Biotecnologie and INSTM
UdR Verona, Università degli Studi di Verona
Verona, Italy

Meiling Tan
School of Chemical Engineering and Technology
Harbin Institute of Technology
Harbin, Heilongjiang province, China

Langping Tu
Changchun Institute of Optics, Fine Mechanics and Physics
Chinese Academy of Sciences
Changchun, China

Fiorenzo Vetrone
Institut National de la Recherche Scientifique—Énergie, Matériaux et Télécommunications
Université du Québec
Varennes, Québec, Canada

D. Wawrzyńczyk
Advanced Materials Engineering and Modelling Group
Faculty of Chemistry, Wroclaw University of Technology
Wybrzeze Wyspianskiego, Wroclaw, Poland

Xiang Wu
Department of Biochemistry and Molecular Pharmacology
University of Massachusetts Medical School
Worcester, Massachusetts

Zhang Yong
Department of Biomedical Engineering
National University of Singapore
Singapore

Nelsi Zaccheroni
Dipartimento di Chimica "G. Ciamician"
Università di Bologna
Bologna, Italy

Hong Zhang
van't Hoff Institute for Molecular Sciences
University of Amsterdam
Amsterdam, The Netherlands

Yuanwei Zhang
Department of Biochemistry and Molecular Pharmacology
University of Massachusetts Medical School
Worcester, Massachusetts

Section I

Syntheses, Mechanism, and Functionalization

1

A Perspective on Lanthanide Chemistry

Simon A. Cotton

CONTENTS

1.1 Introduction 3
1.2 How Many Lanthanides? 3
1.3 High-Coordination Numbers 4
1.4 Complexing Agents for Lanthanides 5
1.5 β-Diketonate Complexes and Luminescence 7
1.6 Low-Coordination Numbers 9
1.7 MRI Agents 10
1.8 Superconductors and Alkoxides 10
1.9 Upconversion and Nanomaterials 11
1.10 Conclusion 12
Abbreviations 12
References 13

1.1 Introduction

Although we now see lanthanides as a mature area of chemistry, lanthanide chemistry was slower to develop than the chemistry of the transition metals, and it is instructive to consider the reasons for this.

1.2 How Many Lanthanides?

Although the first ores of these elements were isolated at the end of the eighteenth century, it was over a century before scientists knew how many lanthanides were there (Evans 1996). In 1787, Arrhenius discovered ytterbite (soon renamed as gadolinite), which was later found to principally contain yttrium, lanthanum, cerium, and neodymium, along with smaller amounts of several other lanthanides. It was Johan Gadolin, who in 1794, isolated yttrium oxide from this ore.

The scene was set for a period of a confused endeavor. The similarity in the properties of the lanthanides made the separations, particularly of adjacent elements, extremely difficult, depending as they did on the techniques such as fractional crystallization. One of the greatest of the early rare-earth researchers, Charles James (a British-born chemist carrying out his research in the United States), reported that "The most important point proved by this work is that the element, giving the characteristic absorption bands of thulium, cannot be separated into simpler substances. After about 15,000 operations the absorption spectrum underwent no change" (James 1911, p. 1342). James was one of the three researchers, who in 1907, discovered lutetium, the last stable lanthanide waiting to be found, and the other two researchers being the Frenchman Georges Urbain and the Austrian Carl Auer von Welsbach, and the priority was usually being given to Urbain.

An application of spectral analysis, following the invention of the spectroscope in 1859, both helped and hindered. The observation of the previous unseen spectral lines not only showed the presence of undiscovered elements, but also led to confusion through an incorrect identification. Mendeleev's periodic classification was severely challenged by the lanthanides (Laing 2005; Scerri 2006). People had no idea on how many lanthanides were there until Moseley's work on x-ray spectroscopy (1913), which led to the concept of an atomic number, which showed that, following the discovery of lutetium, just one rare earth remained to be unearthed, element 61 (Fontani et al. 2014).

In fact, the existence of an element between neodymium and samarium had been postulated in 1902. This suggestion had been advanced by the Czech chemist Bohuslav Brauner, another of the great forgotten figures of lanthanide chemistry, who spent much of his research career working on the exacting business of determining the accurate atomic masses for the lanthanides. The precise positioning of the lanthanides was also problematic. In 1900, "Brauner put forward the view that the rare-earth elements should be regarded as a sort of zone or belt of closely related individuals occupying the place of a single element in the table, using as an analogy the existence of asteroid belts in the solar system" (Levy 1935, p. 1887). (It was, incidentally, Brauner who suggested a test for fluorine, involving inhaling it and noting whether fumes of hydrofluoric acid issued from the nose; he did live up to the age of 79 though.) Claims for the discovery of element 61 in the 1920s ("florentium" and "illinium") rested on a misassignment of spectral lines, and it was eventually realized that the element had no stable isotopes, and it was not until 1947 that promethium was discovered among uranium fission products, and the lanthanide series was complete (Marinsky et al. 1947).

1.3 High-Coordination Numbers

At almost the same time as Moseley's discovery, the formulation of Bragg's law (1912) made it possible to determine the position of atoms in crystals by x-ray diffraction, and this has eventually proved to be vital to our knowledge of lanthanide chemistry. Alfred Werner had relied on the fact that some transition metals—most notably Co(III)—invariably exhibit an octahedral six coordination in their complexes—or as he would explain it, every metal has a fixed number of secondary valencies, that is, a unique coordination number (Constable and Housecroft 2013). This enabled him to rationalize structures of the ammine complexes of $CoCl_3$, where compounds with the formulae $CoCl_3 \cdot nNH_3$ ($n = 3–6$) were known, with the help of conductivity measurements and silver ion precipitation of chloride to confirm the number of ions that are present, enabling him to show that their formulae were $[Co(NH_3)_6]^{3+} (Cl^-)_3$; $[Co(NH_3)_5Cl]^{2+} (Cl^-)_2$; $[Co(NH_3)_4Cl_2]^+ (Cl^-)$; and $[Co(NH_3)_3Cl_3]$. Likewise, he was able to account for the two isomers of $CoCl_3 \cdot 4NH_3$, what we now know as purple *cis*-$[Co(NH_3)_4Cl_2]^+ Cl^-$ and green *trans*-$[Co(NH_3)_4Cl_2]^+ Cl^-$.

Within a few years of Bragg's law, x-ray diffraction studies reported by R. W. G. Wyckoff and R. G. Dickinson in 1921–1922 had confirmed an octahedral coordination in complexes such as $(NH_4)_2[PtCl_6]$ and $[Ni(NH_3)_6]Cl_2$; similarly that $K_2[PtCl_4]$, $(NH_4)_2[PdCl_4]$, and $K_2[PdCl_4]$ all involved a square-planar coordination (Dickinson 1922; Wyckoff 1922). Of course, the isolation of two isomers of $Pt(NH_3)_2Cl_2$ (first made in 1844–1845), showed that these complexes had square-planar geometries, which were not tetrahedral (where only one isomer would exist). Perhaps surprisingly, no structure was reported until much later of any cobalt(III) ammine (Pauling 1994). Lanthanide complexes, of course, exhibit strikingly variable coordination numbers, and the complexes are so labile that very few examples of isolable isomers are known, an exception being the *fac*- and *mer*-$[Ln(OP(NMe_2)_3)_3Cl_3]$ (Petriček et al. 2000). The structures of several hydrated ethylsulfates $[Ln(H_2O)_9] (EtSO_4)_3$ (Ln = Y, La, Ce, Pr, Nd, Sm, Gd, and Dy) were reported in 1937 (Ketelaar 1937) and that of hydrated neodymium bromate, $[Nd(H_2O)_9] (BrO_3)_3$, was reported in 1939 (Helmholz 1939), showing a tricapped trigonal prismatic nine-coordination geometry for the metals. Their significance was not appreciated at the time.

1.4 Complexing Agents for Lanthanides

The Manhattan Project had one particularly important spin-off affecting lanthanide chemistry, with the use of ion-exchange resins for separations. At first, citrate was used to elute individual lanthanides from the resin, but soon, EDTA was recognized as being superior (Choppin and Wong 1994). A 1953 paper showed how the stability constant of the lanthanide–EDTA complexes increases with an increasing atomic number (and decreasing ionic radius). The hexadentate nature of EDTA binding was recognized for the lighter lanthanides at least, but the possibility of water coordination was not discussed (Wheelwright et al. 1953). In a study of lanthanide–EDTA complexes 2 years later, Moeller et al. assumed a six coordination for the lanthanides, but the study of discontinuities in stability-constant data (the "gadolinium break") led to the idea that this was due to a possible change in the denticity of EDTA (Moeller et al. 1955). This point was not clarified for another decade. A logical extension led to the study of the potentially octadentate DTPA, which has complexes with stability constants of at least four orders of magnitude than those of the corresponding EDTA complexes (Harder and Chabarek 1959; Moeller and Thompson 1962). Even at the beginning of the 1960s, when ligand field theory was being successfully applied to the burgeoning area of transition metal coordination chemistry, and new discoveries were being reported daily in transition metal organometallic chemistry—all aided by the application of instrumental methods (nuclear magnetic resonance [NMR], ultraviolet [UV]–visible, and infrared [IR] spectroscopy, as well as magnetic susceptibility measurements)—university-teaching texts had little to say about the lanthanides.

Although the pioneering crystal structures of $[Ln(H_2O)_9]$ $(EtSO_4)_3$ and $[Nd(H_2O)_9]$ $(BrO_3)_3$ had revealed tricapped trigonal prismatic nine coordination as far back as 1937–1939, their significance was not widely appreciated. Right into the 1960s, x-ray diffraction was a relatively slow technique, until the coming of automatic diffractometers and high-speed computing; so, very few structures of lanthanide complexes were determined, reflected in the textbooks of the time. It was generally assumed that yttrium and the lanthanides formed six-coordinate complexes such as the majority of transition metals, despite an increasing evidence to the contrary (Hart and Laming 1963). The presence of bidentate nitrates, leading to 12-coordination in the $[Ce(NO_3)_6]^{3-}$ units in $Ce_2Mg_3(NO_3)_{12}$. $24H_2O$ was recognized in 1963 (Zalkin et al. 1963); the ability of the coordinated nitrate group with its small bite angle to afford high-coordination numbers is now well recognized. However, the first compounds to really challenge the prevailing orthodoxy were the EDTA complexes. J. L. Hoard recognized that while EDTA could wrap itself around small 3d metals and achieve an octahedral coordination of the metal, the Ln^{3+} ions were too large for that to happen. He showed that the compound $La(EDTAH).7H_2O$ contained 10-coordinate $[La(EDTAH)(OH_2)_4]$ units while K $La(EDTA).8H_2O$ proved to contain nine-coordinate $[La(EDTA)(OH_2)_3]^-$ ions. It was also noted that $MLn(EDTA).8H_2O$ (M = K, Ln = La, Nd, or Gd; M = Na, Ln = Nd, Tb, or Er; and M = NH_4, Ln = Nd, or Gd) were isomorphous, arguing for the persistence of nine-coordinate $[La(EDTA)(OH_2)_3]^-$ ions over a range of ionic radii (Hoard et al. 1965; Lind et al. 1965). A subsequent study showed that the later lanthanides could form eight-coordinate $[La(EDTA)(OH_2)_2]^-$ ions and that in some cases the counter-ion could influence which complex crystallized. For example, $Na[Ho(EDTA)(H_2O)_3].5H_2O$ and $K[Ho(EDTA)(H_2O)_3].2H_2O$ contain nine-coordinate $[Ho(EDTA)(H_2O)_3]^-$ ions but $K[Ho(EDTA)(H_2O)_2].3H_2O$ has eight-coordinate

$[Ho(EDTA)(H_2O)_2]^-$ ions. A solid $[C(NH_2)_3]_2[Er(EDTA)(H_2O)_2] ClO_4.6H_2O$ contains an eight-coordinate Er^{3+}, while $Na[Er(EDTA)(H_2O)_3].5H_2O$ contains a nine-coordinate Er^{3+} (Janicki and Mondry 2014).

In the early 1960s, Alan Hart initiated a program of syntheses of the complexes of lanthanide salts with N- and O-donor ligands, including 1,10-phenanthroline, 2,2′-bipyridyl, and triphenylphosphine oxide (Hart and Laming 1964, 1965a, b). IR evidence that the nitrate groups were coordinated in the complexes $[Ln(phen)_2(NO_3)_3]$ and $[Ln(bipy)_2(NO_3)_3]$ indicated coordination numbers greater than six, subsequently confirmed by x-ray diffraction studies (Al-Karaghouli and Wood 1972; Fréchette et al. 1992; Kepert et al. 1996) on $[La(bipy)_2(NO_3)_3]$, $[La(phen)_2(NO_3)_3]$, $[Lu(phen)_2(NO_3)_3]$, and $[Lu(bipy)_2(NO_3)_3]$, showing that the nitrates have a symmetrical bidentate coordination (with rare exceptions, this is the norm for lanthanide complexes). Likewise, the structure of $[Eu(terpy)_3]\ (ClO_4)_3$ (terpy = 2, 2′: 6′, 2″-terpyridine) showed that three terpyridyl ligands might fit around a lanthanide ion, affording a coordination number of 9 (Frost et al. 1969).

Thus, by 1970, as a result of increased activity in lanthanide chemistry, the high-coordination numbers in lanthanide complexes were generally recognized (Choppin 1971).

1.5 β-Diketonate Complexes and Luminescence

At around the same time, β-diketonate complexes of the lanthanides came into prominence (for a review, see Binnemans 2005). Although lanthanide ions are efficient luminescent materials, the overall emission of "inorganic" lanthanide compounds is poor because they are weak absorbers. As long ago as 1942, Weissman recognized that the coordinating organic ligands, including β-diketonates such as benzoylacetone and dibenzoylmethane, resulted in a strong absorption by the organic moiety, and that this could be transferred to a Eu^{3+} ion for subsequent emission, which is the "antenna effect" (Weissman 1942).

Formally similar to the established β-diketonate complexes formed by the transition metals, $[Ln(diketonate)_3]$ had been known since 1897, with the report of hydrated tris acetylacetonate complexes of lanthanum and gadolinium (Urbain 1897). The 3d metals form a number of anhydrous $M(acac)_3$ (e.g., M = Sc, Cr, Mn, Fe, and Co) that contain octahedrally coordinated metals, but several structures reported in the late 1960s showed coordination numbers greater than six for lanthanide compounds. Thus, $Y(acac)_3.3H_2O$ contained eight-coordinate $[Y(acac)_3(H_2O)_2]$ molecules (Cunningham et al. 1967), which was the same coordination number also found in $La(acac)_3.3H_2O$ (which is $[La(acac)_3(H_2O)_2].H_2O$ (Phillips et al. 1968)) and $Cs[M(acac)_4]$ (M = Y, Eu) (Burns and Danford 1969). Using the bulkier 1-phenyl-1,3-butanedionato ligand (Cotton and Legzdins 1968), only one water could coordinate in seven-coordinate[$Y(C_6H_5COCHCOCH_3)_3.H_2O$], and with the even- bulkier tmhd (dpm) ligand, monomeric six-coordinate complexes are obtained (De Villiers and Boeyens 1972) for the later (and smaller) lanthanides, [$Ln(dpm)_3$] or [$Ln(tmhd)_3$] (Ln = Ho–Lu).

A strong red emission from Eu^{3+} in Y_2O_3 had been noted as far back as 1906 by Georges Urbain (Urbain 1906) but it was not for over half a century that this found an application. In 1962, Schimitschek and Schwarz suggested that europium complexes had suitable optical properties for laser materials (Schimitschek and Schwarz 1962), and laser emission was duly observed in 1963–1964 (Brecher et al. 1965; Lempicki and Samelson 1963; Samelson et al. 1964) in Eu(diketonate)$_3$. xH_2O and piperidinium [Eu(diketonate)$_4$] (diketonate, e.g., acac, benzoylacetonate, dibenzoylmethanate, and benzoyltrifluoroacetonate). New europium-based phosphors made mass-produced color television (TV) feasible in the early 1960s, as hitherto the weakness of the available red phosphors had been a limiting factor. Initially the way was led by the discovery that YVO_4: Eu^{3+} was "far superior in both colour and brightness" (Levine and Palilla 1964) to silver-activated zinc cadmium sulfide, hitherto the standard. Within a few years, it in its turn had been superseded by Y_2O_2S: Eu^{3+} (Sovers and Yoshioka 1968) described (Haynes and Brown 1968) as "bright as YVO_4: Eu and distinctly more "red."" The study of Eu^{3+} emission in materials such as $NaInO_2$ and $NaGdO_2$ had shown that the change in site symmetry from centrosymmetric in the former to noncentrosymmetric in the latter was accompanied by pronounced changes in the emission, from predominantly $^5D_0 \rightarrow {}^7F_1$ in the former to $^5D_0 \rightarrow {}^7F_2$ in the latter, resulting in a significant shift in color of the emission (Blasse and Bril 1969).

Although lanthanide chelates made very efficient phosphors, they were not employed in TVs (or in similar commercial applications) as the complexes exhibited poor air stability and also tended to degrade under UV irradiation, where the "inorganic" materials were proving to be superior. Likewise, after the discovery of the Nd:YAG (yttrium aluminum garnet) laser in 1964, it has been materials such as $Y_3Al_5O_{12}$ (YAlG), $Y_3Gd_5O_{12}$ (YGaG), and $Gd_3Ga_5O_{12}$ (GdGaG) lattices that have had the desired properties for a laser host material, being stable, hard, optically isotropic, and accepting substitutionally trivalent ions (Geusic et al. 1964; Reisfeld and Jørgensen 1977). Increasingly, uses are being sought such as phosphors in organic light-emitting diodes (LEDs), and stability problems are being circumvented by dispersing the lanthanide complex in a host matrix. The most popular luminescent rare-earth complex is the strong red emitter [$Eu(tta)_3(phen)$] (Kido and Okamoto 2002; Sano et al. 1995), and ways are continually sought to improve its efficiency (Zhou et al. 2016).

The volatility of the β-diketonate complexes makes them suitable for a number of applications, including the precursors for metal-organic chemical vapor deposition and atomic layer deposition (Binnemans 2005; Hsu et al. 2014) while [$Ln(fod)_3$] complexes are also used as Lewis acid catalysts in organic reactions, such as Diels–Alder syntheses (Bednarski and Danishefsky 1983; Binnemans 2005).

As already noted, the lanthanide β-diketonates form adducts with Lewis bases, usually with one or two donor atoms, thus acting as Lewis acids toward O- and N-donor ligands in particular, and this property was soon to be exploited in NMR-shift reagents (probably the first time that many chemists encountered lanthanides). Originally Hinckley used a pyridine adduct $[Eu(dpm)_3py_2]$ to simplify the NMR spectrum of cholesterol (Hinckley 1969), but it was soon realized (Sanders and Williams 1970) that the base-free europium chelate would be superior, and that $[Pr(dpm)_3]$ would give much-greater shifts to a high field, rather than a low field (Briggs et al. 1970). Soon, the fluorinated complex $[Eu(fod)_3]$ was found to be even better, on account of its greater solubility in non-polar-covalent NMR solvents and for its greater Lewis acidity, due to the electron-withdrawing fluorines (Rondeau and Sievers 1971). Within a few years, the coming of high-frequency NMR spectrometers rendered the common applications of lanthanide-shift reagents redundant, though niche applications remain, like chiral lanthanide-shift reagents, such as $Eu(facam)_3$ and $Eu(hfbc)_3$ (Goering et al. 1974; McCreary et al. 1974; Viswanathan and Toland 1995).

1.6 Low-Coordination Numbers

As already noted, by 1970, it was recognized that lanthanides frequently exhibited high-coordination numbers in their complexes, with the 12-coordination represented by complexes such as $[Ce(NO_3)_6]^{3-}$. Even six coordination personified in compounds such as $Cs_2NaLnCl_6$ (elpasolite structure) was on the low side for the lanthanides (Morss et al. 1970). Up until that time, "low"-coordination numbers of 2 and 3 in transition metal chemistry were confined to certain compounds of the metals with d^{10} electron configurations, such as $[Ag(NH_3)_2]^+$, $[Au(CN)_2]^-$, and $[Hg(CH_3)_2]$, soon to be joined by $[M(PR_3)_2]$ (M = Pd, Pt; PR_3 = bulky phosphine, e.g., $P(^tBu)_3$, PPh^tBu_2) (Matsumoto et al. 1974).

It was found that by using the bulky bis(trimethylsilyl)amide ligand, steric effects could enforce low-coordination numbers of 3d metals. Three coordination was realized in $[Fe\{N(SiMe_3)_2\}_3]$ (Bradley et al. 1969), soon to be joined by the Ti, V, and Cr analogs (Alyea et al. 1972), while it was subsequently determined for M(II) compounds such as $[\{Fe(N(SiMe_3)_2)_2\}_2]$, $[\{Fe(NPh_2))_2\}_2]$, and $[Fe[N(SiMe_3)_2]_2(thf)]$ (Olmstead et al. 1991).

A natural extension of this reasoning led to attempts to make corresponding lanthanide bis(trimethylsilyl)amides, resulting in the successful isolation of the three-coordinated lanthanide complexes $[Ln\{N(SiMe_3)_2\}_3]$, where Ln = Y, La, Ce, Pr, Nd, Sm, Eu, Gd, Ho, Yb, and Lu, which were the first three-coordinate compounds of these metals (Bradley et al. 1973; Ghotra et al. 1973).

Four coordination was first realized in $[Li(thf)_4]$, $[Lu(2,6\text{-dimethylphenyl})_4]$, and the ytterbium analog, by making use of a hindered aryl group (Cotton et al. 1972). In the Ln(II) state, two coordination was more recently achieved, again by using a very bulky ligand, in $[Yb\{C(SiMe_3)_3\}_2]$ (Eaborn et al. 1994; van den Hende et al. 1995). Three-coordinate Yb(II) is similarly found in $[Li(thf)_4][YbR_3]$ ($R = CH(SiMe_3)_2$) (Hitchcock et al. 2002).

1.7 MRI Agents

Following the first experiments on NMR imaging in humans (1977), the strongly paramagnetic Mn(II) was used to enhance tissue discrimination, before it was realized that the even-more paramagnetic Gd^{3+} had the desirable properties for imaging. Linking to research of over 30 years earlier, it was realized that the use of chelating agents would sequester sufficient gadolinium, so that the concentration of free Gd^{3+}(aq) ions would be at manageable levels. In practice, hexadentate EDTA did not form a strong-enough complex (log K ~ 17) but octadentate DTPA formed a sufficiently stable complex for the job (log K ~ 22–23) (Carr et al. 1984a, b; Weinmann et al. 1984). DTPA had originally been studied as a complexing agent for eluting lanthanide ions in rare-earth fractionation (Harder and Chabarek 1959; Moeller and Thompson 1962) and has found uses in sequestering transition metals to stabilize H_2O_2 and to remove Ca^{2+} and Mg^{2+} ions in a wide range of cosmetic products.

The first gadolinium-based magnetic resonance imaging (MRI) agent to come into clinical use was $[Gd(DTPA)(H_2O)]^{2-}$ as Magnevist®, in 1988. An alternative strategy uses a ligand based on a macrocyclic ring with the added donor groups attached, with complexes in commercial use including $[Gd(DOTA)(H_2O)]^-$ (Dotarem®) and $[Gd(HP\text{-}DO3A)(H_2O)]$ (Prohance®) that involve a N_4 macrocycle bearing additional donors (mainly carboxylates) and have slightly greater stability constants than $[Gd(DTPA)(H_2O)]^{2-}$, using a ligand of the same denticity as DOTA. Another type is closely related to $[Gd(DTPA)(H_2O)]^{2-}$, but with two carboxylates replaced by neutral donor groups like amides, affording neutral molecules, such as $[Gd(DTPA\text{-}BMA)(H_2O)]$ (Omniscan) (Bottrill et al. 2006; Caravan 2006; Caravan et al. 1999; Merbach et al. 2013). The comparative stability constants (Kumar et al. 1994) are 17.7 for $[Gd(EDTA]^-$, 22.2 for $[Gd(DTPA]^{2-}$, 21.0 for [Gd(DO3A)], and 25.3 for $[Gd(DOTA]^-$.

1.8 Superconductors and Alkoxides

The 1986 report of superconductivity below ~35 K in $La_{2-x}Ba_xCuO_4$ stimulated a new interest (Bednorz and Müller 1986) and frenetic activity in the rare earths, as this superconducting transition temperature was some 10° higher than found hitherto for any material. Within a year, superconductivity had been described in $YBa_2Cu_3O_{7-x}$ at 93 K (above the boiling point of liquid nitrogen) and "warm" superconductors were thus known (Hazen et al. 1987; Wu et al. 1987; for a recent review, see Keimer et al. 2015).

Attempts to find the best ways of making $YBa_2Cu_3O_{7-x}$ (also known as "YBCO") stimulated research into lanthanide alkoxides as the potential sources of metal oxides by vapor deposition. Compounds such as yttrium isopropoxide had been prepared many years before (Mazdiyasni et al. 1966a, b, 1967) and their formulae were taken to be of the type $[Y(OPr^i)_3]$. The yttrium compound has been used in the sol–gel process to make yttrium-stabilized zirconia. Now, in an age of sophisticated spectroscopic techniques and high-speed crystallographic structural determination, it was shown that yttrium isopropoxide was in fact $[Y_5O(OPr^i)_{13}]$ and had a considerably more complex structure, $[Y_5(\mu_5-O)(\mu_3-OPr^i)_4(\mu_2-OPr^i)_4(OPr^i)_5]$ (Bradley et al. 1990; Poncelet et al. 1989). Subsequently, using the three-coordinate $[Ln\{N(SiMe_3)_2\}_3]$ has afforded a useful route for preparing chloride-free alkoxides, such as trimeric $[La_3(OBu^t)_9(Bu^tOH)_2]$ as well as the neopentoxides $[Ln(ONep)_3]$ (Nep = $OCH_2C(CH_3)_3$; Ln = Sc, Y, and La–Lu except Pm) (Boyle et al. 2007; Bradley et al. 1991). All the neopentoxides have the tetrameric structure $[Ln(\mu-ONep)_2(ONep)]_4$ and afford Ln_2O_3 nanoparticles on hydrolysis. The volatile $[Ce(OCMe_2Pr^i)_4]$, a precursor for CeO_2 films, is a loosely bound dimer, $[(Me_2Pr^iO)_3Ce(\mu_2-OCMe_2Pr^i)_2Ce(OCMe_2Pr^i)_3]$ (Suh et al. 2004). Other oxo alkoxides have been discovered; thus refluxing toluene solutions of $[Ln_3(OBu^t)_9(Bu^tOH)_2]$ (Ln = La, Nd, and Pr) gives compounds $[Ln_5(\mu_5-O)(\mu_3-OBu^t)_4(\mu_2-OBu^t)_4(OBu^t)_5]$, analogous to the isopropoxide (Daniele et al. 2000). Desolvation of $[Ce_2(OPr^i)_8(Pr^iOH)_2]$ gives $[Ce_4O(OPr^i)_{14}]$, with the structure $[Ce_4(\mu_4-O)(\mu_3-OPr^i)_2(\mu-OPr^i)_6(OPr^i)_6]$ (Sirio et al. 1997), while an attempted synthesis of $[Ce(ONep)_4]$ from alcoholysis of $[Ce(OBu^t)_4]$ gives $[Ce_3O(OBu^t)(ONep)_9]$, which has the structure $[Ce_3(\mu_3-O)(\mu_3-OBu^t)(\mu_2-ONep)_3(ONep)_6]$ (Aspinall et al. 2011). Although these compounds have not proved of any great synthetic utility in the synthesis of materials such as YBCO, they nevertheless have provided new insights into the structures of lanthanide compounds.

This brings us to upconversion and nanomaterials, which are the subjects of this book.

1.9 Upconversion and Nanomaterials

Upconversion was first observed, in the 1960s, by Auzel and independently by Ovsyankin and Feofilov (Auzel 1966; Ovsyankin and Feofilov 1966). The phenomenon has been reviewed (Auzel 2004). It was not for another 30 years that it started to come into its own, when nanotechnology began to take off. Luminescent lanthanide nanoparticles have been proved to be extremely useful for bioimaging, biolabels, and bioassays. $NaYF_4:Er^{3+}$, Yb^{3+} was found (Menyuk et al. 1972) to be an efficient upconversion phosphor as early as 1972 but it is only more recently that codoped $NaYF_4:Er^{3+}$, Yb^{3+} (as a green emitter) and $NaYF_4:Tm^{3+}$, Yb^{3+} (as a blue emitter) have been identified as a material with very high upconversion efficiencies (Krämer et al. 2004), which are capable of upconversion multi-color fine-tuning (Wang and Liu 2008). Various syntheses have been developed to synthesize $NaYF_4:Er^{3+}$, Yb^{3+} and similar nanomaterials (Boyer et al. 2006; Liu et al. 2011), notably one involving the thermal decomposition (Boyer et al. 2006) of a mixture of CF_3COONa and *in situ*-synthesized $(CF_3COO)_3Ln$, a method originating in previous studies of the thermal decomposition of $(CF_3COO)_3Ln$ up to some 30 years earlier (Rillings and Roberts 1974; Rüssell 1993).

1.10 Conclusion

We have traveled a long way from the beginnings of lanthanide chemistry to our present sophisticated materials, capable of doing almost anything that we require. Nevertheless, we should always remember that we build on what has gone before; to use the expression attributed to Newton, *"If I have seen further it is by standing on the shoulders of giants."*

Abbreviations

bipy = 2,2′-bipyridyl
DOTA = 1,4,7,10-tetraazacyclododecane-1,4,7,10-tetraacetate
dpm = 2,2,6,6-tetramethylhepta-3,5-dionato (also known as dipivaloylmethane, dpm)
DTPA = diethylene triamine pentaacetate
DTPA–BMA = diethylenetriamine pentaacetic acid–bismethylamide
EDTA = ethylenediaminetetraacetate
facam = 3-trifluoroacetyl-d-camphorato
fod = 7,7,-dimethyl-1,1,2,2,2,3,3-heptafluoroocta-7,7-dimethyl-4,6-dionato
hfbc = 3-heptafluorobutyryl-d-camphorato
HP-DO3A = 10-(2-hydroxypropyl)-1,4,7-tetraazacyclododecane-1,4,7-triacetate
Htta = 2-thenoyltrifluoroacetone (Htta), 4,4,4-trifluoro-1-(2-thienyl)-1,3-butanedione
phen = 1,10-phenanthroline

References

Al-Karaghouli, A. R. and J. S. Wood. 1972. The crystal and molecular structure of trinitratobis(bipyridyl)lanthanum(III). *Inorg. Chem.* 11:2293–2299.

Alyea, E. C., D. C. Bradley, and R. G. Copperthwaite. 1972. Three-co-ordinated transition metal compounds. Part I. The preparation and characterization of tris(bistrimethylsilylamido)-derivatives of scandium, titanium, vanadium, chromium, and iron. *J. Chem. Soc., Dalton Trans.* 1580–1584.

Aspinall, H. C., J. Bacsa, A. C. Jones et al. 2011. Ce(IV) complexes with donor-functionalized alkoxide ligands: Improved precursors for chemical vapor deposition of CeO_2. *Inorg. Chem.* 50:11644–11652.

Auzel, F. 1966. Compteur quantique par transfert d'énergie entre deux ions de terres rares dans un tungstate mixte et dans un verre. *C. R. Acad. Sci., Ser. B* 262:1016–1019.

Auzel, F. 2004. Upconversion and anti-stokes processes with f and d ions in solids. *Chem. Rev.* 104:139–173.

Bednarski, M. and S. Danishefsky. 1983. Mild Lewis acid catalysis: $Eu(fod)_3$-mediated Hetero-Diels-Alder reaction. *J. Am. Chem. Soc.* 105:3716–3717.

Bednorz, J. G. and K. A. Müller. 1986. Possible high T_c superconductivity in the Ba–La–Cu–O system. *Z. Phys. B* 64:189–193.

Binnemans, K. 2005. Rare-earth beta-diketonates, *Handbook on the Physics and Chemistry of Rare Earths*, Elsevier, Amsterdam, vol. 35, pp. 107–272 (comprehensive review on β-diketonates).

Blasse, G. and A. Bril. 1969. On the Eu^{3+} fluorescence in mixed metal oxides. V. The Eu^{3+} fluorescence in the rocksalt lattice. *J. Chem. Phys.* 45:3327–3332.

Bottrill, M., L. Kwok, and N. J. Long. 2006. Lanthanides in magnetic resonance imaging. *Chem. Soc. Rev.* 35:557–571.

Boyer, J.-C., F. Vetrone, L. A. Cuccia, and J. A. Capobianco. 2006. Synthesis of colloidal upconverting $NaYF_4$ nanocrystals doped with Er^{3+}, Yb^{3+} and Tm^{3+}, Yb^{3+} via thermal decomposition of lanthanide trifluoroacetate precursors. *J. Am. Chem. Soc.* 128:7444–7445.

Boyle, T. J., L. A. M. Ottley, S. D. Daniel-Taylor, L. J. Tribby, S. D. Bunge, A. L. Costello, T. M. Alam, J. C. Gordon, and T. M. McCleskey. 2007. Isostructural neo-pentoxide derivatives of group 3 and the lanthanide series metals for the production of Ln_2O_3 nanoparticles. *Inorg. Chem.* 46:3705–3713.

Bradley, D. C., H. Chudzynska, D. M. Frigo, M. E. Hammond, M. B. Hursthouse, and M. A. Mazid. 1990. Pentanuclear oxoalkoxide clusters of scandium, yttrium, indium and ytterbium, x-ray crystal structures of $[M_5(\mu_5\text{-O})(\mu_3\text{-OPr}^i)_4(\mu_2\text{-OPr}^i)_4(\text{OPr}^i)_5]$ (M = In, Yb). *Polyhedron* 9:719–726.

Bradley, D. C., H. Chudzynska, M. B. Hursthouse, and M. Motevalli. 1991. Volatile tris-tertiary-alkoxides of yttrium and lanthanum. The x-ray crystal structure of $[La_3(OBu^t)_9(Bu^tOH)_2]$. *Polyhedron* 10:1049–1059.

Bradley, D. C., J. S. Ghotra, and F. A. Hart. 1973. Low co-ordination numbers in lanthanide and actinide compounds. Part 1. The preparation and characterization of tris{bis(trimethylsilyl)-amido}lanthanides. *J. Chem. Soc., Dalton Trans.* 1021–1023.

Bradley, D. C., M. B. Hursthouse, and P. F. Rodesiler. 1969. The structure of a three-coordinate iron(III) compound. *J. Chem. Soc., Chem. Commun.* 14–15.

Brecher, C., H. Samelson, and A. Lempicki. 1965. Laser phenomena in Europium chelates. III. Spectroscopic effects of chemical composition and molecular structure. *J. Chem. Phys.* 42:1081–1096.

Briggs, J., G. H. Frost, F. A. Hart, G. P. Moss, and M. L. Staniforth. 1970. Lanthanide-induced shifts in nuclear magnetic resonance spectroscopy. Shifts to high field. *Chem. Comm.* 749–750.

Burns, J. H. and M. D. Danford. 1969. The crystal structure of cesium tetrakis(hexafluoroacetylacetona to)europate and -americate. Isomorphism with the yttrate. *Inorg. Chem.* 8:1780–1784.

Caravan, P. 2006. Strategies for increasing the sensitivity of gadolinium based MRI contrast agents. *Chem. Soc. Rev.* 35: 512–523.

Caravan, P., J. J. Ellison, T. J. McMurry, and R. B. Lauffer. 1999. Gadolinium(III) chelates as MRI contrast agents: Structure, dynamics, and applications. *Chem. Rev.* 99:2293–2352.

Carr, D. H., J. Brown, G. M. Bydder, H.-J. Weinmann, U. Speck, D. J. Thomas, and I. R. Young. 1984a. Intravenous chelated gadolinium as a contrast agent in NMR imaging of cerebral tumours. *Lancet* 323:484–486.

Carr, D. H., J. Brown, A. W.-L. Leung, and J. M. Pennock. 1984b. Iron and gadolinium chelates as contrast agents in NMR imaging: Preliminary studies. *J. Comput. Assist. Tomogr.* 8:385–389.

Choppin, G. R. 1971. Structure and thermodynamics of lanthanide and actinide complexes in solution. *Pure Appl. Chem.* 27:23–42.

Choppin, G. R. and P. J. Wong. 1994. Lanthanide aminopolycarboxylates. *ACS Symp. Ser.* 565:346–360 (review).

Constable, E. C. and C. E. Housecroft. 2013. Coordination chemistry: The scientific legacy of Alfred Werner. *Chem. Soc. Rev.* 42:1429–1439.

Cotton, F. A. and P. Legzdins. 1968. An example of the monocapped octahedral form of heptacoordination. The crystal and molecular structure of tris(1-phenyl- 1,3-bu tanedionato)aquoyttrium(III). *Inorg. Chem.* 7:1777–1783.

Cotton, S. A., F. A. Hart, M. B. Hursthouse, and A. J. Welch. 1972. Preparation and molecular structure of a σ-bonded lanthanide phenyl. *J. Chem. Soc. Chem. Comm.* 1225–1226.

Cunningham, J. A., D. E. Sands, and W. F. Wagner. 1967. The crystal and molecular structure of yttrium acetylacetonate trihydrate. *Inorg. Chem.* 6:499–503.

Daniele, S., L. G. Hubert-Pfalzgraf, P. B. Hitchcock, and M. Lappert. 2000. Thermal condensation of trinuclear lanthanide butoxides. Molecular structure of $La_5(\mu_5\text{-}O)(\mu_3\text{-}O^tBu)_4(\mu\text{-}O^tBu)_4(O^tBu)_5$. *Inorg. Chem. Commun.* 3:218–220.

De Villiers, J. P. R. and J. C. A. Boeyens. 1972. Crystal structure of tris-(2,2,6,6-tetramethylheptane-2,5-dionato) erbium(III). *Acta Crystallogr. Sect. B* B28:2335–2340.

Dickinson, R. G. 1922. The crystal structures of potassium chloroplatinite and of potassium and ammonium chloropalladites. *J. Am. Chem. Soc.* 44:2404–2411.

Eaborn, C., P. B. Hitchcock, K. Izod, and J. D. Smith. 1994. A monomeric solvent-free bent lanthanide dialkyl and a lanthanide analogue of a Grignard reagent. Crystal structures of $Yb\{C(SiMe_3)_3\}_2$ and $[Yb\{C(SiMe_3)_3\}I.OEt_2]_2$. *J. Am. Chem. Soc.* 116:12071–12072.

Evans, C.H. 1996. *Episodes from the History of the Rare Earth Elements*, Kluwer, Dordrecht.

Fontani, M., M. Costa, and M. V. Orna. 2014. *The Lost Elements*, Oxford University Press, New York, pp. 118–127, 161–177, 200–218, and 289–309 (the latter for element 61).

Fréchette, M., I. R. Butler, R. Hynes, and C. Detellier. 1992. Structures in solution and in the solid state of the complexes of lanthanum(III) with 1, 10-phenanthroline. X-ray crystallographic and ^{1}H, ^{13}C, ^{17}O, and ^{139}La solution NMR studies. *Inorg. Chem.* 31:1650–1656.

Frost, G. H., F. A. Hart, C. Heath, and M.B. Hursthouse. 1969. The crystal structure of tris-(2,2′,6′,2″-terpyridyl)europium(III) perchlorate. *J. Chem. Soc., Chem. Commun.* 1421.

Geusic, J. E., H. M. Marcos, and L. G. Van Uitert. 1964. Laser oscillations in Nd-doped yttrium aluminium, yttrium gallium and gadolinium garnets. *Appl. Phys. Lett.* 4:182.

Ghotra, J. S., M. B. Hursthouse, and A. J. Welch. 1973. Three-co-ordinate scandium(III) and europium(III); crystal and molecular structures of their trishexamethyldisilylamides. *J. Chem. Soc. Chem. Comm.* 669–670.

Goering, H. L., J. N. Eikenberry, G. S. Koermer, and C. J. Lattimer. 1974. Direct determination of enantiomeric compositions with optically active nuclear magnetic resonance lanthanide shift reagents. *J. Am. Chem. Soc.* 96:1493–1501.

Harder, R. and S. Chabarek. 1959. The interaction of rare earth ions with diethylenetriamminepentaacetic acid. *J. Inorg. Nucl. Chem.* 11:197–209.

Hazen, R., L. Finger, R. Angel, C. Prewitt, N. Ross, H. Mao, C. Hadidiacos, P. Hor, R. Meng, and C. Chu. 1987. Crystallographic description of phases in the Y-Ba-Cu-O superconductor. *Phys. Rev. B* 35:7238–7241.

Hart, F. A. and F. P. Laming. 1963. Some o-phenanthroline complexes of yttrium and the lanthanides. *Proc. Chem. Soc. Lond.* 107.

Hart, F. A. and F. P. Laming. 1964. Complexes of 1,10-phenanthroline with lanthanide chlorides and thiocyanates. *J. Inorg. Nucl. Chem.* 26:579–585.

Hart, F. A. and F. P. Laming. 1965a. Lanthanide complexes-II. Complexes of 1:10-phenanthroline with lanthanide acetates and nitrates. *J. Inorg. Nucl. Chem.* 27:1605–1610.

Hart, F. A. and F. P. Laming. 1965b. Lanthanide complexes-III. Complexes of 2,2′-dipyridyl with lanthanide chlorides, thiocyanates, acetates and nitrates. *J. Inorg. Nucl. Chem.* 27:1825–1829.

Haynes, J. W. and J. J. Brown. 1968. Preparation and luminescence of selected Eu^{3+}-activated rare earth-oxygen-sulfur compounds. *J. Electrochem. Soc.* 115:1060–1066.

Helmholz, L. 1939. The crystal structure of neodymium bromate enneahydrate, $Nd(BrO_3)_3.9H_2O$. *J. Am. Chem. Soc.* 61:1544–1550.

Hinckley, C. C. 1969. Paramagnetic shifts in solutions of cholesterol and the dipyridine adduct of trisdipivalomethanatoeuropium(III). A shift reagent. *J. Am. Chem. Soc.* 91:5160–5162.

Hitchcock, P. B., A. V. Khvostov, and M. F. Lappert. 2002. Synthesis and structures of crystalline bis(trimethylsilyl)methanido complexes of potassium, calcium and ytterbium. *J. Organomet. Chem.* 663:263–268.

Hoard, J. L., B. Lee, and M. D. Lind. 1965. On the structure-dependent behavior of ethylenediaminetetraacetato complexes of the rare earth Ln^{3+} ions. *J. Am. Chem. Soc.* 87:1612–1613.

Hsu, H.-L., K. R. Leong, I.-J. Teng, M. Halamicek, J.-Y. Juang, S.-R. Jian, L. Qian, and N. P. Kherani. 2014. Erbium-doped amorphous carbon-based thin films: A photonic material prepared by low-temperature RF-PEMOCVD. *Materials* 7:1539–1554.

James, C. 1911. Thulium I. *J. Am. Chem. Soc.* 33:1332–1344.
Janicki, R. and A. Mondry. 2014. A new approach to determination of hydration equilibria constants for the case of [Er(EDTA)$(H_2O)_n$] complexes. *Phys. Chem. Chem. Phys.* 16:26823–26831 for a summary of structures of lanthanide EDTA complexes.
Keimer, B., S. A. Kivelson, M. R. Norman, S. Uchida, and J. Zaanen. 2015. From quantum matter to high-temperature superconductivity in copper oxides. *Nature* 518:179–186.
Kepert, D. L., L. I. Semenova, A. N. Sobolev, and A. H. White. 1996. Structural systematics of rare earth complexes. IX. Tris (nitrato- O,O′) (bidentate- N, N′) lutetium(III), N, N′-Bidentate = 2,2′-Bipyridine or 1,10-Phenanthroline. *Aust. J. Chem.* 49:1005–1008.
Ketelaar, J. A. A. 1937. The crystal structures of the ethyl sulphates of the rare earths and yttrium. *Physica* 4:619–630.
Kido, J. and Y. Okamoto. 2002. Organo lanthanide metal complexes for electroluminescent materials. *Chem. Rev.* 102:2357–2368.
Krämer, K. W., D. Biner, G. Frei, H. U. Güdel, M. P. Hehlen, and S. R. Lüthi. 2004. Hexagonal sodium yttrium fluoride based green and blue emitting upconversion phosphors. *Chem. Mater.* 16:1244–1251.
Kumar, K., C. A. Chang, L. C. Francesconi, D. D. Dischino, M. F. Malley, J. Z. Gougoutas, and M. F. Tweedle. 1994. Synthesis, stability, and structure of gadolinium(iii) and yttrium(iii) macrocyclic poly(amino carboxylates). *Inorg. Chem.* 33:3567–3575.
Laing, M. 2005. A revised periodic table: With the lanthanides repositioned. *Found. Chem.* 7:203–233.
Lempicki, A. and H. Samelson 1963. Optical maser action in europium benzoylacetonate. *Phys. Lett.* 4:133–135.
Levine, A. K. and F. C. Palilla. 1964. A new, highly efficient red-emitting cathodoluminescent phosphor (yvo4:eu) for color television. *Appl. Phys. Lett.* 5:118–120.
Levy, S. I. 1935. Brauner memorial lecture. *J. Chem. Soc.* 1876–1890.
Lind, M. D., B. Lee, and J. L. Hoard. 1965. Structure and bonding in a ten-coordinate lanthanum(iii) chelate of ethylenediaminetetraacetic acid. *J. Am. Chem. Soc.* 87:1611–1612.
Liu, D., D. Zhao, D. Zhang, K. Zheng, and W. Qin. 2011. Synthesis and characterization of upconverting $NaYF_4:Er^{3+}$, Yb^{3+} nanocrystals via thermal decomposition of stearate precursor. *J. Nanosci. Nanotechnol.* 11:9770–9773.
Marinsky, J. A., L. E. Glendenin, and C. D. Coryell. 1947. The chemical identification of radioisotopes of neodymium and of element 61. *J. Am. Chem. Soc.* 69:2781–2785.
Matsumoto, M., H. Yoshioka, K. Nakatsu, T. Yoshida, and S. Otsuka. 1974. Two-coordinate palladium(0) complexes, Pd[PPh(t-Bu)$_2$ and Pd[P(t-Bu)$_3$]. *J. Am. Chem. Soc.* 96:3322–3324.
Mazdiyasni, K. S., C. T. Lynch, and J. S. Smith. 1966a. The preparation and some properties of yttrium, dysprosium, and ytterbium alkoxides. *Inorg. Chem.* 5:342–346.
Mazdiyasni, K. S., C. T. Lynch, and J. S. Smith. 1966b. Yttrium, dysprosium, and ytterbium alkoxides and process for making same. U.S. Patent 3278571 (granted Oct. 11, 1966).
Mazdiyasni, K. S., C. T. Lynch, and J. S. Smith. 1967. Yttrium, dysprosium and ytterbium alkoxides. U.S. Patent 3356703 (granted Dec. 5, 1967).

McCreary, M. D., D. W. Lewis, D. L. Wernick, and G. M. Whitesides. 1974. The determination of enantiomeric purity using chiral lanthanide shift reagents. *J. Am. Chem. Soc.* 96:1038–1054.

Menyuk, N., K. Dwight, and J. W. Pierce. 1972. $NaYF_4$: Yb,Er-an efficient upconversion phosphor. *Appl. Phys. Lett.* 21:159–161.

Merbach, A. S., L. Helm, and E. Toth. 2013. *The Chemistry of Contrast Agents in Medical Magnetic Resonance Imaging*, 2nd Edition, John Wiley & Sons, Chichester.

Moeller, T., F. A. J. Moss, and R. H. Marshall. 1955. Observations on the rare earths. LXVI. Some characteristics of ethylenediarninetetraacetic acid chelates of certain rare earth metal ions. *J. Am. Chem. Soc.* 77:3182–3186.

Moeller, T. and L. C. Thompson. 1962. Observations on the rare earths – LXXV. The stabilities of diethylenetriaminepentaacetic acid chelates. *J. Inorg. Nucl. Chem.* 24:499–510.

Morss, L. R., M. Siegal, L. Stenger, and N. Edelstein. 1970. Preparation of cubic chloro complex compounds of trivalent metals : Cs_2NaMCl_6. *Inorg. Chem.* 9:1771–1775.

Olmstead, M. M., P. P. Power, and S. C. Shoner. 1991. Three-coordinate iron complexes: X-ray structural characterization of the amide-bridged dimers $[Fe(NR_2)_2]_2$ (R = $SiMe_3$, C_6H_5) and the adduct $Fe[N(SiMe_3)_2]_2(THF)$ and determination of the association energy of the monomer $Fe(N(SiMe_3)_2]_2$ in solution. *Inorg. Chem.* 30:2547–2551.

Ovsyankin, V. V. and P. P. Feofilov. 1966. Mechanism of summation of electronic excitations in activated crystals. *JETP Lett.* 3:322–323.

Pauling, L. 1994. Early structural coordination chemistry. *ACS Symp. Ser.* 565:68–72.

Petriček, S., A. Demšar, L. Golič, and J. Košmrlj. 2000. New complexes of lanthanide chlorides. Reversible isomerization in octahedral $[LaCl_3(HMPA)_3]$ and the crystal structure of fac-$[SmCl_3(HMPA)_3]$. *Polyhedron* 19:199–204.

Phillips, T., D. E. Sands, and W. F. Wagner. 1968. Crystal and molecular structure of diaquotris(acetylacetonato)lanthanum(III). *Inorg. Chem.* 7:2295–2299.

Poncelet, O., W. J. Sartain, L. G. Hubert-Pfalzgraf, K. Folting, and K. G. Caulton. 1989. Chemistry of yttrium triisopropoxide revisited. Characterization and crystal structure of $[Y_5(m_5\text{-}O)(m_3\text{-}OPr^i)_4(m_2\text{-}OPr^i)_4(OPr^i)_5]$. *Inorg. Chem.* 28:263–267.

Reisfeld, R. and C. H. Jørgensen. 1977. *Lasers and Excited States of Rare Earths*, Springer-Verlag, Berlin.

Rillings, K. W. and J. E. Roberts. 1974. A thermal study of the trifluoroacetates and pentafluoropropionates of praseodymium, samarium and erbium. *Thermochim. Acta* 10:285–298.

Rondeau, R. E. and R. E. Sievers. 1971. New superior paramagnetic shift reagents for nuclear magnetic resonance spectral clarification. *J. Am. Chem. Soc.* 93:1522–1524.

Rüssell, C. 1993. A pyrolytic route to fluoride glasses. I. Preparation and thermal decomposition of metal trifluoroacetates. *J. Non-Cryst. Solids* 152: 161–166.

Samelson, H., A. Lempicki, C. Brecher, and V. A. Brophy. 1964. Room-temperature operation of a europium chelate liquid laser. *Appl. Phys. Lett.* 5:173–174.

Sanders, J. K. M. and D. H. Williams. 1970. A shift reagent for use in nuclear magnetic resonance spectroscopy. A first-order spectrum of n-hexanol. *Chem. Comm.* 422–423.

Sirio, C., L. G. Hubert-Pfalzgraf, and S. Halut. 1997. Facile thermal desolvation of $Ce_2(OPr^i)_8(Pr^iOH)_2$: Characterization and molecular structure of $Ce_4(m_4\text{-}O)(m_3\text{-}OPr^i)_2(m\text{-}OPr^i)_4(OPr^i)_8$. *Polyhedron* 16:1129–1136.

Sano, T., M. Fujita, T. Fujii, Y. Hamada, K. Shibata, and K. Kuroki. 1995. Novel europium complex for electroluminescent devices with sharp red emission. *Jpn. J. Appl. Phys.* 34:1883–1887.

Scerri, E. 2006. *The Periodic Table: Its Story and Its Significance*, Oxford University Press, Oxford.

Schimitschek, E. J. and E. G. K. Schwarz. 1962. Organometallic compounds as possible laser materials. *Nature* 196:832–833.

Sovers, O. J. and T. Yoshioka. 1968. Fluorescence of trivalent-europium-doped yttrium oxysulfide. *J. Chem. Phys.* 49:4945–4954.

Suh, S., J. Guan, L. A. Mîinea, J. M. Lehn, and D. M. Hoffman. 2004. Chemical vapor deposition of cerium oxide films from a cerium alkoxide precursor. *Chem. Mater.* 16:1667–1673.

Urbain, G. 1897. Recherches sur les sables monazités. *Comp. Rend.* 124:618–621.

Urbain, G. 1906. Sur la phosphorescence cathodique de l'europium. *C. R. Acad. Sci. Paris* 142:205–207.

van den Hende, J. R., P. B. Hitchcock, S. A. Holmes, M. F. Lappert, and S. Tian. 1995. Synthesis and characterisation of ytterbium(ii) alkyls. *J. Chem. Soc., Dalton Trans.* 3933–3939.

Viswanathan, T. and A. Toland. 1995. NMR spectroscopy using a chiral lanthanide shift reagent to assess the optical purity of l-phenylethylamine. *J. Chem. Educ.* 72:945–946.

Wang, F. and X. Liu. 2008. Upconversion multicolor fine-tuning: Visible to near-infrared emission from lanthanide-doped $NAYF_4$ nanoparticles. *J. Am. Chem. Soc.* 130:5642–5643.

Weinmann, H.-J., R. C. Brasch, W.-R. Press, and G. W. Wesbey. 1984. Characteristics of gadolinium-DTPA complex: A potential NMR contrast agent. *Am. J. Roentgenol.* 142:619–624.

Weissman, S. I. 1942. Intramolecular energy transfer. The fluorescence of complexes of europium. *J. Chem. Phys.* 10:214–217.

Wheelwright, E. J., F. H. Spedding, and G. Schwarzenbach. 1953. The stability of the rare earth complexes with ethylenediaminetetraacetic acid. *J. Am. Chem. Soc.* 75:4196–4201.

Wu, M. K., J. R. Ashburn, C. J. Torng, P. H. Hor, R. L. Meng, L. Gao, Z. J. Huang, Y. Q. Wang, and C. W. Chu. 1987. Superconductivity at 93 K in a new mixed-phase Y-Ba-Cu-O compound system at ambient pressure. *Phys. Rev. Lett.* 58:908–910.

Wyckoff, R. W. G. 1922. The crystal structures of the hexammoniates of the nickel halides. *J. Am. Chem. Soc.* 44:1239–1245 and references therein.

Zalkin, A., J. D. Forrester, and D. H. Templeton. 1963. Crystal structure of cerium magnesium nitrate hydrate. *J. Chem. Phys.* 39:2881–2891.

Zhou, L., Y. Jiang, R. Cui, Y. Li, X. Zhao, R. Deng, and H. Zhang. 2016. Efficient red organic electroluminescent devices based on trivalent europium complex obtained by designing the device structure with stepwise energy levels. *J. Lumin.* 170:692–696.

2

Principle of Luminescence Upconversion and Its Enhancement in Nanosystems

Hong Zhang, Langping Tu, and Xiao Min Liu

CONTENTS

2.1 Principle of Upconversion Luminescence in Nanosystems 19
2.1.1 Traditional UC Mechanisms 20
2.1.2 EMU Mechanism 23
2.2 UC Dynamics 24
2.3 Challenges and Recent Progress in Improving UC Luminescence in Nanosystems 26
2.3.1 Definition of UC Efficiency and Relevant Measurement Techniques 26
2.3.2 Approaches in Enhancing UC Luminescence of Nanomaterials 26
2.3.2.1 Optimization of Absorption 27
2.3.2.2 Optimization of Energy Transfer 29
2.3.2.3 Optimization of Emission 32
2.4 Future Perspective 34
References 34

2.1 Principle of Upconversion Luminescence in Nanosystems

Most emissions are down conversion (DC) in nature, which means that the emission energy is lower than the excitation energy. In this book, we introduce an opposite concept—upconversion (UC) emission—the emission of one higher energy photon upon the excitation of several lower energy photons. The latter is very attractive for applications in data storage, multi-color displays, photovoltaic devices, biological field, etc. In recent years, the development of nanotechnology has been increasingly inviting scientific interest, especially the interest of the biomedical field, in relevant material systems such as lanthanide (Ln) ions doped materials. In these materials, near-infrared (NIR) photons are converted to higher energy photons ranging from ultraviolet (UV) to NIR. Here, the excitation wavelengths are especially seductive to the biomedical field since it falls in the so-called "optical window" (~650–1300 nm), that is, the spectral range of minimal absorption of human tissue. Besides, auto-fluorescence of the biological background is negligible under the NIR excitation. All these unique features make these materials, once in nanometer sizes, ideal candidates for contrast agents of luminescence imaging. Their application potential is also expanded to solar energy utilization by converting the NIR part of the solar spectrum into visible range in order to match the absorption of commercially available solar cells.

The UC luminescence originates from the peculiar electronic property of lanthanide ions. The lanthanides are a group of elements in the periodic table where the 4f inner shell is filled with electrons. They are mostly stable in trivalent form and the Ln^{3+} ions have the electronic configuration $4f^n5s^25p^6$ where n varies from 0 to 14. The shielding of the 4f electrons of Ln^{3+} by the completed filled $5s^2$ and $5p^6$ subshells results in weak electron–phonon coupling which is responsible for important phenomena such as sharp and narrow f – f transition bands. In addition, the f – f transitions are in principle parity forbidden, but the forbidden can be partially released when the ions locate in a crystals field, where the mixture of states of different parities results in low transition probabilities and substantially long-lived excited states. Moreover, most of the lanthanides have a great many energy levels, some of them are arranged ladder-like, which facilitates UC process according to the successive photon absorption or energy transfer. Consequently, UC emission is theoretically expected for most of lanthanide ions (e.g., Er^{3+}, Tm^{3+}, and Ho^{3+}). The UC emission of these materials is usually termed "upconversion luminescence."

The physics of the UC luminescence has been explored since the 1960s. Thanks to the pioneering work reported by Auzel and others in theoretical and experimental studies, several mechanisms of UC have been proposed (Auzel 2004), including excited-state absorption (ESA), energy transfer upconversion (ETU), photon avalanche (PA), and cooperative sensitization upconversion (CSU). More recently, with the help of specially designed nanostructures, an energy migration-mediated upconversion (EMU) mechanism was proposed (F. Wang et al. 2011). Now, let us review them one by one.

2.1.1 Traditional UC Mechanisms

ESA is the simplest UC mechanism. As seen in Figure 2.1, in this mechanism, sequential absorption within the levels of a given Ln^{3+} ion is responsible for UC emission. Owing to the well ladder-like energy levels of some lanthanide activator ions and the long lifetimes of the intermediate excited energy levels (range from tens of microseconds to milliseconds), the ESA process could theoretically take place in many of the singly doped lanthanide activator ions (e.g., Er^{3+}, Tm^{3+}, Ho^{3+}, Pr^{3+}, and Dy^{3+}). It should be noted that ESA only plays an important role when the doping concentration is relatively low, that

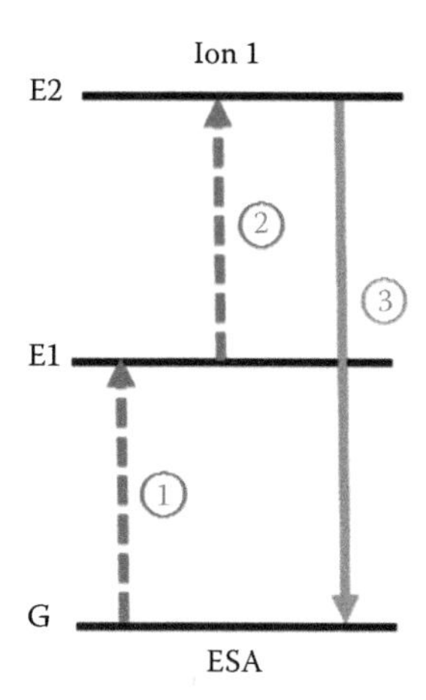

FIGURE 2.1
Schematic diagram of ESA process.

is, energy transfer between two activators is negligible, which leads to insufficient absorption and low UC efficiency.

In the case of ETU, UC emission is induced by the energy transfer between two Ln^{3+} ions, named as activator and sensitizer, respectively. As seen in Figure 2.2a, when both of them are in excited states, UC is assumed to take place to lift the activator onto a higher excited state at the expense of the excitation energy of the sensitizer. The energy transfer probability P_{SA} can generally be written as

$$P_{SA} = \frac{(R_0/R)^S}{\tau_S} \tag{2.1}$$

where τ_S is the actual excited state lifetime of the sensitizer, R_0 is the critical energy transfer distance for which excitation energy transfer and spontaneous deactivation of the sensitizer have equal probabilities, R is the real distance between the activator and sensitizer, and S is a positive integer taking the following values: $S = 6$ for dipole–dipole interactions, $S = 8$ for dipole–quadrupole interactions, and $S = 10$ for quadrupole–quadrupole interactions. It is obvious that the P_{SA} depends heavily on interaction distance, thus the energy transfer is more likely to occur between neighboring Ln^{3+} ions. It should be noticed that the sensitizer and activator can be either the same or different lanthanide ions.

As seen in Figure 2.2b, sometimes one excited ion can transfer part of its energy to another neighboring ion and both end up with some intermediate levels. This process is termed as "cross relaxation" (CR). Compared with ESA, ETU owns much robust UC intensity (one or two orders of magnitude

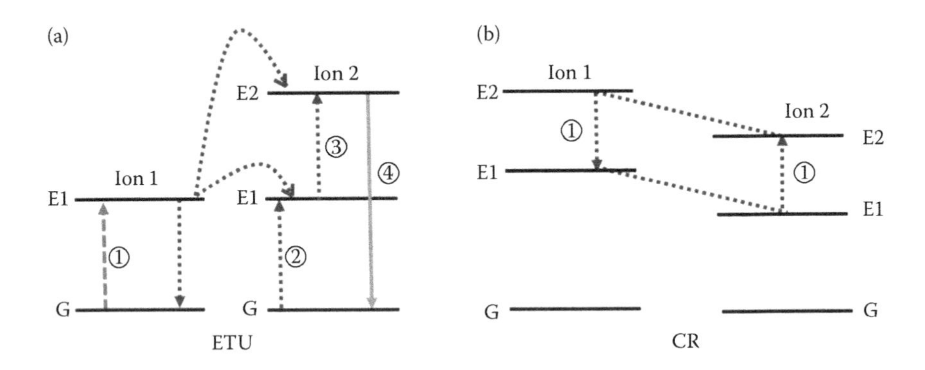

FIGURE 2.2
Schematic diagram of: (a) ETU and (b) CR processes.

stronger than ESA), since it has a larger absorption cross section by introducing sensitizer ions (e.g., Yb^{3+} and Nd^{3+}) into the system. So far, ETU is the most adopted mechanism for UC applications.

The PA process was first proposed by Chivian et al. (1979). As seen in Figure 2.3, with the help of resonant CR between the excited ion and another adjacent ground state ion, the PA-induced UC process could make a feedback loop, which induces a rapid accumulative effect of the intermediate-level population, and finally leading to strong UC emission as an avalanche process. The disadvantage of PA process is that the feedback loop only works well in a very limited UC system, and furthermore, it usually requires a relatively high pump intensity (above a certain threshold) to make the feedback effective. These drawbacks have limited the PA process in practical applications.

CSU, also named as cooperative upconversion (CUC) in some literatures, is a process involving the interaction of three ions. As seen in Figure 2.4, two excited adjacent sensitizer ions cooperatively transfer the contained energy

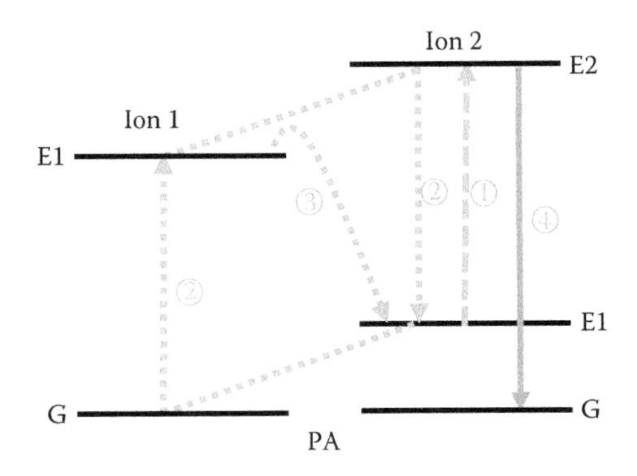

FIGURE 2.3
Schematic diagram of PA processes.

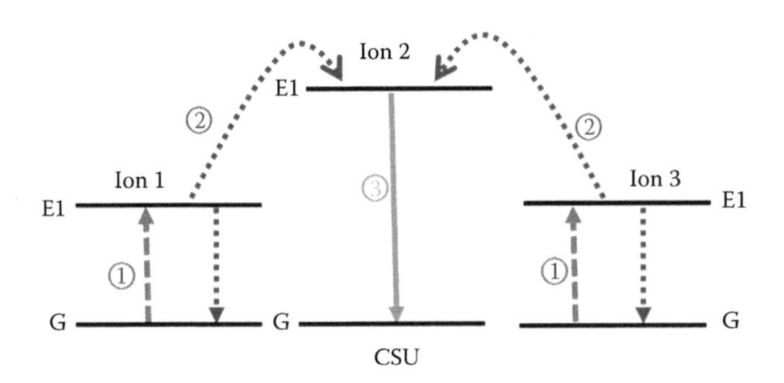

FIGURE 2.4
Schematic diagram of CSU processes.

to one activator ion simultaneously, and the excited activator relaxes back to its ground state by emitting an upconverted photon. Since the emitters have no intermediate energy level matchable to the sensitizer, compared with other methods, CSU reveals a much lower probability. Furthermore, although the CSU mechanism has been proposed for the Yb^{3+}/Tb^{3+} or Yb^{3+}/Eu^{3+} UC system (Maciel et al. 2000; Wang et al. 2008), strictly speaking, no conclusive evidence has been provided to confirm this mechanism.

2.1.2 EMU Mechanism

The above-mentioned traditional UC mechanisms were discovered in bulk materials, where, in order to obtain relatively strong UC emission, sensitizers and activators need to be co-doped to enhance the interaction probability between the two. Therefore, the energy transfer process (from sensitizers to activators) is popularly treated as "short-range" interaction. More specifically, activators are considered to receive energy from neighboring sensitizers directly. The treatment obviously neglects the possibility that the excitation energy of a sensitizer at a distance could in fact migrate via other sensitizers before being transferred to the activators, that is, a relatively "long-range" energy transfer interaction. Unfortunately, due to the limitations of the bulk material system, this "long-range" energy migration process has not been studied well experimentally because it is almost impossible to separate it from other UC mechanisms. Nanotechnology has made it possible to exceed the limitation in recent years. Along with the progress in synthetic technology, more and more complex UC nanostructures become possible, which enables the separation of absorption, transitions, and emission regions in different areas of a nanoparticle. In principle, after photoexcitation, the absorbed energy passes through the transition layer and reaches the activator. The energy migration processes in the transition layer could be readily studied by varying the transition layer thickness and/or the doping concentrations of the mediated ions in the layer. It

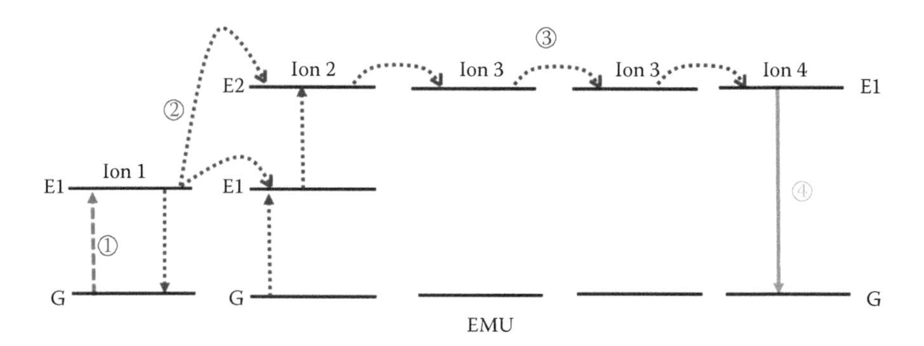

FIGURE 2.5
Schematic diagram of EMU processes.

has been confirmed that the excited energy can travel quite a long distance (e.g., 5 – 10 nm) without significant loss through a Gd^{3+} or Yb^{3+} sublattice formed transition layer (F. Wang et al. 2011; Zhong et al. 2014). The mediated ions (Gd^{3+} or Yb^{3+}) assisted "long-range" energy migration UC phenomenon was named EMU by F. Wang et al. in 2011, as seen in Figure 2.5. The efficient "long-range" EMU process implies that the energy transfer process is actually not a local effect, and the energy could be captured by activators far away (several nanometers) from the sensitizers with the assistance of the energy migration between mediating ions. Based on this understanding, the EMU process can reasonably be assumed to play an important role in the UC process even in the traditional co-doping systems. However, this remains a subject of further study.

2.2 UC Dynamics

The photoluminescence (PL) dynamics is another fundamental aspect of UC emission. Compared with DC materials, the PL dynamics of UC materials has its own characteristics: (i) the measured "lifetime" of the UC emission normally will be much longer than the intrinsic lifetime of the emitting energy level and (ii) in most instances (except ESA), the UC emission usually has a rising component at the initial stage of the time evolution of UC luminescence, in relation with the population process from the long-lived sensitizers to the activators. Theoretically, the UC dynamics could be fully described by a series of differential equations for each individual Ln^{3+} ion. However, these equations are too complex to be solved. One of the most commonly used simplified approaches is taking into account only the population and depopulation processes between different energy levels, and ignoring the multi-step energy migration processes between all identical energy levels. The simplified rate equations could be described as (Chan et al. 2012)

$$\begin{aligned}\frac{dN_i}{dt} &= \sum \text{population rate} - \sum \text{depopulation rate} \\ &= \sum_j \left(N_j A_{ji}^{ED} - N_i A_{ij}^{ED}\right) + \sum_j \left(N_j A_{ji}^{MD} - N_i A_{ij}^{MD}\right) \\ &+ \left(N_{i+1} W_{i+1,i}^{NR} - N_i W_{i,i-1}^{NR}\right) \\ &+ \sum_{ij,kl} \left(N_j N_l C_{ji,lk}^{ET} - N_i N_k C_{ij,kl}^{ET}\right)\end{aligned} \tag{2.2}$$

Here, N_i is the population density of each energy level, A_{ij}^{ED} and A_{ij}^{MD} are the Einstein coefficients for electric dipole (ED) and magnetic dipole (MD) radiative transitions from energy level i to j. $W_{i+1,i}^{NR}$ is the nonradiative multi-phonon relaxation (NMPR) rate constant from energy level $i+1$ to i. $C_{ij,kl}^{ET}$ is the microscopic energy transfer parameter for the transfer of energy via the sensitizer i to j transition and the activator k to l transition. The coefficients ED, MD, and $C_{ij,kl}^{ET}$ can be calculated by Judd–Ofelt theory, while the NMPR rate is treated with a modified energy gap law, and a related description of phonons is used to calculate phonon-assisted energy transfer constants. Finally, the intensity of any given UC emission is proportional to the product of the corresponding N_i of each energy level and its radiative transition rates.

It should be noticed that the theoretical predictions from the rate equation (1.2) usually do not fit the experimental results well. For example, in a simplest UC system with two-level donors and three-level acceptors, it is easy to calculate from the rate equations that the UC PL decay time of the activators for n-photon (the value of n indicates the number of excitation photons required to generate one UC photon) should be $1/n$ of the sensitizer's lifetime (Kingsley et al. 1969), which is, however, not the case in practice. Some modified dynamic models have been proposed in the past few decades. For example, the one suggested by Zusman–Burshteĭn (Artamono et al. 1972; Burshtei 1972) and Yokota–Tanimoto (Yokota and Tanimoto 1967) takes into account energy migration between sensitizers. The Zusman–Burshteĭn model fits the situation where the sensitizer–sensitizer interaction is much stronger than the sensitizer–activator one, whereas the Yokota–Tanimoto model is valid in the opposite case. On the other hand, Grant (1971) and Zubenko et al. (1997) dealt with the problem with time-dependent energy transfer rates instead of constants in the equations. Despite these efforts, the complete understanding of UC dynamics remains a challenge. In our opinion, the deviation mainly caused by (i) the uncertainty of the interaction parameters in the rate equations, (ii) the complex influence of the environment of the nanosystem, and (iii) the energy migration effect. Further investigation is needed.

2.3 Challenges and Recent Progress in Improving UC Luminescence in Nanosystems

The unsatisfactory UC luminescence efficiency (typically <1%) remains one of the main hurdles for application. This scenario has triggered the following questions: what are the responsible channels/steps for the loss of the excitation energy in the nanomaterials? And more interestingly, is it possible to gain even higher UC efficiency in nanomaterials than in macroscopic crystals? In order to obtain answers to these questions, a comprehensive understanding of the UC process in nanostructures is essential.

2.3.1 Definition of UC Efficiency and Relevant Measurement Techniques

The UC efficiency is one of the most important parameters to assess the optical performance of various UC materials. It is usually quantified by a parameter—UC luminescence quantum yield (QY), defined as the ratio of the number of UC emitted photons to the number of absorbed photons

$$QY = \frac{\text{The number of emitted photons}}{\text{The number of absorbed photons}} \tag{2.3}$$

Different from DC emission, the QY of UC luminescence is dependent on the power density of excitation light due to its nature of nonlinearity. Thus, accurate determination of the excitation power density is critical for evaluating the QY of UC luminescence.

One of the popular methods of the UC luminescence QY measurements are implemented by integrating sphere-based equipment, where the number of absorbed/emitted photons can be measured with the aid of photomultiplier tubes. It was reported that, under 150 W/cm^2 excitation power density, the UC QY is only $3 \pm 0.3\%$ in the $NaYF_4$: 20%Yb, 2%Er bulk material, furthermore, the value will be more than one order of magnitude lower in nanosized materials (Boyer and van Veggel 2010).

2.3.2 Approaches in Enhancing UC Luminescence of Nanomaterials

In recent years, numerous efforts have been paid to enhance the luminescence efficiency or controlled spectral modulation of UC materials, especially for the nano-sized UC materials with an eye on the demand of application. Compared with bulk crystals, nano-sized materials exhibit three distinct properties which are important for their UC luminescence.

The first distinct property is the nonnegligible role of the surface properties, which is due to the relatively large surface-to-volume ratio of nanomaterials.

It is well known that the surface can form energy traps which usually quench the UC emission, but it can be beneficial to the luminescence as well. For example, enhancement and/or broadening of the absorption spectra can be realized by anchoring organic molecules or other light harvesting entities onto the surface of nanoparticles.

The second distinct property is that nanomaterials allow tailor-made internal structures. Especially due to the development of nanotechnology, more and more complex nanostructures can be constructed. This property has raised the possibility that the energy transfer paths in nanomaterials can be artificially controlled. For example, although Nd^{3+} ion has a large absorption cross section in the NIR range (~800 nm), it is hardly used as sensitizers directly in bulk UC material because Nd^{3+} will quench the activators' excited state population through a series of harmful CR processes. However, in a tailor-made nanostructure, for example, $NaYF_4$:Yb,Er@$NaYF_4$:Yb@$NaYF_4$:Nd core–shell–shell structure, an intense Er^{3+} UC emission could be observed by restricting Nd^{3+} ions in the outer layer and utilizing Yb^{3+} in the middle layer to bridge energy into the inner emitting layer (Zhong et al. 2014). Another advantage of a nanostructure is that it gives the aspiration that the excitation energy might be "fully liberated" from the negative effects of lattice defects. If there is a tiny defect-free area, it can be "isolated" from the rest of the nanoparticles that might contain defects, and the absorbed energy in this area can in theory be free from nonradiative loss caused by lattice defects (Tu et al. 2015). The concentration quenching effect, which is in relation to the fact that the excitation energy migrates more easily from one ion to another under high doping concentrations which will increase the probability for the energy to be trapped by the defects inside and/or at the surface of the nanoparticles, could thus be suppressed. Therefore, a higher optimal doping concentration, that is, a higher UC efficiency, could be expected in specially designed nanostructures.

The third distinct property of nano-sized materials is that they are susceptible to the environment. Compared with bulk materials, nanomaterials are more susceptible to the environment due to their size limit, which makes external stimuli more effective in modifying UC dynamics. For example, it is possible to modify the radiative transition rate of lanthanides via the metal plasma electric field to enhance the absorption and/or UC emission intensity.

From luminescence dynamics point of view, the UC process of lanthanide ions doped systems can roughly be separated into three stages, including: excitation energy absorption, various energy transfer, and radiative release of the excitation energy, that is, emitting UC photons.

2.3.2.1 Optimization of Absorption

UC emission starts with the absorption of light. A robust UC spectrum relies not only on a high UC emission QY, but also on a large absorption cross section. This is the starting point for developing approaches to improve UC emission. The excitation rate of state *i* can be expressed as

$$R_i \propto I_{exc}\sigma_i N_i \tag{2.4}$$

where σ_i is the absorption cross section of state *i* at the excitation wavelength, N_i is its population density, and I_{exc} is the excitation density. From this relationship, it is obvious that the absorption cross section is key in determining the excitation efficiency. In recent years, optimizing the absorption of UC systems has mainly been following the Yb^{3+}-sensitized approach, Yb^{3+}/Nd^{3+} cooperative-sensitized approach, and dye-sensitized approach.

As mentioned earlier, most of lanthanide activator ions demonstrate insufficient absorption. Furthermore, the concentration of activator ions has to be maintained at a relatively low-level (typical in a range 0.2%–5%) to avoid significant concentration quenching. Therefore, the overall UC efficiency of most activator singly doped nanocrystals is relatively low. Er^{3+} is an exception in this respect, singly doped Er^{3+} ion has a comparatively high UC efficiency since its optimal doping concentration can reach a relatively high level (e.g., 5%–25%) and its ladder-like energy levels are well matched with ~800, ~980, and ~1500 nm excitation. For example, under 1490 nm laser excitation, the UC QY can reach up to ~1.2 ± 0.1% under an excitation density of 150 W/cm^2 in a nano-sized $LiYF_4$ host (Chen et al. 2011).

Since the 1960s, Auzel et al. (1966) found that co-doped with Yb^{3+} ions, the UC emission of some activators (such as Er^{3+}, Tm^{3+}, and Ho^{3+}) could be enhanced over 10 times. The reason is that Yb^{3+} has a relatively large absorption cross section around 980 nm (several times higher than Er^{3+}), and its PL emission resonates well with these typical upconverting Ln^{3+} ions, so the energy transfer from Yb^{3+} to these Ln^{3+} ions occurs effectively. Furthermore, due to the simple energy scheme of Yb^{3+} ion (there is only one excited state $^2F_{5/2}$ in the energy range of interest), the harmful CR processes between Yb^{3+} ions can be avoided, therefore, the "concentration quenching effect" of Yb^{3+} ions is partially suppressed, so the optimal doping concentration of Yb^{3+} ion in nanoparticles can reach up to 20%–40%, which strengthens its advantage. From that time on, the Yb^{3+}-sensitized approach was widely used to enhance the UC efficiency of Er^{3+}, Tm^{3+}, and Ho^{3+} ions.

Another approach to increase the absorption of the UC system is to make better use of its spacial features. For example, a shell coating is a commonly used strategy to enhance the UC emission of a nanoparticle by separating the surface relevant quenching centers and the luminescence centers inside the core. In the majority of reported cases, the shell component is inert, that is, a shell of pure host lattice, and its sole role is to protect the luminescence centers in the core from the surface. Since 2009, a new design of the core–shell structure appeared which contains the sensitizer Yb^{3+} in the shell, that is, an "active shell." The first report was on $NaGdF_4$:Yb^{3+}, Er^{3+} nanoparticles with a shell containing 20% Yb^{3+}-doped $NaGdF_4$ where strong enhancement of the green and red emission bands was realized (Vetrone et al. 2009). Additional energy transfer from excited Yb^{3+} ions in the shell to the Er^{3+} ions in the core was suggested to be responsible for the enhancement. However, there is also anxiety that the Yb^{3+} sensitizers in the shell are harmful for UC emission since they increase the probability of the excitation energy being captured by the surface-related traps. The actual role of the active shell in UC dynamics is not clear yet.

The above-mentioned Yb^{3+}-sensitized UC nanoparticles, regarded as a generation of multimodal bio-probes, have been attracting wide interest in biological applications. However, the Yb^{3+}-sensitized UC materials have only one single narrow absorption band around 980 nm, which restricts its practical application. More recently, Nd^{3+} ions are introduced as an additional NIR absorber and sensitizer in conventional Yb^{3+}-doped UC nanoparticles. The Nd^{3+} has an even larger absorption cross section in the NIR region (1.2×10^{-19} cm^2 at 808 nm) compared to Yb^{3+} (1.2×10^{-20} cm^2 at 980 nm) (Y.-F. Wang et al. 2013) and under the excitation of ~800 nm, the $Nd^{3+} - Yb^{3+}$ energy transfer efficiency is high. The disadvantage that Nd^{3+} will itself quench the emission of activators due to the harmful CR process could be circumvented by a properly designed core–shell nanostructure (Y.-F. Wang et al. 2013; Xie et al. 2013; Zhong et al. 2014). The obvious superiority of the Yb^{3+}/Nd^{3+} cooperative-sensitized approach is the minimization of the overheating effect in biological systems induced by water absorption (the water absorption cross section ~800 nm is much smaller than that at 980 nm).

Despite the fact that Nd^{3+} and Yb^{3+} ions have been excavated to enlarge the absorption ability of UC system, the f – f forbidden transition remains an essential constraint for the lanthanides. Recently, infrared organic dyes were selected as antenna ligands to enlarge the absorption spectrum for UC (Zou et al. 2012). The extinction coefficient of organic dye IR-806 at 806 nm is 390 L/g · cm, which is ~5×10^6 times higher than that of β-$NaYF_4$:Yb^{3+}, Er^{3+} nanoparticles at 975 nm (7×10^{-5} L/g · cm). The overall UC emission of the dye-sensitized nanoparticles is dramatically enhanced about 3300 times as a joint effect of the increase and overall broadening of the absorption spectrum, which was mainly ascribed to the augment of absorption. Nevertheless, most organic molecules suffer from photobleaching, which raises the concern of the photostability of the organic dye-sensitized nanomaterials.

2.3.2.2 Optimization of Energy Transfer

Energy transfer and interactions are critical for UC emission. Generally speaking, the ETU depends not only on the energy transfer between the ions, but also on the initial distribution of the excited states and the boundary conditions of the nanoparticles, for example, surface property, size, and morphology of the nanoparticles. The full description of the ETU process is therefore complex. Here, we will introduce the major factors that affect the ETU process, including the donor–acceptor combination, doping concentration, the excitation power density, and the surface effect.

Typical donor–acceptor combinations are Yb^{3+}/Er^{3+}, Yb^{3+}/Ho^{3+}, Yb^{3+}/Tm^{3+}, etc. It was reported that introducing some new donor–acceptor combinations can manipulate the ETU process, and consequently change the excitation and/or emission spectra of materials. For example, as mentioned earlier, the $Nd^{3+} – Yb^{3+}$ cooperative-sensitized UC materials could shift the excitation of the emission to ~800 nm. On top of that, the UC emission spectrum can also be modulated by the doping elements. Single-band UC emission with high chromatic purity is known to be highly desirable for multicolor imaging, and efforts in this aspect have appeared recently in the literature based on novel donor–acceptor combinations. For example, Er^{3+}/Tm^{3+} (2/2%) co-doped nanoparticles show a spectrally pure red emission (excited by 980 nm) due to the energy transfer between Er^{3+} and Tm^{3+} (Chan et al. 2012). However, because of the insufficient absorption of Er^{3+}, the UC emission is relatively weak. Alternatively, additional doping of Mn^{2+} ions can bring in single-band emission in Yb^{3+}/Er^{3+}, Yb^{3+}/Tm^{3+}, and Yb^{3+}/Ho^{3+} UC systems (Tian et al. 2012; J. Wang et al. 2011). Taking Yb^{3+}/Er^{3+} as an example, the existence of Mn^{2+} ions was considered to disturb the transition possibilities between the green and red emissions of Er^{3+}, with the $Er^{3+} – Mn^{2+}$ energy transfer leading to depopulation of the green emitting $^2H_{11/2}$ and $^4S_{3/2}$ energy levels, and the consequent $Mn^{2+} – Er^{3+}$ back energy transfer increasing the population of the red emitting energy level ($^4F_{9/2}$), resulting in an enhanced red to green emission ratio of Er^{3+}. In addition, doping Ce^{3+} into the Yb^{3+}/Ho^{3+} system could manipulate the red to green emission ratio by tuning the energy transfer process between Ce^{3+} and Ho^{3+} (Chen et al. 2009), and the deep-UV UC emission of Yb^{3+}/Gd^{3+} combination could be enhanced by doping Ho^{3+}, serving as a "bridging ion" in the $Yb^{3+} – Ho^{3+} – Gd^{3+}$ energy transfer process (L. Wang et al. 2013).

Another potential way to improve the UC QY is to optimize the dopant concentrations within the nanoparticles. As noted before, the energy transfer process is usually considered as dipole–dipole, dipole–quadrupole, or quadrupole–quadrupole interactions and is therefore sensitive to the operating distance. The doping concentration thus significantly affects the energy transfer process. According to the reports, increasing the doping concentration of Ln^{3+} ions (either sensitizer or activator) in the nanoparticles could enhance the UC emission to a certain extent. Further increase could make the cascade energy transfer process effective and the concentration quenching phenomenon significant. In practice, the optimal doping concentration of lanthanide ions is usually at a relatively low-level in the range of 0.2%–2% for activators (e.g., Er^{3+}, Tm^{3+}, or Ho^{3+}) with 20%–40% for sensitizer (Yb^{3+}). Over the years, great efforts have been made to elevate the quenching concentration of lanthanide ions in nanoparticles. Recently, there are reports that concentration quenching may be alleviated in some specially designed nanostructures. For example, in ultrasmall (7–10 nm) $NaYF_4$:x%Yb^{3+}, 2%Tm^{3+} nanoparticles, it was demonstrated that the NIR UC emission of Tm^{3+} at 808 nm increases up to 43 times along with an increase in the relative content of Yb^{3+} ions from 20% to 98% (Chen et al. 2010). However, this particularly high quenching concentration of Yb^{3+} is only reported for the Tm^{3+} activator co-doped case and there is no report of similar results for other activators like Er^{3+} or Ho^{3+}. This fact might indicate that the relevant quenching mechanism needs to be further studied. On the other hand, a monotonous increase of the Yb^{3+} concentration up to 98% resulting in about one order of magnitude enhancement of UC intensity in KLu_2F_7: x%Yb^{3+}, 2%Er^{3+} nanoparticle has been reported (Wang et al. 2014). The specificity of the KLu_2F_7 crystal structure is that the doped Yb^{3+} ions are separated as arrays of discrete clusters at the sublattice level and the average distance between the ionic clusters is much larger than the ionic distance within the clusters. In this crystal structure, the excitation energy absorbed by the Yb^{3+} ions tends to be restricted within the discrete cluster rather than migrating a long distance toward other clusters. In this way, the concentration quenching effect can be suppressed significantly if these clusters are quenching center free.

UC emission is a typical nonlinear process. Theoretically, excitation density is directly related to the initial population of the excited states in a photoluminescent system. In the year 2000, Pollnau et al. modeled the relationship of excitation density P with UC emission intensity I, and found that $I \propto P^n$ under low excitation power density. The value of n indicates the number of NIR excitation photons required to generate one UC photon. So far, there is no evidence to suggest that the energy transfer process is dependent on the excitation density if it is relatively low, that is, $<100\ W/cm^2$. Low excitation density is usually applied to the measurement of massive nanoparticles. For single nanoparticle measurements, however, high density excitation is required. It was recently, reported that under high density excitation (e.g., $2.5 \times 10^6\ W/cm^2$) the UC emission is significantly enhanced when the concentration of activator Tm^{3+} is greatly increased from 0.5% to 8% in $NaYF_4$ host (J. Zhao et al. 2013). A similar result was also observed for Er^{3+} ions, co-doped with 20% Yb^{3+} ions, under the low power excitation, its optimal doping concentration is ~2%, where it increases to 20% when the power density is above $3 \times 10^6\ W/cm^2$ (Gargas et al. 2014). The proposed physical picture is based on the variety of the initial distribution of the excited state population in the nanoparticles. The higher density excitation causes more Yb^{3+} ions in the excited state in the nanoparticles, and the critical step in UC emission is the excited state energy transfer from Yb^{3+} to the activator (Tm^{3+} or Er^{3+}). If the number of activators is not enough, these activators will get saturated easily in accepting excitation energy via the sensitizers. From this point of view, under excitation of high density, higher doping levels of the activator could promote the utilization of the excitation energy stored in the sensitizers, and facilitate the UC emission.

Surface characteristics of UC materials are an important issue for the efficiency of UC emission, as they expose numerous lanthanide dopants to surface deactivations (caused by surface defects, lattice strains, surface ligands, and solvents that possess high phonon energy). These processes are strongly manifested in nanoparticles due to the high surface-to-volume ratio. Subsequently, core–shell structures were introduced to improve the UC emission and to study the surface effects of nanoparticles. By controlling the shell thickness, the direct interacting distance of surface effects to the Ln^{3+} ions is confirmed to be around 1.5–5 nm (Gargas et al. 2014). And, it was reported the enhancement factor is from dozens to hundreds when the shell thickness is only 1–2 nm (Zhang et al. 2012). However, the mechanism of surface effect still needs to be clarified. It is assumed that the excited energy states situated on or near the surface can be deactivated directly by neighboring quenching centers. This understanding has loop holes for nanomaterials. For a 20 nm (diameter) sized nanoparticle, even taking the largest surface effect distance (5 nm), there is still ~12.5% area that is inert to the surface effects, which means that the maximal factor of luminescence enhancement induced by shell coating should be around eight, which is not the case. To rationalize the experimental results, an extend explanation based on EMU theory was suggested (Chen et al. 2013; Su et al. 2012): the energy contained in the excited dopants locate in the center of nanophosphors can randomly migrate and travel a long distance to the dopant on/near surface and is quenched by the surface quenching centers. However, this understanding is still half-baked, and the microscopic physical picture, especially the PL dynamics process, needs further study.

2.3.2.3 Optimization of Emission

After the absorption and energy transfer process, the UC emission arises from the depopulation of the emitting energy state, and at this stage, the competition between radiative relaxation and nonradiative relaxation rates is key to the emission intensity. As introduced earlier, for nanosystems, the relevant-doped Ln^{3+} ions are more susceptible to the environment due to the limited space. External stimuli induced radiative/nonradiative rate modification is thus easier to realize in nanomaterials than in macroscopic crystals. We will restrict ourselves here to two aspects: (i) selection of host materials and (ii) plasmonic enhancement.

The phonon-induced energy loss is one of the main reasons for the low efficiency of UC emission, where the excited states' energy converts into phonons of the host via multiphonon-assisted nonradiative relaxation. A proper indicator is the cutoff phonon energy of the host lattice which is an important parameter for the selection of a good host material. Generally speaking, the higher cutoff phonon energy, the lower the UC efficiency. Compared with oxides and oxy-fluorides or oxy-chlorides hosts (lattice phonon energies >500 cm^{-1}), fluorides hosts (e.g., $NaYF_4$, $LiYF_4$, and $NaLuF_4$) have relative lower phonon energies (~350 cm^{-1}) and usually display the highest UC efficiency due to the minimization of nonradiative energy losses in the intermediate/emitting states.

As mentioned earlier, the luminescence of Ln^{3+} ions is mostly due to the ED/quadrupole transitions among the energy levels of the 4f subshell. The radiative transition is in general forbidden due to parity considerations. However, when the rare earth ions are set in an asymmetrical crystal field, the intrinsic wave functions of the 4f subshell mix with other wave functions of opposite parity, such as the wave functions of 5d, 5g, etc. The forbidden nature of the transition is thus partially broken. A highly asymmetrical crystal field is helpful in enhancing the radiative and absorption transition probabilities of rare earth ions. Some methods to change the local crystal fields in macroscopic crystals are also applied to nanosystems. For example, hexagonal $NaYF_4$:Yb^{3+},Er^{3+} microcrystals exhibit visible UC photoluminescence (PL), which is several times higher than their cubic counterparts. On top of that, adding certain ions (e.g., Li^+) into the crystal lattice is sometimes helpful to reduce the crystal symmetry and thus to enhance the UC emission intensity (C. Zhao et al. 2013). Besides tailoring the local crystal field, it was reported that the 2.7 times enhancement of UC PL could be realized in a $BaTiO_3$ (BTO) nanohost by applying a 10 V external field (Hao et al. 2011). In this work, a multilayer film material with a typical parallel plate capacitor was developed, the enhancement was argued to come from the unique crystal structure of the ferroelectric host BTO material. Tetragonal BTO with the point group 4 mm (C_{4v}) at room temperature is noncentrosymmetric. Upon applying an electric field along the direction of spontaneous polarization of the host, the *c*-axis of the lattice elongates and changes the structure symmetry of the BTO host. The UC emission could be enhanced in a controlled manner by simply tuning the applied electric field. The difference in the enhancement of green and red emissions was analyzed based on the Judd–Ofelt (J–O) theory. According to the authors, the green emission of Er^{3+} ions comes from one of the hypersensitive transitions dominated by Ω_2, which is known to be closely associated with the asymmetry of the Ln^{3+} ion sites. This work points to another approach for enhancing UC emission, which could be more robust if better host materials could be explored in the future with higher breakdown voltages.

It is well known that metal nanostructures (i.e., plasmonic substrates) can enhance the emission of a fluorophore due to localized surface plasmon resonance when the distance between the metallic structure and the fluorophore is appropriate. These effects have been widely used for enhancing the downconversion fluorescence of dyes or quantum dots, and were recently introduced to UC nanomaterials. It was found that the UC PL can be enhanced by nanoparticles, nanowires, nanoshells, as well as nanoarrays of Ag and Au with enhancement 5–310-fold by optimizing pertinent experimental parameters. The enhancement could be attributed to (i) the absorption of the UC nanoparticles in relation with the excitation collection effect, (ii) the emission of the activators, and (iii) the nonradiative transition rates of Ln^{3+} ions which can be changed by metal particles.

Accordingly, there are different approaches to enhance the UC emission of nanosystems using a plasmonic field. One scenario is to set the plasmonic resonance with UC emissions. Saboktakin et al. reported 5.2-fold enhancement by Au nanoparticles and 45-fold by Ag nanoparticles in UC luminescence

(Saboktakin et al. 2012). The enhancement was attributed to the increase of both the absorption and the radiative rate of the emission. Another scenario is to set the plasmonic resonance with the excitation wavelength of the UC materials (Saboktakin et al. 2013). In 2013, the plasmonic enhancement of UC PL of nanoparticles in Au nanohole arrays was reported by Saboktakin et al. In this study, Au nanohole arrays were fabricated on transparent glass substrates. By adjusting the size of the apertures, the periodicity of the array, and the thickness of the metallic layer, the plasma band of the metallic nanohole array was tuned to NIR (980 nm), which is resonate with the UC excitation. It was determined that the UC luminescence was intensified 32.6 times for the green emission at ~540 nm and 34.0 times for the red emission at ~650 nm. The authors thus came to the conclusion that the enhancements originated from the absorption improvement due to the resonance between the nanohole arrays and the excitation wavelength of the UC emission.

However, it should be noted that despite a spate of reports of plasmon-enhanced UC PL, plasmon-induced PL quenching also appears (Li et al. 2011). The quenching is attributed to the resonance energy transfer from the Ln^{3+} to the metal particle or the reabsorption of the emitted light by the metal particle. Furthermore, the enhancement or quenching mechanism cannot be discriminated by time-resolved measurements, since both the interactions lead to a shorter PL decay time of Ln^{3+}. Therefore, more investigations are needed on the interaction between plasmonic metal nanoparticles and UC nanostructure.

2.4 Future Perspective

Lanthanide ions doped UC nanoparticles, emerging as a new class of luminescent material, have attracted more and more attention in recent years. However, the brightness of UC emission is still an issue for some applications, mainly caused by (i) the low efficiency of UC and (ii) the narrow and low extinction coefficient of nanoparticles. On the other hand, our comprehension of the UC mechanism is still not sufficient, especially of microscopic UC dynamics. Multidisciplinary efforts, including theoretical modeling and computation, spectroscopy, synthetic chemistry, and chemical engineering, are expected to be the solution to this formidable challenge.

References

Artamono, M., A. Burshtei, C. M. Briskina, A. G. Skleznev, and L. D. Zusman. 1972. Time variation of Nd^{3+} ion luminescence and an estimation of electron excitation migration along ions in glass. *Zh. Eksp. Teor. Fiz.* 62:863.

Auzel, F. 1966. Compteur Quantiquepar Transfert Denergie Entre Deuxionsde Terres Rares DansUn Tungstate MixteEt Dans Un Verre. *C. R. Hebd. Seances Acad. Sci. Ser. B* 262:1016.

Auzel, F. 2004. Upconversion and anti-Stokes processes with f and d ions in solids. *Chem. Rev.* 104:139–173.

Boyer, J.-C. and F. C. J. M. van Veggel. 2010. Absolute quantum yield measurements of colloidal $NaYF_4$: Er^{3+}, Yb^{3+} upconverting nanoparticles. *Nanoscale* 2:1417–1419.

Burshtei, A. 1972. Jump mechanism of energy-transfer. *Zh. Eksp. Teor. Fiz.* 62:1695.

Chan, E. M., G. Han, J. D. Goldberg et al. 2012. Combinatorial discovery of lanthanide-doped nanocrystals with spectrally pure upconverted emission. *Nano Lett.* 12:3839–3845.

Chen, G., H. Liu, G. Somesfalean, H. Liang, and Z. Zhang. 2009. Upconversion emission tuning from green to red in Yb^{3+}/Ho^{3+}-codoped $NaYF_4$ nanocrystals by tridoping with Ce^{3+} ions. *Nanotechnology* 20:385704.

Chen, G., T. Y. Ohulchanskyy, A. Kachynski, H. Agren, and P. N. Prasad. 2011. Intense visible and near-infrared upconversion photoluminescence in colloidal $LiYF_4$: Er^{3+} nanocrystals under excitation at 1490 nm. *ACS Nano* 5:4981–4986.

Chen, G., T. Y. Ohulchanskyy, R. Kumar, H. Agren, and P. N. Prasad. 2010. Ultrasmall monodisperse $NaYF_4$:Yb^{3+}/Tm^{3+} nanocrystals with enhanced near-infrared to near-infrared upconversion photoluminescence. *ACS Nano* 4:3163–3168.

Chen, G., C. Yang, and P. N. Prasad. 2013. Nanophotonics and nanochemistry: Controlling the excitation dynamics for frequency up- and down-conversion in lanthanide-doped nanoparticles. *Acc. Chem. Res.* 46:1474–1486.

Chivian, J. S., W. E. Case, and D. D. Eden. 1979. The photon avalanche—A new phenomenon in Pr^{3+}-based infrared quantum counters. *Appl. Phys. Lett.* 35:124–125.

Gargas, D. J., E. M. Chan, A. D. Ostrowski et al. 2014. Engineering bright sub-10-nm upconverting nanocrystals for single-molecule imaging. *Nat. Nanotechnol.* 9:300–305.

Grant, W. J. C. 1971. Role of rate equations in theory of luminescent energy transfer. *Phys. Rev. B Solid State* 4:648.

Hao, J., Y. Zhang, and X. Wei. 2011. Electric-induced enhancement and modulation of upconversion photoluminescence in epitaxial $BaTiO_3$:Yb/Er thin films. *Angew. Chem. Int. Ed.* 50:6876–6880.

Kingsley, J. D., G. E. Fenner, and S. V. Galginai. 1969. Kinetic and efficiency of infrared-to-visible conversion in LaF_3–Yb, Er. *Appl. Phys. Lett.* 15:115.

Li, Z., L. Wang, Z. Wang, X. Liu, and Y. Xiong. 2011. Modification of $NaYF_4$:Yb,Er@ SiO_2 nanoparticles with gold nanocrystals for tunable green-to-red upconversion emissions. *J. Phys. Chem. C* 115:3291–3296.

Maciel, G. S., A. Biswas, R. Kapoor, and P. N. Prasad. 2000. Blue cooperative upconversion in Yb^{3+}-doped multicomponent sol–gel-processed silica glass for three-dimensional display. *Appl. Phys. Lett.* 76:1978–1980.

Saboktakin, M., X. Ye, U. K. Chettiar et al. 2013. Plasmonic enhancement of nanophosphor upconversion luminescence in Au nanohole arrays. *ACS Nano* 7:7186–7192.

Saboktakin, M., X. Ye, S. J. Oh et al. 2012. Metal-enhanced upconversion luminescence tunable through metal nanoparticle–nanophosphor separation. *ACS Nano* 6:8758–8766.

Su, Q., S. Han, X. Xie et al. 2012. The effect of surface coating on energy migration-mediated upconversion. *J. Am. Chem. Soc.* 134:20849–20857.

Tian, G., Z. Gu, L. Zhou et al. 2012. Mn^{2+} dopant-controlled synthesis of $NaYF_4$:Yb/Er upconversion nanoparticles for *in vivo* imaging and drug delivery. *Adv. Mater.* 24:1226–1231.

Tu, L. P., X. M. Liu, F. Wu, and H. Zhang. 2015. Excitation energy migration dynamics in upconversion nanomaterials. *Chem. Soc. Rev.* 44:1331–1345.

Vetrone, F., R. Naccache, V. Mahalingam, C. G. Morgan, and J. A. Capobianco. 2009. The active-core/active-shell approach: A strategy to enhance the upconversion luminescence in lanthanide-doped nanoparticles. *Adv. Funct. Mater.* 19:2924–2929.

Wang, F., R. Deng, J. Wang et al. 2011. Tuning upconversion through energy migration in core–shell nanoparticles. *Nat. Mater.* 10:968–973.

Wang, H., C.-K. Duan, and P. A. Tanner. 2008. Visible upconversion luminescence from Y_2O_3:Eu^{3+},Yb^{3+}. *J. Phys. Chem. C* 112:16651–16654.

Wang, J., R. Deng, M. A. MacDonald et al. 2014. Enhancing multiphoton upconversion through energy clustering at sublattice level. *Nat. Mater.* 13:157–162.

Wang, J., F. Wang, C. Wang, Z. Liu, and X. Liu. 2011. Single-band upconversion emission in lanthanide-doped $KMnF_3$ Nanocrystals. *Angew. Chem. Int. Ed.* 50:10369–10372.

Wang, L., M. Lan, Z. Liu et al. 2013. Enhanced deep-ultraviolet upconversion emission of Gd^{3+} sensitized by Yb^{3+} and Ho^{3+} in beta-$NaLuF_4$ microcrystals under 980 nm excitation. *J. Mater. Chem. C* 1:2485–2490.

Wang, Y.-F., G.-Y. Liu, L.-D. Sun et al. 2013. Nd^{3+}-sensitized upconversion nanophosphors: Efficient *in vivo* bioimaging probes with minimized heating effect. *ACS Nano* 7:7200–7206.

Xie, X., N. Gao, R. Deng et al. 2013. Mechanistic investigation of photon upconversion in Nd^{3+}-sensitized core–shell nanoparticles. *J. Am. Chem. Society* 135:12608–12611.

Yokota, M. and O. Tanimoto. 1967. Effects of diffusion on energy transfer by resonance. *Journal of the Physical Society of Japan* 22:779.

Zhang, F., R. Che, X. Li et al. 2012. Direct imaging the upconversion nanocrystal core/shell structure at the subnanometer level: Shell thickness dependence in upconverting optical properties. *Nano Lett.* 12:2852–2858.

Zhao, C., X. Kong, X. Liu et al. 2013. Li^+ ion doping: An approach for improving the crystallinity and upconversion emissions of $NaYF_4$:Yb^{3+}, Tm^{3+} nanoparticles. *Nanoscale* 5:8084–8089.

Zhao, J., D. Jin, E. P. Schartner et al. 2013. Single-nanocrystal sensitivity achieved by enhanced upconversion luminescence. *Nat. Nanotechnol.* 8:729–734.

Zhong, Y., G. Tian, Z. Gu et al. 2014. Elimination of photon quenching by a transition layer to fabricate a quenching-shield sandwich structure for 800 nm excited upconversion luminescence of Nd^{3+} sensitized nanoparticles. *Adv. Mater.* 26:2831–2837.

Zou, W., C. Visser, J. A. Maduro, M. S. Pshenichnikov, and J. C. Hummelen. 2012. Broadband dye-sensitized upconversion of near-infrared light. *Nat. Photonics* 6:560–564.

Zubenko, D. A., M. A. Noginov, V. A. Smirnov, and I. A. Shcherbakov. 1997. Different mechanisms of nonlinear quenching of luminescence. *Phys. Rev. B* 55:8881–8886.

3

Synthesis of Upconverting Nanomaterials: Designing the Composition and Nanostructure

Adolfo Speghini, Marco Pedroni, Nelsi Zaccheroni, and Enrico Rampazzo

CONTENTS

3.1 Introduction 37
3.2 Upconverting Nanomaterials: Importance of the Composition 39
3.3 Synthetic Strategies 40
3.3.1 Coprecipitation 40
3.3.1.1 Fluorides 41
3.3.1.2 Oxides 41
3.3.1.3 Oxyfluorides 42
3.3.2 Thermolysis 42
3.3.2.1 Fluorides 43
3.3.2.2 Oxides 45
3.3.2.3 Oxyfluorides 45
3.3.3 Solvo(hydro)thermal 45
3.3.3.1 Fluorides 46
3.3.3.2 Oxides 47
3.3.4 Sol–Gel 49
3.3.5 Combustion 51
3.3.6 Ionic Liquids 52
3.4 Core–Shell Architectures 55
3.4.1 Silica-Shell Formation 57
3.5 Conclusions 60
References 60

3.1 Introduction

Lanthanide-doped upconverting nanostructures are promising materials in the generation of imaging agents for modern biomedical applications, in particular in optical diagnostics (Prodi et al. 2015). Upconversion (UC) is a phenomenon involving optical emission at higher energies than that of the exciting radiation, through sequential absorptions of photons. Lanthanide ions are particularly useful for this process, due to a peculiar energy levels scheme and relatively long-excited states lifetimes (Auzel 2004). In the past decades, many hosts and types of lanthanide ions have been chosen to customize the luminescence properties of the nanosystems, tailoring also their structure to fit efficiently the final application. In this context, some reviews have appeared in the literature describing preparation of lanthanide-doped nanocrystals (DaCosta et al. 2014; Gainer and Romanowski 2014; Hemmer et al. 2013; Y. Liu et al. 2013; Wang and Liu 2009; Yang et al. 2014; Zhou et al. 2012) evidencing the high interest and activity of the field. In these reviews, several factors such as morphology, crystalline phase, size, and components of these nanomaterials have been demonstrated to be crucial parameters acting on their electrical, photophysical, magnetic, and colloidal stability properties (Sun et al. 2014; van Veggel et al. 2012), see Figure 3.1.

The target of this chapter is to briefly illustrate the diverse synthesis of UC nanomaterials, with particular attention to the composition of the host and to the architecture of the nanostructures, tailored to produce efficient luminescent nanomaterials.

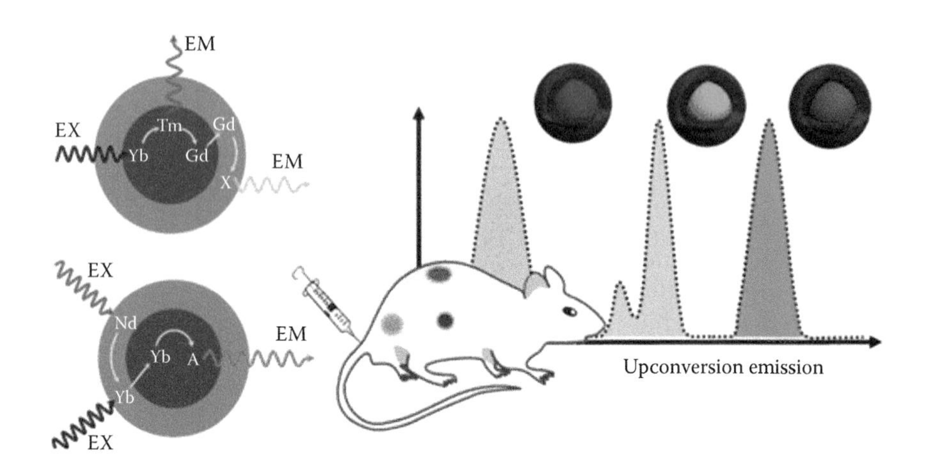

FIGURE 3.1
UC emission obtained from RE nanoparticles with controlled size and structure can be used for many biological applications, thanks to multicolor and tunable emissions. These rationally designed nanostructures and nanocomposites can be engineered to meet various applications, such as imaging, detection, and sensing. (Reprinted with permission from Sun, L. D., Y. F. Wang, and C. H. Yan. 2014. Paradigms and challenges for bioapplication of rare earth upconversion luminescent nanoparticles: Small size and tunable emission/excitation spectra. *Acc. Chem. Res.* 47:1001–1009. Copyright 2014 American Chemical Society.)

3.2 Upconverting Nanomaterials: Importance of the Composition

Among the factors affecting the UC properties of nanomaterials, the choice of the species that absorb (sensitizers) and emit (activators) the radiation is of paramount importance. The mainly exploited sensitizer is Yb^{3+}, characterized by a single transition in the near-infrared (NIR) range, around 980 nm, ($^2F_{7/2} \rightarrow {}^2F_{5/2}$ transition) closely matching the energies of the excited states of several activator ions such as Er^{3+}, Tm^{3+}, and Ho^{3+}. The doping amounts of both sensitizer and activator ions define the luminescence features of the nanomaterials, since they have to be optimized taking into account the delicate balance between the quite high amount required to favor energy transfer toward the activator, and the prevention of detrimental cross-relaxation processes (Mita et al. 1995).

Appropriate selection of suitable host materials for efficient UC emission is also of paramount importance. Important properties of these hosts are

- High chemical stability
- High transparency in the optical range of interest (in ultraviolet (UV), visible, and NIR ranges)
- High optical damage threshold

Moreover, the UC emission efficiency strongly depends on the vibrational properties of the host, influencing nonradiative relaxations for the excited states of the emitting lanthanide ions. The latter processes involve multiphonon assisted deactivation processes, in which the phonons are bridging the energy gap between the emitting level and the next lower lying energy level of the lanthanide ion. As a general rule, the larger is the number of phonons needed to bridge this energy gap and the higher is the efficiency of the radiative emission (Chen et al. 2013). Therefore, to increase the UC emission efficiency, it is desirable to have lanthanide ions embedded in a host for which the phonon energies are as low as possible. The cutoff phonon energy depends on the type of host, and in general, fluoride-based materials have lower phonon energies than oxide compounds (Chen et al. 2014). For instance, cutoff phonon energies for Y_2O_3 (Vetrone et al. 2004) (around 500 cm^{-1}) and ZrO_2 (Patra et al. 2003a) (around 550 cm^{-1}) are much higher than for $NaYF_4$ (Ivaturi et al. 2013) (around 350 cm^{-1}). The efficiency of a UC emission is also strongly dependent on the local symmetry of the site in which the lanthanide ion is accommodated (Peacock 1975). Due to the peculiar character of lanthanide 4f–4f transitions, a lower local symmetry is favorable for increasing the emission efficiency (Krämer et al. 2004; Lin et al. 2014; Schafer et al. 2009).

3.3 Synthetic Strategies

The development of synthetic strategies for efficient luminescent nanomaterials with defined size, shape, composition, and phase is of paramount importance. This section is devoted to the description of the most popular synthetic methods approaches to obtain tailored nanomaterials in a facile and controlled way. We will mainly focus on the advantages (pros) and disadvantages (cons) for each approach and in order to help the reader in the comparison, we have summarized the main features in Table 3.1.

3.3.1 Coprecipitation

The coprecipitation method is characterized by simple protocols, quite short reaction times, simple reaction conditions, and cheap experimental setups.

TABLE 3.1

Overview of the Synthetic Methods Used for the Preparation of Different Hosts with Advantages (Pros) and Disadvantages (Cons)

Synthetic Methods	Hosts (Examples)	Pros	Cons
Coprecipitation	LaF_3, $NaYF_4$, GdF_3 Y_2O_3, Lu_2O_3, $Gd_3Ga_5O_{12}$ $LuPO_4$ $YbPO_4$ GdOF, $Gd_4O_3F_6$ $Lu_6O_5F_8$	• Short preparation time • Relatively cheap and facile procedure	• High T treatment • Poor size control • Aggregation
Thermolysis	LaF_3, $NaYF_4$, $NaGdF_4$ Y_2O_3, Gd_2O_3 LaOF, GdOF	• High quality • Good size and crystallinity control • Low aggregation	• Costly and demanding procedures • High T treatment • Toxic by-products
Solvo(hydro) thermal	$NaYF_4$, $NaLuF_4$ YbF_3, YF_3, GdF_3 CaF_2, SrF_2, $CaGd_3F_{11}$ Gd_2O_3, Er_2O_3	• Relatively mild T • High quality • Good size control • Good dopant control • Low amount of toxic by-products	• Relatively long preparation time
Sol–gel	ZrO_2, $BaTiO_3$, TiO_2 Titanates (nanotubes) Y_2O_3, Gd_2O_3, $Lu_3Ga_5O_{12}$ YVO_4	• Low amount of toxic by-products • Relatively facile procedures	• Final calcination • Aggregation • Relatively long preparation time
Combustion	Y_2O_3, Lu_2O_3, Gd_2O_3 $Gd_3Ga_5O_{12}$	• Short reaction times	• Very high T
ILs	YF_3, GdF_3, EuF_3 $NaYF_4$, $NaGdF_4$	• A "green" approach	• Aggregation

In this procedure, the nucleation of the nanocrystals can be promoted using capping agents (such as polyvinylpyrrolidone (PVP) or polyethyleneimine (PEI)). In some cases, the nanoparticles (NPs) are directly generated without the need of postformation heat treatments, in others cases, annealing is necessary to obtain the desired phase.

3.3.1.1 Fluorides

First examples were published by Van Veggel and coworkers, that synthesized LaF_3 nanoparticles, containing emitters such as Eu, Er, Nd, and Ho, using di-n-octadecyldithiophosphate as a surfactant (Stouwdam and van Veggel 2002) and by Cho et al. (Yi and Chow 2005), who obtained a very small particle size (around 5 nm), with NPs easily dispersible in water.

To prepare the most famous UC phosphor $NaYF_4$, other authors used relatively low reaction temperatures using amines as solvents (Heer et al. 2004). Yi et al. (2004) used and lanthanide–ethylenediaminetetraacetic acid (EDTA) complexes as starting material. The tuning of the molar ratio of EDTA to total lanthanides allowed the control of nanoparticle size from 40 to 170 nm, but samples were affected by low luminescence emission efficiency and rather poor monodispersion. Other approaches succeeded in developing hexagonal-phase $NaYF_4$:Yb,Er/Tm nanocrystals with tunable sizes: these systems were formed starting from small amorphous $NaYF_4$ nanoparticles obtained by precipitation, and by a subsequent treatment at higher temperature (300°C), and showed higher up-conversion emission efficiency (Li and Zhang 2008; Li et al. 2008).

Hollow mesoporous GdF_3 nanoparticles were prepared also by coprecipitation method by Lv et al. (2013) using pH controlled solutions.

3.3.1.2 Oxides

Sesquioxides have been prepared by this method. For instance, Er^{3+}, Yb^{3+} codoped Y_2O_3 upconverting nanoparticles (UCNPs) have been prepared by using a coprecipitation method followed by a postthermal treatment, using a surfactant (cetyltrimethylammonium bromide, CTAB) that had an important role in controlling the size (Lu et al. 2014). Moreover, Er^{3+}, Yb^{3+} codoped Lu_2O_3 nanoparticles with several sizes and shapes (nano-aggregates, submicrometer wires, and nanospheres) were prepared by the coprecipitation technique. It was found that different reactant ratios of lanthanide to urea precipitant produced uniform spherical nanoparticles with sizes of 45, 100, 165, 200, and 250 nm (Zheng et al. 2014).

Er^{3+}-doped Garnet ($Gd_3Ga_5O_{12}$) nanoparticles were also prepared by Daldosso et al. (2008) by coprecipitation, showing quite good UC emission.

Phosphate nanoparticles were also prepared using the coprecipitation technique by Heer et al. (2003), who published a pioneering paper in 2003 about the synthesis of Tm^{3+}, Yb^{3+}-doped $LuPO_4$, and Er^{3+}-doped $YbPO_4$

nanocrystals in the form of transparent colloidal solutions, demonstrating the possibility of obtaining UC blue, green, and red light emission in transparent colloidal solutions by laser excitation in the near-IR region around 980 nm.

3.3.1.3 Oxyfluorides

Few examples of oxyfluoride UC samples prepared by coprecipitation techniques are reported in the literature. Codoped Er^{3+}, Yb^{3+}, Tm^{3+}/Yb^{3+}, Ho^{3+}/Yb^{3+}, and triply doped $Er^{3+}/Tm^{3+}/Yb^{3+}$ single phase GdOF and $Gd_4O_3F_6$ nanoparticles with average particle sizes around 25 and 50 nm, respectively, were prepared in aqueous solution under alkaline conditions by a simple coprecipitation method and a heat treatment at 500°C (Passuello et al. 2011a, b). Due to the heat treatment, a certain degree of agglomeration was observed. In the case of triply doped samples, the nanoparticles show bright white light UC emission upon excitation at 980 nm using a diode laser as the excitation source.

A series of $Er^{3+}/Tm^{3+}/Yb^{3+}$-doped $Lu_6O_5F_8$ nanoparticles have been prepared by a coprecipitation method (Guo et al. 2013). The average size has been tuned from 20 to 320 nm upon increasing Li^+ ion concentration in the host. The detailed crystal structure of $Lu_6O_5F_8$ was analyzed via Rietveld refinement of the powder x-ray diffraction patterns. It is worth mentioning that Li^+ concentration influences the white UC emission. In this case, Li^+ ion behaves as a luminescence intensifier.

3.3.2 Thermolysis

This synthetic approach is often used to form lanthanide-doped nanoparticles, since it provides high quality nanocrystals, with high control over dimension, monodispersity, and photoluminescence (PL) properties. It involves a heat treatment (~300°C) of lanthanide precursors, usually organometallic compounds (e.g., acetate or trifluoroacetate salts) that decompose in nonpolar high boiling organic solvents, such as oleylamine (OM), trioctylphosphine oxide (TOPO), or 1-octadecene (ODE). Surfactants with capping groups, for instance, oleic acid (OA), are in charge to control the nanoparticle size and to prevent their aggregation, due to long hydrocarbon chains. The main parameters affecting this synthetic approach are reaction temperature, metal precursors and their concentration, nature of the solvent, capping agent(s), and reaction time. By careful tuning of these experimental parameters, highly monodispersed nanoparticles with very good crystallinity have been produced using the thermolysis method. Besides these positive features, the thermolysis method presents some serious drawbacks. In particular, quite expensive procedures with demanding aspects such as high temperatures and air sensitive starting materials handling are involved. Moreover, the formation of highly toxic fluorinated by-products could limit the large-scale application of this synthetic approach in bio-related applications.

3.3.2.1 Fluorides

A substantial amount of literature has been published in the last 10 years reporting preparation of fluorides with the thermolysis method. One of the first reports was published by Yan et al. (Zhang et al. 2005), who synthesized LaF3 nanocrystals starting from $La(CF_3COO)_3$ salt, using OA as the capping agent and ODE as the high boiling noncoordinating solvent. This method was extended to other lanthanide-doped nanoparticles, for instance, for the preparation of the famous upconverter material, $NaYF_4$, by several groups (Abel et al. 2009; Boyer et al. 2006; G. Chen et al. 2010; Cuccia and Capobianco 2007; Yi and Chow 2006). Also, the group of Murray and co-workers (Ye et al. 2010), succeeded in the synthesis of $NaYF_4$:Yb, Er nanocrystals with a controlled size and morphology (spherical/nanorod): this approach definitely provided a way to high quality and monodipersed colloids. Also Yin et al. (Yu et al. 2010), synthesized monodisperse β-$NaYF_4$:Yb, Tm nanocrystals with controlled size (25–150 nm), composition, and shape (sphere, hexagonal prism, and hexagonal plate) by thermolysis of metal trifluoroacetates in hot solutions (300–330°C) containing OA, OM, and ODE.

Another interesting host for UC emission, $NaGdF_4$, was also prepared in nanocrystalline form using the thermolysis method (see Figure 3.2), by the group of Capobianco et al. (Boyer et al. 2007; Naccache et al. 2009). Other groups (Cichos et al. 2014; Johnson et al. 2011; Liu et al. 2010; Z.-L. Wang et al. 2010; Zhou et al. 2010) have investigated this host, also for its interesting magnetic resonance imaging (MRI) properties.

A general synthesis of high quality cubic and hexagonal $NaREF_4$(RE: Pr to Lu, Y) nanocrystals (nanopolyhedra, nanorods, nanoplates, and nanospheres) and $NaYF_4$:Yb, Er/Tm nanocrystals (nanopolyhedra and nanoplates) via the co-thermolysis of $Na(CF_3COO)$ and $RE(CF_3COO)_3$ in OA/OM/1-ODE was reported (Mai et al. 2006). By tuning the ratio of Na/RE, solvent composition, reaction temperature and time, control of the phase, shape, and size of the nanocrystals has been achieved. Interesting hosts for UC nanomaterials have demonstrated to be the alkaline earth fluorides (D. Chen et al. 2010; Quan et al. 2008). Uniform alkaline earth metal fluoride MF_2(M = Mg, Ca, and Sr) nanomaterials with various shapes (tetragonal MgF_2 nanoneedles; cubic CaF_2 nanoplates and nanopolyhedra; cubic SrF_2 nanoplates and nanowires) have been synthesized from the thermolysis of alkaline earth metal trifluoroacetate in hot surfactant solutions, with OA, OM, and ODE (Du et al. 2009). The MF_2 nanocrystals were formed by the controlled fluorination of the M–O bond into the M–F bond at the nucleation stage and subsequent growth process. In these cases, the growth of shape-selective MF_2 nanocrystals was likely due to the template direction of micellar structures formed by self-assembly of capping ligands and the so-called "Ostwald ripening" process.

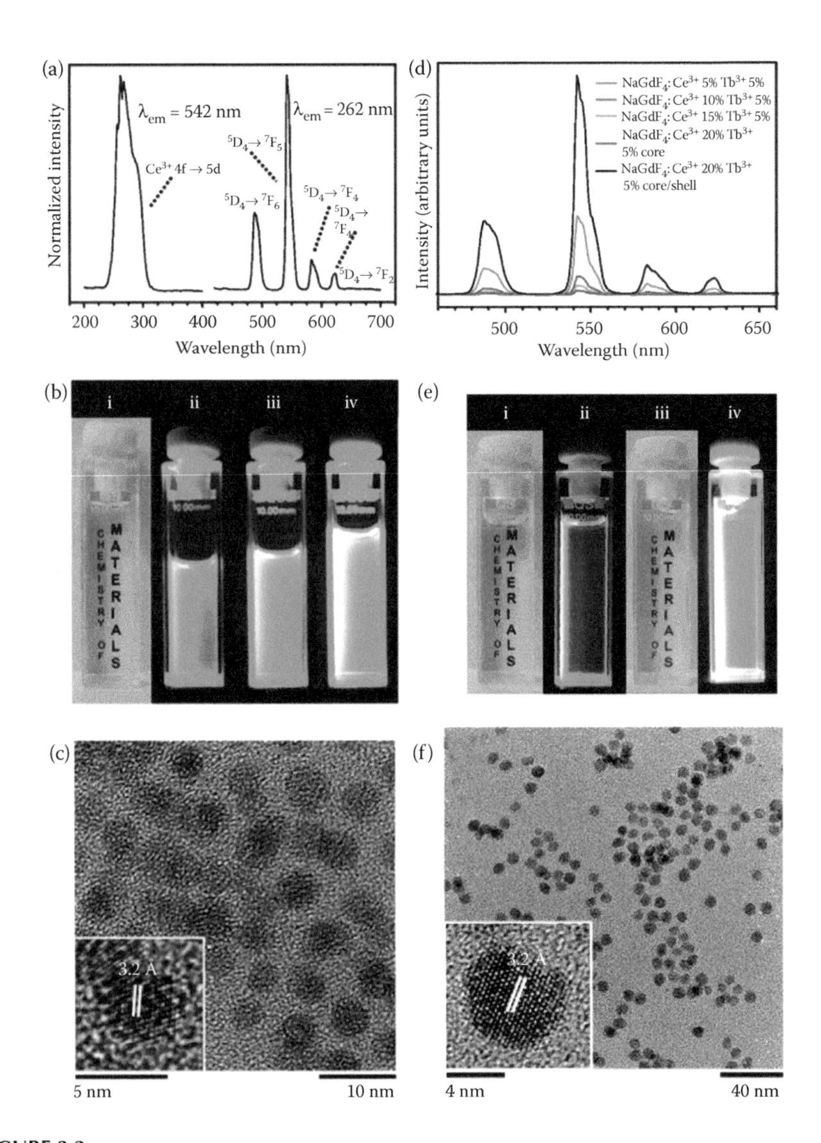

FIGURE 3.2
(a) Excitation and emission spectra of $NaGdF_4$: 15% Ce^{3+}, 5% Tb^{3+} nanoparticles (1 wt% solution in hexane). (b) 1 wt% solution (hexane) of $NaGdF_4$: 10% Ce^{3+}, 5% Tb^{3+} nanoparticles under (i) ambient and (ii) 254 nm UV light. 1 wt% solutions of (iii) $NaGdF_4$: 15% Ce^{3+}, 5% Tb^{3+} and (iv) $NaGdF_4$: 20% Ce^{3+}, 5% Tb^{3+} nanoparticles (hexane) under 254 nm UV light. (c) TEM image of $NaGdF_4$: 20% Ce^{3+}, 5% Tb^{3+}. (d) Emission spectra of 1 wt% solutions (hexane) of $NaGdF_4$:Ce^{3+}, Tb^{3+} core and $NaGdF_4$:Ce^{3+} 20%, Tb^{3+} 5%/$NaYF_4$ core/shell NPs. (e) 1 wt% solution (hexane) of aged $NaGdF_4$: 20% Ce^{3+}, 5%Tb^{3+} nanoparticles under (i) ambient and (ii) 254 nm UV light and $NaGdF_4$: 20% Ce^{3+}, 5% Tb^{3+}/$NaYF_4$ nanoparticles under (iii) ambient and (iv) 254 nm UV light. (f) TEM image of $NaGdF_4$: 20% Ce^{3+}, 5% Tb^{3+}/$NaYF_4$ sample. (Reprinted with permission from Boyer, J. C., J. Gagnon, L. A. Cuccia, and J. A. Capobianco. 2007. Synthesis, characterization, and spectroscopy of $NaGdF_4$: Ce^{3+}, Tb^{3+}/$NaYF_4$ core/shell nanoparticles. *Chem. Mater.* 19 (14):3358–3360. Copyright 2007 American Chemical Society.)

3.3.2.2 Oxides

Some examples are reported in the literature describing the preparation of sesquioxides using a thermolysis method. Well-dispersed Y_2O_3 nanocrystals, with a size less than 10 nm, and self-assembled nanodisks have been synthesized by a simple organometallic route using OM (H. Wang et al. 2005), in which TOPO was added to improve the crystallinity.

Furthermore, monodisperse cubic sesquioxides (from La to Lu and Y) nanomaterials (in the form of ultrathin nanoplates and nanodisks) have been synthesized via a nonhydrolytic approach in OA/OM/ODE (Si et al. 2007). Several lanthanide complexes, such as acetylacetonate, benzoylacetonate, and acetate, have been considered as precursors. The transformation from the complex precursors to the sesquioxides was proposed to occur in two stages: first, the formation of rare earth (RE) oleates by ligand exchange in solution and second, the subsequent decomposition of the oleates into sesquioxides catalyzed by OM.

Gd_2O_3 nanoplates were synthesized by solution-phase decomposition of gadolinium-acetate precursors in the presence of both coordinating and noncoordinating solvents, such as OM, OA, and ODE (Cao 2004).

3.3.2.3 Oxyfluorides

Monodispersed RE oxyfluoride nanocrystals with diverse shapes (cubic RE oxyfluoride nanopolyhedra and nanorods) have been prepared from single-source precursors of metal acetates through controlled fluorination in OA/OM/ODE (Sun et al. 2007). To selectively obtain RE oxyfluoride nanocrystals, the fluorination of the RE–O bond to the RE–F bond at the nucleation stage was controlled by finely tuning the ratio of OA/ODE or OA/OM, and the reaction temperature. Tuning of their shape has been realized by further modifying the reaction conditions. Monodispersed cubic phase LaOF and GdOF nanostructures were prepared by decomposing the lanthanide trifluoroacetate precursors in OA and OM (Du et al. 2008). Nanocrystals from 2 to 7 nm size and various shapes (nanopolyhedra and elongated nanocrystals) have been obtained.

3.3.3 Solvo(hydro)thermal

The hydro-solvothermal strategy requires relatively low temperatures (usually <250°C) taking advantage of experimental setups involving an autoclave reactor. Despite the mild conditions, the method allows one to obtain nanoparticles with a good level of crystallinity, with good control over dimension and morphology by fine tuning the experimental temperature and reaction time, the nature of the solvent and the surfactant and their molar ratios. As an important advantage with respect to the thermolysis method, it produces a lower amount of toxic by-product. Organic solvents as alcohols or amines are used in the presence of surfactant additives such as OA, EDTA, CTAB, or PEI: nonetheless, water is frequently used as well.

3.3.3.1 Fluorides

One of the first examples of production of upconverting $NaYF_4$ nanoparticles by the hydrothermal method is reported by Zeng et al. (2005). The solutions containing the metal precursors were transferred to a Teflon-lined autoclave and heated to 140–200°C for 12–24 h. The author succeeded in controlling the size and morphology of the products by the addition of EDTA and CTAB. $NaYF_4$ nanoparticles were also prepared by this method by several other authors, for instance, by Zhao et al. (2008) who prepared Yb^{3+} and Er^{3+} codoped cubic and hexagonal-phase $NaYF_4$ nanoparticles in different shapes, using a citrate–yttrium-nitrate complex as the precursor and a treatment in autoclave at different temperatures and reaction times. Citric acid was also used as a capping agent by Jiang et al. (2012), that prepared Tm^{3+}, Yb^{3+} codoped α-$NaYF_4$ nanocrystals by a one-step hydrothermal method. The morphology of the nanomaterials maintains a spherical shape when the surfactant amount, hydrothermal time, and hydrothermal temperature were varied. Quite interestingly, under 980 nm excitation, intense UV and blue UC emissions were observed from the α-$NaYF_4$ nanoparticles.

Hao and coworkers synthesized via a hydrothermal method (ethanol–water, OA 170°C, 24 h) fluorescent and magnetic (MRI) $NaLuF_4$:Ln (Ln = Gd^{3+}, Yb^{3+}, Tm^{3+}) nanocrystals with efficient NIR-to-NIR emission, useful for *in vivo* imaging applications. Interestingly, they were able to tune the crystal phase, size, UC properties, and magnetization varying the Gd^{3+} doping degree (Zeng et al. 2012).

One approach proposed as a general strategy to develop nanomaterials, exploits a phase transfer and separation mechanism occurring at the interfaces of the liquid, solid, and solution (LSS) phases present during the synthesis (X. Wang et al. 2005). In the case of lanthanide-doped luminescent nanocrystals with UC emission properties, the reaction between NaF and Ln acetate salts at 180°C was used to obtain approximately round-shaped nanoparticles of $NaYF_4$, YbF_3, LaF_3 (4–12 nm), or oval ones of YF_3 (100 × 500 nm). This strategy was subsequently implemented by Liu and coworkers (F. Wang et al. 2010), that showed how $NaYF_4$ nanocrystals can be rationally tuned in size, phase (cubic/hexagonal), and emission color using trivalent lanthanide dopant ions at controlled concentrations. The reaction time (~2 h) and phase transition occurred at quite moderate reaction temperature (~230°C). The LSS method was also applied to the synthesis of ultra-small (~5 nm) SrF_2 nanocrystals using trivalent lanthanide ions (Ln^{3+}) as doping agents at a concentration up to 40% (mol/mol) (D. Chen et al. 2010). Moreover, with the same wet chemical LSS technique, upconverting monodispersed Yb^{3+}/Er^{3+}-doped CaF_2 nanoparticles have been prepared (G. Wang et al. 2009). The obtained nanomaterials not only can be transparently dispersed in cyclohexane but also can be converted into water-soluble ones by oxidizing OA ligands with the Lemieux-von Rudloff reagent. Very interestingly, upon 980 nm laser excitation, the colloidal dispersion in cyclohexane and water showed bright green UC luminescence (UCL), even slightly stronger that of Yb^{3+}/Er^{3+}-doped $NaYF_4$ nanocrystals.

A facile method for the synthesis of polyethylene glycol (PEG) capped, water dispersible lanthanide-doped UC GdF_3 NPs using a hydrothermal technique was adopted by Passuello et al. (2012). From the investigation, it was found that the layer of PEG coating of the GdF_3 NPs guaranteed a good dispersion of the nanostructure in water and increased the UC emission by decreasing the multiphonon relaxation of the excited states of the Er^{3+} and Tm^{3+} ions due to water phonons.

Er^{3+}/Yb^{3+}, Ho^{3+}/Yb^{3+}, and Tm^{3+}/Yb^{3+} codoped CaF_2 of cubic shape NPs have been prepared by a one-step hydrothermal technique by Pedroni et al. (2011) in water using sodium oleate as surfactant. The obtained NPs were easily dispersed in organic solvents as well as in oleate aqueous solutions, without the need for any postsynthesis reaction. Notably, the Ho^{3+}/Yb^{3+} and Tm^{3+}/Yb^{3+}-doped samples show strong UC emission in the 750–800 nm region upon 980 diode laser excitation, a useful range for biomedical applications. Using the same hydrothermal technique, Tm^{3+}/Yb^{3+}-doped CaF_2 and SrF_2 were directly obtained using citrate anions as capping agents (Pedroni et al. 2013). Colloidal water dispersions of the doped SrF_2 NPs showed a UC emission at 800 nm (due to Tm^{3+} ions) of about two orders of magnitude higher than similarly doped cubic phase $NaYF_4$ NPs prepared with the same hydrothermal technique. It was found that alkali ions (Na^+ or K^+), present as counter cations of the citrate salts used as precursors, can be incorporated in the fluoride host crystals as charge compensators and they have a strong influence of the spectroscopic properties of the lanthanide ions.

A shape-controlled synthesis of monodispersed Yb^{3+} and Er^{3+}-doped $CaGd_3F_{11}$ nanoparticles using a solvothermal method was recently reported by Tian et al. (2014). The morphology of the nanostructures can be tuned to spherical (<10 nm) and one-dimensional nanorods by varying the amounts of the solvents and capping agent in the starting solution. UC emission was found to depend from the shapes of the obtained $CaGd_3F_{11}$ nanoparticles and nanorods, in particular, the nanorods showed a brighter emission with respect to the nanoparticles.

3.3.3.2 Oxides

Hydrothermal synthesis (water, pH = 13, 24 h at 180°C), followed by calcination (800°C) was also used by Qu and coworkers (Z. Liu et al. 2013) to form sesquioxides nanoparticles. In particular, they reported about the synthesis of multimodal PEGylated Gd_2O_3:Yb^{3+}, Er^{3+} nanorods (PEG-UCNPs) for *in vivo* UCL, T_1-enhanced magnetic resonance, and x-ray computed tomography imaging. PEGylation was introduced using a trialkoxysilane-PEG, conferring long blood circulation time, stability *in vivo* and noncytotoxic character, as indicated by small-animal experiments. With two different ligands—OA and aminohexanoic acid (AA)—Li and coworkers (Cao et al. 2011) succeeded in the hydrothermal synthesis of high quality water-soluble UC nanocrystals bearing appropriate functional groups using a one-step synthetic strategy. The OA/AA molar ratio allowed to optimize water dispersibility and provided amino groups for conjugation to folic acid (FA) for targeted bioimaging (Figure 3.3).

An interesting approach for the synthesis of cubic Er_2O_3 nanostructures with high yield and controlled size and shape has been developed via a solvothermal reaction of erbium nitrate in water/ethanol/decanoic acid media (Nguyen et al. 2010). The cubic Er_2O_3 phase was obtained at a temperature lower than 200°C and by tuning experimental parameters (such as the reaction temperature, the concentration of decanoic acid, and erbium precursor), different sizes and a variety of sheaves and brooms can be obtained. Furthermore, a change of the solvent (anhydrous ethanol instead of water/ethanol) had a strong influence on the particle size. Interestingly, at high precursor concentrations, nanorods were formed due to anisotropic

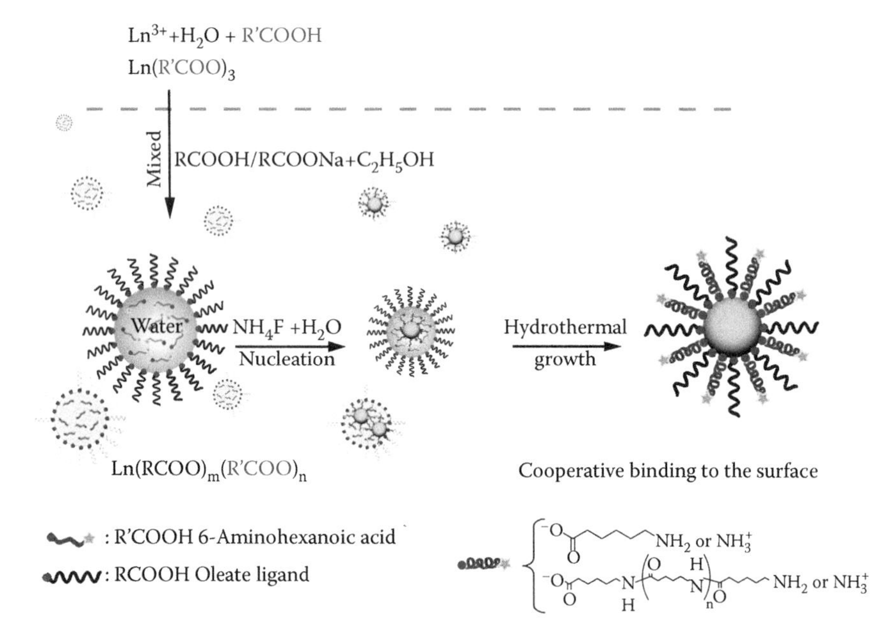

FIGURE 3.3
Scheme of hydrothermal reaction for preparing amino-functionalized UCNPs assisted with binary cooperative ligands: hydrophilic 6-AA and hydrophobic oleate. (Reprinted from *Biomaterials*, 32 (11), Cao, T. Y., Y. Yang, Y. A. Gao, J. Zhou, Z. Q. Li, and F. Y. Li, High-quality water-soluble and surface-functionalized upconversion nanocrystals as luminescent probes for bioimaging, 2959–2968, Copyright 2011, with permission from Elsevier.)

growth. It was found that the UCL properties depend on the particle size of the products.

3.3.4 Sol–Gel

The sol–gel technique has been considered by some groups in the past years to prepare oxide-based UC nanomaterials and it is characterized by hydrolysis and polycondensation of metal alkoxide (or halide)-based precursors. On the other hand, in order to improve the crystallinity of the nanosized materials, a further heat treatment at relatively high temperatures is often carried out. Due to this treatment, the obtained nanoparticles present considerable aggregation, and well-dispersed water solutions are therefore hard to prepare, making their application in the biomedical field difficult. Nonetheless, these NPs can be considered for other applications not requiring very homogeneous dispersions.

An interesting upconverting material is lanthanide-doped ZrO_2. De la Rosa et al. prepared Er^{3+}-doped zirconium oxide using the sol–gel process (De la Rosa-Cruz et al. 2003), followed by an annealing treatment at 1000°C for 10 h. It was found that the crystallite sizes presented dependence from the Er^{3+} concentration in the host, ranging from 28 to 46 nm. UC in the green and red regions were observed with NIR excitation and the intensity of the UC emission bands varied with the Er^{3+}concentration in the host. Moreover, Prasad et al. investigated an interesting modification of the sol–gel technique to generate Er^{3+}-doped ZrO_2 nanoparticles, considering a sol–emulsion–gel technique that used reverse micelles formed in emulsions as reactors for the growth of the nanocrystals (Patra et al. 2002). The authors also studied the effects of the Er^{3+} concentration and different codopants (e.g., Yb^{3+} and Y^{3+}) in the ZrO_2 host on the UC emission. Green and red UC emission at 550 and 670 nm were observed from these oxide nanocrystals upon excitation at 980 nm, the total UC in the green and red regions decreases with increasing concentration of Er^{3+} ions in the host, while it increased with the presence of Y^{3+} and Yb^{3+} ions.

Other binary oxides as TiO_2 have revealed to be interesting as hosts for UC emission. Luo et al. (2011) prepared Er^{3+}-doped anatase TiO_2 nanoparticles via a sol–gel solvothermal method. From emission and excitation spectra as a function on the temperature in the 10–300 K range, a crystal-field (CF) analysis for the Er^{3+} ions assuming a C_{2v} site symmetry revealed a relatively large CF strength (549 cm^{-1}). The UC intensity in Yb^{3+}, Er^{3+} codoped nanoparticles was about five times higher than for Er^{3+} singly doped counterparts, due to efficient Yb^{3+} sensitization and energy transfer UC (ETU).

Patra et al. (2003b) investigated the effects of the Er^{3+} concentration, crystal size and phase, and different processing temperatures on the UC emission for Er^{3+}-doped $BaTiO_3$ and TiO_2 nanocrystals, using a sol–emulsion–gel technique. Using the same experimental setup conditions, and the same Er^{3+} concentration, the observed UC intensity was higher for $BaTiO_3$ than for the TiO_2 host. For the TiO_2 host, the highest UC intensity was observed for samples heat treated at 800°C, where both the anatase and rutile phases were present. From an analysis of UC spectra and power studies, it was confirmed that UC emission was produced by excited-state absorption (ESA) processes.

The preparation of Ho^{3+}, Yb^{3+}-doped titanate nanotubes was carried out by Pedroni et al. (2012) via a two-step procedure: a first sol–gel process to produce the lanthanide-doped titania and a second hydrothermal treatment in alkaline conditions on the obtained doped titania powders to form the titanate nanotubes. A treatment at various temperatures was carried out with the aim of determining the different structural and optical properties of the nanotubes. It was found that on increasing the heat treatment the nanotube UC was stronger, due to reduction of hydroxyl groups and water on the surface of the nanotubes, resulting in changes in the interlayer distances.

Er^{3+} and Yb^{3+}, Er^{3+}-doped Y_2O_3 core–shell particles were synthesized by Dorman et al. (2012) with a two-step process, where the cores were prepared by a molten salt technique and the shell was deposited with a sol–gel process. The authors succeeded inpreparing cores with sizes of 100–150 nm, and shell layers up to 12 nm thick, tunable by controlling the mass ratio between the lanthanide chlorides used as precursors and the core (Er^{3+}-doped Y_2O_3) nanoparticles. A Y_2O_3 shell layer with optimal thickness of 8 nm induced a 53% increase in luminescence lifetimes and visible separation in Stark splitting. Optically active-shell layers, as Yb_2O_3 and Yb^{3+}-doped yttria, also facilitated energy transfer between the lanthanide ions, and it was found that Yb_2O_3 is an interesting host for the shell and it produce an increased lifetime and low pump power needed for UC. Lu et al. (2008) investigated Tm^{3+}, Yb^{3+}-doped Y_2O_3 NPs synthesized with the Pechini sol–gel technique, and coated with SiO_2 or TiO_2 shells using the Stober method. Larger NPs have stronger UC emission than smaller NPs and the core–shell structures are useful to enhance the UCL.

Guo et al. (2004) have investigated the UC properties of other sesquioxides, such as Tm^{3+}, Er^{3+}, and Yb^{3+}-doped cubic phase Gd_2O_3 nanoparticles, prepared by the sol–gel technique. UC emission upon 980 nm laser excitation has been observed in the blue, green, and red regions with a relevant increase of the red emission with respect to the green one.

Nanocrystalline lanthanide-doped $Lu_3Ga_5O_{12}$ garnets have been prepared using a sol–gel technique and subsequent heat treatment at 900°C for 16 h in air, by Venkatramu et al. (2010). The aggregated nanomaterials showed an average particle size of 40 nm. These materials show higher luminescence intensities compared to that found for similarly doped sesquioxides (e.g., Y_2O_3) but also for other nanocrystalline garnets, as $Gd_3Ga_5O_{12}$ and $Y_3Al_5O_{12}$.

An interesting investigation on films composed by nanocrystalline lanthanide-doped YVO_4 nanoparticles was carried out by Yu et al. The films were prepared with a combined use of a Pechini sol–gel process and soft lithography (Yu et al. 2002). From x-ray diffraction analysis, it was found that the films began to crystallize at 400°C and the crystallinity was found to increase on increasing of the annealing temperatures. Starting from nonpatterned phosphor films, mainly consisting of grains with an average size of 90 nm, crystalline films of different thicknesses were obtained. Quite interestingly, the Sm^{3+}-doped YVO_4 films also showed UCL upon laser excitation at 940 nm. In particular, anti-Stokes emissions from the $^4G_{5/2}$ excited state to the lower lying $^6H_{5/2}$, $^6H_{7/2}$, and $^6H_{9/2}$ states have been observed. As the $^6F_{11/2}$ level of Sm^{3+} has a very short lifetime, ESA transitions starting from the $^6F_{11/2}$ energy level have a low probability and the authors proposed that the most probable UC mechanisms could be an energy transfer between Sm^{3+} ions.

3.3.5 Combustion

Combustion synthesis includes controlled explosion reactions and one important advantage of this technique is that the reaction products can be generated in few minutes. Usually, these reactions involve highly exothermic processes that are started by a heat source, to reach temperatures up to 3000°C in the form of a self-sustained combustion propagating through the materials. This technique is most usually considered to synthesize oxide-based upconverting nanoparticles.

Upconverting lanthanide-doped sesquioxides (e.g., Y_2O_3, Lu_2O_3, and Gd_2O_3) prepared by the combustion (or propellant) technique have been investigated by Capobianco et al. in several papers published in the last decade. The first paper on upconverting Er^{3+}-doped nanocrystalline Y_2O_3 appeared in the literature in the year 2000, describing the UC emission of the Er^{3+} ions under laser excitation at 815 nm (Capobianco et al. 2000). The sample was prepared by combustion synthesis, using glycine as a fuel. A further heat treatment at 500°C was needed to decompose the residual nitrate ions. The decay times for the Er^{3+} excited levels obtained for the nanocrystalline sample were found to be in general significantly longer than those observed for the bulk counterpart, due to increased multiphonon relaxation caused by CO_2 and water absorbed onto the surface of the nanosized sample. The same preparation technique was also used to prepare Ho^{3+}-doped Y_2O_3 nanopowders, in order to compare the UC properties with those of the bulk counterpart.

An interesting analysis about the morphological structure of lanthanide-doped Lu_2O_3 powders obtained by propellant synthesis has been carried out by Polizzi et al. (2004). The samples showed a very porous, open morphology with fractal scaling properties. The building blocks of the fractal aggregates are lanthanide-doped cubic Lu_2O_3 crystalline particles with 60–90 nm of average size, which exhibit changes in the lattice parameter proportional to the lanthanide ionic radius. A similar morphological structure was also found for nanocrystalline Y_2O_3 prepared with the same combustion method (Polizzi et al. 2001).

The spectroscopic properties of Er^{3+}, Yb^{3+}-doped Gd_2O_3 nanoparticles prepared by combustion synthesis have been investigated by Singh et al. (2008, 2009), the samples were prepared using urea as the fuel. The solution containing the precursors was heated at 60°C to evaporate water, becoming a gel and this gel was transferred to a furnace and maintained at 500°C until the auto-ignition started and a voluminous structure formed. Changes in the color and the intensity of the UC emission were observed and attributed to the monoclinic to cubic structural transformation in the Gd_2O_3 nanoparticles due to a postsynthesis heat treatment of the sample at 600°C and 900°C. The upconverting Gd_2O_3 nanopowders were found to be also useful for optical thermometry, by considering the Er^{3+} UC emission in the green region from the two thermally coupled excited states $^2H_{11/2}$ and $^4S_{3/2}$ of Er^{3+}, centered at wavelengths of 523 and 548 nm. In the 300–900 K temperature range, the maximum sensitivity derived from the fluorescence intensity ratio technique of the green UC emission is approximately 0.0039 K^{-1}.

Garnets hosts were also considered as hosts for upconverter nanomaterials. Capobianco et al. investigated Er^{3+}-doped and Tm^{3+} and Yb^{3+} codoped $Gd_3Ga_5O_{12}$ (GGG) prepared by the combustion synthesis using glycine as the fuel. This garnet host resulted to be much less prone to incorporate water and CO_2 on the particle surface with respect to sesquioxides, with great benefit to the lanthanide emission properties, increased by a less pronounced nonradiative decay due to water phonons. NIR to visible UC of the nanocrystalline Er^{3+}-doped $Gd_3Ga_5O_{12}$ was studied following excitation of the $^4I_{9/2}$ exited state upon 800 nm laser excitation (Vetrone et al. 2003). It was found that if the Er^{3+} doping is low (around 1%) ESA was the predominant mechanism responsible for populating the upper emitting states. However, as the Er^{3+} concentration was increased to 5%, the decay times for the UC emissions were lengthened and deviated from exponentiality, suggesting the presence of ETU. The 1% each Tm^{3+} and Yb^{3+} codoped $Gd_3Ga_5O_{12}$ nanostructured sample showed strong UC emission in the UV ($^1D_2 \rightarrow {}^3H_6$), blue ($^1D_2 \rightarrow {}^3F_4$), blue-green ($^1G_4 \rightarrow {}^3H_6$), red ($^1G_4 \rightarrow {}^3F_4$), and NIR ($^1G_4 \rightarrow {}^3H_5/{}^3H_4 \rightarrow {}^3H_6$) regions upon excitation of the Yb^{3+} ions with a 980 nm laser radiation (Pandozzi et al. 2005). Due to subsequent energy transfers from the Yb^{3+} ion to the Tm^{3+} ions (energy transfer efficiency about 0.576).

3.3.6 Ionic Liquids

Ionic liquids (ILs) are considered nowadays interesting materials for preparation of inorganic materials as a "green" alternative to the conventional solvents (Lorbeer et al. 2010, 2011). They have unique properties, as thermal and chemical stability, a wide electrochemical window. Moreover, ILs can act as capping agents or surfactants in the inorganic synthesis. Although ILs are very useful for nanoparticles preparation, a certain amount of agglomeration of the prepared nanoparticles is one of the main drawbacks of the technique.

In the past few years, some papers have appeared in the literature to prepare upconverting nanomaterials using the ILs technique, succeeding also to generate various nanocrystal sizes and morphologies. Eu^{3+}-doped GdF_3 nanoparticles have been prepared by the ILs technique by Lorbeer et al. (2010) with a microwave reaction starting from the lanthanide acetates. Fast and efficient synthesis of small, uniform, oxygen-free lanthanide nanofluorides with excellent luminescence has been achieved and a quantum efficiency of up to 145% was determined. Moreover, the same group succeeded in preparing pure, oxygen-free hexagonal EuF_3 nanoparticles by reacting europium acetate hydrate with PF_6 and BF_4 ILs at 120°C using microwave synthesis, with reaction times as short as 30 s (Lorbeer et al. 2011). Extremely small particles (<15 nm) were obtained and the morphology varied from nearly spherical and cuboid to rodlike crystallites forming larger clusters. Very interestingly, the size and shape varied in different ILs. Lanthanide-doped YF_3 nanoparticles were prepared by Nuñez and coworkers using $BmimBF_4$ as the fluoride source (Nuñez and Ocaña 2007). In most cases, highly uniform NPs were obtained and their size could be varied in the nanometer range by adjusting the nature and concentration of the starting lanthanide precursors. Zhang et al. (2008) have synthesized a series of lanthanide fluoride nanocrystals in three ILs (i.e., $OmimPF_6$, $OmimBF_4$, and $BmimPF_6$), utilizing the partial hydrolysis of PF_6^- and BF_4^- to introduce a fluoride source. Lanthanum fluoride nanocrystals can be obtained in a large amount (products up to 0.15 g per 10 mL solvents) with the ILs technique. Interestingly, in these "all-in-one" systems, the ILs acted as solvents, reaction agents, and templates. Regarding different upconverting fluorides, water-soluble and pure hexagonal-phase Yb^{3+} or Er^{3+} and Tm^{3+}-doped $NaYF_4$ nanoparticles were successfully obtained by Liu et al. (2009) with use of a IL, $BmimBF_4$, which acts as solvent, template, and also fluorine source. One interesting advantage of the obtained nanocrystals is that the ILs overlayer on their surface renders them directly dispersed in water.

Spherical $NaYF_4$ nanoclusters have been synthesized using an IL (1-butyl-3-methylimidazolium tetrafluoroborate) based technique using a microwave reaction system. The nanoclusters have diameters ranging from 200 to 430 nm and are formed by the self-assembly of smaller $NaYF_4$ nanoparticles. Quite interestingly, the size of the nanoclusters could be easily controlled by variations of the precursor amounts. It was demonstrated that the ILs have key roles as solvents for the reaction, absorbents of microwave radiation, and the main fluorine sources for the $NaYF_4$ generation. The obtained nanoclusters exhibit excellent UC properties.

An interesting approach combining more synthetic techniques was introduced by He et al. (2011) to prepare lanthanide-doped upconverting $NaGdF_4$ nanocrystals with various crystalline structures. This approach used an OA/IL two-phase system that combined the advantages of the thermal decomposition and ILs techniques, exploiting the two-phase approach in the OA- and IL-phase through a one-step controllable reaction.

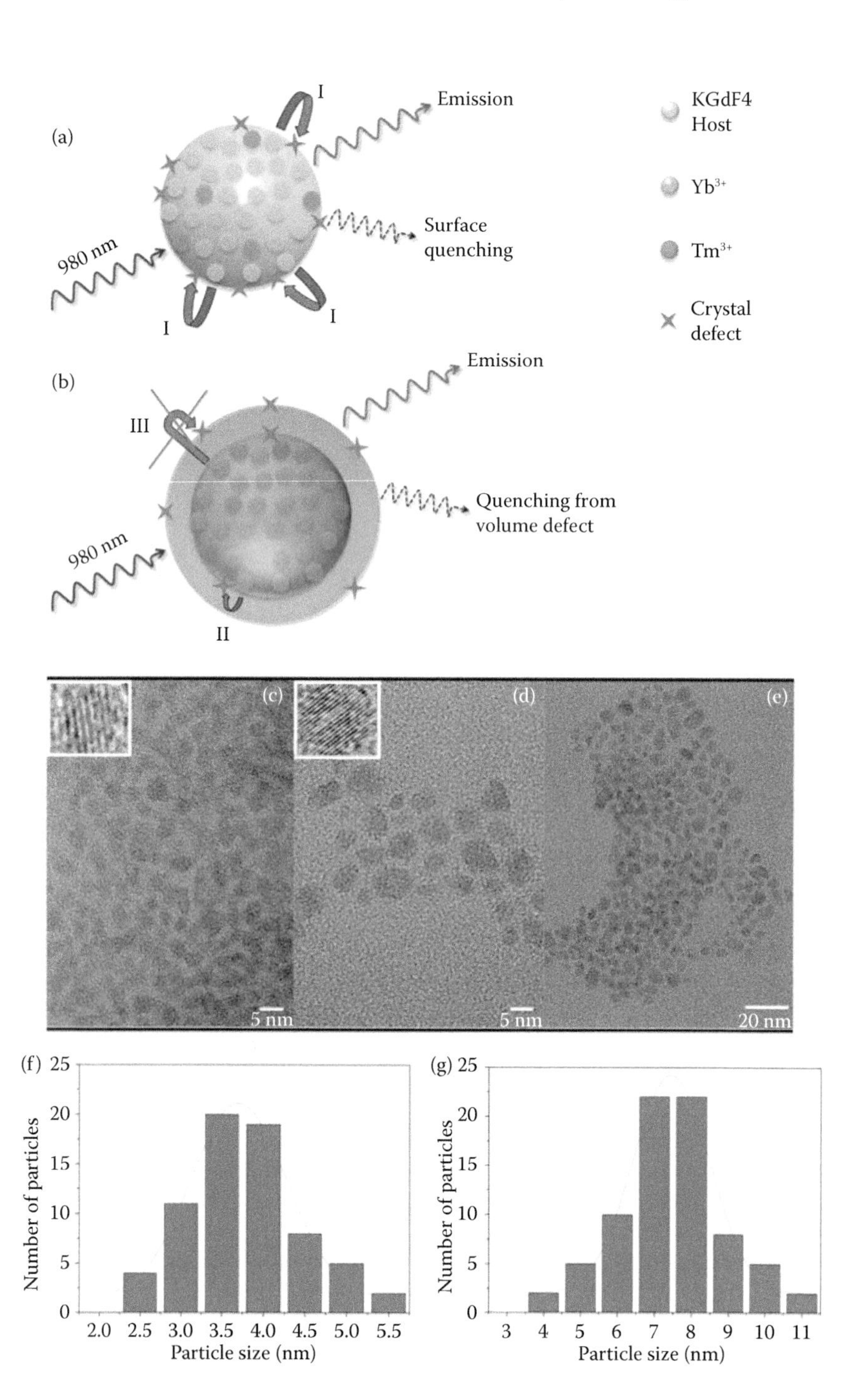

(a)
I
Emission
KGdF4
Host
Yb3+
Tm3+
Crystal
defect
Surface
quenching
980 nm
I
I
(b)
Emission
III
Quenching from
volume defect
980 nm
II
(c)
(d)
(e)
5 nm
5 nm
20 nm
(f)
Number of particles
Particle size (nm)
(g)
Number of particles
Particle size (nm)

3.4 Core–Shell Architectures

"Core-only" lanthanide-doped nanoparticles, prepared with different synthetic approaches, present surface defects that are detrimental for UC emission. Moreover, lanthanide ions on the surface of nanoparticles which have been dispersed in solvents, experience an increased nonradiative multiphonon relaxation (due to the solvent phonons) with respect to those ions located in the interior of the nanoparticle. To improve the emission efficiency, a strategy is to cover the nanoparticle with an additional shell, usually with a second step procedure, using core-only nanoparticles as seeds. Quite recently, this general approach was applied by several research groups to develop NIR-to-NIR emitting nanocrystals, and can also be used to design core–shell (core@shell) nanocrystals that present fine tuning of UC emission—exploiting trapping of the migrating energy by the activators and the elimination of deleterious cross-relaxation phenomena (Wang et al. 2011). With the core@shell approach it is still possible to reach highly monodispersed nanoparticles, but some complications could arise from poorly uniform shell formation or in the purification steps, because of the self-nucleation of the shell precursors.

Capobianco et al. (Vetrone et al. 2009) synthesized hydrophobic $NaGdF_4$:Er^{3+}, Yb^{3+} active-core@$NaGdF_4$ inert-shell nanoparticles and $NaGdF_4$:Er^{3+}, Yb^{3+} active-core@$NaGdF_4$:Yb^{3+} active-shell nanoparticles (average particle size 16 nm) by a modified thermal decomposition synthesis, starting from trifluoroacetate precursors, ODE and OA. It was found that the active shell protects the active Er^{3+} ions from nonradiative decays, and efficiently transfers the incident NIR radiation to the active core. The same research group developed water dispersible ultra-small (<10 nm) multifunctional $KGdF_4$:Tm^{3+}, Yb^{3+} nanoparticles with NIR-to-NIR UC emission properties ($\lambda_{ex} = 980$ nm, $\lambda_{em} = 803$ nm), for which the luminescence efficiency was optimized with the core–shell approach. The $KGdF_4$:Tm^{3+}, Yb^{3+}@$KGdF_4$ core–shell nanoparticles were encapsulated with a PEG-phospholipid shell to obtain a water suspension (Figure 3.4) (Wong et al. 2011).

FIGURE 3.4
(a) Representation showing dopant ions in $KGdF_4$ host and the crystal defects for the core-only nanoparticles. (b) Schematization of the core–shell nanoparticles. (c) TEM image of the $KGdF_4$:Tm^{3+}, Yb^{3+} core-only nanoparticles. (d, e) TEM images of the $KGdF_4$:Tm^{3+}, Yb^{3+}@$KGdF_4$ core–shell nanoparticles showing an increase in the particles size. (f) TEM size distributions for the $KGdF_4$:Tm^{3+}, Yb^{3+} core-only nanoparticles and (g) the $KGdF_4$:Tm^{3+}, Yb^{3+}@$KGdF_4$ core–shell nanoparticles. (Wong, H.-T., F. Vetrone, R. Naccache, H. L. W. Chan, J. Hao, and J. A. Capobianco. 2011. Water dispersible ultra-small multifunctional $KGdF_4$:Tm^{3+}, Yb^{3+} nanoparticles with near-infrared to near-infrared upconversion. *J. Mater. Chem.* 21 (41):16589–16596. Reproduced by permission of The Royal Society of Chemistry.)

Small core@shell oleate-capped $NaGdF_4:Nd^{3+}@NaGdF_4$ nanocrystals (average diameter 15 nm) with efficient NIR-to-NIR downconversion PL ($\lambda_{ex} = 740$ nm, $\lambda_{em} = 850–900$ nm) were developed also by Prasad et al. for *in vitro* and *in vivo* imaging (Chen et al. 2012). They adapted a previously reported synthetic method bearing to hexagonal-phase core@shell $NaYF_4$:Yb, Tm@$NaYF_4$:Yb, Er nanocrystals containing Tm^{3+} and Er^{3+} ions in the core and in the shell, respectively. In this case, a $NaGdF_4$ shell covering a $NaGdF_4:Nd^{3+}$ core suppressed nonradiative recombination processes at the nanoparticle surface, enhancing the PL quantum yield up to 0.40.

Some of the synthetic limitations of these approaches can be by-passed using an epitaxial layer-by-layer growth of the nanocrystals, also called "self-focusing by Ostwald ripening." In this case, sacrificial nanoparticles are injected in the reaction mixture, and upon rapid dissolution (defocusing), they contribute to the formation of a uniform shell surrounding the $NaYF_4:Yb^{3+}@Er^{3+}$ core (self-focusing) (Johnson et al. 2012).

Recently, Zanzoni et al. developed a solvothermal two-step technique to prepare SrF_2 core–shell nanoparticles, in order to investigate the interaction between a protein (ubiquitin) and upconverting NPs. In this work, the $SrF_2:Yb^{3+}@SrF_2:Yb^{3+}$, Tm^{3+} core–shell architecture was adopted to ensure an efficient absorption of the NIR radiation by Yb^{3+} in both the core and shell, and to guarantee that a significant amount of emitting Tm^{3+} ions are located at the NPs surface, permitting a better investigation on the NPs–protein interaction (Zanzoni et al. 2016).

An implementation to the synthesis of lanthanide-doped NPs is based on cation exchange reactions—taking place at the nanoparticle surface—to develop multifunctional nanoprobes.

Following this scheme, a NIR-to-NIR emitting system was developed by Liu et al. (2011). The authors started from $NaYF_4:Yb^{3+}$, Er^{3+} oleate-capped nanoparticles obtained by solvothermal synthesis, to develop multifunctional upconversion nanoparticles ($NaYF_4:Yb^{3+}$, Er^{3+}), combining magnetic (Gd^{3+}), positron emission tomography (PET) (^{18}F), and targeted recognition (FA) properties (Figure 3.5). In particular, Gd^{3+} ions were introduced on the surface of the nanocrystals by cation exchange with Y^{3+} ions, while ^{18}F was introduced for PET imaging by interaction with the RE ions.

In a related example, Van Veggel et al. performed cation exchange with Gd^{3+} ions on upconverting $NaYF_4:Yb^{3+}$, Tm^{3+} nanoparticles (Dong et al. 2012). They started from oleate-stabilized $NaYF_4:Yb^{3+}$, Tm^{3+} nanoparticles (average diameter 19–20 nm) synthesized using a coprecipitation method in organic media at 300°C (Figure 3.6).

Ligand exchange of the as-prepared oleate-stabilized nanoparticles with PVP (Johnson et al. 2010) provided a water dispersible system that was exposed to Gd^{3+} to obtain the bimodal $NaYF_4:Yb^{3+}$, Tm@$NaGdF_4$ core@shell nanoparticles. The sub-nanometer $NaGdF_4$ shell (ca. 0.6 nm) conferred very high proton relaxivity to the nanoparticle for targeted MRI applications.

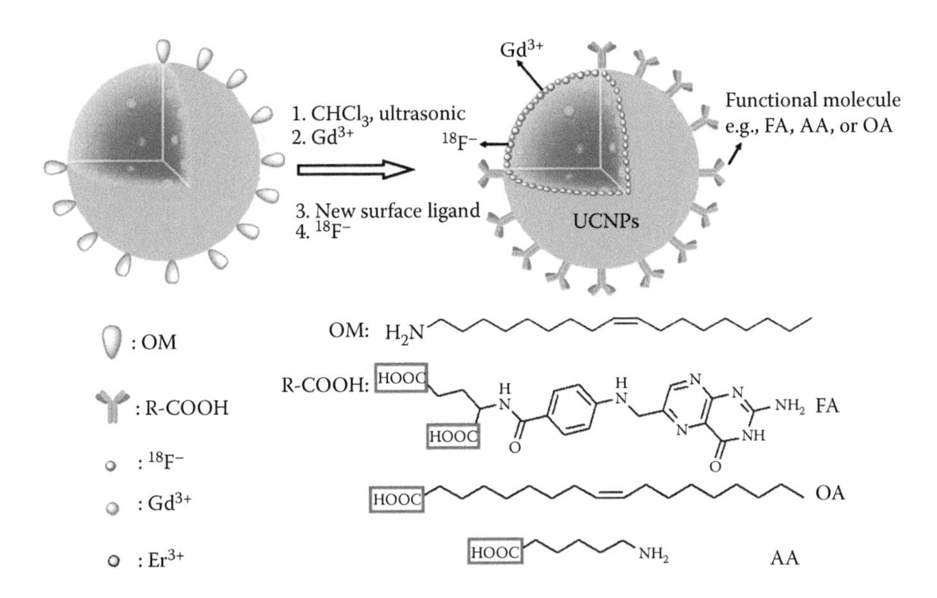

FIGURE 3.5
Schematic representation of ^{18}F-labeled magnetic-UC functional nanocrystals. OM: oleyl amine; FA: folic acid; OA: oleic acid; and AA: aminocaproic acid. (Reprinted with permission from Liu, Q., Y. Sun, C. Li, J. Zhou, C. Li, T. Yang, X. Zhang, T. Yi, D. Wu, and F. Li. 2011. F-18-labeled magnetic-upconversion nanophosphors via rare-earth cation-assisted ligand assembly. *ACS Nano* 5 (4):3146–3157. Copyright 2011 American Chemical Society.)

3.4.1 Silica-Shell Formation

Nanoparticles tailored for biological imaging applications need to provide stable suspension in water and buffered solutions, and often the possibility to introduce chemical functionalities for biomolecule targeting. Most of the "as-prepared" lanthanide-doped nanoparticles are capped with hydrophobic molecules such as oleate (Li et al. 2008) or OM (Chan et al. 2012), and for this reason, several approaches were developed to increase the polarity of the nanoparticle surface to obtain water dispersibility. An overview of these synthetic approaches has been recently reviewed in focused publications (Muhr et al. 2014; Sedlmeier and Gorris 2014). Silica encapsulation offers the possibility to endow lanthanide-doped nanocrystals with a surface material that is hydrophilic, transparent to radiation and photophysically inert, simple to functionalize, and intrinsically nontoxic (Arap et al. 2013; Bonacchi et al. 2011; Genovese et al. 2014). The shell usually does not affect the emission efficiency of the nanocrystals, and actually in some cases improves it. Beside this behavior, silica can be independently doped with other fluorophores or contrast agents (Qian et al. 2009), or drugs to develop multimodal imaging or theranostic tools: this possibility is fostered by mesoporous silica shells (Li et al. 2013; P. Yang et al. 2012), formed

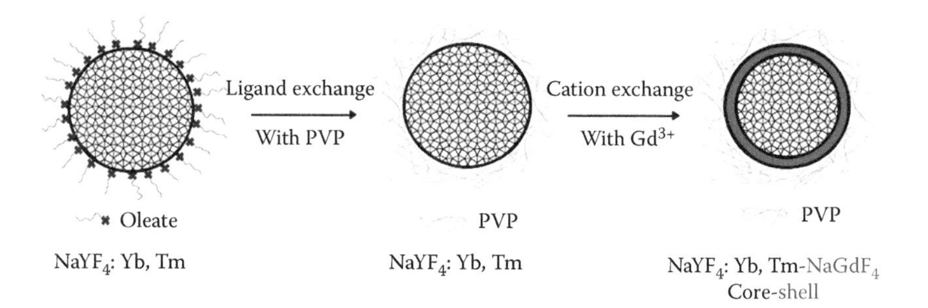

FIGURE 3.6
Schematization of the cation exchange process to form $NaYF_4$:Yb, Tm@$NaGdF_4$ core–shell nanoparticles. (Dong, C., A. Korinek, B. Blasiak, B. Tomanek, and F. C. J. M. van Veggel. 2012. Cation exchange: A facile method to make $NaYF_4$:Yb^{3+}, Tm^{3+}@$NaGdF_4$ core@shell nanoparticles with a thin, tunable, and uniform shell. *Chem. Mater.* 24 (7):1297–1305. Reproduced by permission of The Royal Society of Chemistry.)

adding to the synthetic mixture surfactant agents like CTAB (Liu et al. 2012; Qian et al. 2009).

A few synthetic approaches are available to cover lanthanides-doped nanosystems: nanoparticles presenting hydrophobic capping agents are frequently coated using a reverse microemulsion approach (water-in-oil) (Jalil and Zhang 2008). Silica coating involves an increase of the nanoparticles average diameter, a process that may cause some loss of monodispersity of the starting material: for this reason, most of the reports adopt this approach, starting from very monodisperse OA-capped nanocrystals and using the microemulsion method for silica shell formation, that guarantees nanocrystals core confinement within the reverse micelles of the microemulsion (Li et al. 2008). Shell thickness can be varied mainly acting on the amount of tetrathoxysilane introduced in the microemulsion and/or on the surfactant-to-water ratio, the main parameter affecting the number of reverse micelles within the microemulsion.

In a recent example, Li et al. (Zhu et al. 2014) developed a NIR-to-NIR emitting imaging (Figure 3.7).

Lanthanide-doped core was synthesized by a solvothermal method (average diameter 20 nm), and during the silica coating step ethoxysilane functionalized PEG (PEG–siloxane) was added into the microemulsion to achieve PEGylation (final hydrodynamic diameter ~65 nm).

Hydrophilic nanocrystals are usually coated in water–alcohol mixtures using a Stöber approach (Stöber et al. 1968). With this strategy, Wolfbeis et al. (Mader et al. 2010) faced the coating of lanthanide-doped nanoparticles and microparticles with average diameters spanning from 50 nm to 15 μm. They adopted a click chemistry-based strategy to functionalize the surface of the nano- and microparticles with a few "click reagents" and alkyne-modified fluorescent dyes.

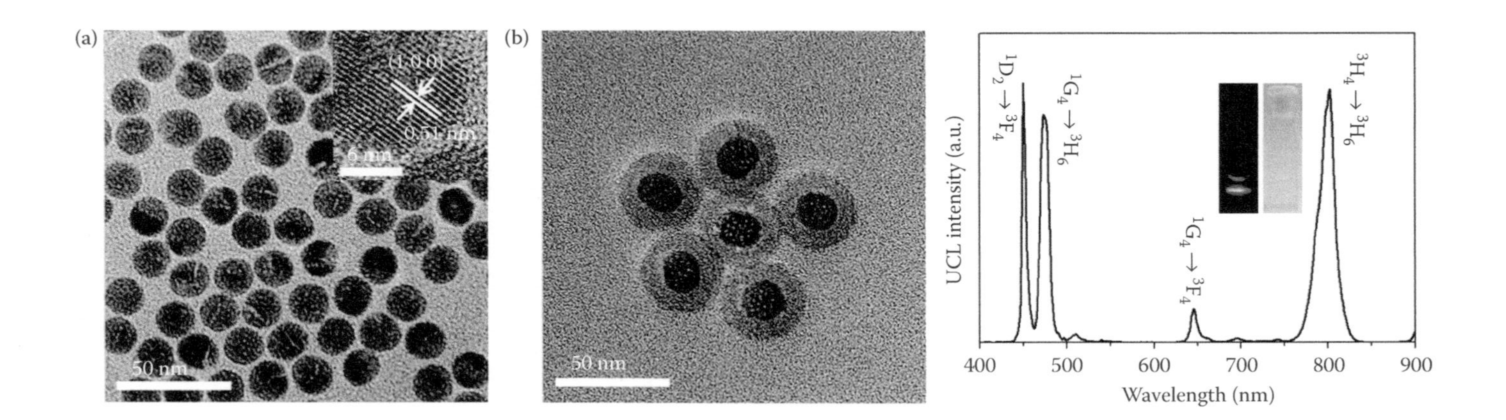

FIGURE 3.7
TEM images of (a) $NaYF_4$:Yb, Tm UCNPs and (b) $NaYF_4:Yb^{3+}$, $Tm^{3+}@SiO_2$ nanoparticles modified with PEG (PEG-UCNPs), and room temperature UCL spectrum of PEG-UCNPs in water (5 mg/mL, $\lambda_{ex} = 980$ nm). (Zhu, X., B. Da Silva, X. Zou, B. Shen, Y. Sun, W. Feng, and F. Li. 2014. Intra-arterial infusion of PEGylated upconversion nanophosphors to improve the initial uptake by tumors *in vivo*. *RSC Adv.* 4 (45):23580–23584. Reproduced by permission of The Royal Society of Chemistry.)

The silica modification quite often produces systems that are prone to aggregation processes and unspecific binding, and for this reason just the formation of an ultrathin silica shell of few nanometers (Li and Zhang 2006; F. Liu et al. 2013) or proper functionalization treatments (Bagwe et al. 2006) were developed.

Polymer-coated lanthanide-doped nanoparticles can also be coated by a silica shell, since some polymers such as PVP (Li and Zhang 2006) favor the condensation of a silica layer. This approach moves toward more direct coating protocols since PVP is often used as a stabilizing agents in many hydrothermal syntheses. Surface modification can be carried out during the growth of the silica shell or after its formation, and besides colloidal stabilization endow nanoparticles with moieties bearing $-NH_2$ (M. Wang et al. 2009), –COOH (F. Liu et al. 2013), –SH (Y. Yang et al. 2012), or functional groups for click chemistry (Mader et al. 2010).

3.5 Conclusions

Despite the great attention devoted to UCNPs presenting unique optical and chemical properties, their practical application is still limited. This is mainly due to a few demanding features including their nontrivial and, sometimes, poorly reproducible synthetic procedures. In this chapter, we have summarized the recent achievements in the development of more and more efficient preparation methods, with emphasis on the pros and cons of each strategy (Table 3.1). Presently, with a careful control and tuning of the reaction conditions, it is possible to obtain monodispersed and customized UCNPs with many different components. It has to be underlined, however, that the quantum yield of these materials is still generally not very high and it dramatically depends not only on the components but also on the architecture of the nanoparticle (local environment around the lanthanide ions and structure). Consequently, the research on new and optimized synthetic and surface modification methods is a fundamental step toward the application of these species. Great efforts are now devoted to investigate nondemanding and reproducible approaches for large-scale preparations of multifunctional, multi-shell UCNPs able to fulfill practical needs.

References

Abel, K. A., J.-C. Boyer, and F. C. J. M. van Veggel. 2009. Hard proof of the $NaYF_4$/$NaGdF_4$ nanocrystal core/shell structure. *J. Am. Chem. Soc.* 131 (41):14644–14645.

Arap, W., R. Pasqualini, M. Montalti, L. Petrizza, L. Prodi, E. Rampazzo, N. Zaccheroni, and S. Marchio. 2013. Luminescent silica nanoparticles for cancer diagnosis. *Curr. Med. Chem.* 20 (17):2195–2211.

Auzel, F. 2004. Upconversion and anti-Stokes processes with f and d ions in solids. *Chem. Rev.* 104 (1):139–173.

Bagwe, R. P., L. R. Hilliard, and W. Tan. 2006. Surface modification of silica nanoparticles to reduce aggregation and nonspecific binding. *Langmuir* 22 (9):4357–4362.

Bonacchi, S., D. Genovese, R. Juris, M. Montalti, L. Prodi, E. Rampazzo, and N. Zaccheroni. 2011. Luminescent silica nanoparticles: Extending the frontiers of brightness. *Angew. Chem. Int. Ed.* 50 (18):4056–4066.

Boyer, J. C., J. Gagnon, L. A. Cuccia, and J. A. Capobianco. 2007. Synthesis, characterization, and spectroscopy of $NaGdF_4$: Ce^{3+}, $Tb^{3+}/NaYF_4$ core/shell nanoparticles. *Chem. Mater.* 19 (14):3358–3360.

Boyer, J.-C., F. Vetrone, L. A. Cuccia, and J. A. Capobianco. 2006. Synthesis of colloidal upconverting $NaYF_4$ nanocrystals doped with Er^{3+}, Yb^{3+} and Tm^{3+}, Yb^{3+} via thermal decomposition of lanthanide trifluoroacetate precursors. *J. Am. Chem. Soc.* 128 (23):7444–7445.

Cao, T. Y., Y. Yang, Y. A. Gao, J. Zhou, Z. Q. Li, and F. Y. Li. 2011. High-quality water-soluble and surface-functionalized upconversion nanocrystals as luminescent probes for bioimaging. *Biomaterials* 32 (11):2959–2968.

Cao, Y. C. 2004. Synthesis of square gadolinium-oxide nanoplates. *J. Am. Chem. Soc.* 126 (24):7456–7457.

Capobianco, J. A., F. Vetrone, T. D'Alesio, G. Tessari, A. Speghini, and M. Bettinelli. 2000. Optical spectroscopy of nanocrystalline cubic Y_2O_3:Er^{3+} obtained by combustion synthesis. *Phys. Chem. Chem. Phys.* 2 (14):3203–3207.

Chan, E. M., G. Han, J. D. Goldberg, D. J. Gargas, A. D. Ostrowski, P. J. Schuck, B. C. Cohen, and D. J. Milliron. 2012. Combinatorial discovery of lanthanide-doped nanocrystals with spectrally pure upconverted emission. *Nano Lett.* 12:3839–3845.

Chen, D., Y. Yu, F. Huang, P. Huang, A. Yang, and Y. Wang. 2010. Modifying the size and shape of monodisperse bifunctional alkaline-earth fluoride nanocrystals through lanthanide doping. *J. Am. Chem. Soc.* 132 (29):9976–9978.

Chen, G., T. Y. Ohulchanskyy, R. Kumar, H. Agren, and P. N. Prasad. 2010. Ultrasmall monodisperse $NaYF_4$:Yb^{3+}/Tm^{3+} nanocrystals with enhanced near-infrared to near-infrared upconversion photoluminescence. *ACS Nano* 4 (6):3163–3168.

Chen, G., H. Qiu, P. N. Prasad, and X. Chen. 2014. Upconversion nanoparticles: Design, nanochemistry, and applications in theranostics. *Chem. Rev.* 114 (10):5161–5214.

Chen, G., C. Yang, and P. N. Prasad. 2013. Nanophotonics and nanochemistry: Controlling the excitation dynamics for frequency up- and down-conversion in lanthanide-doped nanoparticles. *Acc. Chem. Res.* 46 (7):1474–1486.

Chen, G. Y., T. Y. Ohulchanskyy, S. Liu, W. C. Law, F. Wu, M. T. Swihart, H. Agren, and P. N. Prasad. 2012. Core/shell $NaGdF_4$:$Nd^{3+}/NaGdF_4$ nanocrystals with efficient near-infrared to near-infrared downconversion photoluminescence for bioimaging applications. *ACS Nano* 6 (4):2969–2977.

Cichos, J., L. Marciniak, D. Hreniak, W. Strek, and M. Karbowiak. 2014. The effect of surface ligand, solvent and Yb^{3+} co-doping on the luminescence properties of Er^{3+} in colloidal $NaGdF_4$ nanocrystals. *J. Mater. Chem. C* 2 (39):8244–8251.

Cuccia, L. A. and J. A. Capobianco. 2007. Synthesis of colloidal upconverting $NaYF_4$: Er^{3+}/Yb^{3+} and Tm^{3+}/Yb^{3+} monodisperse nanocrystals. *Nano Lett.* 7 (3):847–852.

DaCosta, M. V., S. Doughan, Y. Han, and U. J. Krull. 2014. Lanthanide upconversion nanoparticles and applications in bioassays and bioimaging: A review. *Anal. Chim. Acta* 832:1–33.

Daldosso, M., D. Falcomer, A. Speghini, M. Bettinelli, S. Enzo, B. Lasio, and S. Polizzi. 2008. Synthesis, structural investigation and luminescence spectroscopy of nanocrystalline $Gd_3Ga_5O_{12}$ doped with lanthanide ions. *J. Alloy. Compd.* 451 (1–2):553–556.

De la Rosa-Cruz, E., L. A. Diaz-Torres, R. A. Rodriguez-Rojas, M. A. Meneses-Nava, O. Barbosa-Garcia, and P. Salas. 2003. Luminescence and visible upconversion in nanocrystalline ZrO_2:Er^{3+}. *Appl. Phys. Lett.* 83 (24):4903–4905.

Dong, C., A. Korinek, B. Blasiak, B. Tomanek, and F. C. J. M. van Veggel. 2012. Cation exchange: A facile method to make $NaYF_4$:Yb, Tm-$NaGdF_4$ core–shell nanoparticles with a thin, tunable, and uniform shell. *Chem. Mater.* 24 (7):1297–1305.

Dorman, J. A., J. H. Choi, G. Kuzmanich, and J. P. Chang. 2012. Elucidating the effects of a rare-earth oxide shell on the luminescence dynamics of Er^{3+}:Y_2O_3 nanoparticles. *J. Phys. Chem. C* 116 (18):10333–10340.

Du, Y.-P., X. Sun, Y.-W. Zhang, Z.-G. Yan, L.-D. Sun, and C.-H. Yan. 2009. Uniform alkaline earth fluoride nanocrystals with diverse shapes grown from thermolysis of metal trifluoroacetates in hot surfactant solutions. *Cryst. Growth Des.* 9 (4):2013–2019.

Du, Y.-P., Y.-W. Zhang, L.-D. Sun, and C.-H. Yan. 2008. Luminescent monodisperse nanocrystals of lanthanide oxyfluorides synthesized from trifluoroacetate precursors in high-boiling solvents. *J. Phys. Chem. C* 112 (2):405–415.

Gainer, C. F. and M. Romanowski. 2014. A review of synthetic methods for the production of upconverting lanthanide nanoparticles. *J. Innov. Opt. Heal. Sci.* 07 (02):1330007.

Genovese, D., E. Rampazzo, S. Bonacchi, M. Montalti, N. Zaccheroni, and L. Prodi. 2014. Energy transfer processes in dye-doped nanostructures yield cooperative and versatile fluorescent probes. *Nanoscale* 6 (6):3022–3036.

Guo, H., N. Dong, M. Yin, W. P. Zhang, L. R. Lou, and S. D. Xia. 2004. Visible upconversion in rare earth ion-doped Gd_2O_3 nanocrystals. *J. Phys. Chem. B* 108 (50):19205–19209.

Guo, L., Y. Wang, Y. Wang, J. Zhang, P. Dong, and W. Zeng. 2013. Structure, enhancement and white luminescence of multifunctional $Lu_6O_5F_8$:20%Yb^{3+}, 1%Er^{3+}(Tm^{3+}) nanoparticles via further doping with Li^+ under different excitation sources. *Nanoscale* 5 (6):2491–2504.

He, M., P. Huang, C. Zhang, H. Hu, C. Bao, G. Gao, R. He, and D. Cui. 2011. Dual phase-controlled synthesis of uniform lanthanide-doped $NaGdF_4$ upconversion nanocrystals via an OA/ionic liquid two-phase system for *in vivo* dual-modality imaging. *Adv. Funct. Mater.* 21 (23):4470–4477.

Heer, S., K. Kömpe, H. U. Gudel, and M. Haase. 2004. Highly efficient multicolour upconversion emission in transparent colloids of lanthanide-doped $NaYF_4$ nanocrystals. *Adv. Mater.* 16 (23–24):2102–2105.

Heer, S., O. Lehmann, M. Haase, and H. U. Gudel. 2003. Blue, green, and red upconversion emission from lanthanide-doped $LuPO_4$ and $YbPO_4$ nanocrystals in a transparent colloidal solution. *Angew. Chem. Int. Ed.* 42 (27):3179–3182.

Hemmer, E., N. Venkatachalam, H. Hyodo, A. Hattori, Y. Ebina, H. Kishimoto, and K. Soga. 2013. Upconverting and NIR emitting rare earth based nanostructures for NIR-bioimaging. *Nanoscale* 5 (23):11339–11361.

Ivaturi, A., S. K. W. MacDougall, R. Martín-Rodríguez, M. Quintanilla, J. Marques-Hueso, K. W. Krämer, A. Meijerink, and B. S. Richards. 2013. Optimizing infrared to near infrared upconversion quantum yield of β-$NaYF_4$:Er^{3+} in fluoropolymer matrix for photovoltaic devices. *J. Appl. Phys.* 114 (1):013505.

Jalil, R. A. and Y. Zhang. 2008. Biocompatibility of silica coated $NaYF_4$ upconversion fluorescent nanocrystals. *Biomaterials* 29:4122–4128.

Jiang, T., W. Qin, W. Di, R. Yang, D. Liu, X. Zhai, and G. Qin. 2012. Citric acid-assisted hydrothermal synthesis of α-$NaYF_4$:Yb^{3+}, Tm^{3+} nanocrystals and their enhanced ultraviolet upconversion emissions. *CrystEngComm* 14 (6):2302–2307.

Johnson, N. J. J., A. Korinek, C. Dong, and F. C. J. M. van Veggel. 2012. Self-focusing by Ostwald ripening: A strategy for layer-by-layer epitaxial growth on upconverting nanocrystals. *J. Am. Chem. Soc.* 134 (27):11068–11071.

Johnson, N. J. J., W. Oakden, G. J. Stanisz, R. Scott Prosser, and F. C. J. M. van Veggel. 2011. Size-tunable, ultra small $NaGdF_4$ nanoparticles: Insights into their T1 MRI contrast enhancement. *Chem. Mater.* 23 (16):3714–3722.

Johnson, N. J. J., N. M. Sangeetha, J.-C. Boyer, and F. C. J. M. van Veggel. 2010. Facile ligand-exchange with polyvinylpyrrolidone and subsequent silica coating of hydrophobic upconverting [small beta]-$NaYF_4$:Yb^{3+}/Er^{3+} nanoparticles. *Nanoscale* 2 (5):771–777.

Krämer, K. W., D. Biner, G. Frei, H. U. Güdel, M. P. Hehlen, and S. R. Lüthi. 2004. Hexagonal sodium yttrium fluoride based green and blue emitting upconversion phosphors. *Chem. Mater.* 16 (7):1244–1251.

Li, C., Z. Hou, Y. Dai, D. Yang, Z. Cheng, P. Ma, and J. A. Lin. 2013. A facile fabrication of upconversion luminescent and mesoporous core–shell structured β-$NaYF_4$:Yb^{3+}, Er^{3+}@mSiO_2 nanocomposite spheres for anti-cancer drug delivery and cell imaging. *Biomater. Sci.* 1:213–223.

Li, Z. and Y. Zhang. 2006. Monodisperse silica-coated polyvinylpyrrolidone/$NaYF_4$ nanocrystals with multicolor upconversion fluorescence emission. *Angew. Chem. Int. Ed.* 45 (46):7732–7735.

Li, Z. and Y. Zhang. 2008. An efficient and user-friendly method for the synthesis of hexagonal-phase $NaYF_4$: Yb, Er/Tm nanocrystals with controllable shape and upconversion fluorescence. *Nanotechnology* 19 (34):345606.

Li, Z., Y. Zhang, and S. Jiang. 2008. Multicolor core/shell-structured upconversion fluorescent nanoparticles. *Adv. Mater.* 20 (24):4765–4769.

Lin, M., Y. Zhao, M. Liu, M. S. Qiu, Y. Q. Dong, Z. F. Duan, Y. H. Li, B. Pingguan-Murphy, T. J. Lu, and F. Xu. 2014. Synthesis of upconversion $NaYF_4$:Yb^{3+}, Er^{3+} particles with enhanced luminescent intensity through control of morphology and phase. *J. Mater. Chem. C* 2 (19):3671–3676.

Liu, F., Q. Zhao, H. You, and Z. Wang. 2013. Synthesis of stable carboxy-terminated $NaYF_4$: Yb^{3+}, Er^{3+}@SiO_2 nanoparticles with ultrathin shell for biolabeling applications. *Nanoscale* 5 (3):1047–1053.

Liu, J., W. Bu, S. Zhang, F. Chen, H. Xing, L. Pan, L. Zhou, W. Peng, and J. Shi. 2012. Controlled synthesis of uniform and monodisperse upconversion core/mesoporous silica shell nanocomposites for bimodal imaging. *Chem. Eur. J.* 18 (8):2335–2341.

Liu, Q., Y. Sun, C. Li, J. Zhou, C. Li, T. Yang, X. Zhang, T. Yi, D. Wu, and F. Li. 2011. F-18-labeled magnetic-upconversion nanophosphors via rare-earth cation-assisted ligand assembly. *ACS Nano* 5 (4):3146–3157.

Liu, X. M., J. W. Zhao, Y. J. Sun, K. Song, Y. Yu, C. A. Du, X. G. Kong, and H. Zhang. 2009. Ionothermal synthesis of hexagonal-phase $NaYF_4$:Yb^{3+}, Er^{3+}/Tm^{3+} upconversion nanophosphors. *Chem. Commun.* 6628–6630.
Liu, Y., D. Tu, H. Zhu, and X. Chen. 2013. Lanthanide-doped luminescent nanoprobes: Controlled synthesis, optical spectroscopy, and bioapplications. *Chem. Soc. Rev.* 42:6924–6958.
Liu, Y., D. Tu, H. Zhu, R. Li, W. Luo, and X. Chen. 2010. A strategy to achieve efficient dual-mode luminescence of Eu^{3+} in lanthanides doped multifunctional $NaGdF_4$ nanocrystals. *Adv. Mater.* 22 (30):3266–3271.
Liu, Z., F. Pu, S. Huang, Q. Yuan, J. Ren, and X. Qu. 2013. Long-circulating Gd_2O_3:Yb^{3+}, Er^{3+} up-conversion nanoprobes as high-performance contrast agents for multimodality imaging. *Biomaterials* 34 (6):1712–1721.
Lorbeer, C., J. Cybińska, and A.-V. Mudring. 2010. Facile preparation of quantum cutting GdF_3:Eu^{3+} nanoparticles from ionic liquids. *Chem. Commun. (Cambridge, England)* 46 (4):571–573.
Lorbeer, C., J. Cybińska, and A.-V. Mudring. 2011. Europium(III) fluoride nanoparticles from ionic liquids: Structural, morphological, and luminescent properties. *Cryst. Growth Des.* 11 (4):1040–1048.
Lu, Q., F. Y. Guo, L. Sun, A. H. Li, and L. C. Zhao. 2008. Silica-/titania-coated Y_2O_3: Tm^{3+}, Yb^{3+} nanoparticles with improvement in upconversion luminescence induced by different thickness shells. *J. Appl. Phys.* 103 (12):123533.
Lu, Q. P., Y. B. Hou, A. W. Tang, Y. Z. Lu, L. F. Lv, and F. Teng. 2014. Controlled synthesis and defect dependent upconversion luminescence of Y_2O_3: Yb, Er nanoparticles. *J. Appl. Phys.* 115 (7):074309.
Luo, W. Q., C. Y. Fu, R. F. Li, Y. S. Liu, H. M. Zhu, and X. Y. Chen. 2011. Er^{3+}-doped anatase TiO_2 nanocrystals: Crystal-field levels, excited-state dynamics, upconversion, and defect luminescence. *Small* 7 (21):3046–3056.
Lv, R., S. Gai, Y. Dai, N. Niu, F. He, and P. Yang. 2013. Highly uniform hollow GdF_3 spheres: Controllable synthesis, tuned luminescence, and drug-release properties. *ACS Appl. Mater. Inter.* 5 (21):10806–10818.
Mader, H. S., M. Link, D. E. Achatz, K. Uhlmann, X. Li, and O. S. Wolfbeis. 2010. Surface-modified upconverting microparticles and nanoparticles for use in click chemistries. *Chem. Eur. J.* 16 (18):5416–5424.
Mai, H. X., Y. W. Zhang, R. Si, Z. G. Yan, L. Sun, L. P. You, and C. H. Yan. 2006. High-quality sodium rare-earth fluoride nanocrystals: Controlled synthesis and optical properties. *J. Am. Chem. Soc.* 128 (19):6426–6436.
Mita, Y., H. Yamamoto, K. Katayanagi, and S. Shionoya. 1995. Energy transfer processes in Er^{3+}- and Yb^{3+}-doped infrared upconversion materials. *J. Appl. Phys.* 78 (2):1219–1223.
Muhr, V., S. Wilhelm, T. Hirsch, and O. S. Wolfbeis. 2014. Upconversion nanoparticles: From hydrophobic to hydrophilic surfaces. *Acc. Chem. Res.* 47 (12):3481–3493.
Naccache, R., F. Vetrone, V. Mahalingam, L. A. Cuccia, and J. A. Capobianco. 2009. Controlled synthesis and water dispersibility of hexagonal phase $NaGdF_4$:Ho^{3+}/Yb^{3+} nanoparticles. *Chem. Mater.* 21 (4):717–723.
Nguyen, T. D., C. T. Dinh, and T. O. Do. 2010. Shape- and size-controlled synthesis of monoclinic ErOOH and cubic Er_2O_3 from micro- to nanostructures and their upconversion luminescence. *ACS Nano* 4 (4):2263–2273.
Nuñez, N. O. and M. Ocaña. 2007. An ionic liquid based synthesis method for uniform luminescent lanthanide fluoride nanoparticles. *Nanotechnology* 18 (45):455606.

Pandozzi, F., F. Vetrone, J. C. Boyer, R. Naccache, J. A. Capobianco, A. Speghini, and M. Bettinelli. 2005. A spectroscopic analysis of blue and ultraviolet upconverted emissions from $Gd_3Ga_5O_{12}$:Tm^{3+}, Yb^{3+} nanocrystals. *J. Phys. Chem. B* 109 (37):17400–17405.

Passuello, T., M. Pedroni, F. Piccinelli et al. 2012. PEG-capped, lanthanide doped GdF_3 nanoparticles: Luminescent and T_2 contrast agents for optical and MRI multimodal imaging. *Nanoscale* 4 (24):7682–7689.

Passuello, T., F. Piccinelli, M. Pedroni, M. Bettinelli, F. Mangiarini, R. Naccache, F. Vetrone, J. A. Capobianco, and A. Speghini. 2011a. White light upconversion of nanocrystalline Er/Tm/Yb doped tetragonal $Gd_4O_3F_6$. *Opt. Mater.* 33 (4):643–646.

Passuello, T., F. Piccinelli, M. Pedroni, S. Polizzi, F. Mangiarini, F. Vetrone, M. Bettinelli, and A. Speghini. 2011b. NIR-to-visible and NIR-to-NIR upconversion in lanthanide doped nanocrystalline GdOF with trigonal structure. *Opt. Mater.* 33 (10):1500–1505.

Patra, A., C. S. Friend, R. Kapoor, and P. N. Prasad. 2002. Upconversion in Er^{3+}:ZrO_2 nanocrystals. *J. Phys. Chem. B* 106 (8):1909–1912.

Patra, A., C. S. Friend, R. Kapoor, and P. N. Prasad. 2003a. Effect of crystal nature on upconversion luminescence in Er^{3+}:ZrO_2 nanocrystals. *Appl. Phys. Lett.* 83 (2):284.

Patra, A., C. S. Friend, R. Kapoor, and P. N. Prasad. 2003b. Fluorescence upconversion properties of Er^{3+}-doped TiO_2 and $BaTiO_3$ nanocrystallites. *Chem. Mater.* 15 (19):3650–3655.

Peacock, R. D. 1975. The intensities of lanthanide f–f transitions. *Struct. Bond.* 22:83–122.

Pedroni, M., F. Piccinelli, T. Passuello, M. Giarola, G. Mariotto, S. Polizzi, M. Bettinelli, and A. Speghini. 2011. Lanthanide doped upconverting colloidal CaF_2 nanoparticles prepared by a single-step hydrothermal method: Toward efficient materials with near infrared-to-near infrared upconversion emission. *Nanoscale* 3 (4):1456–1460.

Pedroni, M., F. Piccinelli, T. Passuello et al. 2013. Water (H_2O and D_2O) dispersible NIR-to-NIR upconverting Yb^{3+}/Tm^{3+} doped MF_2 (M = Ca, Sr) colloids: Influence of the host crystal. *Cryst. Growth Des.* 13 (11):4906–4913.

Pedroni, M., F. Piccinelli, S. Polizzi, A. Speghini, M. Bettinelli, and P. Haro-González. 2012. Upconverting Ho–Yb doped titanate nanotubes. *Mater. Lett.* 80:81–83.

Polizzi, S., S. Bucella, A. Speghini, F. Vetrone, R. Naccache, J. C. Boyer, and J. A. Capobianco. 2004. Nanostructured lanthanide-doped Lu_2O_3 obtained by propellant synthesis. *Chem. Mater.* 16 (7):1330–1335.

Polizzi, S., G. Fagherazzi, M. Battagliarin, M. Bettinelli, and A. Speghini. 2001. Fractal aggregates of lanthanide-doped Y_2O_3 nanoparticles obtained by propellant synthesis. *J. Mater. Res.* 16 (1):146–154.

Prodi, L., E. Rampazzo, F. Rastrelli, A. Speghini, and N. Zaccheroni. 2015. Imaging agents based on lanthanide doped nanoparticles. *Chem. Soc. Rev.* 44 (14):4922–4952.

Qian, H. S., H. C. Guo, P. C.-L. Ho, R. Mahendran, and Y. Zhang. 2009. Mesoporous-silica-coated up-conversion fluorescent nanoparticles for photodynamic therapy. *Small* 5 (20):2285–2290.

Quan, Z., D. Yang, P. Yang, X. Zhang, H. Lian, X. Liu, and J. Lin. 2008. Uniform colloidal alkaline earth metal fluoride nanocrystals: Nonhydrolytic synthesis and luminescence properties. *Inorg. Chem.* 47 (20):9509–9517.

Schafer, H., P. Ptacek, H. Eickmeier, and M. Haase. 2009. Synthesis of hexagonal Yb^{3+}, Er^{3+}-doped $NaYF_4$ nanocrystals at low temperature. *Adv. Funct. Mater.* 19 (19):3091–3097.

Sedlmeier, A. and H. H. Gorris. 2014. Surface modification and characterization of photon-upconverting nanoparticles for bioanalytical applications. *Chem. Soc. Rev.* 44:1526–1560.

Si, R., Y. W. Zhang, H. P. Zhou, L. D. Sun, and C. H. Yan. 2007. Controlled-synthesis, self-assembly behavior, and surface-dependent optical properties of high-quality rare-earth oxide nanocrystals. *Chem. Mater.* 19 (1):18–27.

Singh, S. K., K. Kumar, and S. B. Rai. 2008. Multifunctional Er^{3+}–Yb^{3+} codoped Gd_2O_3 nanocrystalline phosphor synthesized through optimized combustion route. *Appl. Phys. B* 94 (1):165–173.

Singh, S. K., K. Kumar, and S. B. Rai. 2009. Er^{3+}/Yb^{3+} codoped Gd_2O_3 nano-phosphor for optical thermometry. *Sensor. Actuat. A Phys.* 149 (1):16–20.

Stöber, W., A. Fink, and E. Bohn. 1968. Controlled growth of monodisperse silica spheres in the micron size range. *J. Colloid Interf. Sci.* 26 (1):62–69.

Stouwdam, J. W. and F. C. J. M. van Veggel. 2002. Near-infrared emission of redispersible Er^{3+}, Nd^{3+}, and Ho^{3+} doped LaF_3 nanoparticles. *Nano Lett.* 2 (7):733–737.

Sun, L. D., Y. F. Wang, and C. H. Yan. 2014. Paradigms and challenges for bioapplication of rare earth upconversion luminescent nanoparticles: Small size and tunable emission/excitation spectra. *Acc. Chem. Res.* 47:1001–1009.

Sun, X., Y.-W. Zhang, Y.-P. Du, Z.-G. Yan, R. Si, L.-P. You, and C.-H. Yan. 2007. From trifluoroacetate complex precursors to monodisperse rare-earth fluoride and oxyfluoride nanocrystals with diverse shapes through controlled fluorination in solution phase. *Chem. Eur. J.* 13 (8):2320–2332.

Tian, Y., H. Y. Yang, S. S. Yu, and X. Li. 2014. Shape-controlled synthesis of monodispersed ultrasmall $CaGd_3F_{11}$ nanocrystals for potential dual-modal probes. *ChemPlusChem* 79 (11):1584–1589.

van Veggel, F. C. J. M., C. Dong, N. J. J. Johnson, and J. Pichaandi. 2012. Ln^{3+}-doped nanoparticles for upconversion and magnetic resonance imaging: Some critical notes on recent progress and some aspects to be considered. *Nanoscale* 4 (23):7309–7321.

Venkatramu, V., M. Giarola, G. Mariotto, S. Enzo, S. Polizzi, C. K. Jayasankar, F. Piccinelli, M. Bettinelli, and A. Speghini. 2010. Nanocrystalline lanthanide-doped $Lu_3Ga_5O_{12}$ garnets: Interesting materials for light-emitting devices. *Nanotechnology* 21 (17):175703.

Vetrone, F., J. C. Boyer, J. A. Capobianco, A. Speghini, and M. Bettinelli. 2003. Luminescence spectroscopy and near-infrared to visible upconversion of nanocrystalline $Gd_3Ga_5O_{12}$:Er^{3+}. *J. Phys. Chem. B* 107 (39):10747–10752.

Vetrone, F., J. C. Boyer, J. A. Capobianco, A. Speghini, and M. Bettinelli. 2004. A spectroscopic investigation of trivalent lanthanide doped Y_2O_3 nanocrystals. *Nanotechnology* 15 (1):75–81.

Vetrone, F., R. Naccache, V. Mahalingam, C. G. Morgan, and J. A. Capobianco. 2009. The active-core/active-shell approach: A strategy to enhance the upconversion luminescence in lanthanide-doped nanoparticles. *Adv. Funct. Mater.* 19 (18):2924–2929.

Wang, F., R. R. Deng, J. Wang, Q. X. Wang, Y. Han, H. M. Zhu, X. Y. Chen, and X. G. Liu. 2011. Tuning upconversion through energy migration in core–shell nanoparticles. *Nat. Mater.* 10 (12):968–973.

Wang, F., Y. Han, C. S. Lim, Y. Lu, J. Wang, J. Xu, H. Chen, C. Zhang, M. Hong, and X. Liu. 2010. Simultaneous phase and size control of upconversion nanocrystals through lanthanide doping. *Nature* 463 (7284):1061–1065.

Wang, F. and X. Liu. 2009. Recent advances in the chemistry of lanthanide-doped upconversion nanocrystals. *Chem. Soc. Rev.* 38 (4):976–989.

Wang, G., Q. Peng, and Y. Li. 2009. Upconversion luminescence of monodisperse CaF_2:Yb^{3+}/Er^{3+} nanocrystals. *J. Am. Chem. Soc.* 131 (40):14200–14201.

Wang, H., M. Uehara, H. Nakamura, M. Miyazaki, and H. Maeda. 2005. Synthesis of well-dispersed Y_2O_3:Eu nanocrystals and self-assembled nanodisks using a simple non-hydrolytic route. *Adv. Mater.* 17 (20):2506–2509.

Wang, M., C. Mi, Y. Zhang, J. Liu, F. Li, C. Mao, and S. Xu. 2009. NIR-responsive silica-coated $NaYbF_4$:Er/Tm/Ho upconversion fluorescent nanoparticles with tunable emission colors and their applications in immunolabeling and fluorescent imaging of cancer cells. *J. Phys. Chem. C* 113 (44):19021–19027.

Wang, X., J. Zhuang, Q. Peng, and Y. D. Li. 2005. A general strategy for nanocrystal synthesis. *Nature* 437 (7055):121–124.

Wang, Z.-L., J. H. Hao, and H. L. W. Chan. 2010. Down- and up-conversion photoluminescence, cathodoluminescence and paramagnetic properties of $NaGdF_4$: Yb^{3+}, Er^{3+} submicron disks assembled from primary nanocrystals. *J. Mater. Chem.* 20 (16):3178–3185.

Wong, H.-T., F. Vetrone, R. Naccache, H. L. W. Chan, J. Hao, and J. A. Capobianco. 2011. Water dispersible ultra-small multifunctional $KGdF_4$:Tm^{3+}, Yb^{3+} nanoparticles with near-infrared to near-infrared upconversion. *J. Mater. Chem.* 21 (41):16589–16596.

Yang, D., P. Ma, Z. Hou, Z. Cheng, C. Li, and J. Lin. 2014. Current advances in lanthanide ion (Ln^{3+})-based upconversion nanomaterials for drug delivery. *Chem. Soc. Rev.* 44:1416–1448.

Yang, P., S. Gai, and J. Lin. 2012. Functionalized mesoporous silica materials for controlled drug delivery. *Chem. Soc. Rev.* 41 (9):3679–3698.

Yang, Y., Q. Shao, R. Deng et al. 2012. *In vitro* and *in vivo* uncaging and bioluminescence imaging by using photocaged upconversion nanoparticles. *Angew. Chem. Int. Ed.* 51 (13):3125–3129.

Ye, X., J. E. Collins, Y. Kang, J. Chen, D. T. N. Chen, A. G. Yodh, and C. B. Murray. 2010. Morphologically controlled synthesis of colloidal upconversion nanophosphors and their shape-directed self-assembly. *Proc. Natl. Acad. Sci. USA* 107 (52):22430–22435.

Yi, G.-S. and G.-M. Chow. 2005. Colloidal LaF_3:Yb, Er, LaF_3:Yb, Ho and LaF_3:Yb, Tm nanocrystals with multicolor upconversion fluorescence. *J. Mater. Chem.* 15 (41):4460–4464.

Yi, G. S. and G. M. Chow. 2006. Synthesis of hexagonal-phase $NaYF_4$:Yb, Er and NaYF4:Yb, Tm nanocrystals with efficient up-conversion fluorescence. *Adv. Funct. Mater.* 16 (18):2324–2329.

Yi, G. S., H. C. Lu, S. Y. Zhao, G. Yue, W. J. Yang, D. P. Chen, and L. H. Guo. 2004. Synthesis, characterization, and biological application of size-controlled nanocrystalline $NaYF_4$: Yb, Er infrared-to-visible up-conversion phosphors. *Nano Lett.* 4 (11):2191–2196.

Yu, M., J. Lin, Z. Wang, J. Fu, S. Wang, H. J. Zhang, and Y. C. Han. 2002. Fabrication, patterning, and optical properties of nanocrystalline YVO_4:A (A = Eu^{3+}, Dy^{3+}, Sm^{3+}, Er^{3+}) phosphor films via sol–gel soft lithography. *Chem. Mater.* 14 (5):2224–2231.

Yu, X., M. Li, M. Xie, L. Chen, Y. Li, and Q. Wang. 2010. Dopant-controlled synthesis of water-soluble hexagonal $NaYF_4$ nanorods with efficient upconversion fluorescence for multicolor bioimaging. *Nano Res.* 3 (1):51–60.

Zanzoni, S., M. Pedroni, M. D'Onofrio, A. Speghini, and M. Assfalg. 2016. Paramagnetic nanoparticles leave their mark on nuclear spins of transiently adsorbed proteins. *J. Am. Chem. Soc.* 138 (1):72–75.

Zeng, J. H., J. Su, Z. H. Li, R. X. Yan, and Y. D. Li. 2005. Synthesis and upconversion luminescence of hexagonal-phase $NaYF_4$:Yb, Er^{3+} phosphors of controlled size and morphology. *Adv. Mater.* 17 (17):2119–2123.

Zeng, S., J. Xiao, Q. Yang, and J. Hao. 2012. Bi-functional $NaLuF_4$:Gd^{3+}/Yb^{3+}/Tm^{3+} nanocrystals: Structure controlled synthesis, near-infrared upconversion emission and tunable magnetic properties. *J. Mater. Chem.* 22 (19):9870–9874.

Zhang, C., J. Chen, Y. C. Zhou, and D. Q. Li. 2008. Ionic liquid-based "all-in-one" synthesis and photoluminescence properties of lanthanide fluorides. *J. Phys. Chem. C* 112 (27):10083–10088.

Zhang, Y. W., X. Sun, R. Si, L. P. You, and C. H. Yan. 2005. Single-crystalline and monodisperse LaF_3 triangular nanoplates from a single-source precursor. *J. Am. Chem. Soc.* 127 (10):3260–3261.

Zhao, J., Y. Sun, X. Kong, L. Tian, Y. Wang, L. Tu, J. Zhao, and H. Zhang. 2008. Controlled synthesis, formation mechanism, and great enhancement of red upconversion luminescence of $NaYF_4$:Yb^{3+}, Er^{3+} nanocrystals/submicroplates at low doping level. *J. Phys. Chem. B* 112 (49):15666–15672.

Zheng, K. Z., W. Y. Song, C. J. Lv, Z. Y. Liu, and W. P. Qin. 2014. Controllable synthesis and size-dependent upconversion luminescence properties of Lu_2O_3:Yb^{3+}/Er^{3+} nanospheres. *CrystEngComm* 16 (20):4329–4337.

Zhou, J., Z. Liu, and F. Li. 2012. Upconversion nanophosphors for small-animal imaging. *Chem. Soc. Rev.* 41:1323–1349.

Zhou, J., Y. Sun, X. Du, L. Xiong, H. Hu, and F. Li. 2010. Dual-modality *in vivo* imaging using rare-earth nanocrystals with near-infrared to near-infrared (NIR-to-NIR) upconversion luminescence and magnetic resonance properties. *Biomaterials* 31 (12):3287–3295.

Zhu, X., B. Da Silva, X. Zou, B. Shen, Y. Sun, W. Feng, and F. Li. 2014. Intra-arterial infusion of PEGylated upconversion nanophosphors to improve the initial uptake by tumors *in vivo*. *RSC Adv.* 4 (45):23580–23584.

4

Functionalization Aspects of Water Dispersible Upconversion Nanoparticles

**Markus Buchner, Verena Muhr,
Sandy-Franziska Himmelstoß, and Thomas Hirsch**

CONTENTS

4.1 Introduction....69
4.2 Synthesis of UCNPs....71
4.3 Surface Modifications of Hydrophobic UCNPs....72
 4.3.1 Amphiphilic Coatings....72
 4.3.2 Encapsulation with Silica....74
 4.3.3 Ligand Exchange....75
4.4 Protein Conjugation....76
4.5 Conjugation to Nucleic Acids....79
4.6 Conjugation to Dyes....87
4.7 Conclusion....94
Acknowledgment....95
References....95

4.1 Introduction

Upconversion nanoparticles (UCNPs)/nanomaterials have the unique property to absorb near-infrared (NIR) light which leads to an anti-*Stokes* emission from the ultraviolet (UV) over the visible to the NIR range (Zhang 2014). The NIR light is absorbed by sensitizer ions (e.g., Yb^{3+} and Nd^{3+}), and then transferred to so-called activator ions (e.g., Tm^{3+}, Er^{3+}, and Ho^{3+}) via a nonradiative, resonant energy transfer process. The lanthanide ions are preferably embedded in host lattices with low phonon energy and similar ionic radii to reduce nonradiative deactivation processes (Haase and Schäfer 2011; Suyver et al. 2006; Wang and Liu 2009). Fluorides combine low phonon energies (~300–400 cm^{-1}) with high chemical stability compared to other halide ions or oxide materials. The doping ratio of the sensitizer and activator ions is very critical, in order to generate highly fluorescent nanomaterials and to avoid self-quenching of the excited states. A $NaYF_4$ hexagonal host lattice

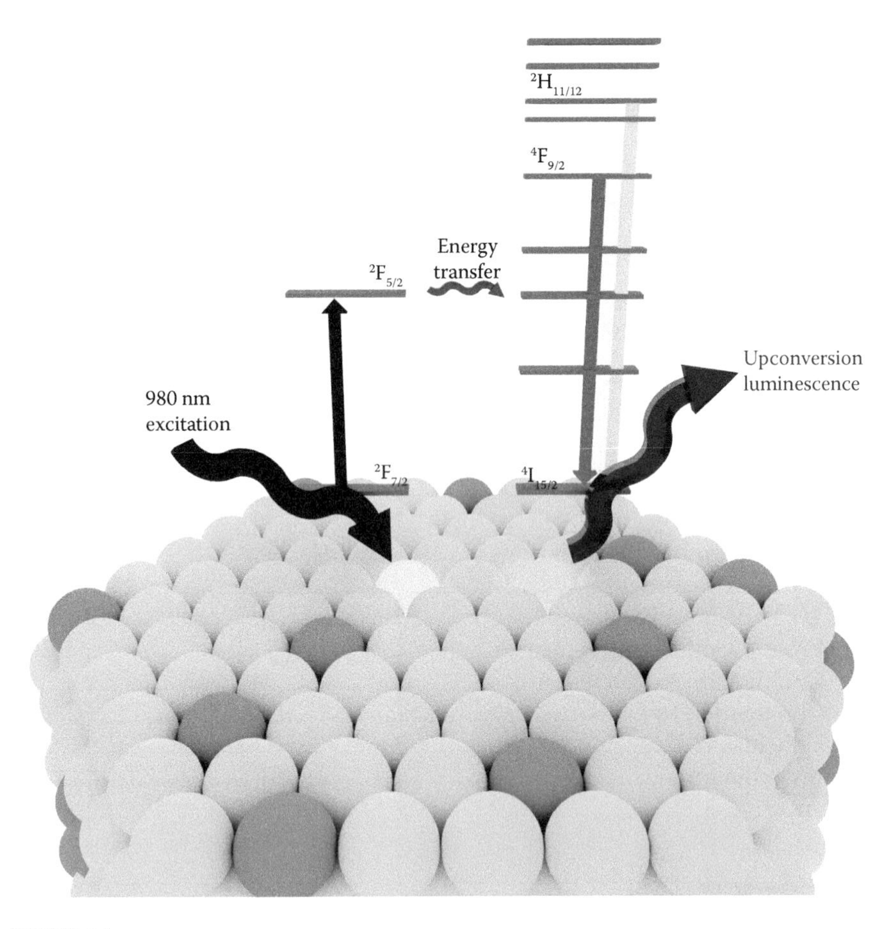

FIGURE 4.1
Hexagonal crystal lattice of $NaYF_4$ doped with 20% Yb ions and 2% Er ions. The Yb^{3+} ions are excited at 980 nm ($^2F_{7/2} \rightarrow {}^2F_{5/2}$) and transfer their energy to Er^{3+}, which can then either emit from the $^2H_{11/2}$ state (green emission) or the $^4F_{9/2}$ state after internal relaxation (red emission).

doped with Yb^{3+} and Er^{3+} ions and a simplified overview of the energy transfers inside the crystal is depicted in Figure 4.1.

The use of conventional fluorescent dyes in bioassays, biosensors, and imaging has been well established, but suffers from several drawbacks, such as high background fluorescence in biological matrices, photobleaching, small *Stokes* shifts, and short luminescence lifetimes. These disadvantages can be circumvented by the replacement of the dyes by UCNPs. The NIR excitation within the so-called optical window of biological material allows for deep tissue penetration without triggering autofluorescence of the matrix and reduced light scattering, leading to significantly higher signal-to-noise ratios. Beneficial properties, like high photostability, narrow emission bands, long luminescence lifetimes, and low cytotoxicity, additionally qualify UCNPs for their use in the design of biosensing applications. Both homogenous and heterogeneous assays have been developed, facing different challenges when dealing with the optimization of the surface modification process regarding stability, sensitivity, and selectivity. Nucleic acids, proteins, and fluorescent organic dyes represent the most popular signal molecules for detection schemes using UCNPs.

4.2 Synthesis of UCNPs

The three main strategies for the synthesis of UCNPs are the thermolysis, the Ostwald ripening, and the hydrothermal strategy (Chen et al. 2014; Zhou et al. 2012). The hydrothermal strategy is usually performed in autoclaves with Teflon inserts. The precursors are mixed with surfactants like ethylenediaminetetraacetic acid (EDTA) or polyethylenimine. Hydrothermal syntheses typically have reaction times from 5 up to 24 h under high-pressure and temperatures up to 200°C (Wang et al. 2012). The diameter of the nanoparticles is in the range of 100–500 nm. It is extremely difficult to obtain smaller particles, especially diameters of about 20 nm, which is important for biological applications. The size distribution of the synthesized particles has been improved over the last years, resulting in almost monodisperse UCNPs. Another drawback is that the progress of the reaction cannot be controlled and the synthesis of uniform core–shell particles has not been realized via the hydrothermal strategy so far. The main advantage of the hydrothermal synthesis is the preparation of directly water dispersible nanoparticles.

In contrast, particles produced by thermal decomposition are not water dispersible after the synthesis (Boyer et al. 2006, 2007). The surface of the nanocrystal needs to be stabilized with organic surfactants. The most commonly used organic surfactants are oleate, oleylamine, and trioctylphosphine oxide. However, this method is superior regarding particle-size distributions and crystallinity.

The Ostwald ripening strategy relies on the fact that the growth of larger particles is energetically favored compared to smaller nanoparticles (Johnson et al. 2012; Rinkel et al. 2014). The synthesis is carried out in high boiling solvents (e.g., oleic acid and octadecene). First, lanthanide–surfactant complexes are formed as precursors at room temperature. In the next step, the precursors are heated up to approximately 300°C to grow the upconversion nanocrystals by Ostwald ripening. The advantages of the method are the narrow size distribution of the monodisperse nanoparticles, the purity in the crystal phase, and their enhanced optical properties. Wilhelm et al. (2015) showed that the progress of the reaction can be monitored by laser excitation allowing also a rough estimation of the size of the nanoparticles. Furthermore, the synthesis of uniform core–shell nanoparticles can be achieved by the Ostwald ripening method. Wang and Liu (2014) summarized the synthesis of core–shell nanoparticles for multicolor imaging. Moreover, Wang et al. (2010) showed that the luminescent properties of small nanoparticles (<20 nm) increase dramatically by the surface passivation effect when growing a nondoped shell consisting of $NaGdF_4$ around Yb^{3+} and Tm^{3+}-doped $NaGdF_4$ core particles. Jayakumar et al. (2014) reported new strategies to shift the excitation wavelength of the UCNPs from 980 to 808 nm by incorporating Nd^{3+} ions in core–shell nanoparticles. In comparison with the excitation wavelength at 980 nm, water absorption is minimized under 808 nm laser excitation, yielding bright, luminescent nanoparticles in aqueous systems after successful surface modification. At the same time, the water heating effect caused by the 980 nm excitation is avoided, which is favorable for biological samples.

One drawback of the Ostwald ripening method is the hydrophobic character of the nanoparticles, which requests further surface modification for bioanalytical applications.

4.3 Surface Modifications of Hydrophobic UCNPs

The development of increasingly advanced, reproducible, and controllable synthesis protocols for efficient UCNPs resulted in interesting possibilities for the investigation of novel, powerful biosensing applications. The most efficient way to synthesize small, monodisperse UCNPs up to date is the Ostwald ripening method (Muhr et al. 2014). As mentioned before, the particles are covered with surfactants like oleic acid or oleylamine. The polar headgroup of the surfactants points toward the surface of the nanoparticles and coordinates the metal ions at the surface of the nanoparticles. The hydrophobic tails point outwards and restrict the colloidal stability of the nanoparticles to organic solvents like chloroform or cyclohexane. Consequently, the need for effective surface modification strategies to transfer hydrophobic nanocrystals into aqueous media is increased. Over the years, a variety of methods to render the particles water dispersible have been developed. The most common and effective strategies can be categorized into three main groups: amphiphilic coatings, silica shell formation, and ligand exchange. These three methods are shown in Figure 4.2 and are described as follows.

4.3.1 Amphiphilic Coatings

One straightforward and very powerful technique to warrant water dispersibility is the formation of an additional layer on top of oleate-capped UCNPs via the deposition of amphiphilic molecules. Their long alkyl

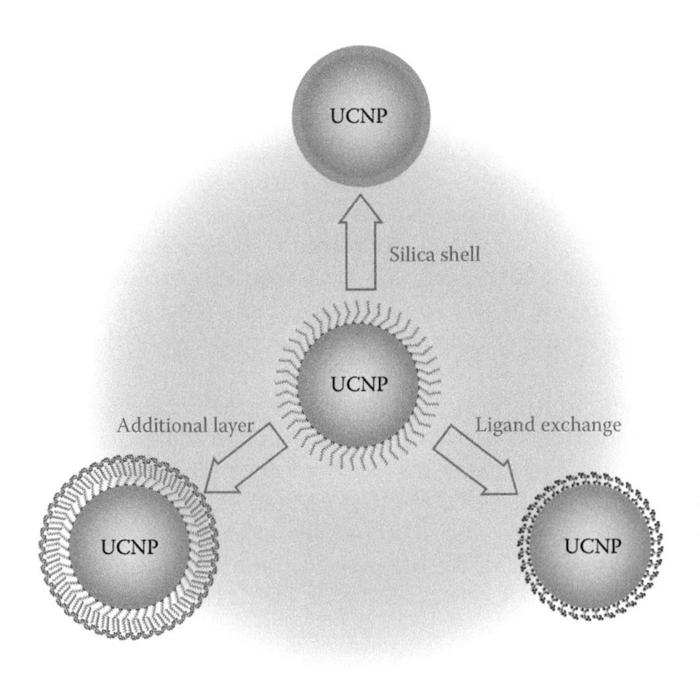

FIGURE 4.2
Overview of surface modification methods for transferring hydrophobic UCNPs into hydrophilic media by ligand exchange, additional layer deposition, and silica shell formation.

chains are able to intercalate between the hydrophobic oleate molecules based on van-der-Waals (vdW) interactions, thus forming a bilayer around the particle surface. However, the maximum length of the alkyl chain leading to a successful surface coating is limited by the length of the oleate itself. If the chain length exceeds the length of the oleate, the vdW interaction becomes too weak to stabilize the bilayer. The hydrophilic head groups of the amphiphiles are directed toward the solvent, which consequently renders the particles dispersible in aqueous media. The choice of the head group provides a simple way to alter and adjust the resulting surface charge according to the desired application. This first charged layer on the particle surface also enables the possibility for layer-by-layer coatings, that is, the deposition of several layers with alternating charges. However, the head groups of the amphiphilic molecules do not only ensure water dispersibility. They also represent functional groups allowing for the coupling of the coated UCNPs to receptor molecules or the attachment of, for example, dyes and stabilizing molecules.

In this context, phospholipids represent a group of frequently used amphiphiles for the surface modification of UCNPs. Phospholipids are commercially available with a great variety of different head groups, which ensures flexibility regarding the final surface coating. Particles modified with phospholipids are generally taken up by cells quite easily, and exhibit low cytotoxicity and biocompatibility, which is usually achieved by the addition of poly(ethylene glycol) (PEG) units (Nam et al. 2011; Park et al. 2009).

However, phospholipids are quite challenging regarding their synthesis and consequently expensive. Therefore, surface modification using much less-expensive surfactants has also been investigated. For example, TWEEN-80 (Ren et al. 2012) and several long-chain alkylammonium derived surfactants (Liang et al. 2012) have been successfully applied as amphiphilic surface coatings on UCNPs. Nevertheless, the colloidal stability of the resulting particles in aqueous systems was not as good as of the phospholipid-coated ones.

The use of amphiphilic polymers instead of small surfactants or phospholipids is one approach to further increase colloidal stability (Pellegrino et al. 2004). As these polymers contain a large number of hydrophobic alkyl chains per molecule, they are better stabilized against ligand detachment by a chelating effect. Poly(maleic anhydride-*alt*-1-octadecene) (PMAO) is the most widely used representative for the group of amphiphilic polymers (Li et al. 2014; Wang et al. 2011). UCNPs coated with PMAO display excellent colloidal stability in buffers over a wide pH range and in cell culture media for weeks (Jiang et al. 2012).

4.3.2 Encapsulation with Silica

The deposition of a silica shell on UCNPs is one of the most common techniques to achieve water dispersibility and to introduce functional groups to the particle surface. Hydrophobic UCNPs can be coated with a uniform layer of silica based on a reverse microemulsion method, which involves the polymerization of silicate precursors in the presence of a surfactant that forms and stabilizes the reverse microemulsion. The precursor tetraethyl orthosilicate (TEOS) and the nonionic surfactant Igepal CO-520 are most often used. Usually, a small amount of ammonia is added in order to keep the concentration of silicic acid above the nucleation concentration, which leads to the steady growth of a uniform silica shell. The thickness of the resulting shell depends on the amount of both precursor material and surfactant present during the shell growth phase. Particles encapsulated within silica are dispersible in water, are taken up by cells fast, and display low cytotoxicity (Jalil and Zhang 2008).

Yet, the colloidal stability of such UCNPs@SiO_2 in aqueous dispersions is quite poor, which leads to their aggregation and consequently precipitation within hours (Idris et al. 2012; Wang et al. 2009a). This problem arises through the centrifugation of the particles during purification. After centrifugation, it is often completely impossible to redisperse the particles in water. One solution for this problem is the stabilization of the silica-coated UCNPs by an increased surface charge in order to reduce the aggregation tendency (Bagwe et al. 2006). The desired high surface charge can be achieved by the introduction of negatively or positively charged functional groups on the silica surface. This can be done either by silanizing the already coated particles or by the addition of functional organosilanes during the polymerization process, for example, carboxyethylsilanetriol (F. Liu et al. 2013) or aminopropyltrimethoxysilane (Wang et al. 2009b). By optimizing the surface chemistry on UCNPs@SiO_2, it was possible to stabilize them in various kinds of buffers, including phosphate buffer and cell culture media.

It is also possible to synthesize mesoporous silica shells around nanoparticles. They exhibit an extremely large surface area with tunable pore sizes. Molecules can be easily trapped inside the pores of this material. These properties make them especially interesting for applications in drug delivery (J. Liu et al. 2013; K. Li et al. 2015). Another possibility is to load the mesoporous shell with dyes or receptor molecules that otherwise would not be able to penetrate the cell membrane (C. Li et al. 2013).

4.3.3 Ligand Exchange

Surface modification by ligand exchange is a third possibility to obtain water dispersible UCNPs. Here, the original ligand, for example, oleate is replaced either by small hydrophilic molecules or polymers. In a typical one-step procedure, the hydrophobic UCNPs are treated with a solution containing an excess of the desired new surface ligand at elevated temperatures for several hours. Usually, the protocol has to be adjusted for every single ligand regarding optimum reaction conditions. Commonly used hydrophilic molecules for the modification of nanoparticles via ligand exchange are citrate, poly(acrylic acid) (PAA) (J. Wang et al. 2014), poly(vinylpyrrolidone) (Johnson et al. 2010), and PEGylated compounds (Tong et al. 2015).

Other protocols for ligand exchange are based on two-step processes. The first step is the complete removal of the hydrophobic ligand oleate. One possibility to strip off the oleate is the addition of strong acids, for example, HCl (Bogdan et al. 2011). The oleate is protonated by the acid and thus released from the particle surface, leading to the generation of more or less ligand free particles. Another way to remove the oleate is the treatment with nitrosyl tetrafluoroborate ($NOBF_4$) (Dong et al. 2011). In this case, BF_4^- ions provide an electrostatic stabilization of the ligand free particles, consequently preventing particle aggregation during the exchange process, as it is usually the case for most one-step methods. In the second step, the desired new capping molecule is attached. After isolation and purification, the bare nanoparticles are stirred together with a solution containing the new surface ligand for passivation of the nanocrystal surface. Here, usually heating and protective gas are not required, and reaction times are considerably shorter compared to the more common one-step procedures (Ni et al. 2014; Y.-F. Wang et al. 2013).

In general, ligand exchange is an extremely versatile technique, as there are almost no limitations to their applicability. The only requirement for the new ligand molecule is the presence of a functional group capable of coordinating to the surface of the UCNPs. Negatively charged groups, such as carboxylates, sulfonates, phosphonates, and readily available free electron pairs, like in amino groups, fulfill this requirement. Thus, almost every molecule containing at least one of the mentioned functional groups can be successfully applied for the ligand exchange process. In contrast to surface modification by amphiphilic coatings and silica shells, ligand exchange provides a way to attach the new ligand directly to the surface of the UCNPs. The distance between the particle and ligand is thus minimized, which is, for example, especially important for applications relying on energy transfer processes between UCNPs and organic dyes immobilized on their surface.

When working with multifunctional molecules, the concentration of the UCNPs and the desired ligand has to be carefully adjusted in order to prevent cross-linking between several particles. Nevertheless, crosslinking and thus aggregation of the particles during the ligand exchange cannot be prevented completely. Another drawback of this method is the poor colloidal stability of the water dispersible UCNPs in solutions with high ionic strength and especially in phosphate buffer. Here, the use of strongly coordinating phosphates as new capping molecules provides increased stability in buffered systems (Ma et al. 2014; Zhao et al. 2014). Also, the application of water-soluble polymers instead of small molecules as the new ligand, improves the temporal colloidal stability in these systems due to the higher stability of the polymer against its detachment from the particle surface. Additional to the water dispersibility, a great number of possible functions suitable for further coupling are provided at the surface, since not every single functional group of one polymer chain is directed toward the particle. The first charged polymer layer can also function as first charged layer for successive layer-by-layer coatings (C. Wang et al. 2013).

To introduce biofunctionality to UCNPs, the surface of the water dispersible particles needs to be further engineered. The next chapters give an overview of strategies for coupling proteins, nucleic acids, and dyes to UCNPs as well as their bioanalytical applications.

4.4 Protein Conjugation

Antibodies, a special class of proteins and best known for their unique recognition ability, have been attached to UCNPs, for example, for the establishment of an immune sandwich assay of carcinoembryonic antigen (CEA) (Y. Li et al. 2015). The CEA in test samples was first captured with magnetic beads (MBs) coated with anti-CEA antibodies. After washing steps assisted by magnetic separation, the addition of PAA-capped UCNP with anti-CEA antibodies led to the formation of sandwich structures on the MBs connected by the CEA. Excess UCNP composites were removed and the luminescence emission of the UCNPs was recorded, resulting in a linear behavior for concentrations of the CEA between 0.05 and 20 ng mL^{-1}. The anti-CEA coated UCNPs exhibited good stability in 4-(2-hydroxyethyl)-1-piperazineethanesulfonic acid (HEPES) buffer.

Ong et al. (2014) proved the high photostability of the UCNPs in comparison with the green fluorescent protein (GFP). First, they performed a ligand exchange to remove oleate from the surface of the UCNPs and introduce citrate as a new ligand. As example, an anti-*E. coli* antibody was bound via classical 1-ethyl-3-(3-dimethylaminopropyl)carbodiimide (EDC) chemistry to the citrate-coated nanoparticles. After binding to *E. coli*, the halftime of GFP was determined to 75 s by fluorescence imaging, while the signal of the UCNPs stayed nearly constant over 30 min. Special focus was put on the colloidal stability and monodispersity of the nanoparticles. After the binding of the antibody, the hydrodynamic diameter of the particles increased about 200 nm with a low PdI of 0.162. Consequently, also with reference to the transmission electron microscopy (TEM) images, agglomeration was not critical. The binding of the particles to the *E. coli* cells showed only vulnerability against the growth medium lysogeny broth.

Binding of antibodies to silica-modified UCNPs generates conjugates of UCNPs and antibodies colloidally stable in phosphate buffer. Zhang et al. (2007) embedded a photosensitizer into silica shells around UCNPs and an anti-MUC1/episialin antibody was bound to the amino functionalized silica shell through cyanogen bromide.

Qiao et al. (2015) presented a very elegant technique to bind antibodies to the surface of UCNPs. They removed oleic acid from the core–shell $NaGdF_4$:Yb,Er @$NaGdF_4$ particles by heating to 50°C and adding an asymmetric PEG containing a maleimide group at one end and a diphosphate on the other end. The disulfide groups of a monoclonal antibody were reduced with tris(2-carboxyethyl)phosphine hydrochloride and the obtained thiol groups were bound to the maleimide groups on the surface of the nanoparticles by click reaction. Due to the coordinating phosphate groups on the surface, the modified particles were even stable in phosphate buffered saline (PBS) and can be used for a large area of applications.

Wang et al. (2009b) grew a silica shell around $NaYF_4$:Yb,Er nanoparticles using the classical *Stöber* synthesis. In a second step, 3-aminopropyltrimethoxysilane was used to functionalize the surface of the nanoparticles with amino groups and a rabbit anti-CEA8 antibody was bound to the nanoparticles via EDC chemistry.

The unique recognition properties and the high catalytic activities of enzymes are commercially used in several bioassays and sensors. In combination with UCNPs, the most critical point is the crosslinking of the particles and enzymes after immobilization to the surface of the nanoparticles. Crosslinking and as a result changes in the tertiary structure lead to lower catalytic activities. Gao et al. (2015) immobilized a caged protein kinase (PKA) on the surface of UCNPs, which can be activated by NIR light inside cells. $NaYF_4$:Yb,Tm nanoparticles were synthesized in high boiling solvents and a silica shell was formed by a reverse microemulsion method by adding Igepal CO-520 and TEOS. Considering the corresponding TEM images and dynamic light scattering (DLS) measurements, the particles were partially aggregated. The caged protein was bound to the surface of the silica-coated nanoparticles via electrostatic interactions. With the Bradford assay, the amount of the protein in the supernatant was calculated to roughly 4.5 nmol of PKA per 1 mg nanoparticle. The thiol groups of the enzymes were caged with o-nitrobenzyl bromide. After NIR excitation of the UCNPs, o-nitrobenzyl bromide was cleaved from the active center as shown in Figure 4.3 and the enzyme was used for the monitoring of intracellular signal transduction.

Chien et al. (2013) used a similar method to target cancer cells by creating a photoresponsive molecule with folic acid and o-nitrobenzyl bromide. The photosensitive protecting group was cleaved from folic acid by NIR excitation. Wang et al. (2015) developed a successful detection method for hyaluronidase. Hyaluronic acid-capped $NaYF_4$:Yb,Er UCNPs were bound with EDC chemistry to poly(m-phenylenediamine) nanospheres. Due to the electron rich properties of the nanospheres, the green emission of the UCNPs was fully quenched. In presence of hyaluronidase, the UCNPs were released from the surface and the green luminescence of the particles was recovered.

In relationship with UCNPs, enzymes are very often used in combination with organic molecules and dyes, which are able to absorb the emitted upconversion luminescence. Wilhelm et al. (2014) showed that they could monitor the production of nicotinamide adenine dinucleotide (NADH) and flavin adenine dinucleotide (FAD) due to the spectral overlap with Yb^{3+}, Tm^{3+}-doped core–shell $NaYF_4$ nanoparticles.

Wilhelm et al. (2013) presented a general approach for biomodification of silica-coated UCNPs with proteins. First, a silica shell was formed by the reverse microemulsion method. Subsequently, the silica surface was

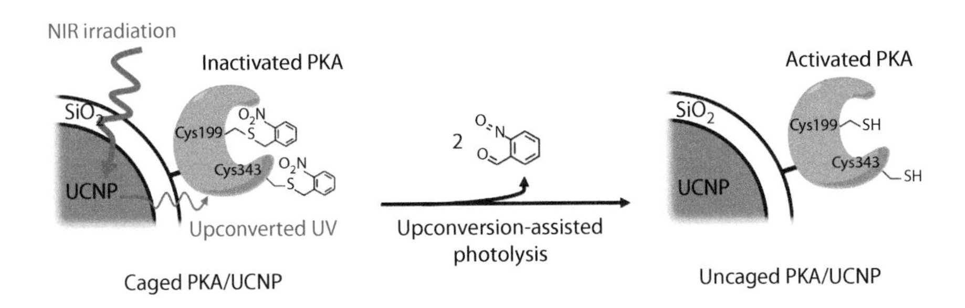

FIGURE 4.3
Caged PKA/UCNP complex design and the process of upconversion-assisted PKA uncaging. (Reprinted with permission from Gao, H.-D., P. Thanasekaran, C.-W. Chiang et al. 2015. Construction of a near-infrared-activatable enzyme platform to remotely trigger intracellular signal transduction using an upconversion nanoparticle. *ACS Nano* 9, 7041–7051. Copyright 2015 American Chemical Society.)

modified with a silane-PEG2000-N-hydroxysuccinimide (NHS) for biocompatibility and reactivity toward amino groups. The functionalized particles were conjugated to streptavidin-coated MBs demonstrating the high reactivity toward proteins.

Min et al. (2014) used photoactivation through UCNPs to release a platinum antitumor drug. An amino functionalized silica shell was formed around $NaYF_4$:Yb,Tm particles, using TEOS and (3-aminopropyl)triethoxysilane. Mal-dPEG$^{TM}{}_6$-NHS was used as a linker to bind trans, trans, trans-$[Pt(N_3)_2(OH)_2(py)_2]$ to a thiol group of a peptide linker to the silica surface of the particles. By NIR excitation, the cell toxic Pt(IV) complex was cleaved from the peptide linker by the blue Tm^{3+} emission. The triggered cell death was monitored by an energy transfer of Tm^{3+}-doped particles to Cy5^{TM}, also attached to the surface of the UCNPs. The energy transfer was first blocked by the quencher Qsy21, which was cleaved from the surface of the nanoparticles by caspase-3 by entering apoptosis.

4.5 Conjugation to Nucleic Acids

Nucleic acids play an irreplaceable role in the detection of pathogens and toxins. Countless DNA hybridization assays and aptasensors have been developed for quantification and imaging purposes *in vitro* and *in vivo*. While common colorimetric and fluorescence-based techniques usually suffer from high background signals, photobleaching or cytotoxicity, the application of UCNPs as fluorescent labels circumvents these drawbacks due to their NIR excitation, large anti-*Stokes* shifted emission, and exceptional photostability. The conjugation of nucleotides to the surface of UCNPs is the most critical step during the design of a sensor or imaging platform. Extensive crosslinking due to multifunctional ligands, control over the amount of receptors on the particle surface, and the stability of the modified UCNPs in buffers and cell culture media represent the most important factors to consider when facing the decision on the type of surface modification strategy.

Ligand exchange is by far the most popular technique to functionalize UCNPs with nucleic acids due to its abundant versatility and simplicity. Either the oligonucleotide itself is used for the ligand replacement and is directly bound to the particle surface, or water-soluble molecules or polymers containing functional groups are used as new surface ligands to which subsequently the nucleotides can be coupled. The direct attachment of the nucleic acids represents the most straightforward approach, as it does not require any precedent chemical modification of the DNA for coupling chemistry. The nucleotides are hereby efficiently bound to the surface of the UCNPs via their strongly coordinating phosphate moieties. Due to the high binding affinity of those phosphate groups on the particle surface, the

DNA-coated UCNPs exhibit excellent colloidal stability in phosphate buffer and cell culture media.

For example, DNA was directly attached to $NaYF_4$:Yb,Tm@$NaYF_4$ nanoparticles aimed to determine S1 endonuclease activity (Huang et al. 2015a). The UCNPs were first modified with DNA by stirring the oligonucleotide together with oleate-capped particles in a two-phase system consisting of water and cyclohexane. The oleate was completely replaced by the ssDNA and the UCNPs were transferred into the aqueous phase. These DNA-modified UCNPs were able to adsorb to the surface of graphene oxide due to π-stacking interactions. Graphene oxide is capable of quenching the luminescence of the UCNPs adsorbed on its surface. When the DNA on the UCNPs was cleaved by the addition of the endonuclease, the particles were released from the graphene oxide. Consequently, the quenching effect decreased, leading to stronger emission intensities of the UCNPs. By monitoring the increase of the luminescence intensity, S1 nuclease activity of 1×10^{-4} units mL^{-1} was detected selectively.

The same group used the identical modification strategy with DNA via direct exchange at the liquid–liquid interface for the development of a detection scheme for Hg^{2+} ions based on nonradiative electron/hole recombination annihilation through an effective electron transfer process (Huang et al. 2015b). The Hg^{2+} ions could be selectively detected in a dynamic range between 10 nM and 10 μM.

In another example, the oleate ligand of $NaYF_4$:Yb,Er UCNPs was directly replaced by ssDNA by vigorous stirring in a water–chloroform mixture, leading to the formation of water dispersible DNA-coated particles (L.-L. Li et al. 2013). The UCNP–DNA conjugate showed stability against ligand detachment after hybridization with the complementary DNA strand labeled with gold nanoparticles (AuNPs), illustrating their potential for applications in hybridization assays as shown in Figure 4.4. The possibility for cell imaging and cancer cell targeting was also demonstrated by the hybridization of the UCNP–DNA conjugates with Cy3 labeled complementary DNA or the immobilization of a nucleolin-specific aptamer on the UCNPs, respectively. The DNA-modified UCNPs showed exceptional transfection capabilities, making them an excellent material of choice for targeted bioimaging and gene delivery.

Despite its simplicity, the direct attachment of the nucleotides is prone to several problems. Although the high binding affinity of the phosphate groups to the surface of the UCNPs leads to colloidal stability, the odds are that several UCNPs are crosslinked due the presence of multiple phosphate functionalities on one DNA strand. This can cause the formation of large networks and aggregates that are no longer colloidally stable in dispersion and precipitate. There is also the risk that the DNA strands wrap themselves around the particle. The nucleotides are then no longer available for any hybridization with the target DNA, which then leads to reduced detection sensitivities.

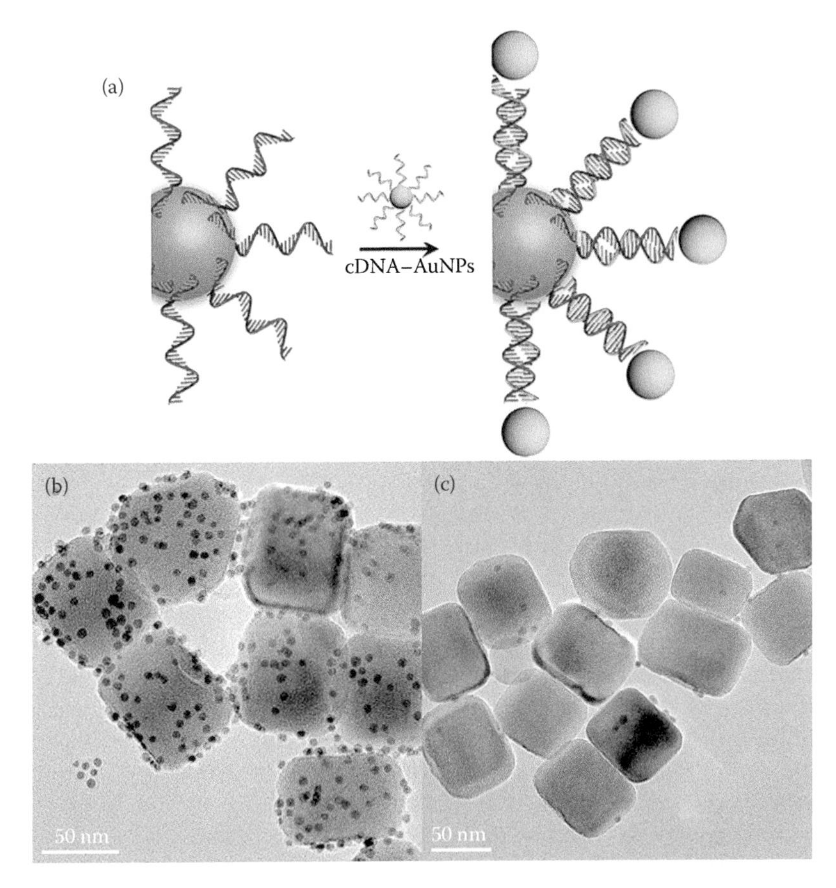

FIGURE 4.4
(a) DNA-directed assembly of UCNPs and AuNPs. TEM images of T30-UCNPs assembled with AuNPs bearing (b) complementary DNA and (c) noncomplementary DNA. (Reprinted with permission from Li et al. 2013. Copyright 2013 American Chemical Society.)

This can be prevented by the use of linker molecules, which are first bound to the UCNPs by ligand exchange, followed by the coupling of nucleic acids in a second step. These linker molecules have to carry both, a functional group with high binding affinity to the surface of the UCNPs and carboxyl or amino moieties, for binding to correspondingly functionalized nucleic acids.

Small bifunctional molecules such as dimercaptosuccinic acid (Lan et al. 2015) and aminoethanephosphonic acid (Song et al. 2012) have been successfully used as linkers for the conjugation of DNA to UCNPs. After ligand exchange, the carboxyl and amino groups were activated with EDC/NHS or glutaraldehyde, respectively, in order to couple amino-functionalized DNA or avidin for subsequent attachment of biotin-modified DNA. Both conjugates were used for the design of energy transfer-based sensors for the detection of either microRNA or adenosine triphosphate (ATP).

Ligands containing single binding sites, like in the two examples mentioned above, are prone to detachment from the particle surface, particularly in buffers containing phosphate ions and high ionic strength in general. In contrast to that, the use of polymers comprising multiple binding sites provides better stabilization against ligand detachment and functional groups for subsequent bioconjugation. The most frequently applied polymer is PAA.

In order to conjugate hydrophobic $NaYF_4$:Yb,Er,Mn to ssDNA for the detection of *staphylococcal enterotoxin B* (SEB), the oleate was first replaced by PAA in a one-step ligand exchange process in diethylene glycol (S. Wu et al. 2013). The ssDNA was then attached to the PAA-coated particles via carbodiimide chemistry and afterwards hybridized with the complementary DNA strand labeled with the dark quencher BHQ_3. After the hybridization, no red luminescence from the Mn^{2+}-doped UCNPs occurred. The degradation of the DNA by treatment with SEB led to the release of the quencher and consequently an increase of the UCNP luminescence intensity proportional to the SEB concentration. The detection limit for SEB was determined to be as low as 0.3 pg mL^{-1} with high reproducibility and specificity.

A similar approach was used for the development of a method for the simultaneous detection of three pathogenic bacteria (S. Wu et al. 2014). Three different kinds of UCNPs were synthesized, each exhibiting a different emission wavelength in the blue (Tm), green (Ho), or red (Er, Mn) region. After exchanging the oleate with PAA in a one-step procedure in diethylene glycol, three different amino-functionalized DNA strands (aptamers) were attached to the carboxyl groups on the particle surface by activation with EDC/NHS. Magnetic nanoparticles (MNPs) were modified equally with the complementary DNA, and both particle types were connected by hybridization. When the target bacteria were present, the UCNP-aptamer conjugates preferably bound to the specific binding sites on the bacteria. The luminescence signal of the remaining UCNP–MNPs conjugates was evaluated after magnetic separation from the UCNPs bound to the bacteria. The more bacteria present, the lower was the residual upconversion luminescence intensity. The whole assay principle is displayed in Figure 4.5. The limits of detection for the three target bacteria were found to be between 10 and 25 cfu mL^{-1}.

Ligand exchange is also popular for the design of heterogeneous immunoassays and aptasensors. Ulrich Krull et al. have especially performed pioneering research in the area of solid phase-based assay formats.

They performed a hybridization assay based on energy transfer on cellulose paper (Zhou et al. 2014). The UCNPs were first coated with citrate via a two-step ligand exchange assisted by HCl, conjugated to streptavidin by carbodiimide activation, and suspended in borate buffer. The modified UCNPs were immobilized on cellulose paper in small spots and biotin-tagged probe

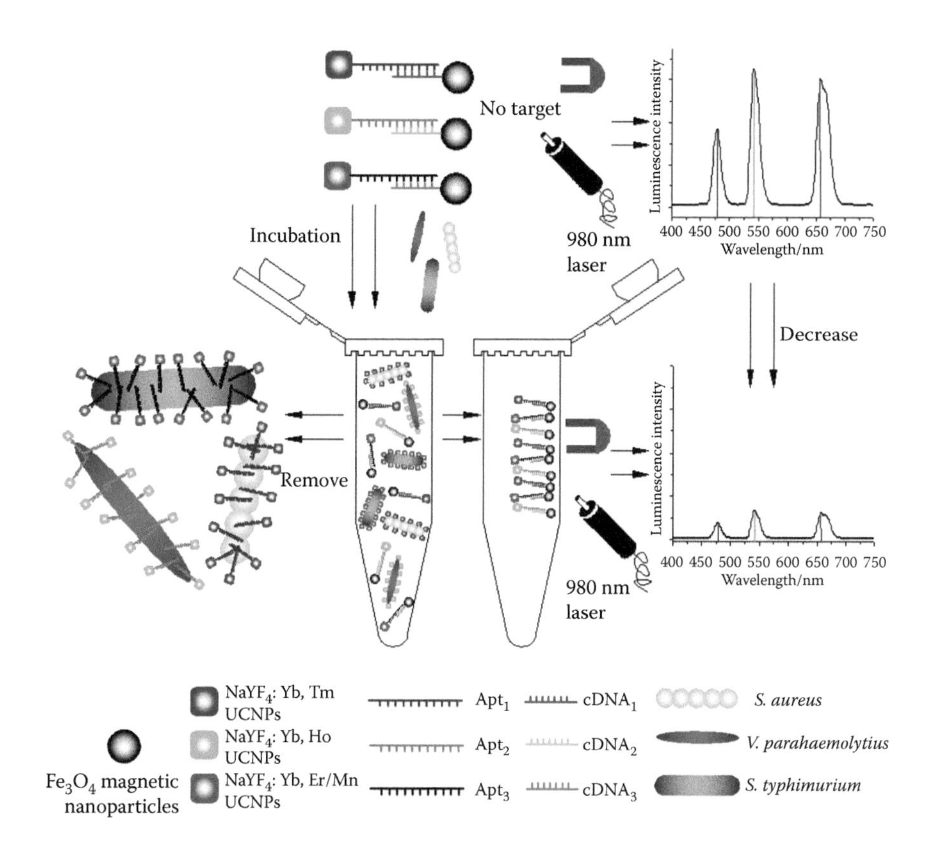

FIGURE 4.5
Schematic illustration of the multiplexed luminescence bioassay based on aptamer-modified UCNPs for the simultaneous detection of various pathogenic bacteria. (Reprinted with permission from Wu et al. 2014. Copyright 2013 American Chemical Society.)

DNA was bound to the particles on the paper. After treatment with the Cy3-labeled target DNA solution, the luminescence intensity of the UCNPs was read out using an epifluorescence microscope. The more target DNA was added, the lower was the remaining luminescence intensity due to a more efficient energy transfer from the UCNPs to the bound Cy3, achieving an limit of detection (LOD) of 34 fmol of target DNA. With this assay format, it was even possible to reliably discriminate between the complementary DNA and one base pair mismatch targets.

In another approach, UCNPs bound on cover slips have been used for the detection of thrombin (Doughan et al. 2014). $NaYF_4$:Yb,Tm@$NaYF_4$ UCNPs were first modified with o-phosphorylethanolamine (PEA) by ligand exchange. These particles were then immobilized on aldehyde-functionalized cover slips forming a densely packed layer on the glass surface. The immobilized nanoparticles showed stability against washing at three different pH

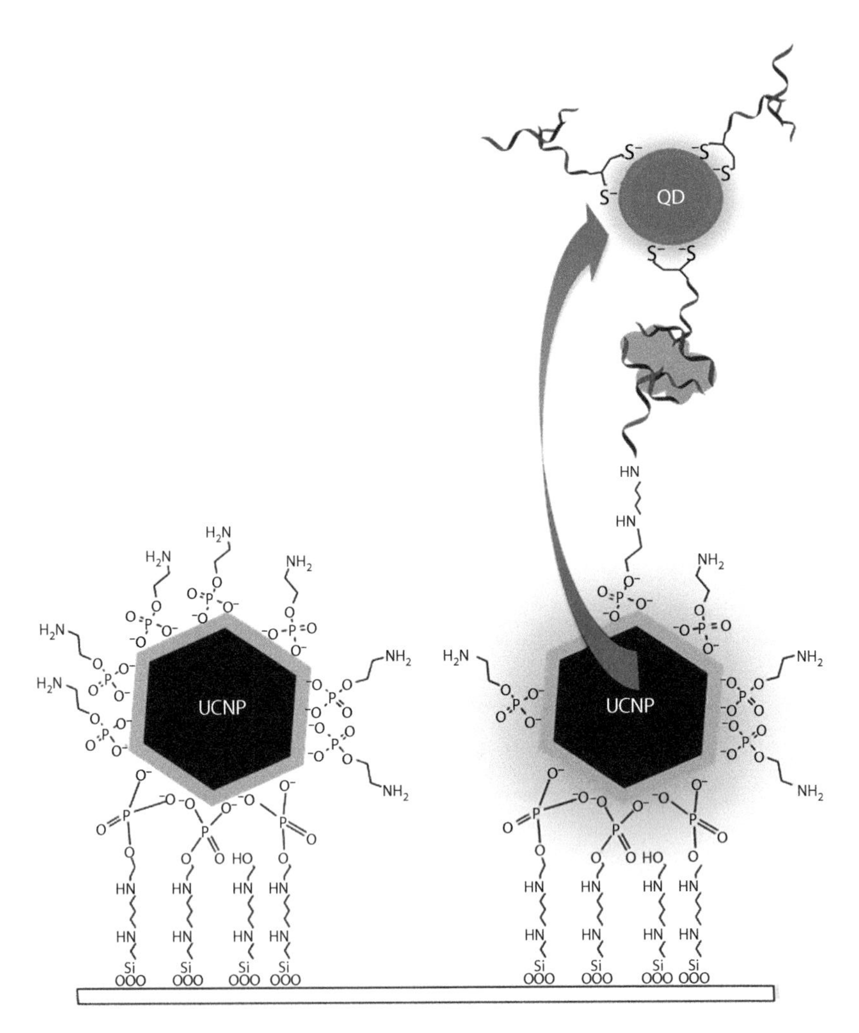

FIGURE 4.6
Schematic showing immobilization of PEA-UCNP on aldehyde-functionalized cover slip and use in the detection of thrombin using two different thrombin-specific aptamers. (Reprinted with permission from Doughan, S., Y. Han, U. Uddayasankar, and U. J. Krull. 2014. Solid-phase covalent immobilization of upconverting nanoparticles for biosensing by luminescence resonance energy transfer. *ACS Appl. Mater. Inter.* 6, 14061–14068. Copyright 2014 American Chemical Society.)

values of 5.5, 7.4, and 8.5; no leakage was observed. Next, a thrombin-specific aptamer was conjugated to the UCNPs on the cover slips. After thrombin was captured, quantum dots (QDs) also tagged with a thrombin-specific aptamer were added. The resulting conjugate is depicted in Figure 4.6. The overlap of the UCNP emission and the QD absorption led to a reduction of the observed upconversion luminescence intensity of the UCNPs and to an increase of the luminescence of the QDs. It was possible to detect down to 230 fmol thrombin by evaluation of the intensity ratios of all involved emission peaks.

Generally, UCNP–DNA conjugates obtained via ligand replacement are prone to aggregation during the exchange process caused by crosslinking between multiple particles and control over the loading capacity of the nucleic acids on the particle surface is rather difficult.

The formation of a bilayer on top of oleate-coated UCNPs with the help of amphiphilic molecules is much less commonly used to prepare particles for conjugation to nucleic acids, but does not suffer from the same drawbacks as the ligand exchange strategy mentioned earlier.

In one example, such UCNPs were used for the determination of human immunodeficiency virus (HIV) antibodies (Y.-M. Wu et al. 2014). Here, the particles were simultaneously modified with two different phospholipids in order to control the receptor density on the particle surface. One lipid carried a PEG chain imparting increased colloidal stability and biocompatibility, the other one an HIV specific aptamer. The luminescence of these water dispersible UCNPs was completely quenched by GOx due to π-stacking interactions. In the presence of HIV, the aptamer immobilized on the particles specifically bound to the target virus, which prevented the adsorption to the GOx and led to linearly increasing luminescence intensity for HIV concentrations between 5 and 150 nM in diluted human blood serum.

Coating with phospholipids enables efficient control over surface loading with the DNA probe in one step by adjusting the ratio of two or more different phospholipids used for the surface modification, while at the same time imparting excellent colloidal stability in a variety of buffers. However, phospholipids are expensive and their synthesis and purification is fairly elaborate.

A second scarcely employed strategy for the immobilization of nucleotides to UCNPs is the prevenient growth of a functional silica shell suitable for bioconjugation.

For the design of a DNA sensor, the polymerization of TEOS around the particles in a reverse microemulsion was followed by silanization with 3-aminopropyl triethoxysilane, resulting in amino-functionalized $NaYF_4$:Yb,Er@SiO_2 (Alonso-Cristobal et al. 2015). The amino groups were then reacted with succinic anhydride and the resulting carboxyl moieties were used for the attachment of amino-functionalized ssDNA via carbodiimide coupling, as illustrated in Figure 4.7. Due to electrostatic repulsion between the charged functional groups, the reaction steps did not cause aggregation, precipitation of the particles, and no crosslinking was observed. The hybridization of the target DNA to the modified UCNPs was monitored by the evaluation of the quenching efficiency of GOx caused by π–π interactions with the remaining ssDNA on the particles. The LOD of this sensor setup was calculated to be 5 nM.

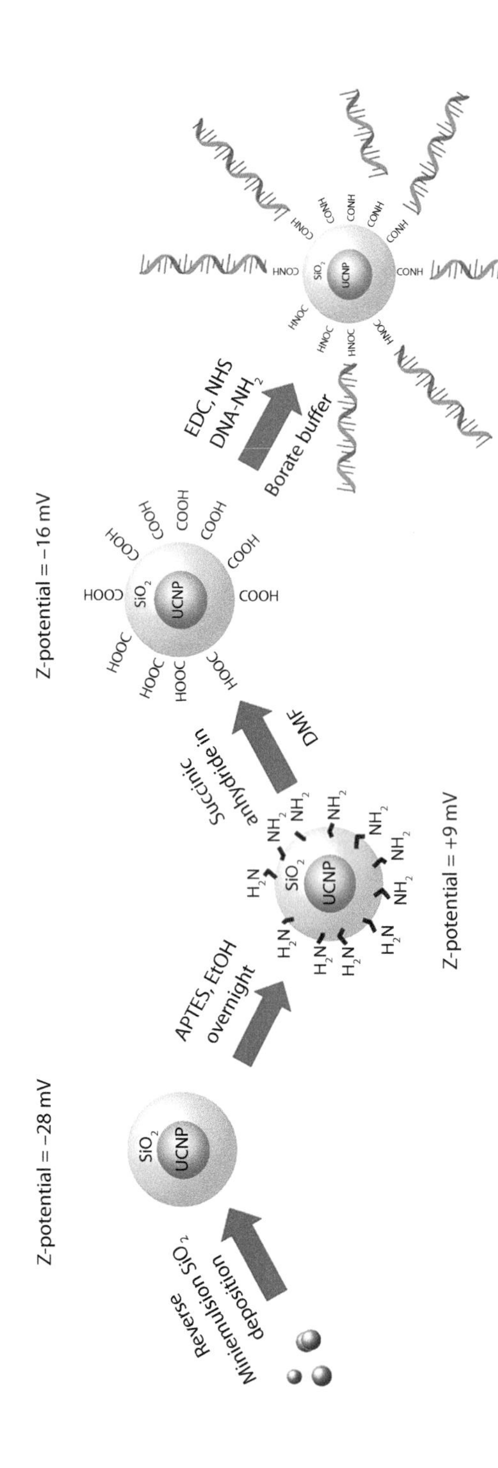

FIGURE 4.7
Schematic illustration of the chemical route for the functionalization of UCNPs@SiO_2 nanoparticles. (Reprinted with permission from Alonso-Cristobal, P., P. Vilela, A. El-Sagheer et al. 2015. Highly sensitive DNA sensor based on upconversion nanoparticles and graphene oxide. *ACS Appl. Mater. Inter.* 7, 12422–12429. Copyright 2015 American Chemical Society.)

4.6 Conjugation to Dyes

Fluorescent dyes are the most important tools for cellular imaging, sensing of intracellular parameters, and bioassays. Yet, many applications are limited by photobleaching and low signal-to-noise ratios caused by strong scattering of UV light, high autofluorescence, and unspecific binding. NIR excitable UCNPs can act as nanocarriers for those established dyes and eliminate these shortcomings by the design of conjugates relying on energy transfer mechanisms between the particles and the respective dyes. All applications share the necessity to select dyes that feature an overlap of their absorption spectrum with any emission band of the UCNPs. With this requirement in mind, numerous sensing schemes based on Förster resonance energy transfer (FRET) and inner filter effects have been reported. Almost as many different surface modification strategies for the attachment of fluorescent dyes to the particles have been described, all of which vary regarding stability, surface loading, distance between donor and acceptor, and sensitivity. All these parameters have to be carefully evaluated in respect to the final application before choosing one method.

Ligand exchange facilitates the attachment of dyes directly to the surface of UCNPs, which is especially favorable for efficient FRET processes due to the minimization of the distance between the donor–acceptor pair. In respect to applications in biological systems, this method only works for dyes which are water soluble and carry at least one functional group capable of binding to the surface of the UCNPs. Consequently, the colloidal stability in aqueous media solely depends on the binding affinity and stabilizing properties of the dye.

The direct immobilization of the dye by ligand exchange was used for the *in vivo* detection of hydroxyl radicals (Z. Li et al. 2015). The oleate was first removed from the surface of hydrophobic $NaYF_4$:Yb,Tm@$NaYF_4$ UCNPs by the addition of HCl. Subsequently, the surface of the obtained ligand free particles was passivated with a carboxyl-modified azo dye that absorbed the blue luminescence around 475 nm of the UCNPs and efficiently quenched the intensity of the affected UCNP emission. Hydroxyl radicals oxidized the azo dye, thus altering its absorption properties and preventing the energy transfer from the UCNPs. This process was monitored by the corresponding increase of the luminescence intensity of the blue UCNP emission. The colloidal stability of the nanoprobes under physiological conditions was sufficient for the successful performance of *in vivo* measurements.

If the active dye does not possess any functional groups or only weakly coordinating ones, a linker molecule can be first attached to the UCNPs by ligand replacement. This first layer can then be used to immobilize the respective dyes covalently or electrostatically. Suitable linkers in these cases are hydrophilic polymers, like PAA and poly(allyl amine) (PAAm). They simultaneously impart water dispersibility, a high surface charge for

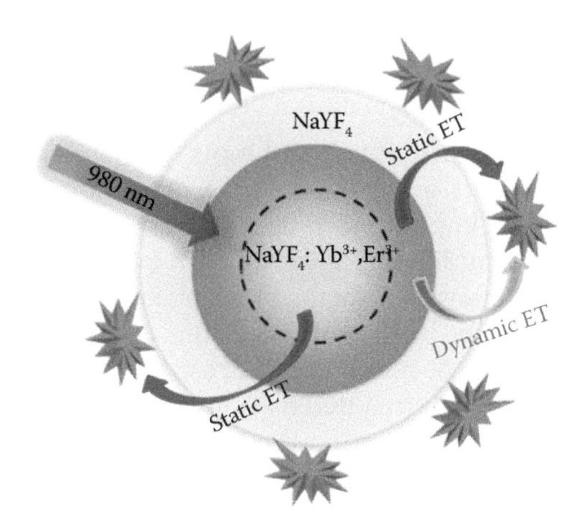

FIGURE 4.8
Energy transfer model from Er^{3+} in core–shell $NaYF_4$:Yb^{3+},Er^{3+}@$NaYF_4$ UCNPs to surface-bound acceptor rose bengal. (Reprinted with permission from Ding et al. 2015. Copyright 2015 American Chemical Society.)

electrostatic adsorption, and functional groups for covalent attachment of the dyes.

This strategy was used for the development of a UCNP nanocarrier for the photosensitizer rose bengal (Y. Ding et al. 2015). First, the oleate was exchanged with PAAm assisted by the addition of HCl. Afterwards, rose bengal was coupled to the amino groups by carbodiimide activation. Figure 4.8 shows the modified core–shell nanoparticles with the covalently bound dye molecules. The resulting system was used to determine the effect of the distance between UCNP and dye on the energy transfer efficiency.

In another example, PAA was introduced as the first polymer layer after the ligand exchange (Peng et al. 2015). A positively charged Zn^{2+} sensitive dye was adsorbed to the polymer-coated UCNPs via electrostatic interactions. In the absence of Zn^{2+} ions, the absorption of the dye nicely overlaps with the blue emission of $NaYF_4$:Yb,Tm. Increasing Zn^{2+} concentrations cause a blue shift of the dye absorption and the recovery of the UCNP emission. The modified particles were stable in cell culture media and were successfully applied for *in vivo* detection of Zn^{2+} in zebrafish.

Protease activities have also been monitored based on energy transfer between UCNPs and the fluorescent dye carboxytetramethylrhodamine (TAMRA) (Zeng et al. 2015). The C-terminus of an oligopeptide was modified with TAMRA and the peptide was immobilized on the surface of UCNPs by ligand exchange, resulting in a reduction of the luminescence intensity of the green upconversion emission. Cleavage of the peptide by trypsin and caspase-3 released the TAMRA dye from the particle surface and the emission intensity gradually increased. The particles displayed great colloidal stability when transferred into water under high peptide concentrations to ensure sufficient surface coverage and charge. They were used for the sensitive determination of protease activities and tumor cell imaging.

Hydrophobic, not water-soluble dyes cannot be attached to the UCNPs via ligand exchange for use in bioapplications, as they do not render the particles water dispersible. One alternative is the coating of oleate-capped UCNPs with amphiphilic molecules. This technique can be used to trap small dyes (and other molecules) inside the hydrophobic bilayer. At the same time, the particles are transferred into the aqueous phase and the choice of the amphiphilic molecule provides additional possibilities for further conjugation steps. Modifications with amphiphilic molecules generally display low aggregation tendencies during the coating process and lead to excellent colloidal stability in various buffers and cell culture media.

The photosensitizer merocyanine 540 was loaded into the bilayer formed by the addition of various phospholipids (H. Wang et al. 2014) or amphiphilic polymers, such as the triblock copolymer Pluronic®F-127 (H. Ding et al. 2015) to oleate-capped UCNPs. Furthermore, the particles were decorated with targeting moieties for tumor cells. The successful entrapment of the photosensitizer within the hydrophobic bilayer was demonstrated by the efficient production of singlet oxygen upon NIR excitation of the UCNPs and energy transfer to the MC540 inside breast cancer cells.

Molecules inside the hydrophobic bilayer were shown to retain their responsive properties. For example, the methylmercury sensitive cyanine dye hCy7 was immobilized within a bilayer formed by the amphiphilic blockpolymer PMAO–PEG and the oleate ligands (Y. Liu et al. 2013). In absence of methylmercury, the red emission at 660 nm of the UCNPs was quenched due to an efficient energy transfer to the dye. The addition of methylmercury induced a shift of the hCy7 absorption and the upconversion luminescence was recovered. Figure 4.9 shows the surface modification and the detection principle of the nanoparticles against $MeHg^+$. The sensor did not show cross sensitivity to numerous metal ions while methylmercury could be monitored *in vivo* with a detection limit of about 0.8 ppb.

The same group used a very similar technique to detect hypochlorite (Zou et al. 2015). They incorporated the cyanine dye hCy3 with long alkyl side chains inside the bilayer formed by an amphiphilic polymer around oleate-capped $NaYF_4$:Yb,Nd,Er@$NaYF_4$:Nd nanoparticles. The hCy3 dye was able to absorb the green emission of the UCNPs. The fluorescent dye reacted irreversibly with ClO^- leading to an increase of the luminescence intensity of the upconversion nanocrystals. Hypochlorite pretreated HeLa cells were incubated with the 808 nm excitable nanoparticles to detect the pathogen using confocal microscopy.

Aside from entrapment within the surfactant bilayer, the dyes can also be covalently attached to the hydrophilic part of the amphiphilic molecules. This is required if the dyes are too hydrophilic for efficient vdW interactions with the alkyl chains or too large for stable incorporation in the bilayer in order to prevent leakage.

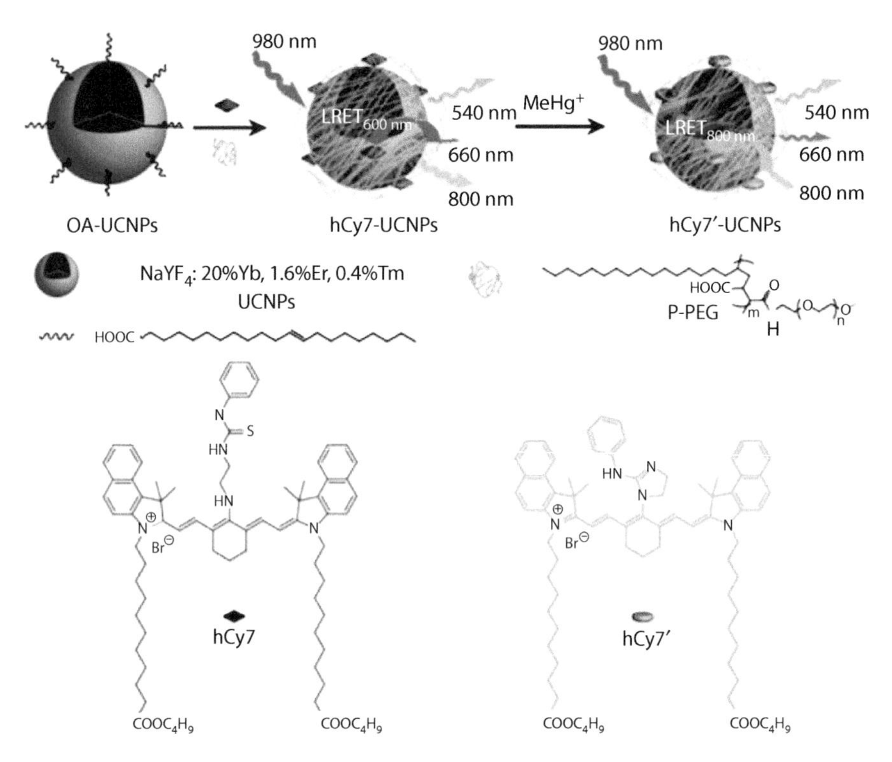

FIGURE 4.9
Schematic illustration of the synthesis of UCNPs-hCy7 and its sensing principle to $MeHg^{+}$ with a change in upconversion emission. (Reprinted with permission from Liu et al. 2013. Copyright 2013 American Chemical Society.)

With this strategy, UCNPs coated with both a porphyrin-functionalized phospholipid and a PEG–phospholipid suitable for multimodal imaging were obtained (Rieffel et al. 2015). Despite only two imaging components were used, the UCNP itself and the porphyrin, the particles were active in six different imaging modalities, including positron emission tomography, X-ray computed tomography, and photoacoustic imaging. This was demonstrated by *in vivo* lymphatic imaging in mice and revealed significant contrast enhancement.

High stability against leaching of the dyes from the surface of the UCNPs is provided by silica coating with subsequent covalent attachment of the dye. The conjugation can be performed by either surface silanization with silane-modified dyes or by coupling to functional groups on the silica surface.

The former method was used for the design of core–shell nanocomposites for NIR imaging and photothermal therapy (Shan et al. 2013). $NaYF_4$:Yb,Er UCNPs were encapsulated inside a silica shell by the polymerization of TEOS via the reverse microemulsion method. Secondly, the silane-modified NIR canine dye CyTE-777-triethoxysilane was covalently bound to the silica shell. These dye carrying nanoparticles were colloidally stable in water for over 1 year. Incubation of macrophage cells with the nanocomposites enabled upconversion imaging upon excitation at 980 nm. Excitation of the organic dye loaded on the surface of the UCNPs led to efficient cell death and a cell viability of only 50% after 1 h, proving the photothermal treatment capabilities of the particles.

Functional groups can be integrated into the silica shell by the addition of functional organosilanes during the growth period of the silica shell. Amino groups were generated by the application of, for example, (N-(3-trimethoxysilyl)propyl)ethylene diamine (Arppe et al. 2014) and 3-(triethoxysilyl)propyl isocyanate (Ma et al. 2015) in the polymerization phase. Afterwards, pH-sensitive organic dyes, pHrodo™ Red and xylene orange, displaying a spectral overlap with one emission band of the UCNPs were coupled to those amino groups via EDC/NHS chemistry. The ratiometric readout of pH-induced changes in the emission intensities caused by shifts in the absorption spectra of the dyes revealed a resolution of 0.3 pH units. Furthermore, as the UCNP–dye composites were colloidally stable under physiological conditions, they were used for the imaging of intracellular pH values inside HeLa cells.

Dyes can also be directly and covalently integrated into the silica shell (Zhou et al. 2015). In one example, first a silica shell was formed around UCNPs with three different dopants by the polymerization of TEOS. Then, either aminopropyl triethoxysilane-modified rhodamine B isothiocyanate or nickel (II) phthalocyanine-tetrasulfonic acid was added and polymerized on top of the previous silica shell. This dye-containing layer served as an emission filter to generate single-band emitting UCNPs and provided dispersibility and stability in physiological buffers. This way, single green, red, and blue emitting UCNPs were obtained by filtering either the red or the green emission in Yb,Er- and Yb,Tm-doped particles, respectively, with the dye layer. The conjugation to specific antibodies was performed by coupling to the amino groups of the dye-doped silica shell. The surface engineering leading to these single-band emitting conjugates is shown in Figure 4.10. The excellent properties of the modified UCNPs were demonstrated by multiplexed imaging of cancer biomarkers.

Although intentional in the last example, the downside of the modification with a silica shell regarding energy transfer efficiency is the significant increase of the distance between the donor UCNPs and the acceptor dyes. Mesoporous or rattle-structured (J. Liu et al. 2014) silica at least partly bypasses this problem, since the attachment of the dye does not only occur at the outer surface of the silica shell, but also within the whole shell itself. This also significantly increases the loading capacity of the whole system due to a greatly enlarged surface area available for the immobilization.

Mesoporous silica ($mSiO_2$) is usually synthesized by removing the template for the porous structure after the polymerization of the silicate precursors. In a first step, the silica shell is completely formed around UCNPs coated with surfactants, such as hexadecyl trimethylammonium bromide (CTAB).

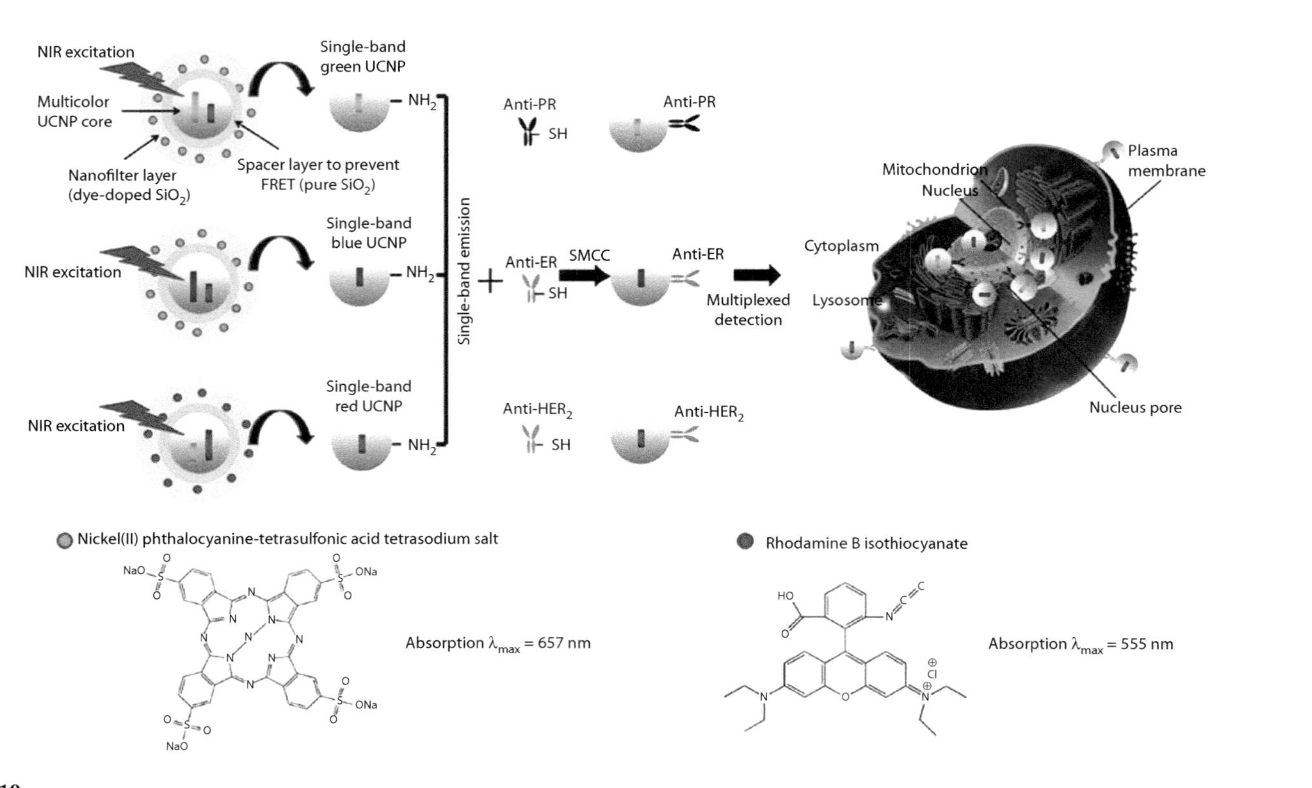

FIGURE 4.10
Surface amino modifications of the multilayer structure of green, blue, and red single-band emitting UCNPs and conjugates with antibodies for multiplexed *in situ* molecular mapping of breast cancer biomarkers. (Reprinted with permission from Zhou, L., R. Wang, C. Yao et al. 2015. Single-band upconversion nanoprobes for multiplexed simultaneous *in situ* molecular mapping of cancer biomarkers. *Nat. Commun.* 6, 6938, distributed under a Creative Commons CC-BY license.)

Subsequently, the surfactant can be removed by ion exchange, resulting in the formation of disordered, wormhole-like structures inside the silica shell (S. Liu et al. 2014). Particles covered with $mSiO_2$ displayed colloidal stability in PBS. The pores were, for example, loaded with a merocyanine (MC) dye, which is reactive against hydrogen sulfide. Its oxidized form possesses a spectral overlap with the green UCNP emission, while the reduced structure formed in presence of HS^- does not, as illustrated in Figure 4.11. This allowed

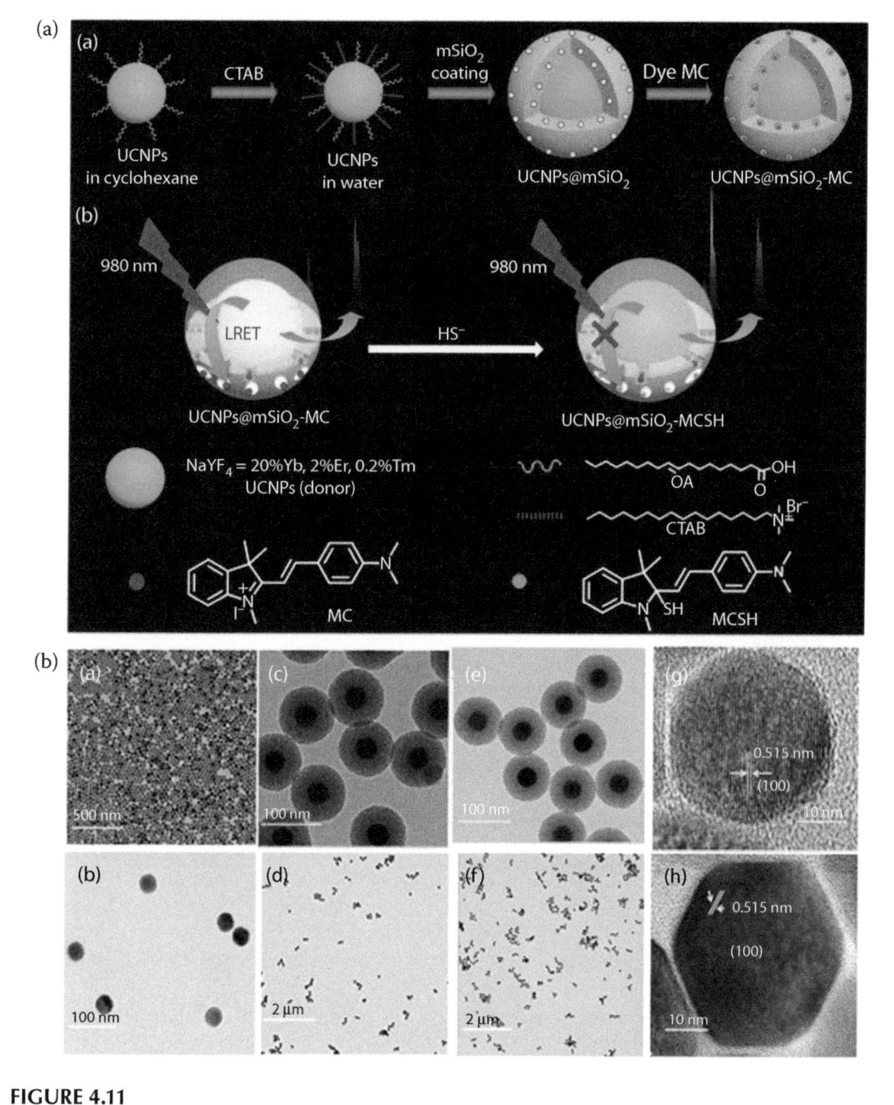

FIGURE 4.11
(a) Design strategy and synthetic route of nanoprobe UCNPs@SiO_2-MC. (b) Sensing mechanism and energy transfer process of UCNPs@SiO_2-MC toward HS^-. (Adapted with permission from Liu et al. 2014. Copyright 2014 American Chemical Society.)

the ratiometric evaluation of the luminescence intensities with a detection limit of 0.58 μM.

4.7 Conclusion

The most common surface modifications of hydrophobic UCNPs are the ligand exchange method and the encapsulation with silica to obtain water stable dispersions. For the ligand exchange strategy, PAA is the most popular hydrophilic ligand, due to its high binding affinity to the surface of the nanoparticles utilizing the chelate effect. Direct attachment of the signal molecules is also a frequently performed method requiring only functional groups capable of coordinating to the lanthanide ions on the surface. Encapsulation with inorganic silica shells creates colloidally stable particles in aqueous media. One drawback of the silica coating is the tendency for agglomeration, especially after centrifugation, which can be avoided by smart surface engineering.

The modified nanoparticles are often colloidally stable in HEPES and tris(hydroxymethyl)aminomethane (TRIS) buffer; however, they suffer from agglomeration in phosphate-buffered solutions. To achieve stability in PBS, coating with phospholipids and ligand exchange with phosphonic acid groups have proven their applicability in biological samples. For further (bio)modification, almost exclusively classical EDC chemistry or maleimide chemistry are used to bind DNA, dyes, and proteins to the surface of the particles. The overall well-established click chemistry is scarcely applied for bioconjugation of UCNPs (C. Liu et al. 2014; T. Wu et al. 2013). Very often the aspects of crosslinking and agglomeration are not further discussed. Table 4.1 gives a general overview of aspects to consider

TABLE 4.1

Suitability of Surface Modification Strategies Regarding the Final Application

Application	Surface Modification Strategy	Advantages
Sensors (via energy transfer)	Ligand exchange, mesoporous silica shell	Minimum distance between donor and acceptor
Electrophoresis	Amphiphilic polymers, silica coating	Stability against ligand detachment
Cellular imaging	Silica coating, amphiphilic polymers	Stability in cell culture media
Bioconjugation	Amphiphilic molecules, silica coating, ligand exchange	Variety of head groups available, stability in buffers
Cellular uptake	Amphiphilic polymers, silica shell	Mimicking of biological systems, tunable surface charge

when choosing the appropriate surface modification method for different applications.

One general issue of all surface modifications is the difficulty to compare individual methods and the absence of a standard for the surface characterization. The amount of surface ligands and signal molecules is often not characterized resulting in a low reproducibility of the techniques. For example, the amount of dye molecules loaded onto the surface of UCNPs nanoparticles is rarely specified or proofed. Due to the high dye loading capacities of the UCNPs, absorbance and fluorescence measurements are only conditionally suitable.

Therefore, inductively coupled plasma measurements, nuclear magnetic resonance spectroscopy, and thermogravimetric analysis should be generally established to characterize the surface of the modified UCNPs.

Despite the challenges in characterization and control of the surface properties of functionalized UCNPs, numerous bioapplications of UCNPs clearly demonstrate the outstanding advantage of almost background free measurements upon NIR excitation. Nevertheless, the strong water absorption at the typical excitation wavelength of 980 nm causes heating of biological samples and low upconversion efficiency. This drawback has recently been overcome by the design of core–shell architectures with additional Nd^{3+} ions as sensitizers, shifting the excitation wavelength to 808 nm. It is expected that this class of UCNPs will gain further interest in upconversion technologies, especially for bioanalytical applications.

Acknowledgment

We acknowledge the COST action CM1403 "The European Upconversion Network: From the Design of Photon-Upconverting Nanomaterials to Biomedical Applications."

References

Alonso-Cristobal, P., P. Vilela, A. El-Sagheer et al. 2015. Highly sensitive DNA sensor based on upconversion nanoparticles and graphene oxide. *ACS Appl. Mater. Inter.* 7:12422–12429.

Arppe, R., T. Nareoja, S. Nylund et al. 2014. Photon upconversion sensitized nanoprobes for sensing and imaging of pH. *Nanoscale* 6:6837–6843.

Bagwe, R. P., L. R. Hilliard, and W. Tan. 2006. Surface modification of silica nanoparticles to reduce aggregation and nonspecific binding. *Langmuir* 22:4357–4362.

Bogdan, N., F. Vetrone, G. A. Ozin, and J. A. Capobianco. 2011. Synthesis of ligand-free colloidally stable water dispersible brightly luminescent lanthanide-doped upconverting nanoparticles. *Nano Lett.* 11:835–840.

Boyer, J.-C., L. A. Cuccia, and J. A. Capobianco. 2007. Synthesis of colloidal upconverting $NaYF_4$: Er^{3+}/Yb^{3+} and Tm^{3+}/Yb^{3+} monodisperse nanocrystals. *Nano Lett.* 7:847–852.

Boyer, J.-C., F. Vetrone, L. A. Cuccia, and J. A. Capobianco. 2006. Synthesis of colloidal upconverting $NaYF_4$ nanocrystals doped with Er^{3+}, Yb^{3+} and Tm^{3+}, Yb^{3+} via thermal decomposition of lanthanide trifluoroacetate precursors. *J. Am. Chem. Soc.* 128:7444–7445.

Chen, G., H. Qiu, P. N. Prasad, and X. Chen. 2014. Upconversion nanoparticles: Design, nanochemistry, and applications in theranostics. *Chem. Rev.* 114:5161–5214.

Chien, Y.-H., Y.-L. Chou, S.-W. Wang et al. 2013. Near-infrared light photocontrolled targeting, bioimaging, and chemotherapy with caged upconversion nanoparticles *in vitro* and *in vivo*. *ACS Nano* 7:8516–8528.

Ding, H., Y. Lv, D. Ni et al. 2015. Erythrocyte membrane-coated NIR-triggered biomimetic nanovectors with programmed delivery for photodynamic therapy of cancer. *Nanoscale* 7:9806–9815.

Ding, Y., F. Wu, Y. Zhang et al. 2015. Interplay between static and dynamic energy transfer in biofunctional upconversion nanoplatforms. *J. Phys. Chem. Lett.* 6:2518–2523.

Dong, A., X. Ye, J. Chen et al. 2011. A generalized ligand-exchange strategy enabling sequential surface functionalization of colloidal nanocrystals. *J. Am. Chem. Soc.* 133:998–1006.

Doughan, S., Y. Han, U. Uddayasankar, and U. J. Krull. 2014. Solid-phase covalent immobilization of upconverting nanoparticles for biosensing by luminescence resonance energy transfer. *ACS Appl. Mater. Inter.* 6:14061–14068.

Gao, H.-D., P. Thanasekaran, C.-W. Chiang et al. 2015. Construction of a near-infrared-activatable enzyme platform to remotely trigger intracellular signal transduction using an upconversion nanoparticle. *ACS Nano* 9:7041–7051.

Haase, M. and H. Schäfer. 2011. Upconverting nanoparticles. *Angew. Chem. Int. Ed.* 50:5808–5829.

Huang, L.-J., X. Tian, J.-T. Yi, R.-Q. Yu, and X. Chu. 2015a. A turn-on upconversion fluorescence resonance energy transfer biosensor for ultrasensitive endonuclease detection. *Anal Methods* 7:7474–7479.

Huang, L.-J., R.-Q. Yu, and X. Chu. 2015b. DNA-functionalized upconversion nanoparticles as biosensors for rapid, sensitive, and selective detection of Hg^{2+} in complex matrices. *Analyst* 140:4987–4990.

Idris, N. M., M. K. Gnanasammandhan, J. Zhang, P. C. Ho, R. Mahendran, and Y. Zhang. 2012. *In vivo* photodynamic therapy using upconversion nanoparticles as remote-controlled nanotransducers. *Nat. Med.* 18:1580–1585.

Jalil, R. A. and Y. Zhang. 2008. Biocompatibility of silica coated $NaYF_4$ upconversion fluorescent nanocrystals. *Biomaterials* 29:4122–4128.

Jayakumar, M. K. G., N. M. Idris, K. Huang, and Y. Zhang. 2014. A paradigm shift in the excitation wavelength of upconversion nanoparticles. *Nanoscale* 6:8441–8443.

Jiang, G., J. Pichaandi, N. J. J. Johnson, R. D. Burke, and F. C. J. M. van Veggel. 2012. An effective polymer cross-linking strategy to obtain stable dispersions of upconverting $NaYF_4$ nanoparticles in buffers and biological growth media for biolabeling applications. *Langmuir* 28:3239–3247.

Johnson, N. J. J., A. Korinek, C. Dong, and F. C. J. M. van Veggel. 2012. Self-focusing by Ostwald ripening: A strategy for layer-by-layer epitaxial growth on upconverting nanocrystals. *J. Am. Chem. Soc.* 134:11068–11071.

Johnson, N. J. J., N. M. Sangeetha, J.-C. Boyer, and F. C. J. M. van Veggel. 2010. Facile ligand-exchange with polyvinylpyrrolidone and subsequent silica coating of hydrophobic upconverting β-$NaYF_4$:Yb^{3+}/Er^{3+} nanoparticles. *Nanoscale* 2:771–777.

Lan, J., F. Wen, F. Fu et al. 2015. A photoluminescent biosensor based on long-range self-assembled DNA cascades and upconversion nanoparticles for the detection of breast cancer-associated circulating microRNA in serum samples. *RSC Adv.* 5:18008–18012.

Li, C., Z. Hou, Y. Dai et al. 2013. A facile fabrication of upconversion luminescent and mesoporous core–shell structured β-$NaYF_4$: Yb^{3+} Er^{3+}@$mSiO_2$ nanocomposite spheres for anti-cancer drug delivery and cell imaging. *Biomater. Sci.* 1:213–223.

Li, K., Q. Su, W. Yuan et al. 2015. Ratiometric monitoring of intracellular drug release by an upconversion drug delivery nanosystem. *ACS Appl. Mater. Inter.* 7:12278–12286.

Li, L.-L., P. Wu, K. Hwang, and Y. Lu. 2013. An exceptionally simple strategy for DNA-functionalized up-conversion nanoparticles as biocompatible agents for nanoassembly, DNA delivery, and imaging. *J. Am. Chem. Soc.* 135:2411–2414.

Li, X., Y. Wu, Y. Liu, X. Zou, L. Yao, F. Li, and W. Feng. 2014. Cyclometallated ruthenium complex-modified upconversion nanophosphors for selective detection of Hg^{2+} ions in water. *Nanoscale* 6:1020–1028.

Li, Y., Z. Wu, and Z. Liu. 2015. An immune sandwich assay of carcinoembryonic antigen based on the joint use of upconversion phosphors and magnetic beads. *Analyst* 140:4083–4088.

Li, Z., T. Liang, S. Lv, Q. Zhuang, and Z. Liu. 2015. A rationally designed upconversion nanoprobe for *in vivo* detection of hydroxyl radical. *J. Am. Chem. Soc.* 137:11179–11185.

Liang, S., X. Zhang, Z. Wu et al. 2012. Decoration of up-converting $NaYF_4$:Yb,Er(Tm) nanoparticles with surfactant bilayer. A versatile strategy to perform oil-to-water phase transfer and subsequently surface silication. *CrystEngComm* 14:3484–3489.

Liu, C., W. Ma, Z. Gao et al. 2014. Upconversion luminescence nanoparticles-based lateral flow immunochromatographic assay for cephalexin detection. *J. Mater. Chem. C* 2:9637–9642.

Liu, F., Q. Zhao, H. You, and Z. Wang. 2013. Synthesis of stable carboxy-terminated $NaYF_4$: Yb^{3+}, Er^{3+}@SiO_2 nanoparticles with ultrathin shell for biolabeling applications. *Nanoscale* 5:1047–1053.

Liu, J., W. Bu, L. Pan, and J. Shi. 2013. NIR-triggered anticancer drug delivery by upconverting nanoparticles with integrated azobenzene-modified mesoporous silica. *Angew. Chem. Int. Ed.* 52:4375–4379.

Liu, J., Y. Liu, W. Bu et al. 2014. Ultrasensitive nanosensors based on upconversion nanoparticles for selective hypoxia imaging *in vivo* upon near-infrared excitation. *J. Am. Chem. Soc.* 136:9701–9709.

Liu, S., L. Zhang, T. Yang et al. 2014. Development of upconversion luminescent probe for ratiometric sensing and bioimaging of hydrogen sulfide. *ACS Appl. Mater. Inter.* 6:11013–11017.

Liu, Y., M. Chen, T. Cao et al. 2013. A cyanine-modified nanosystem for *in vivo* upconversion luminescence bioimaging of methylmercury. *J. Am. Chem. Soc.* 135:9869–9876.

Lu, S., D. Tu, P. Hu et al. 2015. Multifunctional nano-bioprobes based on rattle-structured upconverting luminescent nanoparticles. *Angew. Chem. Int. Ed.* 54:7915–7919.

Ma, C., T. Bian, S. Yang et al. 2014. Fabrication of versatile cyclodextrin-functionalized upconversion luminescence nanoplatform for biomedical imaging. *Anal. Chem.* 86:6508–6515.

Ma, T., Y. Ma, S. Liu et al. 2015. Dye-conjugated upconversion nanoparticles for ratiometric imaging of intracellular pH values. *J. Mater. Chem. C* 3:6616–6620.

Min, Y., J. Li, F. Liu, E. K. Yeow, and B. Xing. 2014. Near-infrared light-mediated photoactivation of a platinum antitumor prodrug and simultaneous cellular apoptosis imaging by upconversion luminescent nanoparticles. *Angew. Chem. Int. Ed.* 53:1012–1016.

Muhr, V., S. Wilhelm, T. Hirsch, and O. S. Wolfbeis. 2014. Upconversion nanoparticles: From hydrophobic to hydrophilic surfaces. *Acc. Chem. Res.* 47:3481–3493.

Nam, S. H., Y. M. Bae, Y. I. Park et al. 2011. Long-term real-time tracking of lanthanide ion doped upconverting nanoparticles in living cells. *Angew. Chem.* 123:6217–6221.

Ni, D., J. Zhang, W. Bu et al. 2014. Dual-targeting upconversion nanoprobes across the blood-brain barrier for magnetic resonance/fluorescence imaging of intracranial glioblastoma. *ACS Nano* 8:1231–1242.

Ong, L. C., L. Y. Ang, S. Alonso, and Y. Zhang. 2014. Bacterial imaging with photostable upconversion fluorescent nanoparticles. *Biomaterials* 35:2987–2998.

Park, Y. I., J. H. Kim, K. T. Lee et al. 2009. Nonblinking and nonbleaching upconverting nanoparticles as an optical imaging nanoprobe and T1 magnetic resonance imaging contrast agent. *Adv. Mater.* 21:4467–4471.

Pellegrino, T., L. Manna, S. Kudera et al. 2004. Hydrophobic nanocrystals coated with an amphiphilic polymer shell: A general route to water soluble nanocrystals. *Nano Lett.* 4:703–707.

Peng, J., W. Xu, C. L. Teoh et al. 2015. High-efficiency *in vitro* and *in vivo* detection of Zn^{2+} by dye-assembled upconversion nanoparticles. *J. Am. Chem. Soc.* 137:2336–2342.

Qiao, R., C. Liu, M. Liu et al. 2015. Ultrasensitive *in vivo* detection of primary gastric tumor and lymphatic metastasis using upconversion nanoparticles. *ACS Nano* 9:2120–2129.

Ren, W., G. Tian, S. Jian et al. 2012. TWEEN coated $NaYF_4$:Yb,Er/$NaYF_4$ core/shell upconversion nanoparticles for bioimaging and drug delivery. *RSC Adv.* 2:7037–7041.

Rieffel, J., F. Chen, J. Kim et al. 2015. Hexamodal imaging with porphyrin–phospholipid-coated upconversion nanoparticles. *Adv. Mater.* 27:1785–1790.

Rinkel, T., J. Nordmann, A. N. Raj, and M. Haase. 2014. Ostwald-ripening and particle size focusing of sub-10 nm $NaYF_4$ upconversion nanocrystals. *Nanoscale* 6:14523–14530.

Shan, G., R. Weissleder, and S. A. Hilderbrand. 2013. Upconverting organic dye doped core–shell nano-composites for dual-modality NIR imaging and photothermal therapy. *Theranostics* 3:267–274.

Song, K., X. Kong, X. Liu et al. 2012. Aptamer optical biosensor without bio-breakage using upconversion nanoparticles as donors. *Chem. Commun.* 48:1156–1158.

Suyver, J. F., J. Grimm, M. K. van Veen, D. Biner, K. W. Krämer, and H. U. Güdel. 2006. Upconversion spectroscopy and properties of $NaYF_4$ doped with Er^{3+}, Tm^{3+} and/or Yb^{3+}. *J. Lumin.* 117:1–12.

Tong, L., E. Lu, J. Pichaandi, P. Cao, M. Nitz, and M. A. Winnik. 2015. Quantification of surface ligands on $NaYF_4$ nanoparticles by three independent analytical techniques. *Chem. Mater.* 27:4899–4910.

Wang, C., L. Cheng, Y. Liu et al. 2013. Imaging-guided pH-sensitive photodynamic therapy using charge reversible upconversion nanoparticles under near-infrared light. *Adv. Funct. Mater.* 23:3077–3086.

Wang, C., H. Tao, L. Cheng, and Z. Liu. 2011. Near-infrared light induced *in vivo* photodynamic therapy of cancer based on upconversion nanoparticles. *Biomaterials* 32:6145–6154.

Wang, F. and X. Liu. 2009. Recent advances in the chemistry of lanthanide-doped upconversion nanocrystals. *Chem. Soc. Rev.* 38:976–989.

Wang, F. and X. Liu. 2014. Multicolor tuning of lanthanide-doped nanoparticles by single wavelength excitation. *Acc. Chem. Res.* 47:1378–1385.

Wang, F., J. Wang, and X. Liu. 2010. Direct evidence of a surface quenching effect on size-dependent luminescence of upconversion nanoparticles. *Angew. Chem. Int. Ed.* 49:7456–7460.

Wang, H., Z. Liu, S. Wang et al. 2014. MC540 and upconverting nanocrystal coloaded polymeric liposome for near-infrared light-triggered photodynamic therapy and cell fluorescent imaging. *ACS Appl. Mater. Inter.* 6:3219–3225.

Wang, J., T. Wei, X. Li et al. 2014. Near-infrared-light-mediated imaging of latent fingerprints based on molecular recognition. *Angew. Chem.* 126:1642–1646.

Wang, L., W. Qin, Z. Liu et al. 2012. Improved 800 nm emission of Tm^{3+} sensitized by Yb^{3+} and Ho^{3+} in β-$NaYF_4$ nanocrystals under 980 nm excitation. *Opt. Express* 20:7602.

Wang, M., C.-C. Mi, W.-X. Wang et al. 2009a. Immunolabeling and NIR-excited fluorescent imaging of HeLa cells by using $NaYF_4$:Yb,Er upconversion nanoparticles. *ACS Nano* 3:1580–1586.

Wang, M., C. Mi, Y. Zhang, J. Liu, F. Li, C. Mao, and S. Xu. 2009b. NIR-responsive silica-coated $NaYbF_4$:Er/Tm/Ho upconversion fluorescent nanoparticles with tunable emission colors and their applications in immunolabeling and fluorescent imaging of cancer cells. *J. Phys. Chem. C* 113:19021–19027.

Wang, Y.-F., G.-Y. Liu, L.-D. Sun, J.-W. Xiao, J.-C. Zhou, and C.-H. Yan. 2013. Nd^{3+}-sensitized upconversion nanophosphors: Efficient *in vivo* bioimaging probes with minimized heating effect. *ACS Nano* 7:7200–7206.

Wang, Z., X. Li, Y. Song, L. Li, W. Shi, and H. Ma. 2015. An upconversion luminescence nanoprobe for the ultrasensitive detection of hyaluronidase. *Anal. Chem.* 87:5816–5823.

Wilhelm, S., M. Barrio, J. Heiland et al. 2014. Spectrally matched upconverting luminescent nanoparticles for monitoring enzymatic reactions. *ACS Appl. Mater. Inter.* 6:15427–15433.

Wilhelm, S., T. Hirsch, W. M. Patterson, E. Scheucher, T. Mayr, and O. S. Wolfbeis. 2013. Multicolor upconversion nanoparticles for protein conjugation. *Theranostics* 3:239–248.

Wilhelm, S., M. Kaiser, C. Würth et al. 2015. Water dispersible upconverting nanoparticles: Effects of surface modification on their luminescence and colloidal stability. *Nanoscale* 7:1403–1410.

Wu, S., N. Duan, X. Ma, Y. Xia, H. Wang, and Z. Wang. 2013. A highly sensitive fluorescence resonance energy transfer aptasensor for staphylococcal enterotoxin B detection based on exonuclease-catalyzed target recycling strategy. *Anal. Chim. Acta* 782:59–66.

Wu, S., N. Duan, Z. Shi, C. Fang, and Z. Wang. 2014. Simultaneous aptasensor for multiplex pathogenic bacteria detection based on multicolor upconversion nanoparticles labels. *Anal. Chem.* 86:3100–3107.
Wu, T., M. Barker, K. M. Arafeh, J.-C. Boyer, C.-J. Carling, and N. R. Branda. 2013. A UV-blocking polymer shell prevents one-photon photoreactions while allowing multi-photon processes in encapsulated upconverting nanoparticles. *Angew. Chem. Int. Ed.* 52:11106–11109.
Wu, Y.-M., Y. Cen, L.-J. Huang, R.-Q. Yu, and X. Chu. 2014. Upconversion fluorescence resonance energy transfer biosensor for sensitive detection of human immunodeficiency virus antibodies in human serum. *Chem. Commun.* 50 4759–4762.
Zeng, T., T. Zhang, W. Wei et al. 2015. Compact, programmable, and stable biofunctionalized upconversion nanoparticles prepared through peptide-mediated phase transfer for high-sensitive protease sensing and *in vivo* apoptosis imaging. *ACS Appl. Mater. Inter.* 7:11849–11856.
Zhang, F. 2014. *Photon Upconversion Nanomaterials. Nan Sci Tec.*, Springer, New York.
Zhang, P., W. Steelant, M. Kumar, and M. Scholfield. 2007. Versatile photosensitizers for photodynamic therapy at infrared excitation. *J. Am. Chem. Soc.* 129:4526–4527.
Zhao, G., L. Tong, P. Cao, M. Nitz, and M. A. Winnik. 2014. Functional PEG–PAMAM-tetraphosphonate capped $NaLnF_4$ nanoparticles and their colloidal stability in phosphate buffer. *Langmuir* 30:6980–6989.
Zhou, F., M. O. Noor, and U. J. Krull. 2014. Luminescence resonance energy transfer-based nucleic acid hybridization assay on cellulose paper with upconverting phosphor as donors. *Anal. Chem.* 86:2719–2726.
Zhou, J., Z. Liu, and F. Li. 2012. Upconversion nanophosphors for small-animal imaging. *Chem. Soc. Rev.* 41:1323–1349.
Zhou, L., R. Wang, C. Yao et al. 2015. Single-band upconversion nanoprobes for multiplexed simultaneous *in situ* molecular mapping of cancer biomarkers. *Nat. Commun.* 6:6938.
Zou, X., Y. Liu, X. Zhu et al. 2015. An Nd^{3+} sensitized upconversion nanophosphor modified with a cyanine dye for the ratiometric upconversion luminescence bioimaging of hypochlorite. *Nanoscale* 7:4105–4113.

5

Synergistic Effects in Organic-Coated Upconversion Nanoparticles

Laura Francés-Soriano, María González-Béjar, and Julia Pérez-Prieto

CONTENTS

5.1 Introduction 102
5.2 A Brief Explanation of Upconverting Nanomaterials 104
5.2.1 The Upconversion Phenomenon 104
5.2.2 Upconversion Materials 106
5.2.3 UC Material Multicolor Emission and Its Tuning 107
5.2.4 The Ligand Anchoring Group 108
5.3 Synergistic Effects between the UC Nanomaterial and the Organic Ligand 109
5.3.1 The Role of the Ligand in the Preparation of the Nanomaterial 109
5.3.2 The Role of the Ligand in the Colloidal Stability of UCNPs in Aqueous and Nonaqueous Solvents, as Well in the UCNP Phase Transfer 111
5.3.3 The Role of the Ligand in the Optical Properties of the Nanomaterial 114
5.3.4 The Role of the NPs in the Organic Ligand Functionality 115
5.3.5 Synergism in the Nanohybrid Functionality 115
5.3.5.1 Functionality 115
5.3.5.2 Synergistic Effect between the UCNP and the Organic Ligand in the Nanosystem Functionality: Encapsulation versus Ionic and Covalent Binding 115
5.3.5.3 Targeting 117
5.3.5.4 Sensing 118
5.3.5.5 Bioimaging 121
5.3.5.6 Therapy 122
5.3.5.7 Drug Delivery and Chemotherapy 126
5.4 Summary 129
References 130

5.1 Introduction

Lanthanide-doped upconversion nanoparticles (UCNPs), such as $NaYF_4$ doped with Yb^{3+} and Er^{3+}, exhibit narrow and high-intensive emissions in the visible spectrum when excited by a low power energy, continuous-wave (CW) near-infrared (NIR) laser, that is, a large anti-Stokes shift (see Figure 5.1). Of special relevance for biomedical applications, NIR light exhibits a remarkably deep penetration into tissues, and UCNPs present luminescence with a high signal-to-noise ratio, they are very resistant to photobleaching and photoblinking, they exhibit excellent chemical and thermal stability and relatively low toxicity, and they are suitable for long-time particle tracking (G. Chen et al. 2014; Dacosta et al. 2014; Gnach et al. 2015; Gu et al. 2013; Y. Liu et al. 2013; M. Wang et al. 2011). These unique properties not only make UCNPs particularly ideal in nanomedicine, but they are also promising in nanotechnology, since they can be used for photovoltaic cells (as NIR absorbers), photocatalysis (as sensitizers), and security applications (anti-counterfeiting materials) (G. Chen et al. 2015; Gnach and Bednarkiewicz 2012; Ramasamy et al. 2014; Won Jin et al. 2009).

A general feature common to all nanoparticles (NPs) is that they possess a high surface-to-volume ratio. By making use of this property, NPs can be capped with a considerable number of ligands, which have an anchoring atom at one end with affinity for the NP surface and the other end provides the NP periphery with the hydrophobicity or hydrophilicity needed to give rise to stable organic or aqueous NP colloidal solutions, respectively (See Figure 5.2) (Sedlmeier and Gorris 2015).

A highly relevant role that ligands can also play is that of introducing additional functionality to the nanosystem. In this case, the NP acts as a 3D-scaffold which makes it possible to provide a high local concentration of a functional moiety, such as fluorophores and photosensitizers. The UCNP combined with its organic capping can act as a nanohybrid with properties that differ from those of the individual components (Wolfbeis 2015). This symbiotic effect arises not only from the possibility of placing a high local concentration of a functional group on the NP surface, but

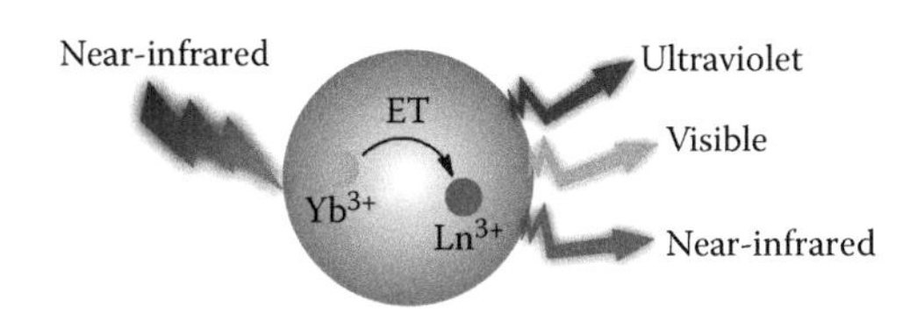

FIGURE 5.1
UCNP emission after excitation with NIR light.

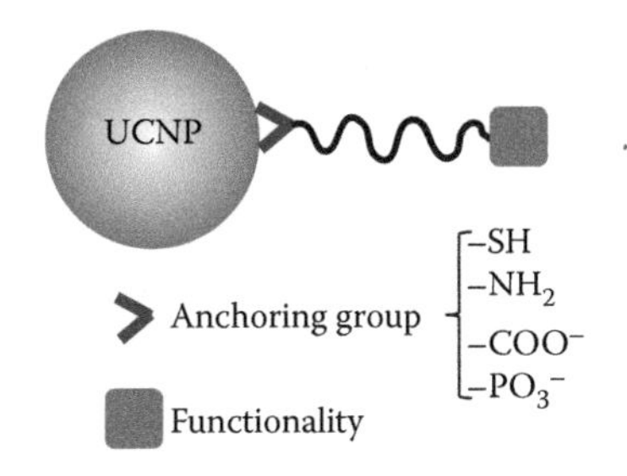

FIGURE 5.2
General representation of organic ligand coated UCNPs.

also from the potential incorporation of simultaneous functionalities on the NP capping, the high encapsulating capacity of the nanohybrid, the capacity of the ligand to produce an efficient interdigitation of amphiphilic molecules and the potential of the ligand to be used for establishing specific interactions, among others (Sedlmeier and Gorris 2015; Wolfbeis 2015). The symbiosis between NPs and their ligands enhances the unique properties of the upconversion (UC) material and the nanohybrid can be applied in molecular recognition, drug-delivery, phototherapy, and so on. Figure 5.3 shows a scheme of possible modifications which will be discussed in the following section.

In this chapter, strategies recently devised for the preparation of smart photoactive UCNPs will be analyzed in terms of the symbiotic effect between the organic and the inorganic components.

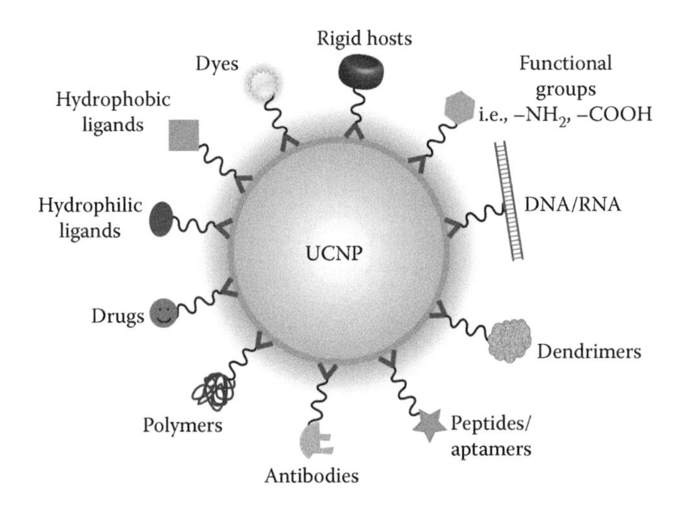

FIGURE 5.3
Possible modifications of UCNPs with organic ligands.

5.2 A Brief Explanation of Upconverting Nanomaterials

5.2.1 The Upconversion Phenomenon

NIR light can be used to excite UCNPs in complex mixtures that contain chromophores which absorb ultraviolet–visible (UV–Vis) light powerfully and this makes them relevant for biological and nanotechnological applications (Idris et al. 2015).

The UC process in lanthanides involves the sequential absorption of two or more NIR photons and it can take place via three competitive mechanisms: excited state absorption (ESA), energy transfer upconversion (ETU), and photon avalanche (PA), Figure 5.4a (Auzel 2004; Tu et al. 2015). These mechanisms differ in how the multiphoton absorption process occurs. In the ESA process, there is a successive absorption of two photons by a single ion: (1) the first photon is resonant with the transition from the ground state (GS) level to the excited metastable (ES1) level and (2) the second photon possesses energy that is resonant with the transition from ES1 to the higher excited state ES2. Then, the photon drops from ES2 to GS and the UC emission

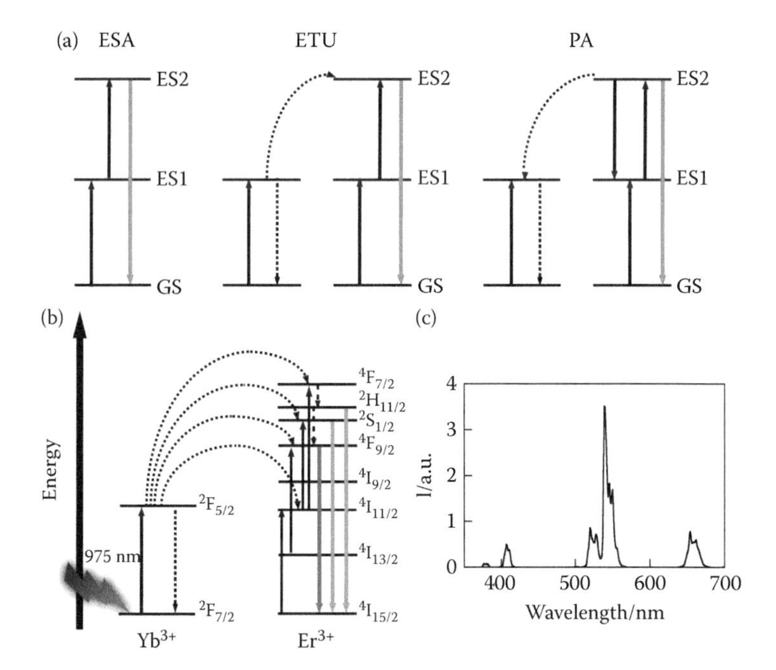

FIGURE 5.4
(a) UC processes for lanthanide-doped crystals: ESA, ETU, and PA. (b) Schematic representation of the ETU mechanism for a host matrix doped with Yb and Er ions; the dashed/dotted, dashed, and full arrows represent photon excitation, energy transfer, and emission processes, respectively, and (c) Emission spectrum (λ_{ex} = 975 nm) of $NaYF_4$:Yb, Er UCNPs.

occurs. This mechanism requires the concentration of the active ion to be low to avoid transfer loss through cross-relaxation between the luminescent centers and it is suitable for single-doped materials.

The ETU mechanism is based on a sequential absorption of two photons to populate the ES2 level. The main difference between ETU and ESA is that the excitation in ETU occurs through energy transfer between two adjacent ions. In the ETU process, each of two adjacent ions can absorb a photon of the same energy, thereby populating their excited level. A nonradiative energy transfer process promotes one of the ions to the upper ES2 and eventually leads to ES2–GS emission, while the other ion relaxes back to GS (Auzel 2004). In this type of UC process, the concentration of ions, which determines their average distance, has a strong influence on the UCNP emission efficiency. This is the most frequently observed and developed mechanism in UCNPs (Naccache et al. 2015).

In the PA process, the ES1 level of an ion is reached by a nonresonant excitation and then a cross-relaxation energy transfer occurs between the excited ion and the adjacent ion at the GS, thus resulting in the population of the intermediate level ES1 of both ions after a number of excitation cycles; the ES1 level acts as a storage reservoir for pumping energy. Subsequently, the two ions readily populate the ES2 level to initiate further cross-relaxation and the ES2 population increases exponentially, finally resulting in a strong UC emission as an avalanche process. The PA mechanism requires the pump intensity to reach a threshold value to generate a robust photocycle (Auzel 2004). This process suffers from several drawbacks, such as a strong dependence on the pump power and a slow response to excitation (up to several seconds) due to numerous looping cycles of cross-relaxation processes. This mechanism is hardly ever observed.

Figure 5.4b depicts the ETU mechanism for an UCNP consisting of a host matrix doped with Yb and Er ions. For this system, an electron of the Yb^{3+} is excited from the $^2F_{7/2}$ to $^2F_{5/2}$ level, and then the Yb^{3+} excited state transfers its energy to the Er^{3+} leading to the $^4I_{11/2}$ excited state. During the population of $^4I_{11/2}$, a second Yb^{3+}excited state transfers its energy to this level, thus populating a higher $^4F_{7/2}$ excited state of the Er^{3+} ion. The Er^{3+} ion can then relax nonradiatively to the $^2H_{11/2}$ and $^4S_{3/2}$ levels, leading to green emissions at 525 and 543 nm (Figure 5.4b). Competitively, the electron can further relax and populate the $^4F_{9/2}$ level, thus leading to a red emission at 659 nm. These are two-photon processes, but three-photon processes can also be observed, mainly when using a high pumping power. In the case of the Yb/Tm pair, two-, three-, and four-photon processes can be observed with emissions at 695 and 800 nm (two-photon), 450 and 650 nm (three-photon), and 360 and 450 nm (four-photon).

Whereas two-photon absorption processes are observed in some organic chromophores and NPs, such as CdSe quantum dots, that occur via "virtual states," the ETU in UCNPs occurs via the real long-lived excited states of the rare-earth ions in the UCNPs. As a consequence, UC luminescence can be observed by using inexpensive CW diode lasers with a moderate excitation power density of ca. 10 W cm^{-2}. By contrast, the simultaneous two-photon absorption processes in dyes require expensive ultra-short-pulsed lasers with high power density excitation (10^5–10^9 W cm^{-2}).

5.2.2 Upconversion Materials

The UC material is composed of a host matrix doped with an activator and a sensitizer, which are both rare earth (RE) ions (Gai et al. 2014). Whereas the sensitizer only possesses one excited state, the activator has ladder-like energy levels and the energy difference between three or more subsequent levels is close to that of the energy separation between the GS of the sensitizer and its single excited state in order to make the multiple photon absorption and energy transfer possible. The sensitizer possesses a higher absorption cross-section at the NIR region than the activator (e.g., 10 $M^{-1}cm^{-1}$ for Yb^{3+} vs. about 1 $M^{-1}cm^{-1}$ for Er^{3+}). The concentration of the sensitizer in the host matrix is larger than that of the activator to enhance the energy transfer from the sensitizer to the activator and to avoid self-quenching effects of the activator. Er^{3+}, Tm^{3+}, and Ho^{3+} are generally selected as activators for UC photoluminescence (Gai et al. 2014; Schäfer et al. 2007).

The host matrix plays a crucial role in the UC process. It should be taken into account that the $4f^n$ transitions of the RE ions (emissive species) are strongly forbidden by the parity selection rule (Naccache et al. 2015). As a consequence, the lanthanide ions usually exhibit low molar absorption coefficients and their excited states are long-lived (from μs to ms).

These transitions are perturbed by static crystalline fields and therefore they become partially allowed in the host matrix. This host lattice effect explains the higher emission of the bulk UC material than that of the nanoparticle, as well as the decrease of the emissive efficiency of a nanoparticle as its size decreases, specifically, the ratio of RE ions near the NP surface increases and the perturbation by the crystalline field decreases. The ions at the NP surface behave as "dormant ions."

In addition, the ideal host materials should offer a dense lattice to the dopant and should possess low lattice phonon energy. As the RE^{3+} ions and alkaline ions exhibit similar ionic sizes, their inorganic compounds are a good choice to reduce lattice impurities, which would quench the UCNP luminescence (Schäfer et al. 2007). A matrix with low phonon energy minimizes the highly competitive nonradiative processes, thus resulting in more emissive UCNPs. The most popular inorganic host materials for UCNPs are RE fluorides, since they have low phonon energies (about 350 cm^{-1}) and high chemical stability (Haase and Schäfer 2011).

In addition, the crystal structure phase of the host material is highly relevant for the emissive properties of the UCNP. Thus, the thermodynamically stable hexagonal (β) phase $NaYF_4$ (β-$NaYF_4$) is more suitable than the cubic (α) phase (α-$NaYF_4$), since the former is of lower symmetry. As a result, the intensity of the green emission of β-$NaYF_4$: Yb, Er is 10 times higher than that of the cubic phase.

There are several strategies to increase the luminescence efficiency of UCNPs. For instance, the dormant ions near the UCNP surface can become active ions by formation of an undoped shell of the same material as the host matrix (e.g., $NaYF_4$:Yb, Er@$NaYF_4$). This strategy has resulted in up to 30-fold enhanced emission compared to that of the core; this is ascribed to the shell crystalline field effect. In addition, the growth of a silica shell also results in more emissive UCNPs (e.g., $NaYF_4$:Yb, Er@SiO_2) (Xie et al. 2013). The enhanced emission due to the noncrystalline shell can be attributed to its low phonon energy and/or to an increased perturbation due to the ligand field effect. The improved spontaneous emission rate is also consistent with an enhanced local density of states brought about by the change of the refractive index at the interface between the core and the shell.

Moreover, it has also been demonstrated that the use of a fluoride host matrix with ytterbium instead of yttrium is beneficial for the UCNP efficiency. Thus, the NIR-to-NIR photoluminescence of $NaYbF_4$:Tm (2%) was four times higher than that of $NaYF_4$:Yb (20%), Tm (2%) (Chen et al. 2010).

In summary, UCNPs can be capped with an inorganic shell, whose role is to increase their luminescence (by reducing the amount of surface defects and/or isolating them from interactions with emission quenchers).

Furthermore, the organic capping can also be beneficial for the emissive properties of the UCNP although high energy C–H and C–C vibrational oscillators reduce the emission through nonradiative relaxation processes (this quenching effect is more important for long chains) (Heer et al. 2004; Suyver et al. 2005).

Although there are several strategies to enhance the emission of UCNPs, the quantum yield (Φ_f) of an UC process, defined as the ratio of photons absorbed by a fluorophore to those emitted via fluorescence, is quite low; two or more photons are required for UC emission and, therefore, its efficiency depends not only on the absorption cross-section of the sensitizer at the excitation wavelength but also on the power density of the excitation source, and as mentioned above it also depends on the particle size. Thus, the Φ_f of the green emission of $NaYF_4$:Er,Yb UCNPs at 150 W cm^{-2} power density changes from 0.005% to 0.1% to 0.3% when the nanoparticle size changes from 8–10 nm to 30 nm to 100 nm (Boyer and Van Veggel 2010). In addition, at a considerably lower power density (20 W cm^{-2}), the Φ_f of bulk $NaYF_4$:Er,Yb is 3.0%.

5.2.3 UC Material Multicolor Emission and Its Tuning

The relative intensities of the different emission peaks of UCNPs can be changed accurately by varying the dopant ion concentration, mainly that of the sensitizer (Li and Lin 2010). Thus, the green-to-red emission of $NaYF_4$:Yb, Er decreases by increasing Yb^{3+} concentrations. This has been ascribed to the decrease of the Yb–Er interatomic distance, which facilitates the back energy transfer from Er^{3+} to Yb^{3+}. Thus, the output color of these NPs has been changed from green to yellow (combination of red and green) and then to red by increasing the concentrations of Yb^{3+} ion (Guo et al. 2012).

The output emission of the UCNP also depends on the nature of the dopant ions. Thus, $NaYF_4$:Yb, Tm UCNPs emit blue since they possess a dominant emission in the visible at ca. 470 nm together with a weak one at ca. 650 nm, as well as a stronger emission in the NIR (at 800 nm).

In addition, the UCNP emission output can also be manipulated by using inorganic perovskites as host matrices (such as $KMnF_3$:Yb, Er), thus leading to UCNPs with a high-purity single-band emission (red emission in this case) (Lei et al. 2015). This is ascribed to an efficient nonradiative energy transfer from the level responsible for the green emission ($^2H_{11/2}$ and $^4S_{3/2}$ of Er) to the ^1T level of Mn^{2+}, followed by back energy transfer to the level responsible for the red emission ($^4F_{9/2}$ level of Er).

Finally, a white emission can be obtained by using a three-component dopant system. For example the $NaYF_4$:Yb,Er, Tm combines the red and green emissions of Er with the blue of Tm. The output emission of the system can be varied from blue to white by gradually increasing the Er concentration, that is, by increasing the intensity of the green and red emissions.

5.2.4 The Ligand Anchoring Group

Though the UCNPs can also be prepared as "naked" systems (Bogdan et al. 2011; A. Dong et al. 2011), they are usually covered with an organic capping (UCNP@organic capping), whose role is to keep them isolated and make them dispersible (in either organic and/or aqueous media) or even to provide them with an additional functionality. The organic ligand can play a key role in different stages of a nanoparticle, such as in its preparation, dispersibility in a medium, optical performance, and application in different areas.

Obviously, core and core–shell UCNPs (shell refers to inorganic materials) will eventually be capped by an organic capping, either by surface modification, for example, silica, or by noncovalent binding (Muhr et al. 2014).

Therefore, several strategies can be combined to obtain the best synergy possible between the organic coating and the UCNPs. In fact, the role of the organic capping is not only to determine the dispersibility of a nanoparticle in a specific medium, but also to provide the nanosystem with additional functionality and/or impede the UCNPs from causing any undesirable interference with the environment (e.g., cellular) or their aggregation. Therefore, the organic capping should remain firmly anchored to the UCNP.

Organic ligands are attached to the UCNP surface via an anchoring group (Figure 5.2). The most typical groups are NH_2 (e.g., oleylamine (OM)), SH (e.g., polyethyleneglycol [PEG]-SH (Voliani et al. 2013)), COOH (e.g., oleic acid (OA)), and PO_3H (e.g., PEG-PO_3H (Boyer et al. 2010)). However, gradual desorption of the ligands from the NP surface occurs due to the noncovalent linkage. This desorption can be decreased by using multichelating ligands, such as polymers and dendrimers. Among multichelating polymers are (i) polyacrylic acid (PAA) and polyanionic dendrimers with carboxylate groups, (ii) poly (amidoamine) and branched polyethylenimine (PEI) with amino groups, (iii) p(MEO_2MA-co-SEMA) copolymer (where MEO_2MA is 2-(2-methoxyethoxy)ethyl methacrylate and SEMA is thiolate-containing methacrylate) with thiolate groups, polyphosphoric acid with phosphonic groups, and (iv) polyvinylpyrrolidone (PVP) with amide groups (Cheng and Lin 2015; Liras et al. 2014).

In addition, rigid dendrimers have been used as the organic capping of UCNPs. These dendrimers are anchored to the UCNP surface via carboxylate groups, but their globular shape permits a large number of the carboxylates to be at the nanoparticle periphery, thus enhancing the water-dispersibility of the UCNPs as well as offering the possibility of further conjugation (Esipova et al. 2012). The most typical anchoring group of UCNPs is the carboxylate. However, the carboxylic acid pKa is ca. 5 (e.g., in OA) and therefore the carboxylate anchoring group becomes progressively protonated in acidic media and consequently the carboxylic ligand eventually separates from the UCNP surface. In fact, this is the strategy (protonation of oleate-capped UCNP (UCNP@OA) at pH below 4) used to prepare "naked" UCNPs (Bogdan et al. 2011). Leaching does not only apply to carboxylate ligands but also to ligands anchored via other groups.

5.3 Synergistic Effects between the UC Nanomaterial and the Organic Ligand

The *organic ligand* can be crucial at different stages of a nanoparticle, from the nanoparticle preparation to its dispersibility in a medium, from its optical performance to its application in different areas. *The nanoparticle* can act as a carrier of functional organic molecules, making it possible to increase the local concentration of a functional molecule. In addition, *the nanoparticle and its organic ligand* can cooperate to encapsulate functional molecules close to the nanoparticle surface. Examples of the synergistic effect between the inorganic component (the nanoparticle) and the organic component (the organic ligand) as well as its relevance to provide smart nanosystems will be presented in the following sections.

5.3.1 The Role of the Ligand in the Preparation of the Nanomaterial

Upconverting nanomaterials can be prepared by several methods: coprecipitation, thermal decomposition, hydro(solvo)thermal synthesis, sol–gel processing, and combustion synthesis (Chen et al. 2012; Wang and Liu 2009).

The reaction temperature, time, and composition of the reaction mixture can determine the size, phase, or morphology of the UCNPs (Chen et al. 2012; Guo et al. 2012; Naccache et al. 2015; Pan et al. 2013; Voliani et al. 2014; Wei, et al. 2006; Yang et al. 2012; Ye et al. 2010).

However, the anchoring ligand can also play a crucial role on such features. These NPs are usually synthesized in the presence of a noncoordinating solvent with a high boiling point (octadecene, ODE) and a capping ligand (OA) that prevents their agglomeration (Chan 2015; Li and Lin 2010; M. Wang et al. 2011) and makes them dispersible in organic solvents (F. Zhang). Alternatively, other solvents/ligands, for example, trioctylphosphine (TOP) (Shan and Ju 2009), OM (Yi and Chow 2006), and trioctylphosphine oxide (TOPO) (Zhuravleva et al. 2005) can be used for the synthesis of UCNPs.

The amount of ligand employed during the synthesis can determine the shape of the resulting UCNPs (Guo et al. 2012; Li and Zhang 2008; Mai et al. 2006; Zhang et al. 2005). Li and Zhang showed that the ligand concentration in the reaction mixture affects the growth rate of the nanocrystals in the [100] and [001] directions. It is faster in the [100] direction at low ligand concentration, both rates are similar when the ligand concentration increases, and it is faster in the [001] direction at high concentrations. Therefore, hexagonal nanoplates, spherical nanocrystals, and elliptic nanocrystals can be obtained by changing the ligand concentration. Similar results have been obtained by J. Guo et al. when increasing the concentration of OA.

H.-X. Mai et al. have found that addition of small amounts of OM to the reaction mixture containing OA/ODE increases the energy barrier between α-$NaReF_4$ and β-$NaReF_4$ phases and consequently, the presence of large amounts of OM decreases the growth rate of β-$NaReF_4$.

In addition, OA/OM/ODE mixtures can lead to ultrasmall UCNPs (7–10 nm) which are attractive for bioimaging, since they are easily eliminated from the body (Chen et al. 2010). Alternatively, ultrasmall hexagonal $NaYF_4$:Yb,Er(Tm) UCNPs can be obtained by using OM as both ligand and solvent (Yi and Chow 2006). The coordination of lanthanide cations with amines is weaker than with carboxylate groups and OM facilitates the phase transition of $NaYF_4$:Yb,Er(Tm) from cubic to hexagonal in comparison with OA.

Also, TOP has been used as capping ligand in combination with OA (Shan and Ju 2009; Shan et al. 2007) to lead to OA–TOP co-capped $NaYF_4$ NPs. TOP is crucial in stopping the NPs from aggregating and allowing the three-dimensional growth of NPs to obtain hexagonal nanoprisms. The cooperative action between TOP and OA decreases the energy barrier of the α-to-β-phase transition, thus allowing the synthesis of β-UCNPs at lower temperatures (Shan and Ju 2007).

Yang et al. (2012) have demonstrated the involvement of the organic capping ligand in the final shape of the particles prepared by hydrothermal synthesis by using bifunctional carboxylic compounds, namely malonic acid (MA), oxalic acid (OXA), succinic acid (SA), and tartaric acid (TA). One of the carboxylic group anchors to the particle surface, whereas the other can be used for further functionalization of the particle periphery. The shape of the material changes from rods to hexagonal prisms, hollow prisms, and tubes by increasing the amount of MA. Differences in acidity result in differences in the anchoring capacity of the carboxylic acids to the crystal surface and in the growth direction. Thus, disks, polygonal columns, and hexagonal tablets are obtained by using OXA, SA, and TA, respectively.

Therefore, though an exhaustive study about the influence of the ligand in the preparation of UCNPs is still lacking, the above-mentioned data are consistent with the role of the ligand anchoring group in the phase, size, and morphology of the resulting UCNP.

5.3.2 The Role of the Ligand in the Colloidal Stability of UCNPs in Aqueous and Nonaqueous Solvents, as Well in the UCNP Phase Transfer

The ligand on the UCNP surface avoids their aggregation thanks to steric or electrostatic repulsions and makes the UCNPs dispersible in organic or aqueous media (Sedlmeier and Gorris 2015). Apolar ligands provide lipophilicity to the nanoparticle surface, whereas polar ligands, which after anchoring to the UCNP surface provide polarity to the UCNPs surface, make the UCNPs hydrophilic.

In general, the UCNPs are synthesized in the presence of hydrophobic ligands (OA, TOP, and OM), but they can be become water-dispersible or amphiphilic through surface modification by ligand exchange, ligand oxidation, electrostatic interactions, layer by layer assembly, bio-conjugation, polymerization, silanization…(Muhr et al. 2014; M. Wang et al. 2011).

Moreover, these modifications can be beneficial for the optical properties and/or functionality (see Figure 5.3) of the NPs but these points will be discussed in Sections 5.3.3; 5.3.4; and 5.3.5. One of the most common strategies to obtain the UCNP with the desired functionality is ligand exchange, that is, the displacement of the original ligand by another ligand which can provide new properties or functionalities. The new organic ligand has to bind more strongly to the surface of the nanoparticle to allow a fast and effective displacement. Some examples are discussed below.

Poly(allylamine) (PAAm) replaces the ligand of oleate-capped $NaYF_4$:Yb, Er NPs at room temperature, thus leading to hydrophilic NPs. However, it requires prolonged reactions times (about 36 h) due to the strong interaction between the carboxylate group of the oleate and the UCNP surface (Xia et al. 2014). An increase of the reaction temperature can shortcut the reaction time as it has been shown in the exchange using PEG-SH to produce water-dispersible, PEG-capped UCNPs (heating at 60°C overnight) (Voliani et al. 2013). The use of a tetraprotonic acid, namely 1-hydroxyethane-1,1-diphosphonic acid (HEDP), shortens further the reaction time and hydrophilic UCNPs can be successfully prepared after stirring the reaction mixture at

60°C during 4 h (Schäfer et al. 2007). Therefore, the direct ligand exchange strategy needs the optimization of parameters such as concentration, stirring time, and temperature for each ligand. Moreover, NPs can aggregate during the ligand exchange (B. Dong et al. 2011). Hence, the direct ligand exchange can be a relatively difficult strategy with a tedious work-up and be time-consuming (from 4 h to several days) (Muhr et al. 2014).

Ligand exchange reactions involving two steps can be advantageous: first, the OA ligand is removed via acid treatment or by reacting with $NOBF_4$ and then new ligands, such as OM, PVP, cucurbit[n]uril (CB), and polyglutamic dendrimers, are easily attached to the NPs (Esipova et al. 2012; Francés-Soriano et al. 2015; Muhr et al. 2014). This method reduces significantly the aggregation of the UCNPs during the ligand exchange and allows for the attachment to the nanoparticle surface of any organic capping agent with an appropriate anchoring group (Muhr et al. 2014).

While ligand exchange involves a replacement of the original ligand by a new ligand, the colloidal stability of UCNPs in water can be improved by adding a new layer of organic coating via ligand attraction with the original ligand. There are three possibilities: (i) hydrophobic interactions with amphiphilic polymers or surfactants, (ii) electrostatic interactions, or (iii) host–guest interactions.

The use of amphiphilic polymers or surfactants is the most common strategy. The long alkyl chains of the polymers or surfactants intercalate between the chains of OA-capped UCNPs via van der Waals interactions, while the hydrophilic groups or chains of the polymer are in the nanosystem periphery and therefore the NPs are dispersible in water (Muhr et al. 2014).

Amphiphilic polymers, such as PEI (Yan et al. 2013), PEG (C. Wang et al. 2011), PAA (Yi and Chow 2007), N-succinyl-N′-octyl chitosan (SOC) (Cui et al. 2012) can transfer UCNPs from nonpolar solvents to water via hydrophobic interactions while maintaining the original ligand. PEI-modified graphene oxide (PEI-GO) was used as nanocarrier of UCNPs to prepare water-dispersible NPs (Yan et al. 2013). A water solution of PEI-GO was added to a chloroform dispersion of UCNP@OA forming a biphasic system. Authors observed that, after stirring this system, UCNPs were transferred to aqueous phase due to the hydrophobic interaction between the hydrophobic plane of PEI-GO and oleate alkyl chains on the surface of UCNPs, whereas the hydrophilic plane made the dispersibility of UCNPs possible in aqueous media and physiological solutions.

In addition, it has been shown that the hydrophobic chains of the PEG-grafted poly(maleic anhydride-alt-1-octadecene) can interact via van der Waals with the lipophilic chain of oleate on the UCNP surface, whereas the hydrophobic outer groups allow the dispersibility of UCNPs in water (See Figure 5.5) (C. Wang et al. 2011). Analogous interactions were observed between alkyl chains of OM-capped $NaYF_4$:Yb, Tm, and isopropyl groups in the modified PAA polymer, thus resulting in hydrophilic NPs due to

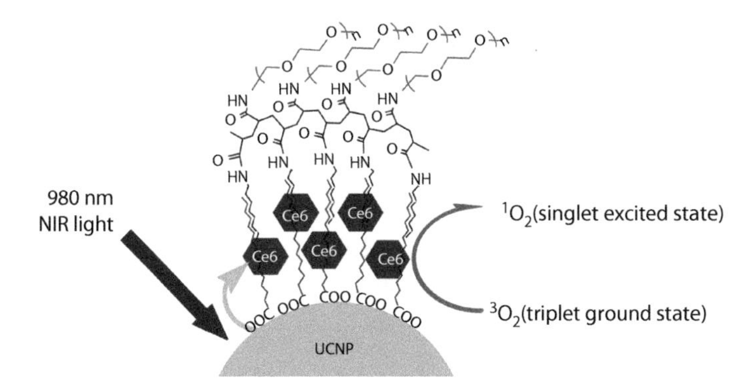

FIGURE 5.5
Schematic drawing showing NIR-induced singlet oxygen generation using of Ce6 encapsulated in PEGylated UCNPs (J. Liu et al. 2013). (Reprinted from *Biomaterials*, 32/26, Wang, C., H. Tao, L. Cheng, Z. Liu, Near-infrared light induced *in vivo* PDT of cancer based on UCNPs, 6145–6154, Copyright 2011, with permission from Elsevier.)

carboxyl groups of the modified PAA extending outward (Yi and Chow 2007).

Furthermore, surfactants, such as TWEEN 80 or cetyltrimethylammonium bromide (CTAB) can be used to convert hydrophobic UCNPs NPs into hydrophilic ones (Ren et al. 2012). The TWEEN surfactant is very cheap and it can prevent the nonspecific adsorption of proteins. Hydrophobic interactions between the lipophilic tails of TWEEN and the oleate layer on the surface of UCNPs allow ligand interdigitation and the UCNPs are transferable from an apolar solvent to water, biological buffers, and culture media.

Also, cationic CTAB, an ionic sodium dodecyl sulfate and nonanionic PEG tert-octylphenyl ether (C_8PhE_{10}) surfactants were successfully interdigitated to the nanoparticle ligand (Liang et al. 2012). High concentrations of surfactant gave rise to bilayer-modified individual UCNPs while at lower concentrations the aggregation of the UCNPs was observed. Interestingly, the bilayer-modified UCNPs can facilitate the nanoparticle silication, thus improving the stability of the nanosystem and enhancing their application.

In addition, hydrophobic UCNPs can be transferred to aqueous media via electrostatic interactions between organic capping layers. For example, multilayer nanohybrids can be built via electrostatic attractions between different layers of positively charged poly-(allylamine hydrochloride) (PAH) and negatively charged poly(sodium 4-styrenesulfonate) (PSS). Thus, UCNPs were negatively charged and could interact via electrostatic attraction with positively charged PAH, forming a layer of PAH around the NPs. After that, a negatively charged layer of PSS was added and subsequently assembled with PAH again. This method is known as layer-by-layer surface modification and makes it easy to control the shell thickness (Wang et al. 2005).

Another efficient and simple strategy to transfer hydrophobic NPs into water is by using host–guest interactions, such as that between β-cyclodextrin (CD) and adamantaneacetic acid (ADAA) (Liu et al. 2010). Thus, UCNP@ADAA NPs were transferred to a water solution containing CD by forming the UCNP@ADAA@CD nanohybrid.

5.3.3 The Role of the Ligand in the Optical Properties of the Nanomaterial

The emission ion of UCNPs is usually quenched to a higher extent in the presence of water molecules or alcohols than in organic solvents (Arppe et al. 2015; Wilhelm et al. 2015). Therefore, the way in which the modification of the UCNP is carried out to make them water dispersible is crucial for their final emission efficiency (Wilhelm et al. 2015).

A useful strategy to obtain nanohybrids dispersible in water while keeping or enhancing their emission is to use multi-anchoring polymers that meet two requirements: they must be firmly attached to the UCNP surface, thus avoiding the direct contact with water molecules, and they must provide hydrophilicity to final UCNP@polymer nanohybrids.

For instance, β-$NaYF_4$:Yb,Er, UCNPs have been capped with a thin polymer shell by replacing the oleate ligand of UCNPs by multidentate thiolate-grafting of P(MEO_2MA-*co*-SEMA) copolymers (Liras et al. 2014). The presence of the 2-(2-methoxyethoxy)ethyl side chains of MEO_2MA extending out of the nanohybrid made them water-dispersible, whereas the part anchoring to the surface was hydrophobic and prevented water from approaching the UCNP surface as opposed to ligands with a polar chain, such as PEG (Voliani et al. 2013). Interestingly, the amphiphilic nature of the UCNP@P(MEO_2MA-*co*-SEMA) nanohybrids made it possible to compare the emission in dichloromethane of the UCNP@polymer nanohybrids with that of UCNP@OA, thus showing that the nanohybrids exhibit enhanced emission by a factor of up to 10 compared with that of UCNP@OA. Moreover, the emission of the UCNP@polymer nanohybrids in water is even twofold greater than that of UCNP@OA in chloroform. The emission enhancement provided by the polymer can be explained by the reduction of surface defects and/or the isolation the UCNP from interactions with emission quenchers.

In addition, it has recently been reported that both the emission performance and the intensity ratio of the red and green UC emission bands of water-dispersible β-$NaYF_4$:Yb, Er UCNPs depend on the surface modification strategy chosen for their preparation from hydrophobic UCNPs (in particular, UCNP@OA) (Wilhelm et al. 2015). The deposition of an amphiphilic polymer on the top of the UCNP@OA exhibited reduced nonradiative quenching by water as compared to UCNPs prepared via ligand exchange. Moreover, the bilayer UCNPs showed a brighter green luminescence compared to the intensity of the red emission, whereas ligand exchange led to the opposite.

This trend was hardly affected by the chemical nature of the ligand (Wilhelm et al. 2015).

5.3.4 The Role of the NPs in the Organic Ligand Functionality

As mentioned above, the organic ligand can modulate important features of the UCNP, such as dispersibility, emission efficiency, as shown in UCNP@P(MEO$_2$MA-*co*-SEMA) nanohybrids, where the polymer provides the nanoparticle with amphiphilic and thermosensitive properties due to the lateral ethylene glycol chains. Consequently, they exhibit reversible aggregation/disaggregation above and below the lower critical solution temperature, respectively (Liras et al. 2014).

Interestingly, the nanoparticle can modulate the properties of the organic capping. Thus, the lower critical solution temperature value changed drastically for these polymers covalently bound to the nanoparticle surface and it was about that of human body temperature in some of them. This can be ascribed to the flexibility of the polymers to adopt suitable conformations to anchor the UCNP surface.

5.3.5 Synergism in the Nanohybrid Functionality

5.3.5.1 Functionality

It is important to highlight that the large surface/volume ratio of the NPs makes it possible to decorate them with a high number of functional groups at the periphery by choosing the proper organic capping. For example, UCNPs coated with PEGs with an anchoring group in one end and amino and/or carboxylic groups at the other can be further conjugated with other functional molecules by performing simple coupling reactions (either coupling of amine-functionalized molecules to the free carboxylates on the nanohybrid periphery or coupling of carboxylic-functionalized groups to the amine groups on the nanohybrid periphery). This strategy has been applied to the preparation of water-dispersible UCNPs with two different payloads (fluorescein and rhodamine derivatives) by using two orthogonally functionalized PEGs (Voliani et al. 2013). More examples of the symbiosis of the UCNP and the organic capping to provide additional functionality will be discussed in the following sections, since they have been designed for specific applications.

5.3.5.2 Synergistic Effect between the UCNP and the Organic Ligand in the Nanosystem Functionality: Encapsulation versus Ionic and Covalent Binding

The nanoparticle and the organic capping can work together to endow the nanohybrid new capacities as compared to that of the individual components, that is, the naked nanoparticle and the ligand. In principle, the ligand

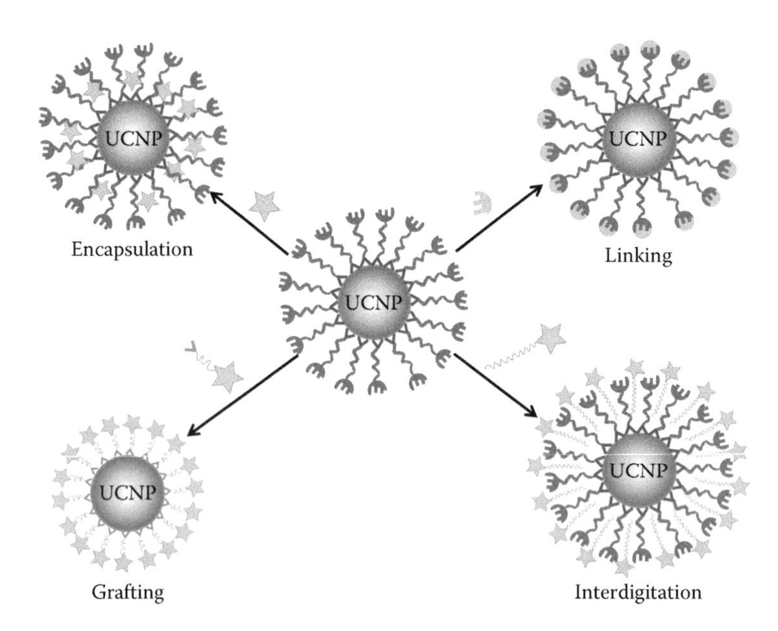

FIGURE 5.6
Schematic representation of the mechanisms to add functionality to the UCNP.

can be a molecule whose primary target can be to decorate the nanoparticle periphery in such a way that permits the further binding of functional molecules, such as, biomolecule, photosensitizer (PS), and drug.

Figure 5.6 illustrates different strategies to add functionality to the UCNPs via the organic capping: (i) covalent or ionic linkage of functional organic molecules to the ligand functional group; (ii) interdigitation between the chains of the organic ligand and functional molecule, either surfactant or polymer; and (iii) encapsulation of the functional molecule within the nanoparticle organic capping. This chapter is devoted to the synergistic effects between UCNPs and the organic ligands; UCNP refers to both NPs with and without an inorganic shell. Thus, core UCNPs capped with inorganic shells, such as silica and titanium dioxide, will be mentioned throughout the chapter, but the emphasis will be placed in the role of the organic capping.

Most of the examples reported in the literature benefit from the covalent bonding between groups, such as carboxylic, amines, and thiols, to locate functional (bio)molecules at the periphery of organic-capped UCNPs. However, there are also examples of noncovalent bonding, such as dipole-ionic and van der Waals interactions.

With regard to ionic binding, UCNPs capped with cucurbiturils (UCNP@CB hybrids) have been used as scaffolds for cationic dyes (such as methylene blue and pyronin Y), due to their anchoring to the CB free carbonyl portal (a charge–dipole interaction) (Francés-Soriano et al. 2015). This made it possible to locate a high concentration of the dyes close to the UCNP and lead to efficient energy transfer from the UCNP to the dye. These UCNP@CB@dye supramolecular systems may be highly advantageous in photodynamic therapy (PDT) taking into account the interest of these tricyclic basic dyes in PDT.

It is also possible to use bi-layer water-dispersible nanohybrids to encapsulate nonpolar organic molecules, for example, to load chlorin e6 (Ce6) within the hydrophobic part of the organic capping (see Figure 5.5) (J. Liu et al. 2013).

5.3.5.3 Targeting

Target ability refers to the selective delivery of UCNPs to diseased tissues and is of high relevance for biomedical applications. A specific interaction between the surface ligand and cells or biomolecules is needed in order to minimize damage to normal tissues. Small molecules, peptides, proteins, aptamers, dendrimers, carbohydrates, and antibodies anchored on the UCNP surface or linked to the UCNP ligand can be recognized by specific cells (Friedman et al. 2013). Targeting can be combined with other applications of UCNPs, such as drug delivery, sensing, and bioimaging. These applications will be discussed in the following sections.

Small molecules such as carbohydrates (Bogdan et al. 2010) and folic acid (Ma et al. 2012; Pan et al. 2013; Xing et al. 2015; Yang et al. 2013) have been extensively studied as targeting ligands (Chen et al. 2014). Folic acid is suitable for tumor cells targeting because folate receptors are highly over-expressed on the surface of many tumor types. Xing et al. developed NPs coated with a folate-conjugated, light-responsive, and amphiphilic copolymer consisting of 1′-(2-methacryloxyethyl)-3′,3′-dimethyl-6-nitro-spiro (2H-1-benzopyran-2′,2′-in-doline) (SPMA), poly(ethylene glycol) methacrylate (MAPEG), and N-acryloxysuccinimide (NSA) units (Xing et al. 2015). Hydrophobic interactions between the alkyl chains of the copolymer and the lipophilic chains of the UCNP ligand permit the addition of a new capping layer on the UCNP surface placing the folate receptor at the nanohybrid periphery. The nanohybrid shows high emission efficiency, good dispersibility, stability in aqueous media, and is suitable for tumor targeting.

Furthermore, targeting can be through internalization of UNCPs by cells. The NPs ligand can play a crucial role, because it can control the UCNPs–cell interactions (Jin et al. 2011). Jin et al. have studied the effect of the charge of $NaYF_4$:Yb, Er UCNPs on the cellular uptake efficiency (determined by multiphoton microscopy), using PVP, PEI, and PAA as the organic capping. At the same conditions, positively charged UCNP@PEI NPs exhibited greatly enhanced cellular uptake in comparison with that of UCNP@PVP and UCNP@PAA (neutral and negative charged polymers, respectively) NPs, because the cell surface is usually negatively charged. So, the charge of the organic capping can improve the efficiency of the NPs internalization. Analogous results were obtained in the comparative study of $NaGdF_4$:Yb, Er UCNPs capped with PEG, PEI, and 6-aminocapronic acid (6AA) (Tsang et al. 2015b). Thus, the electrostatic interaction between $-NH_2$ groups on the UCNP@PEI periphery and the cell membrane increased the efficiency in the cell uptake of the nanohybrid.

5.3.5.4 Sensing

UCNPs can be used to sense a physical stimulus or (bio)chemical species in their surroundings by the combination of the UCNP emission and organic capping functionality.

To date, studies of the suitability of UCNPs in physical sensing has mainly focused in temperature sensing (Chen et al. 2014; Hao et al. 2013). Although there are studies of their response to strain (Wisser et al. 2015), electrical (Hao et al. 2011; Huang and Hsieh 2009) and magnetic fields (Tikhomirov et al. 2009; G. Wang et al. 2011; Valiente et al. 2009), high radiation energy, and NIR-light, but this sensing makes use of their core intrinsic properties. Here we make special emphasis on the synergic effects between the organic capping and the UCNP in sensing. See recent reviews for more information regarding physical and (bio)chemical sensing with UCNPs (Christ and Schäferling 2015; González-Béjar and Pérez-Prieto 2015; Tsang et al. 2015a).

In thermometry of solids, there is no need to perform a modification on the organic coating of the synthesized NPs. However, it becomes crucial to prepare water-dispersible UCNPs with a nontoxic and biocompatible capping if they are going to be used for nanothermometry *in vivo* (Vetrone et al. 2010b). To that purpose *Vetrone* et al. used PEI-capped α-$NaYF_4$:Yb, Er NPs (Vetrone et al. 2010a). We have mentioned above an example of the cooperation between an UCNP and a polymer to sense temperature at values close to those of human body (Liras et al. 2014).

The organic ligand can play a key role in (bio)chemical sensing. Chemical sensing is usually based on UCNPs coated with a large number of photoactive molecules whose absorbance overlaps with some of the UCNP emission bands. Consequently, the sensing capacity is due to either inner filter effects or energy transfer processes from the UCNP to the capping molecule (e.g., a fluorophore within the polymeric coating). However, in biochemical sensing, the UCNPs are usually coated with receptors.

The starting point for (bio)chemical sensing with UCNPs is to meet the dispersibility requirement (either in organic or aqueous solvents) by making use of one of the strategies mentioned throughout this chapter (ligand exchange, layer by layer assembly, etc). In addition, the organic ligand must include either an organic molecule with sensing capabilities to the desired analyte (Hao et al. 2013) and the optical response must be proportional to the concentration of the analyte if the purpose is its quantification.

Figure 5.7 illustrates a nanohybrid where the on/off energy transfer between the UCNP (donor) and the fluorophore (acceptor) can be activated or deactivated upon recognition of the analyte. Many analytes (metal ions, gas molecules…) (Chen et al. 2014) have been recognized by using UCNP nanohybrids (Christ and Schäferling 2015; Hao et al. 2013).

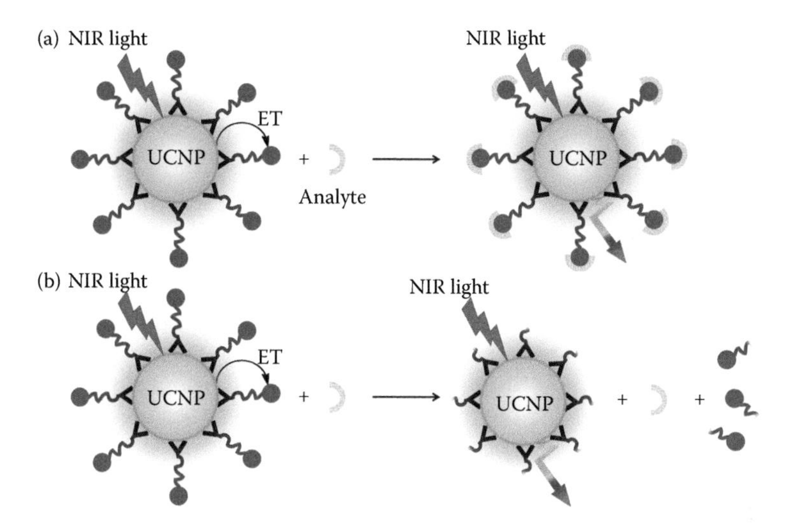

FIGURE 5.7
Analyte recognition via on/off energy transfer between UCNP (donor) and the fluorophore (acceptor). (a) Inhibition of the energy transfer through linking of the analyte to the acceptor. (b) Inhibition of the energy transfer because the analyte (usually an enzyme) cut off the bond between UCNP and fluorophore (acceptor).

Undoubtedly, pH reflects the concentration of protons in the medium and is a key target parameter in a broad range of applications such as life sciences, food and beverage processing, soil examination, and marine and pharmaceutical research (Wisser et al. 2015). Among the several detection methods available to detect pH changes, luminescent methods are recommended and most widely used (Wencel et al. 2014). Therefore, in this chapter we have selected pH sensing to illustrate possible designs of chemical sensors.

The aim is to avoid the drawbacks of commonly used pH indicators based on organic fluorophores, such as their photodegradation, photobleaching, and autofluorescence. In particular, water-dispersible UCNPs have been recently proven to be either pH-responsive (Wencel et al. 2014) or useful for pH measurement and sensing (Arppe et al. 2014; Esipova et al. 2012; Sun et al. 2009; Xie et al. 2012). The useful pH ranges reported for these systems are 6–11 for bromothymol blue/UCNPs (Esipova et al. 2012) and 6–10 for Nile blue derivative (ETH5418)/UCNPs (Sun et al. 2009) thin films, 2.5–7.5 for pHrodoTM Red dye/UCNPs (Xie et al. 2012), 5–9 for porphyrin/UCNPs (Arppe et al. 2014) and 5–8 for graphene oxide/UCNPs (Esipova et al. 2012). In addition, UCNP-based nanohybrids have been used for controlled release (Yan et al. 2014) and imaging guided PDT (Chen et al. 2014) using pH variations. Hexagonal $NaYF_4$:Yb, Er NPs ($NaYF_4$:Er^{3+}, Yb^{3+}) have been used for these purposes due to their superior UC efficiency (C. Wang et al. 2013).

Studies on the pH effect on UCNPs capped with azelaic acids ($HOOC(CH_2)_7COOH$) have proved that their emission is independent of the pH in the range from 2 to 11 (Xie et al. 2012). By contrast, *N. Bogdan* et al. (Bogdan et al. 2011) have reported the influence of the pH on the red emission of naked $NaYF_4$:Yb, Er UCNPs. Nevertheless, two strategies have been proposed to make UCNPs pH sensitive based on the cooperativity between the organic capping and the UCNP. One of them is based on the stabilization of emissive NPs by multidentate pH-responsive polymers (Bogdan et al. 2012). The other one (useful for pH measurements) is based on the combination of UCNPs with a pH-probe (fluorescent or not; it can also be a nanomaterial) embedded in a silica shell or polymer matrix. The pH-dependent probe suffers a spectral shift according to the pH, and as a consequence the degree of overlap between its absorption and the UCNPs emission wavelengths varies, which results in a change in the ratio between the UCNPs emission bands and may also influence the energy transfer from the UCNPs to the acceptor (if the probe used can act as acceptor and not just as absorption filter) (Arppe et al. 2014; Liras et al. 2014; Sun et al. 2009). Gold nanoparticles (AuNPs) and UCNP can lead to pH sensitive systems (Arppe et al. 2014; Luo et al. 2013; Zhang et al. 2009). In some examples, the synergistic effect is due to the ligand on the AuNPs and it will not be discussed here (Zhang et al. 2009).

An example that perfectly shows the synergistic effect between the organic ligand and the UCNP for pH detection has been reported by *Zhao* et al. (Yan et al. 2014). They have designed a biocompatible pH sensitive film based on energy transfer between positively charged PEI-coated $NaYF_4$:Yb, Er UCNPs and negatively charged UCNPs and graphene oxide (GO). The interaction between them becomes stronger with increasing pH, since at acidic pHs values (e.g., pH = 5.00) the negative charge of the GO nanosheets decreases. Interestingly, the sensor film has been applied in real urine measurements.

In the case of biochemical sensing (or bioassay) the nanohybrids allow detection of biomolecules (antibody, immunoglobulin, etc.) in a solution and it can also be designed for the quantification of the biomolecule. The bioassay can be homogeneous or heterogeneous. The first assay consists in simply mixing the analytes and the sample and then carrying out an optical measurement, whereas the heterogeneous assay requires multiple steps in which the analytes are added usually in solution to a solid where the nanohybrid has been fixed. Later on, the solid is washed to ensure that only those analytes that have been recognized remain attached and separated at different points in the assay (Chen et al. 2014).

DNA/RNA analysis is of great interest in molecular biology, genetics, and molecular medicine (Zhang et al. 2006). *P. Zhang* et al. have synthesized UCNPs using a sandwich-type hybridization format composed by two shorter oligonucleotides with specific designed sequence to recognize the longer target oligonucleotide. One of the shorter oligonucleotides is covalently anchored to the nanoparticle surface and the other is used to attach a fluorophore which absorption spectrum overlaps with the emission spectrum of the NPs. Upon irradiation with NIR light and in the presence of the longer target oligonucleotide, emission of the fluorophore is observed due to energy transfer from the UCNP to the fluorophore. This study demonstrates the high capacity of the nanohybrid to recognize DNA with a detection limit of 2.3 nM.

Aptamers are DNA/RNA molecules that are analogous to antibodies in their range of target recognition and in their high specificity (Wu et al. 2014). *Wu* et al. developed a highly stable, specific, and sensitive nanohybrid combining the multicolor properties of UCNPs with aptamers. Three different aptamers with specificity to *Staphylococcus aureus, Vibrio parahaemolyticus,* and *Salmonella typhimurium* bacteria were used. The specific recognition and high affinity of each nanohybrid for the target bacterium permit the detection of various types of bacteria coexisting in the same sample.

In addition, UCNPs conjugated with glycodendrimers have proved to recognize lectins via lanthanide resonance energy transfer (Bogdan et al. 2010). Poly(amidoamine) dendrimers were grafted to the surface of $NaGdF_4$:Yb, Er NPs, thus allowing the subsequent linking of carbohydrates to the dendrimers via a thiourea bond. The carbohydrates have an aromatic group in their structure, which plays two important roles. Whereas one of them recognizes proteins, such as lectin, via hydrophobic interactions, the other avoids the contact of water with the nanoparticle surface, which can quench the UCNP luminescence. The resulting UCNP nanohybrids possess excellent emission properties, which can be used to follow the recognition process of fluorophore-labeled Con A lectin by imaging.

5.3.5.5 Bioimaging

Fluorescence imaging is a noninvasive technique for real-time monitoring cells, biological processes, and morphological details in tissue and animals (Wu and Butt 2015). This technique is relevant for studying biological and pathological processes to provide important clues to the progression of several diseases (Dong et al. 2015). Bioimaging is applied in toxicity assessment, cellular and whole-body imaging, optical tomography, multimodal imaging, etc. (Ma and Ni 2015).

The NIR-response of UCNPs makes them particularly suitable for bioimaging applications due to the elimination of autofluorescence from biosamples (Kobayashi et al. 2009; Prodi et al. 2015), thus enabling high signal-to-noise ratio. In addition, UCNPs possess narrow emission bands that are advantageous for multicolor imaging (Cheng et al. 2011b), they show high photostability as compared to quantum dots and organic dyes, and they have low toxicity (Gnach et al. 2015). Interestingly, NIR light exhibits a remarkably deep penetration in tissues (Idris et al. 2009; Yu et al. 2009).

The organic capping plays a key role in the application of the UCNP in this field, since it can tune the emission properties of the nanoparticle and improve its stability and dispersibility in water or physiological media for *in vivo* studies (Chen et al. 2014). As mentioned in Section 5.3.5.4, surface properties can control the UCNP–cell interactions. Therefore, improved cellular uptake without losing UC emission is desirable in cellular imaging. Three different polymers have been used to bind to the nanoparticle surface: PEG, PEI, and 6AA (Tsang et al. 2015b). UCL bioimaging in HeLa cells showed that PEI-capped UCNPs exhibited enhanced luminescence due to the presence of PEI amino groups, which have a high affinity for the cell membrane.

Lanthanide ions of the UCNPs can form complexes with cellular phosphate residues and this could be a biological hazard and interfere in the fluorescence of nanomaterials.

Ethylene diamine-tetra methylene phosphonic acid (EDTMP) is a compound with four C-PO$(OH)_2$ groups that can anchor to the surface of UCNPs with high affinity (R. Li et al. 2015). Consequently, EDMPT can be used to protect the UCNP and prevent its complexation to cellular phosphates, thus preserving its fluorescence.

AA-capped $NaGdF_4$:Yb, Tm, Er (UCNP@AA) NPs were developed to achieve whole body bioimaging of a mouse (Zhou et al. 2010). After 40 min of an intravenous injection of a solution containing UCNP@AA, an upconvertion luminescence (UCL) signal was detected and an image of the whole body was obtained. The presence of AA on the nanoparticle surface conferred good water dispersibility to the UCNP as well as low cytotoxicity for the biological system, therefore the UCNP@AA nanohybrid was suitable for bioimaging.

Another interesting example of the ligand role is tuning the multicolor UC emission of $NaYF_4$:Yb, Er by adjusting the molar ratio of two coordinating ligands on the surface of UCNPs (Niu et al. 2010). NPs were synthesized using a mixture of octadecylamine and oleamide via a thermolysis procedure. The increase of the oleamide/octadecylamine molar ratio caused the color emission output to change from green to red. Multicolored UCNPs are important for multiplexed bioimaging because they enable simultaneous monitoring of the interaction between multiple proteins or cells within an organism and a panel of sub-level parts of an organ (Dong et al. 2015; Zhou et al. 2015).

5.3.5.6 Therapy

As explained in Section 5.3.5.5, UCNPs seem to be biologically friendly (Gnach et al. 2015) and are of a noninvasive nature (Chen et al. 2014). These qualities make them suitable for theranostic applications, since they can integrate imaging and therapy capacities into a single platform and therefore they are useful for diagnosis, drug delivery, and monitoring of therapeutic response (Xie et al. 2010). The key features to consider when selecting a luminescent particle for therapy are the absorption wavelength, the emission wavelength, and the size of the particle and the capping ligand. UCNPs are used for two purposes in therapeutic applications. One, they convert NIR to UV–Vis, which extends the operating range of therapeutic agents thanks to energy transfer processes or induced photorelease (Zhou et al. 2015). And two, as discussed in the previous section, they can be used as probes to monitor (bio)chemical changes and they can be employed to monitor the distribution and metabolism of drugs (Zhou et al. 2015). In this section, we discuss a number of notable therapies in the field, which are: PDT, photo thermal therapy (PTT), and chemotherapy, and a combination of them all.

5.3.5.6.1 Photodynamic Therapy

PDT is a noninvasive cancer treatment based on selective delivery/administration of a PS, which upon selective excitation is capable of generating singlet oxygen (type II mechanism) and/or other reactive oxygen species (ROS) (type I mechanism), causing tumoral cells apoptosis and/or necrosis exclusively in the illuminated area and its surroundings (Brown et al. 2004; Dougherty 1993). To date, most PSs used for this purpose are organic molecules or complexes that absorb visible light and are soluble in aqueous solutions. Unfortunately, they tend to aggregate and they lose their photochemical activity upon prolonged illumination (Ideat et al. 2005). Even worse, the penetration of visible light through tissues is very limited as compared to NIR. So it is advantageous to use UCNPs to absorb NIR light and be able to transfer this energy to a PS. With regard to the excitation wavelength, 975 and 920 nm are the most used to excite Yb^{3+} in the UCNPs (Chen et al. 2013; Vetrone et al. 2010b), though other designs using 808 nm have recently been reported (e.g., rose bengal (RB)-capped $NaYF_4$:Yb,Ho@$NaYF_4$:Nd@$NaYF_4$ UCNPs (Dong et al. 2015; Wang et al. 2015). As illustrated in previous sections (e.g., sensing), a pre-requisite for the Energy transfer (ET) process is that the surrounding chromophore (a PS in this particular case) absorbs light where the UCNPs emit. Here we emphasize the photophysical and photochemical events that take place after excitation of the PS through the ET process to lead eventually to reactive species that cause tumoral cells death. Figure 5.8 shows a simplified representation of the process: energy transfer from the UCNP to the PS leads to the PS singlet-excited state (^{1}PS) that would lead to its triplet excited state (^{3}PS) by intersystem crossing. Then, singlet oxygen (and/or other ROS such as hydroxyl radicals or superoxide anion radical) would be generated by triplet–triplet energy transfer (or electron transfer) from the (^{3}PS) to oxygen in the GS.

The high surface/volume ratio of the NIR-responsive UCNPs combined with the capacity of their surface and ligands to establish stabilizing binding interactions with the PDT PS evidences the cooperative effect in the organic-coated UCNP system. Undoubtedly, an ideal UCNP–PS nanohybrid for PDT would be a nontoxic and biocompatible, stable, water-dispersible, small-sized (able to permeate the cell membrane) nanohybrid with a large loading of a PS. It is also desirable to avoid leaching of the PS into the living system.

There are several designs to load water-dispersible UCNPs, usually $NaYF_4$:Yb(Er and/or Tm), with a PS or combination of PSs to maximize the overlap with the UCNPs emission (Idris et al. 2012). Some PSs have been

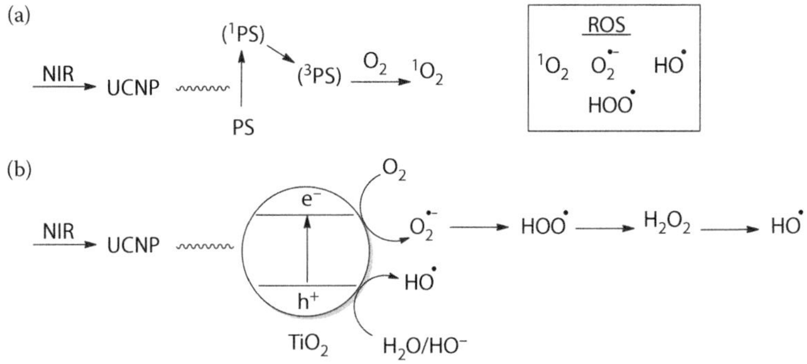

FIGURE 5.8
Photosensitized generation of singlet oxygen (a) and other ROS (b).

embedded in polymeric shells around the UCNP core (e.g., ZnPc (Cui et al. 2013), pyropheophorbide a (Zhou et al. 2012), monomalonic fullerene (X. Liu et al. 2013), and Ce6 (Cheng et al. 2011a; Park et al. 2012)).

In addition, PSs, such as rose bengal (RB), Zn(II) phthalocyanine (ZnPc), and Bodipy derivatives have been covalently bonded to $NaYF_4$:RE UCNP scapped with 2-aminoethyl dihydrogen phosphate (Liu et al. 2012), PAAm (Xia et al. 2014), and PEI (Topel et al. 2014), respectively.

Moreover, it has been recently demonstrated that carboxy-functionalized PSs, such as a carboxybodipy (Bodipy-COOH), (González-Béjar et al. 2014) and carboxyphthalocyanine zinc (ZnPc-COOH) (M. Wang et al. 2014) can be grafted to the UCNP surface to lead to nanohybrids able to generate singlet oxygen efficiently. Singlet oxygen is one of the possible ROS that can lead to tumor cell killing.

A novel series of $NaYF_4$:Yb,Tm@TiO_2 nanosystems has been recently designed for *in vivo* tumor PDT (Lucky et al. 2015). In order to make this nanosystem water-dispersible, the nanosystem was coated with a polymeric shell (PEG derivative). The mechanism associated to ROS for TiO_2 (Hoffmann et al. 1995; Linsebigler et al. 1995; Oppenlander 2003) is activated after NIR excitation of the UCNP followed by energy transfer to TiO_2. Figure 5.8 illustrates the charge separation (electron/holepair) induced upon TiO_2 irradiation (Anpo and Kamat 2010; García et al. 2009; Khataee and Fathinia 2013). These electron–hole pairs can recombine and the charge carriers can competitively migrate to the TiO_2 surface, eventually leading to hydroxyl free radicals in the presence of water and oxygen (Furube et al. 1999). The *in vivo* tumor PDT treatment with $NaYF_4$:Yb,Tm@TiO_2 proved efficient in delaying the tumor growth.

Most of the just described nanosystems have been tested for PDT *in vitro* with diverse tumor cell lines, but a few of them have been successfully tested *in vivo* (mice have been used for this purpose). Generally, the tumor of the mice is treated with the nanosystem and NIR-light and its evolution is compared with that of other tumors under control treatment conditions (i.e., saline solution and NIR-irradiation; no irradiation and the nanosystem) in order to make sure that the volume of the tumor decreases exclusively due to UCNP/NIR-light combination. In some cases, the nanosystem is also loaded with folate to increase the tumor-selectivity since cancer cells overexpress folate receptor. The above-mentioned PDT nanocomposites that comprise one type of PS and have been tested *in vivo* are discussed below.

The first demonstration of NIR light-induced *in vivo* PDT of cancer with $NaYF_4$:Yb,Er UCNPs loaded with Ce6 as PS was reported by Liu et al. (Furube et al. 1999). In this nanosystem, Ce6 absorbs red emission of Er^{3+} and it is intercalated between the lipophilic chains of the interdigitated area between the oleate capping the UCNP surface and a polymeric coating (Figure 5.5). Two important findings were proven in this report. First, the tumor volume of mice treated with Ce6-functionalized UCNPs and NIR light irradiation was much smaller (about 60%) than that of mice treated with saline and visible light irradiation. Secondly, tissue penetration depth observed by NIR irradiation was much deeper than that of visible illumination. Interestingly, later on the authors developed charge-reversible $NaYF_4$:Mn,Yb,Er UCNPs loaded with Ce6 for pH-sensitive *in vivo* PDT (Park et al. 2012). These UCNPs emit strong red light emission at ca. 660 nm under 980 nm NIR that can be absorbed by the PS to generate singlet oxygen. The novelty here is due to the use of a pH-sensitive polymer that improves the cancer cell killing efficacy in a slightly acidic environment; it has to be taken into account that tumor extracellular environment is slightly acidic (pH 6.5–6.8). The nanosystem is negatively charged in neutral and alkaline environments but becomes positively charged under slightly acidic conditions. This enhances the cellular uptake of the NPs due to a highly favorable interaction with negatively charged cell membranes. This example combines several synergistic effects that allow pH sensing and singlet oxygen generation *in vivo* under NIR illumination.

Cui et al. designed $NaYF_4$:Yb,Er UCNPs coated with folate-modified chitosan to anchor PS molecules of ZnPc in order to realize *in vivo* targeted deep-tissue PDT and a tumor inhabitation ratio of 50% was achieved (Cui et al. 2012, 2013).

In 2014, ZnPc-$LiYF_4$:Yb, Er (Xia et al. 2014) and 5-aminolevulinic acid-capped $NaYF_4$:Yb,Er@ CaF_2 NPs (M. Wang et al. 2014) were also constructed and proven to be successful.

Another example of application in tumor targeted imaging and PDT is that of UCNPs co-loaded with aminophenylboronic acid (APBA) and hyaluronated fullerene (HAC_{60}) (X. Wang et al. 2014). APBA was used as a stabilizer in the synthesis and as a functional ligand for targeting bioimaging, whereas HAC_{60} was used as an acceptor to generate singlet oxygen. The two ligands on the surface of the NPs made possible the quenching of UC emission via energy transfer processes. Also, uptake of the nanohybrid by PC12 cancer cells was dramatically enhanced compared with HAC_{60} without UCNPs.

Therefore, the synergistic effect between the core NPs and the capping ligands led to a nanohybrid with two color fluorescent signals, high specificity, high cell uptake, and improved PDT.

5.3.5.6.2 Photothermal Therapy

In PTT treatment, the tumor loaded with the proper agent (e.g., NPs) is exposed to an excitation source. This light is absorbed by the agent and converted it into heat (local hyperthermia), thus killing cancer cells (e.g., the temperature of the tissue raises and, as a consequence, cancerous cells are ablated) (Philipp et al. 1995).

To date, AuNPs (Kennedy et al. 2011), organic dyes, graphene oxides, and quantum dots have shown to be effective agents for PTT. Although UCNPs are relevant due to their capacity to absorb NIR light, they require the presence of the above-mentioned agents to be used in PTT and the only role of the organic capping is to provide hydrophilicity to the nanosystem. For example, PEG-modified $NaYF_4$:Yb,Er@ Fe_3O_4@Au nanohybrids have been used for dual-mode imaging guided PTT (Cheng et al. 2011a; Cheng et al. 2012). In this nanocomposite, PEG provides water dispersibility, $NaYF_4$:Yb,Er generates UC emissions, Fe_3O_4 is responsible for the T2-weighted Magnetic resonance imaging (MRI), and the Au shell converts NIR light into heat. After intravenous injection of the nanocomposite, the temperature at the tumors surface increases to ~50°C, as compared to ~38°C for irradiated tumors in the absence of the nanocomposites.

However, nanohybrids comprising UCNPs and functional organic ligands are of interest for combined PTT/PDT therapy. An interesting system is that in which PAAm-capped $NaYF_4$:Yb,Er,Tm@$NaYF_4$ UCNPs (UCNPs-NH_2) was coupled to functionalized graphene oxide (NGO-COOH) and ZnPc molecules were adsorbed on the GO surface (Y. Wang et al. 2013). The efficiency in the cancer cell death increased by using two lasers (808 nm for PTT, 630 nm for PDT) simultaneously to irradiate the nanocomposites that had been up-taken by the cancer cells (Figure 5.9).

5.3.5.7 Drug Delivery and Chemotherapy

Chemotherapy is the main technique to cure diseases like cancer with the disadvantage to control the local effective therapeutic concentration, causing cytotoxicity at high concentrations (Dong et al. 2015). In general, UCNP-assisted chemotherapeutic processes fall into two categories: (a) UCL-guided monitoring of the degree of drug release (Dong et al. 2015) and (b) NIR-triggered drug release based on UCNPs (Figure 5.10) (Wu and Butt 2015).

In addition, UCNPs act as versatile nano-transducers converting NIR light to light of shorter wavelengths and, consequently, NIR-light excitation can eventually induce changes in close-lying molecules which absorb at shorter wavelengths and/or can induce their release by photo-cleavage. These NIR light-induced processes are of particular interest in biological samples for

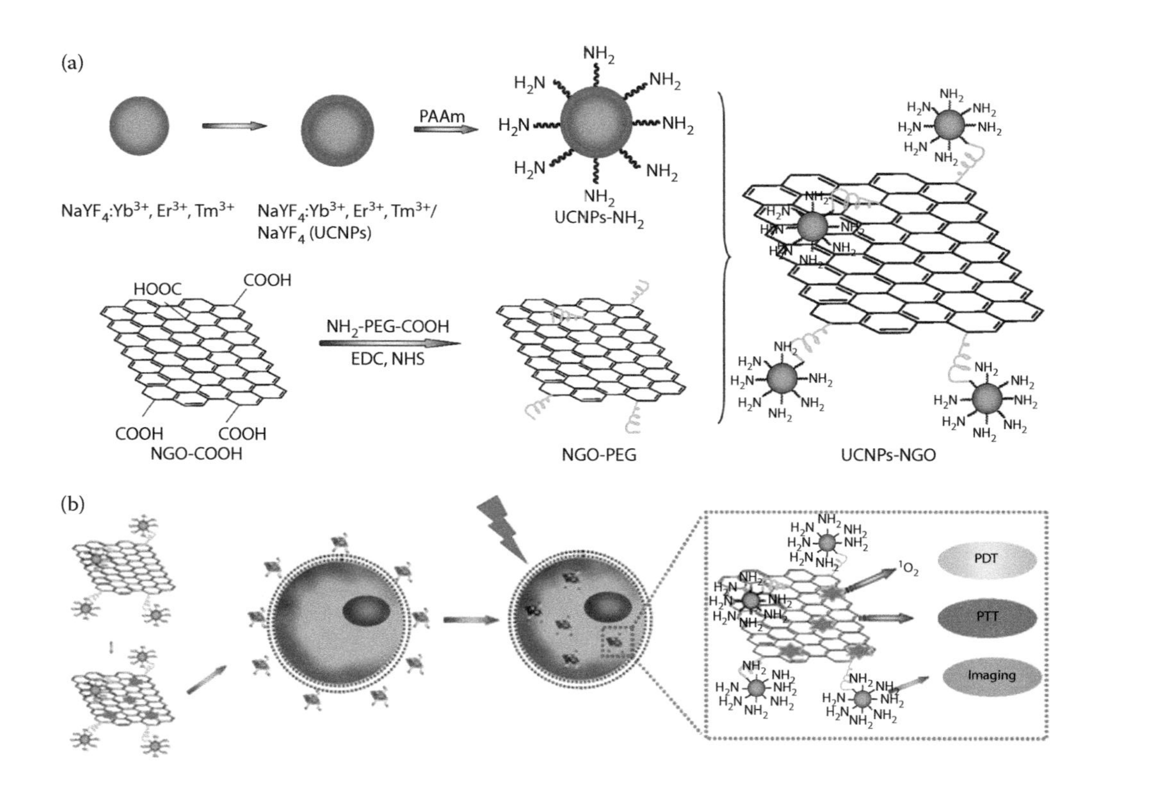

FIGURE 5.9
(a) Schematic illustration of the synthetic procedure for UCNPs–NGO: Numbers of core–shell structured UCNPs being covalently grafted with NGO via bifunctional PEG; (b) Schematic illustration of UCNPs–NGO/ZnPc as a multifunctional theranostic nanoplatform for cancer treatment. (Reprinted from *Biomaterials*, 34/31, Wang, Y., H. Wang, D. Liu, S. Song, X. Wang, H. Zhang, Graphene oxide covalently grafted UCNPs for combined NIR mediated imaging and photothermal/photodynamic cancer therapy, 7715–7724, Copyright 2013, with permission from Elsevier.)

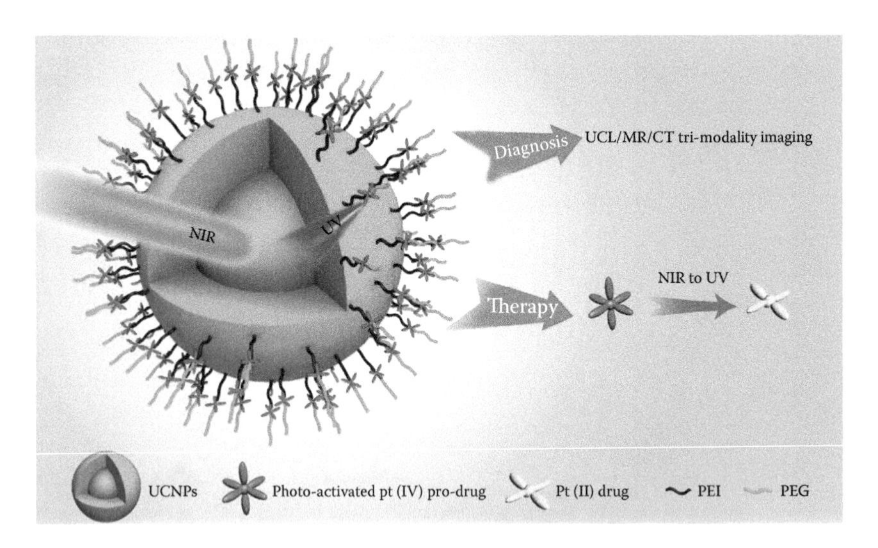

FIGURE 5.10
Schematic illustration of the characterization of UCNP–DPP–PEG NPs and their use for diagnosis and photorelease. (Adapted from Dai, Y., H. Xiao, J. Liu et al. 2013. In vivo multimodality imaging and cancer therapy by NIR light-triggered trans-platinum pro-drug-conjugated upconverison NPs. Journal of the American Chemical Society 135 (50):18920–18929.). Copyright 2015 American Chemical Society.)

remote-control release of caged compounds, such as drugs (Yang et al. 2015), since their light-dependence determines where and when a process is initiated (Idris et al. 2015).

An example is that occurring after NIR excitation of $NaYF_4$:Yb, Er NPs capped with PAA in the presence of eosin Y (photoinitiator that exhibits a strong absorption band between 460 and 540 nm), *N*-vinylpyrrolidone (NVP), and Poly(ethylene-glycol)-diacrylate (PEGDA). Energy transfer from the UCNP to eosin Y generates free radicals which initiate the polyaddition reaction of the carbon–carbon double bond of NVP and PEGDA (Xiao et al. 2013) to form cross-linked hydrogels. Progressive light irradiation of the system eventually leads to UCNP–PEGDA hybrid microspheres due to the aggregation of the NPs via hydrogen bonding between the PAA-capping and the PEGDA hydrogel. These hybrid microspheres can be loaded with ZnPc by hydrophobic interactions to give rise to a system with a high efficiency in singlet oxygen generation (Xiao et al. 2013).

In addition, UCNP luminescence can be absorbed by close-lying molecules, eventually inducing drug-release by photocleavage. For example, NIR laser light excitation of UCNPs encapsulated in UV light-degradable polymer particles causes the degradation of the polymer into its monomers by inducing cleavage of a protective group (Viger et al. 2013). This strategy also proved successful to control the release of coumarin 153 (C153) remotely by irradiating at 980 nm.

Moreover, PEG-capped (Ruggiero et al. 2015) and PEI-capped (Dai et al. 2013) β-$NaYF_4$:Yb,Tm@$NaGdF_4$:Yb NPs were covalently conjugated to a platinum (IV) antitumor drug, namely the *trans,trans,trans*-[$Pt(N_3)_2(NH_3)(py)(O_2CCH_2CH_2COOH)_2$] (DPP) pro-drug (Figure 5.10). An added layer of PEG improved the stability and biocompatibility of the nanohybrid. The synergistic effect between the organic capping and the nanoparticle core allowed the selective drug delivery induced by NIR excitation (30 min of irradiation led to 48% of the drug delivery) (Dai et al. 2013). These UCNP nanohybrids were used as multifunctional drug delivery systems which combine NIR-activated platinum pro-drug delivery and UCL/magnetic resonance/computer tomography trimodality imaging.

The activity of proteins in protein-based therapies with UCNPs can be modified as a consequence of their covalent attachment to the UCNPs and therefore, it would be advantageous if the delivery of the nanohybrid was followed by the protein release. In this way, a DNA-mediated hierarchical hollow UCNP has been developed (Zhou et al. 2014). Hollow UCNPs were functionalized with 3-aminopropyltriethoxysilane (APTES) to afford UCNPs-NH_2, and then a spiropyran dye (SP-COOH) was conjugated. When the SP molecule is irradiated with UV-light it isomerizes to the merocyanine (MC) form. This isomerization is reversible with visible light. The authors demonstrated that positively charged MC-functionalized UCNPs allow the electrostatic interaction with negatively charged proteins, such as β-galactose. So, after irradiation at 980 nm, hollow UCNPs transform the NIR light into visible light, which is absorbed by MC, reverting back to the neutral SP form. Hence, the trapped proteins are released to the solution (70 wt% in 720 min after NIR exposure) (Zhou et al. 2014).

5.4 Summary

The UCNPs can be capped with a variety of organic compounds which possess an anchoring group to attach to the nanoparticle surface and can also contain functional molecules and/or functional groups. The combination of both components can be used to create smart NIR-responsive nanosystems with many biological and nanotechnological applications. In this chapter, we have tried to select typical examples to visualize how the nanoparticle and the organic capping work together to provide the desired functionality; additional examples can be found in related literature reviews. (Auzel 2004), (Wolfbeis 2015), (Idris et al. 2015), (Wu et al. 2015), (Zhou et al. 2015), (Yang et al. 2015), (Li et al. 2015), (Sedlmeier and Gorris 2015), (Prodi et al. 2015), (Naccache et al. 2015), (González-Béjar and Pérez-Prieto 2015), (Gnach et al. 2015), (Dong et al. 2015), (Christ and Schäferling 2015), (Cheng and Lin 2015), (Chan et al. 2015), (Tsang et al. 2015a), (G. Chen et al.

2015); (Park et al. 2015), (Chan 2015), (H. Chen et al. 2015), (Bünzli 2015), (Kerr De La Rica 2015), (Tu et al. 2015), (Wu and Butt 2015), (Ramasamy et al. 2014), (Muhr et al. 2014), (Gai et al. 2014), (Dacosta et al. 2014), (Chen et al. 2014), (Gu et al. 2013), (Chen et al. 2013), (Y. Liu et al. 2013), (Gnach and Bednarkiewicz 2012), (Chen et al. 2012), (Wang et al. 2011), (Haase and Schäfer 2011), (Fischer et al. 2011), (Li and Lin 2010).

References

Anpo, M. and P. V. Kamat. 2010. *Enviromentally Begin Photocatalysts. Applications of Titanium Oxide-Based Materials.* New York: Springer Science Business Media, LCC.

Arppe, R., I. Hyppanen, N. Perala et al. 2015. Quenching of the upconversion luminescence of $NaYF_4$:Yb^{3+},Er^{3+} and $NaYF_4$:Yb^{3+},Tm^{3+} nanophosphors by water: The role of the sensitizer Yb^{3+} in non-radiative relaxation. *Nanoscale* 7 (27):11746–11757.

Arppe, R., T. Nareoja, S. Nylund et al. 2014. Photon upconversion sensitized nanoprobes for sensing and imaging of pH. *Nanoscale* 6 (12):6837–6843.

Auzel, F. 2004. Upconversion and anti-stokes processes with f and d ions in solids. *Chem. Rev.* 104 (1):139–174.

Bogdan, N., E. M. Rodriguez, F. Sanz-Rodriguez et al. 2012. Bio-functionalization of ligand-free upconverting lanthanide doped nanoparticles for bio-imaging and cell targeting. *Nanoscale* 4 (12):3647–3650.

Bogdan, N., F. Vetrone, G. A. Ozin et al. 2011. Synthesis of ligand-free colloidally stable water dispersible brightly luminescent lanthanide-doped upconverting nanoparticles. *Nano Lett.* 11 (2):835–840.

Bogdan, N., F. Vetrone, R. Roy et al. 2010. Carbohydrate-coated lanthanide-doped upconverting nanoparticles for lectin recognition. *J. Mater. Chem.* 20 (35):7543.

Boyer, J.-C., M.-P. Manseau, J. I. Murray et al. 2010. Surface modification of upconverting $NaYF_4$ nanoparticles with PEG-Phosphate ligands for NIR (800 nm) biolabeling within the biological window. *Langmuir* 26 (2):1157–1164.

Boyer, J.-C. and F. C. J. M. Van Veggel. 2010. Absolute quantum yield measurements of colloidal $NaYF_4$: Er^{3+}, Yb^{3+} upconverting nanoparticles. *Nanoscale* 2 (8):1417–1419.

Brown, S. B., E. A. Brown, and I. Walker. 2004. The present and future role of photodynamic therapy in cancer treatment. *Lancet Oncol.* 5 (8):497–508.

Bünzli, J.-C. G. 2015. Lanthanide light for biology and medical diagnosis. *J. Lumin.* doi: http://dx.doi.org/10.1016/j.jlumin.2015.07.033.

Cui, S., H. Chen, H. Zhu et al. 2012. Amphiphilic chitosan modified upconversion nanoparticles for in vivo photodynamic therapy induced by near-infrared light. *J. Mater. Chem.* 22:4861–4873.

Cui, S., D. Yin, Y. Chen et al. 2013. In vivo targeted deep-tissue photodynamic therapy based on near-infrared light triggered upconversion nanoconstruct. *ACS Nano* 7 (1):676–688.

Chan, E. M. 2015. Combinatorial approaches for developing upconverting nanomaterials: High-throughput screening, modeling, and applications. *Chem. Soc. Rev.* 44 (6):1653–1679.

Chan, E. M., E. S. Levy, and B. E. Cohen. 2015. Rationally designed energy transfer in upconverting nanoparticles. *Adv. Mater.* 27 (38):5753–5761.

Chen, G., H. Agren, T. Y. Ohulchanskyy et al. 2015. Light upconverting core–shell nanostructures: Nanophotonic control for emerging applications. *Chem. Soc. Rev.* 44 (6):1680–1713.

Chen, G., T. Y. Ohulchanskyy, R. Kumar et al. 2010. Ultrasmall monodisperse $NaYF_4$:Yb^{3+}/Tm^{3+}nanocrystals with enhanced near-Infrared to near-infrared upconversion photoluminescence. *ACS Nano* 4 (6):3163–3168.

Chen, G., H. Qiu, P. N. Prasad et al. 2014. Upconversion nanoparticles: Design, nanochemistry, and applications in theranostics. *Chem. Rev.* 114 (10):5161–5214.

Chen, G., C. Yang, and P. N. Prasad. 2013. Nanophotonics and nanochemistry: Controlling the excitation dynamics for frequency up- and down-conversion in lanthanide-doped nanoparticles. *Acc. Chem. Res.* 46 (7):1474–1486.

Chen, H., D. Shi, Y. Wang et al. 2015. The advances in applying inorganic fluorescent nanomaterials for the detection of hepatocellular carcinoma and other cancers. *RSC Adv.* 5 (97):79572–79584.

Chen, J. and J. X. Zhao. 2012. Upconversion nanomaterials: Synthesis, mechanism, and applications in sensing. *Sensors* 12 (3):2414–2435.

Cheng, L., K. Yang, Y. Li et al. 2011a. Facile preparation of multifunctional upconversion nanoprobes for multimodal imaging and dual-targeted photothermal therapy. *Angew. Chem. Int. Ed.* 50:7385–7390.

Cheng, L., K. Yang, Y. Li et al. 2012. Multifunctional nanoparticles for upconversion luminescence/MR multimodal imaging and magnetically targeted photothermal therapy. *Biomaterials* 33 (7):2215–2222.

Cheng, L., K. Yang, M. Shao et al. 2011b. Multicolor in vivo imaging of upconversion nanoparticles with emissions tuned by luminescence resonance energy transfer. *J. Phys. Chem.* C 115 (6):2686–2692.

Cheng, Z. and J. Lin. 2015. Synthesis and application of nanohybrids based on upconverting nanoparticles and polymers. *Macromol. Rapid Comm.* 36 (9):790–827.

Christ, S. and M. Schäferling. 2015. Chemical sensing and imaging based on photon upconverting nano- and microcrystals: A review. *Methods Appl. Fluoresc.* 3 (3):034004.

Dacosta, M. V., S. Doughan, Y. Han et al. 2014. Lanthanide upconversion nanoparticles and applications in bioassays and bioimaging: A review. *Anal. Chim. Acta* 832:1–33.

Dai, Y., H. Xiao, J. Liu et al. 2013. In vivo multimodality imaging and cancer therapy by near-infrared light-triggered trans-platinum pro-drug-conjugated upconversion nanoparticles. *J. Am. Chem. Soc.* 135 (50):18920–18929.

Dong, A., X. Ye, J. Chen et al. 2011. A generalized ligand-exchange strategy enabling sequential surface functionalization of colloidal nanocrystals. *J. Am. Chem. Soc.* 133 (4):998–1006.

Dong, B., S. Xu, J. Sun et al. 2011. Multifunctional $NaYF_4$:Yb^{3+},Er^{3+} Integration of upconversion imaging and photothermal therapy. *J. Mater. Chem.* 21 (17):6193–6200.

Dong, H., S. R. Du, X. Y. Zheng et al. 2015. Lanthanide nanoparticles: From design toward bioimaging and therapy. *Chem. Rev.* 115 (19):10725–10815.

Dougherty, T. J. 1993. Photodynamic therapy. *Photochem. Photobiol.* 58 (6):895–900.

Esipova, T. V., X. Ye, J. E. Collins et al. 2012. Dendritic upconverting nanoparticles enable in vivo multiphoton microscopy with low-power continuous wave sources. *Proc. Natl. Acad. Sci.* 109 (51):20826–20831.

Friedman, A. D., S. E. Claypool, and R. Liu. 2013. The smart targeting of nanoparticles. *Curr. Pharm. Des.* 19 (35):6315–6329.

Fischer, L. H., G. S. Harms, and O. S. Wolfbeis. 2011. Upconverting nanoparticles for nanoscale thermometry. Angew. *Chem., Int. Ed.* 50 (20):4546–4551.

Francés-Soriano, L., M. González-Béjar, and J. Pérez-Prieto. 2015. Cucurbit[n]uril-capped upconversion nanoparticles as highly emissive scaffolds for energy acceptors. *Nanoscale* 7 (12):5140–5146.

Furube, A., T. Asahi, H. Masuhara et al. 1999. Charge carrier dynamics of standard TiO_2 catalysts revealed by femtosecond diffuse reflectance spectroscopy. *J. Phys. Chem.* B 103 (16):3120–3127.

Gai, S., C. Li, P. Yang et al. 2014. Recent progress in rare earth micro/nanocrystals: Soft chemical synthesis, luminescent properties, and biomedical applications. *Chem. Rev.* 114 (4):2343–2389.

García, J. C., J. I. Simionato, A. E. C. Da Silva et al. 2009. Solar photocatalytic degradation of real textile effluents by associated titanium dioxide and hydrogen peroxide. *Sol. Energy* 83:316–322.

Gnach, A. and A. Bednarkiewicz. 2012. Lanthanide-doped up-converting nanoparticles: Merits and challenges. *Nano Today* 7 (6):532–563.

Gnach, A., T. Lipinski, A. Bednarkiewicz et al. 2015. Upconverting nanoparticles: Assessing the toxicity. *Chem. Soc. Rev.* 44 (6):1561–1584.

González-Béjar, M., M. Liras, L. Francés-Soriano et al. 2014. NIR excitation of upconversion nanohybrids containing a surface grafted Bodipy induces oxygen-mediated cancer cell death. *J. Mater. Chem.* B 2 (28):4554–4563.

González-Béjar, M. and J. Pérez-Prieto. 2015. Upconversion luminescent nanoparticles in physical sensing and in monitoring physical processes in biological samples. *Methods Appl. Fluoresc.* 3 (4):042002.

Gu, Z., L. Yan, G. Tian et al. 2013. Recent advances in design and fabrication of upconversion nanoparticles and their safe theranostic applications. *Adv. Mater.* 25 (28):3758–3779.

Guo, J., F. Ma, S. Gu et al. 2012. Solvothermal synthesis and upconversion spectroscopy of monophase hexagonal $NaYF_4$:Yb^{3+}/Er^{3+} nanosized crystallines. *J. Alloy. Compd.* 523:161–166.

Haase, M. and H. Schäfer. 2011. Upconverting nanoparticles. *Angew Chem Int Ed* 50 (26):5808–5829.

Hao, J., Y. Zhang, and X. Wei. 2011. Electric-induced enhancement and modulation of upconversion photoluminescence in epitaxial $BaTiO_3$:Yb/Er thin films. *Angew. Chem., Int. Ed.* 50 (30):6876–6880.

Hao, S., G. Chen, and C. Yang. 2013. Sensing using rare-earth-doped upconversion nanoparticles. *Theranostics* 3 (5):331–345.

Heer, S., K. Kömpe, H. U. Güdel et al. 2004. Highly efficient multicolour upconversion emission in transparent colloids of lanthanide-doped $NaYF_4$ nanocrystals. *Adv. Mater.* 16 (23–24):2102–2105.

Hoffmann, M. R., S. T. Martin, W. Choi et al. 1995. Environmental applications of semiconductor photocatalysis. *Chem. Rev.* 95 (1):69–96.

Huang, T. C. and W. F. Hsieh. 2009. Er-Yb codoped ferroelectrics for controlling visible upconversion emissions. *J. Fluoresc.* 19:511–516.

Ideat, R., F. Tasaka, W.-D. Jang. et al. 2005. Nanotechnology-based photodynamic therapy for neovascular disease using a supramolecular nanocarrier loaded with a dendritic photosensitizer. *Nano Lett.* 5 (12):2426–2431.

Idris, N. M., M. K. Gnanasammandhan, J. Zhang et al. 2012. In vivo photodynamic therapy using upconversion nanoparticles as remote-controlled nanotransducers. *Nature Med.* 18 (10):1580–1585.

Idris, N. M., M. K. G. Jayakumar, A. Bansal et al. 2015. Upconversion nanoparticles as versatile light nanotransducers for photoactivation applications. *Chem. Soc. Rev.* 44 (6):1449–1478.

Idris, N. M., Z. Li, L. Ye et al. 2009. Tracking transplanted cells in live animal using upconversion fluorescent nanoparticles. *Biomaterials* 30 (28):5104–5113.

Jin, J., Y-Juan Gu, Cornelia Wing-Yin Man et al. 2011. Polymer-coated $NaYF_4$:Yb^{3+}, Er^{3+} upconversion nanoparticles for charge-dependent cellular imaging. *ACS Nano* 5 (10):7838–7847.

Kennedy, L. C., Bickford L. R., Lewinski N. A. et al. 2011. A new era for cancer treatment: Gold-nanoparticle-mediated thermal therapies. *Small* 7 (2):169–183.

Kerr, C. A. and R. De La Rica. 2015. Photoluminescent nanosensors for intracellular detection. *Anal. Methods* 7 (17):7067–7075.

Khataee, A. R. and M. Fathinia. 2013. New and future developments in catalysis: Nanoparticle catalysis by surface plasmon. In *Recent Advances in Photocatalytic Processes by Nanomaterials*, edited by Steven L. Sun, 267–288. United Kingdom: Elsevier Science Ltd.

Kobayashi, H., N. Kosaka, M. Ogawa et al. 2009. In vivo multiple color lymphatic imaging using upconverting nanocrystals. *J. Mater. Chem.* 19 (36):6481.

Lei, L., J. Zhou, J. Zhang et al. 2015. The use of zinc ions to control the size of Yb/Er:$KMnF_3$ nanocrystals with single band emission. *CrystEngComm* 17 (44):8457–8462.

Li, C. and J. Lin. 2010. Rare earth fluoride nano-/microcrystals: Synthesis, surface modification and application. *J. Mater. Chem.* 20 (33):6831.

Li, R., Z. Ji, J. Dong et al. 2015. Enhancing the imaging and biosafety of upconversion nanoparticles through phosphonate coating. *ACS Nano* 9 (3):3293–3306.

Li, X., F. Zhang, and D. Zhao. 2015. Lab on upconversion nanoparticles: Optical properties and applications engineering via designed nanostructure. *Chem. Soc. Rev.* 44 (6):1346–1378.

Li, Z. and Y. Zhang. 2008. An efficient and user-friendly method for the synthesis of hexagonal-phase $NaYF_4$:Yb, Er/Tm nanocrystals with controllable shape and upconversion fluorescence. *Nanotechnology* 19 (34):345606.

Liang, S., X. Zhang, Z. Wu et al. 2012. Decoration of up-converting NaYF4:Yb,Er(Tm) nanoparticles with surfactant bilayer. A versatile strategy to perform oil-to-water phase transfer and subsequently surface silication. *CrystEngComm* 14 (10):3484.

Linsebigler, A. L., G. Lu, and J. T. Yates Jr. 1995. Photocatalysis on TiO_2 Surfaces: Principles, mechanisms, and selected results. *Chem. Rev.* 95:735–758.

Liras, M., M. González-Béjar, E. Peinado et al. 2014. Thin amphiphilic polymer-capped upconversion nanoparticles: Enhanced emission and thermoresponsive properties. *Chem. Mater.* 26 (13):4014–4022.

Liu, J., W. Bu, L. Pan et al. 2013. NIR-triggered anticancer drug delivery by upconverting nanoparticles with integrated azobenzene-modified mesoporous silica. *Angew. Chem. Int. Ed.* 52 (16):4375–4379.

Liu, K., X. Liu, Q. Zeng et al. 2012. Covalently assembled NIR nanoplatform for simultaneous fluorescence imaging and photodynamic therapy of cancer cells. *ACS Nano* 6 (5):4054–4062.

Liu, Q., C. Li, I. Yang et al. 2010. "Drawing" upconversion nanophosphors into water through host-guest interaction. *Chem. Commun.* 46 (30):5551–5553.
Liu, X., M. Zheng, X. Kong et al. 2013. Separately doped upconversion-C_{60} nanoplatform for NIR imaging-guided photodynamic therapy of cancer cells. *Chem. Commun.* 49 (31):3224–3226.
Liu, Y., D. Tu, H. Zhu et al. 2013. Lanthanide-doped luminescent nanoprobes: Controlled synthesis, optical spectroscopy, and bioapplications. *Chem. Soc. Rev.* 42 (16):6924–6958.
Lucky, S. S., N. Muhammad Idris, Z. Li et al. 2015. Titania coated upconversion nanoparticles for near-infrared light triggered photodynamic therapy. *ACS Nano* 9 (1):191–205.
Luo, M., Q. Li, M. Chen et al. 2013. Controlled assembly of gold and rare-earth upconversion nanoparticles for ratiometric sensing applications. *Wuhan Univ. J. Nat. Sci.* 18 (4):277–282.
Ma, J., P. Huang, M. He et al. 2012. Folic acid-conjugated LaF_3:Yb,Tm@SiO_2 nanoprobes for targeting dual-modality imaging of upconversion luminescence and X-ray computed tomography. *J. Phys. Chem.* B 116 (48):14062–14070.
Ma, X. and X. Ni. 2015. Fabrication of polythiophene-TiO2 heterojunction solar cells coupled with upconversion nanoparticles. *J. Mater. Sci.: Mater. Electron.* 26 (2):1129–1135.
Mai, H.-X., Y.-W. Zhang, R. Si et al. 2006. High-quality sodium rare-earth fluoride nanocrystals: Controlled synthesis and optical properties. *J. Am. Chem. Soc.* 128 (19):6426–6436.
Muhr, V., S. Wilhelm, T. Hirsch et al. 2014. Upconversion nanoparticles: From hydrophobic to hydrophilic surfaces. *Acc. Chem. Res.* 47 (12):3481–3493.
Naccache, R., Yu Q., and Capobianco J. A. 2015. The fluoride host: Nucleation, growth, and upconversion of lanthanide-doped nanoparticles. *Adv. Opt. Mat.k.* 3 (4):482–509.
Niu, W., S. Wu, and S. Zhang. 2010. A facile and general approach for the multicolor tuning of lanthanide-ion doped $NaYF_4$ upconversion nanoparticles within a fixed composition. *J. Mater. Chem.* 20 (41):9113.
Oppenlander, T. 2003. *Photochemical Purification fo Water and Air.* Wiley-VCH Verlag GmbH & Co. KGaA, Weinheim.
Pan, L., M. He, J. Ma et al. 2013. Phase and size controllable synthesis of $NaYbF_4$ nanocrystals in oleic acid/ionic liquid two-phase system for targeted fluorescent imaging of gastric cancer. *Theranostics* 3 (3):210–222.
Park, Y. I., H. M. Kim, J. H. Kim et al. 2012. Theranostic probe based on lanthanide-doped nanoparticles for simultaneous in vivo dual-modal imaging and photodynamic therapy. *Adv. Mater.* 24 (42):5755–5761.
Park, Y. I., K. T. Lee, Y. D. Suh et al. 2015. Upconverting nanoparticles: A versatile platform for wide-field two-photon microscopy and multi-modal in vivo imaging. *Chem. Soc. Rev.* 44 (6):1302–1317.
Philipp, C. M., E. Rohde, and H. P. Berlien. 1995. Nd:YAG laser procedures in tumor treatment. *Semin. Surg. Oncol.* 11 (4):290–298.
Prodi, L., E. Rampazzo, F. Rastrelli et al. 2015. Imaging agents based on lanthanide doped nanoparticles. *Chem. Soc. Rev.* 44 (14):4922–4952.
Ramasamy, P., P. Manivasakan, and J. Kim 2014. Upconversion nanophosphors for solar cell applications. *RSC Adv.* 4 (66):34873–34895.
Ren, W., G. Tian, S. Jian et al. 2012. TWEEN coated $NaYF_4$:Yb,Er/$NaYF_4$ core/shell upconversion nanoparticles for bioimaging and drug delivery. *RSC Adv.* 2 (18):7037.

Ruggiero, E., J. Hernandez-Gil, J. C. Mareque-Rivas et al. 2015. Near infrared activation of an anticancer PtIV complex by Tm-doped upconversion nanoparticles. *Chem. Commun.* 51 (11):2091–2094.

Schäfer, H., P. Ptacek, K. Kömpe et al. 2007. Lanthanide-doped $NaYF_4$ nanocrystals in aqueous solution displaying strong up-conversion emission. *Chem. Mater.* 19 (6):1396–1400.

Sedlmeier, A. and H. H. Gorris. 2015. Surface modification and characterization of photon-upconverting nanoparticles for bioanalytical applications. *Chem. Soc. Rev.* 44 (6):1526–1560.

Shan, J. and Y. Ju. 2007. Controlled synthesis of lanthanide-doped $NaYF_4$ upconversion nanocrystals via ligand induced crystal phase transition and silica coating. *Appl. Phys. Lett.* 91 (12):123103.

Shan, J. and Y. Ju. 2009. A single-step synthesis and the kinetic mechanism for monodisperse and hexagonal-phase $NaYF_4$:Yb, Er upconversion nanophosphors. *Nanotechnology* 20:275603.

Shan, J., X. Qin, N. Yao et al. 2007. Synthesis of monodisperse hexagonal $NaYF_4$:Yb, Ln (Ln = Er, Ho and Tm) upconversion nanocrystals in TOPO. *Nanotechnology* 18 (44):445607.

Sun, L.-N., H. Peng, M. I. J. Stich et al. 2009. pH sensor based on upconverting luminescent lanthanide nanorods. *Chem. Commun.* (33):5000–5002.

Suyver, J. F., A. Aebischer, D. Biner et al. 2005. Novel materials doped with trivalent lanthanides and transition metal ions showing near-infrared to visible photon upconversion. *Opt. Mater.* 27 (6):1111–1130.

Tikhomirov, V. K., L. F. Chibotaru, D. Saurel et al. 2009. Er^{3+}-doped nanoparticles for optical detection of magnetic field. *Nano Lett.* 9 (2):721–724.

Topel, S. D., G. T. Cin, and E. U. Akkaya. 2014. Near IR excitation of heavy atom free Bodipy photosensitizers through the intermediacy of upconverting nanoparticles. *Chem. Commun.* 50 (64):8896–8899.

Tsang, M.-K., G. Bai, and J. Hao. 2015a. Stimuli responsive upconversion luminescence nanomaterials and films for various applications. *Chem. Soc. Rev.* 44 (6):1585–1607.

Tsang, M.-K., C.-F. Chan, K.-L. Wong et al. 2015b. Comparative studies of upconversion luminescence characteristics and cell bioimaging based on one-step synthesized upconversion nanoparticles capped with different functional groups. *J. Lumin.* 157:172–178.

Tu, L., X. Liu, F. Wu et al. 2015. Excitation energy migration dynamics in upconversion nanomaterials. *Chem. Soc. Rev.* 44 (6):1331–1345.

Valiente, R., M. Millot, F. Rodríguez et al. 2009. Er^{3+} luminescence as a sensor of high pressure and strong external magnetic fields. *High Pressure Res.* 29 (4):748–753.

Vetrone, F., R. Naccache, A. Juarranz De La Fuente et al. 2010a. Intracellular imaging of HeLa cells by non-functionalized $NaYF_4$:Er^{3+}, Yb^{3+} upconverting nanoparticles. *Nanoscale* 2 (4):495–498.

Vetrone, F., R. Naccache, A. Zamarrón et al. 2010b. Temperature sensing using fluorescent nanothermometers. *ACS Nano* 4 (6):3254–3258.

Viger, M. L., M. Grossman, N. Fomina et al. 2013. Low power upconverted near-IR light for efficient polymeric nanoparticle degradation and cargo release. *Adv. Mater.* 25 (27):3733–3738.

Voliani, V., M. Gemmi, L. Francés-Soriano et al. 2014. Texture and phase recognition analysis of β-$NaYF_4$ nanocrystals. *J. Phys. Chem. C* 118 (21):11404–11408.

Voliani, V., M. González-Béjar, V. Herranz-Pérez et al. 2013. Orthogonal functionalisation of upconverting $NaYF_4$ nanocrystals. *Chem. Eur. J.* 19 (40):13538–13546.
Wang, C., L. Cheng, Y. Liu et al. 2013. Imaging-guided pH-sensitive photodynamic therapy using charge reversible upconversion nanoparticles under near-infrared light. *Adv. Funct. Mater.* 23 (24):3077–3086.
Wang, C., H. Tao, L. Cheng et al. 2011. Near-infrared light induced in vivo photodynamic therapy of cancer based on upconversion nanoparticles. *Biomaterials* 32 (26):6145–6154.
Wang, D., B. Xue, X. Kong et al. 2015. 808 nm driven Nd^{3+}-sensitized upconversion nanostructures for photodynamic therapy and simultaneous fluorescence imaging. *Nanoscale* 7 (1):190–197.
Wang, F. and X. Liu 2009. Recent advances in the chemistry of lanthanide-doped upconversion nanocrystals. *Chem. Soc. Rev.* 38 (4):976–989.
Wang, G., Q. Peng, and Y. Li 2011. Lanthanide-doped nanocrystals: Synthesis, optical-magnetic properties, and applications. *Acc. Chem. Res.* 44 (5):322–332.
Wang, L., R. Yan, Z. Huo et al. 2005. Fluorescence resonant energy transfer biosensor based on upconversion-luminescent nanoparticles. *Angew. Chem. Int. Ed.* 44 (37):6054–6057.
Wang, M., G. Abbineni, A. Clevenger et al. 2011. Upconversion nanoparticles: synthesis, surface modification and biological applications. *Nanomedicine* 7 (6):710–729.
Wang, M., Z. Chen, W. Zheng et al. 2014. Lanthanide-doped upconversion nanoparticles electrostatically coupled with photosensitizers for near-infrared-triggered photodynamic therapy. *Nanoscale* 6 (14):8274–8282.
Wang, X., C. X. Yang, J. T. Chen et al. 2014. A dual-targeting upconversion nanoplatform for two-color fluorescence imaging-guided photodynamic therapy. *Anal. Chem.* 86 (7):3263–3267.
Wang, Y., H. Wang, D. Liu et al. 2013. Graphene oxide covalently grafted upconversion nanoparticles for combined NIR mediated imaging and photothermal/photodynamic cancer therapy. *Biomaterials* 34 (31):7715–7724.
Wei, Y., F. Lu, X. Zhang et al. 2006. Synthesis of oil-dispersible hexagonal-phase and hexagonal-shaped $NaYF_4$:Yb,Er nanoplates. *Chem. Mater.* 18 (24):5733–5737.
Wencel, D., T. Abel, and C. Mcdonagh. 2014. Optical chemical pH sensors. *Anal. Chem.* 86 (1):15–29.
Wilhelm, S., M. Kaiser, C. Wurth et al. 2015. Water dispersible upconverting nanoparticles: effects of surface modification on their luminescence and colloidal stability. *Nanoscale* 7 (4):1403–1410.
Wisser, M. D., M. Chea, Y. Lin et al. 2015. Strain-induced modification of optical selection rules in lanthanide-based upconverting nanoparticles. *Nano Lett.* 15 (3):1891–1897.
Wolfbeis, O. S. 2015. An overview of nanoparticles commonly used in fluorescent bioimaging. *Chem. Soc. Rev.* 44 (14):4743–4768.
Won Jin, K., N. Marcin, and N. P. Paras. 2009. Color-coded multilayer photopatterned microstructures using lanthanide (III) ion co-doped $NaYF_4$ nanoparticles with upconversion luminescence for possible applications in security. *Nanotechnology* 20 (18):185301.
Wu, S. and H.-J. Butt. 2015. Near-infrared-sensitive materials based on upconverting nanoparticles. *Adv. Mater.* doi: 10.1002/adma.201502843.
Wu, S., N. Duan, Z. Shi et al. 2014. Simultaneous aptasensor for multiplex pathogenic bacteria detection based on multicolor upconversion nanoparticles labels. *Anal. Chem.* 86 (6):3100–3107.

Wu, X., G. Chen, J. Shen et al. 2015. Upconversion nanoparticles: A versatile solution to multiscale biological imaging. *Bioconjug. Chem.* 26 (2):166–175.

Xia, L., X. Kong, X. Liu et al. 2014. An upconversion nanoparticle-Zinc phthalocyanine based nanophotosensitizer for photodynamic therapy. *Biomaterials* 35 (13):4146–4156.

Xiao, Q., Y. Ji, Z. Xiao et al. 2013. Novel multifunctional $NaYF_4:Er^{3+},Yb^{3+}$/PEGDA hybrid microspheres: NIR-light-activated photopolymerization and drug delivery. *Chem. Commun.* 49:1527–1529.

Xie, D., H. Peng, S. Huang et al. 2013. Core–shell structure in doped inorganic nanoparticles: Approaches for optimizing luminescence properties. *J. Nanomater.* 2013: 1–10.

Xie, J., Lee S., and Chen X. 2010. Nanoparticle-based theranostic agents. *Adv. Drug Deliver. Rev.* 62 (11):1064–1079.

Xie, L., Y. Qin, and H.-Y. Chen. 2012. Polymeric optodes based on upconverting nanorods for fluorescent measurements of pH and metal ions in blood samples. *Anal. Chem.* 84 (4):1969–1974.

Xing, Q., N. Li, Y. Jiao et al. 2015. Near-infrared light-controlled drug release and cancer therapy with polymer-caged upconversion nanoparticles. *RSC Adv.* 5 (7):5269–5276.

Yan, L., Y.-N. Chang, W. Yin et al. 2014. Biocompatible and flexible graphene oxide/upconversion nanoparticle hybrid film for optical pH sensing. *Phys. Chem. Chem. Phys.* 16 (4):1576–1582.

Yan, L., Y.-N. Chang, L. Zhao et al. 2013. The use of polyethylenimine-modified graphene oxide as a nanocarrier for transferring hydrophobic nanocrystals into water to produce water-dispersible hybrids for use in drug delivery. *Carbon* 57:120–129.

Yang, D., P. A. Ma, Z. Hou et al. 2015. Current advances in lanthanide ion (Ln^{3+})-based upconversion nanomaterials for drug delivery. *Chem. Soc. Rev.* 44 (6):1416–1448.

Yang, J., D. Shen, X. Li et al. 2012. One-step hydrothermal synthesis of carboxyl-functionalized upconversion phosphors for bioapplications. *Chemistry* 18 (43):13642–13650.

Yang, Y., B. Velmurugan, X. Liu et al. 2013. NIR photoresponsive crosslinked upconverting nanocarriers toward selective intracellular drug release. *Small* 9 (17):2937–2944.

Ye, X., J. E. Collins, Y. Kang et al. 2010. Morphologically controlled synthesis of colloidal upconversion nanophosphors and their shape-directed self-assembly. *PNAS* 107 (52):22430–22435.

Yi, G.-S. and G.-M. Chow. 2007. Water-soluble $NaYF_4$:Yb,Er(Tm)/$NaYF_4$/polymer core/shell/shell nanoparticles with significant enhancement of upconversion fluorescence. *Chem. Mater.* 19 (3):341–343.

Yi, G. S. and G. M. Chow. 2006. Synthesis of hexagonal-phase $NaYF_4$:Yb,Er and $NaYF_4$:Yb,Tm nanocrystals with efficient up-conversion fluorescence. *Adv. Funct. Mater.* 16 (18):2324–2329.

Yu, M., F. Li, Z. Chen et al. 2009. Laser scanning up-conversion luminescence microscopy for imaging cells labeled with rare-earth nanophosphors. *Anal. Chem.* 81 (3):930–935.

Zhang, S.-Z., L.-D. Sun, H. Tian et al. 2009. Reversible luminescence switching of $NaYF_4$:Yb,Er nanoparticles with controlled assembly of gold nanoparticles. *Chem. Commun.* (18):2547–2549.

Zhang, P., S. Rogelj, K. Nguyen et al. 2006. Design of a highly sensitive and specific nucleotide sensor based on photon upconverting particles. *J. Am. Chem. Soc.* 128 (38):12410–12411.
Zhang, Y.-W., X. Sun, R. Si et al. 2005. Single-crystalline and monodisperse LaF_3 triangular nanoplates from a single-source precursor. *J. Am. Chem. Soc.* 127 (10):3260–3261.
Zhou, A., Y. Wei, B. Wu et al. 2012. Pyropheophorbide a and c (RGDyK) comodified chitosan-wrapped upconversion nanoparticle for targeted near-infrared photodynamic therapy. *Mol. Pharmaceutics* 9:1580–1589.
Zhou, J., Q. Liu, W. Feng et al. 2015. Upconversion luminescent materials: Advances and applications. *Chem. Rev.* 115 (1):395–465.
Zhou, J., Y. Sun, X. Du et al. 2010. Dual-modality in vivo imaging using rare-earth nanocrystals with near-infrared to near-infrared (NIR-to-NIR) upconversion luminescence and magnetic resonance properties. *Biomaterials* 31 (12):3287–3295.
Zhou, L., Z. Chen, K. Dong et al. 2014. DNA-mediated construction of hollow upconversion nanoparticles for protein harvesting and near-infrared light triggered release. *Adv. Mater.* 26 (15):2424–2430.
Zhuravleva, N. G., A. A. Eliseev, N. A. Sapoletova et al. 2005. The synthesis of EuF_3/TOPO nanoparticles. *Mater. Sci. Eng.* C 25 (5–8):549–552.

6

Tuning Optical Properties of Lanthanide Upconversion Nanoparticles

Yuanwei Zhang, Zhanjun Li, Xiang Wu, and Gang Han

CONTENTS

6.1 Introduction 139
6.2 Tuning Excitation Wavelengths 141
  6.2.1 Nd^{3+}-Sensitized 800 nm Laser Excitation 142
  6.2.2 Nd^{3+}-Sensitized 745 nm LED Excitation 147
  6.2.3 Organic Dye-Sensitized NIR Excitation 147
6.3 Tuning Emission Properties 149
  6.3.1 Multicolor Arrays 150
    6.3.1.1 By Combination of Lanthanide Emitters 150
    6.3.1.2 Through the Pulse Width of IR Laser Beams 152
    6.3.1.3 By Tuning Laser Power 153
  6.3.2 Single-Band Upconversion Emission 155
  6.3.3 Tuning Emission Lifetime 156
6.4 Conclusion and Outlook 158
References 158

6.1 Introduction

Upconversion nanoparticles (UCNPs) have received profound interest in recent years as promising agents for biological studies with several advantages over traditional down conversion fluorophores. They are nonblinking, nonphotobleaching, and able to emit short-wavelength light after sequentially absorbing two or more photons of near-infrared (NIR) light. Thanks to the advancements in nanotechnology, the development of UCNPs entered a higher level with many approaches to push UCNPs into the next generation, which are small in size with high upconversion emission intensities and tunable absorption/emission properties. This chapter focuses mainly on the recent progress in engineering excitation and emission properties of UCNPs systems.

Upconversion materials are commonly excited under NIR irradiation to exhibit unique large anti-Stokes shifts (Gamelin and Gudel 1998; Haase and

Schafer 2011; Wermuth and Gudel 1999). In particular, upon continuous wave (CW) excitation at 980 nm, lanthanide-doped UCNPs display broad emissions from UV to NIR with sharp emission lines and long lifetimes (Mader et al. 2010). UCNPs applied for luminescent imaging and analytical assays are well acknowledged (Feng et al. 2013; Liu et al. 2013). Compared with conventional photoluminescence materials such as organic dyes and quantum dots, lanthanide-doped UCNPs are advantageous (Gu et al. 2013; Wang et al. 2011a,b) because of their NIR absorption (a region with low absorption and self-emission of the bio-environment), low-toxicity, sharp emissions, long lifetime, excellent photostability, and large anti-Stokes shift (Naczynski et al. 2013). Due to these fascinating properties, lanthanide-doped UCNPs have numerous potential applications, such as upconversion lasers (Joubert et al. 1993; Lenth and Macfarlane 1990), solar cells spectrum modifiers (Khan et al. 2011; Shalav et al. 2005), photodynamic therapy remote controllers (Idris and Gnanasammandhan 2012), optical waveguide amplifiers (Kik and Polman 2003; Strohhofer and Polman 2001), and temperature sensors (Fisher and Harms 2011).

The whole field of photon upconversion can be traced back to 1959, with a proposition that infrared (IR) light could be detected through sequential absorption by a solid-state ion (Bloembergen 1959). Since then, upconversion luminescence has been chiefly investigated through bulk material, primarily rare-earth metals due to their characteristic fluorescent properties (Chen et al. 2014). These systems, however, were far too big for most biological applications. The emergence of nanotechnology in the late 1900s quickly sparked interest in the potential of ion-doped nanoparticles. In early 2000, Güdel and Haase give the first examples of transparent colloidal solution of UCNPs to emit visible light under NIR irradiation, which subsequently propagated intensive research and the use of UCNPs as biomolecular agents (Heer et al. 2003, 2004). Since then, crystalline fluoride-based host lattices $M(RE)F_4$ doped with lanthanides were the most popular means of studying UC emission. In particular, $NaYF_4$ coupled with 20%Yb^{3+}, 2%Er^{3+} (0.5%Tm^{3+}) were long used and considered to be the optimal doping conditions.

These first-generation UCNPs have been synthesized from nearly all upconversion materials known to date, which include oxides, phosphates, vanadates, yttrium fluoride-based, and various mixed systems (Chen et al. 2014). Several well-established fabrication routes exist, such as thermal decomposition, hydro-thermal synthesis, co-precipitation, and a sol–gel process. The choice of methods depends not only on the composition of the nanoparticle, but also its desired properties. Among them, wet-chemical strategies of thermal decomposition and hydro-thermal synthesis are most applicable for bio-related studies, as nanoparticles with narrow size distribution and high dispersibility can be conducted. In 2005, Yan et al. reported the first preparation of highly monodispersed LaF_3 nanoparticles (16 nm) through the pyrolysis of $La(CF_3COO)_3$ in a mixed solvent of oleic acid and 1-octadecene at 280°C (Zhang et al. 2005). In 2008, Zhang et al. developed a unique temperature approach based on replacement reactions to synthesize monodispersed β-$NaYF_4$:Yb,Er/Tm nanocrystals with a 21 nm spherical size (Li et al. 2008). These advancements in synthetic routes have triggered the fabrication of a new generation of nanoparticles, with the aim of improving the properties of UCNPs for functional applications.

However, constraints and challenges remain in developing UCNPs that are suitable for studies in multifaceted bio-environments and varied assays, such as relatively poor upconversion efficiency and large particle sizes. In addition, for studies of complicated bio-assays, multi-channel detections are favorable. This requires versatile imaging agents with tunable excitation wavelengths and emission colors. In general, the excitation and emission wavelength/intensities of UCNPs are highly dependent on the doped lanthanide ions. For instance, the most widely employed sensitizer Yb^{3+} only has a narrow absorption band around 980 nm, and the optional lanthanide emitters are quite restricted according to the energy matching.

This chapter focuses on the problems of tuning excitation/emission properties of UCNPs, by examining the recent design and engineering strategies of modern nanoparticles. Rather than providing a complete survey of applications like most recent review articles (Dong et al. 2015; Idris and Jayakumar 2015; Zheng et al. 2015), our emphasis lies in the construction of more wavelength versatile nanoparticles that offer future opportunities for the functional enhancement of UCNPs.

6.2 Tuning Excitation Wavelengths

Traditional wavelengths used to excite Yb^{3+}-based UCNPs centered around 980 nm. However, this excitation window poses a significant problem for biological applications, as can be seen in Figure 6.1, 980 nm light has a heavy optical absorption by water (Shen et al. 2013). Because cells and tissues withhold 980 nm radiation and can potentially induce heat damages (McNichols et al. 2004), this becomes the major obstacle for the application of UCNPs in water abundant live systems (Kou et al. 1993; Nam et al. 2011). In particular, such heating effect is likely to be more severe in certain experiments that require high power density and long-time irradiation, such as single nanoparticle imaging or longitudinal deep tissue imaging.

In order to minimize the heating effect of biological specimens upon irradiation by 980 nm laser, He et al. employed a 915 nm laser to excite Yb^{3+} sensitized α-$NaYF_4$:Ln nanoparticles, which showed appreciably less heating and better imaging depth (Zhan et al. 2011). In theory, lanthanide-doped UCNPs are excitable by light from 900 to 1000 nm due to the electron transition absorption between $^2F_{7/2}$ and $^2F_{5/2}$ of Yb^{3+} ions. In the absorption spectrum of UCNPs, there is a strong and steep absorption band located

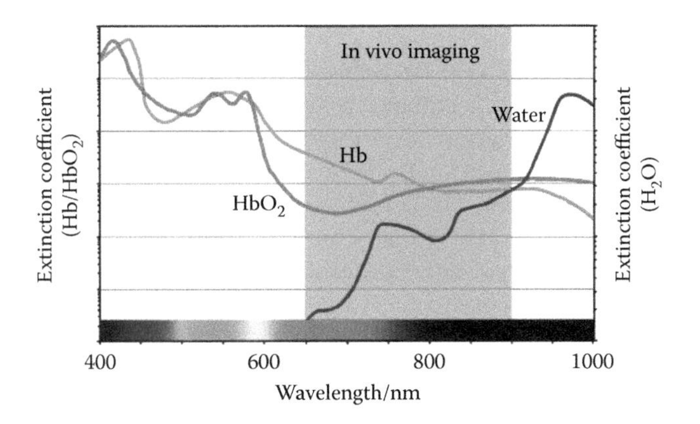

FIGURE 6.1
Spectra profiles of tissue optical window. (Hb: hemoglobin; HbO_2: oxyhemoglobin). (Adapted from Shen, J., G. Y. Chen et al., *Adv. Optical Mater.*, 1, 644–650, 2013. With permission.)

in the range of 900–1000 nm. Although the maximum absorption is around the wavelength of 974 nm, it is also capable of being excited by light around 920 nm. The 915 nm laser excited UCNPs have been used for HeLa cancer cell imaging and nude mouse *in vivo* imaging, and exhibited good experimental outcomes. However, this excitation window around 915 nm still overlapped with water absorption quite a lot.

6.2.1 Nd^{3+}-Sensitized 800 nm Laser Excitation

In contrast, as can be seen in the tissue optical window of Figure 6.1, the local minima of water absorption is at ca. 800 nm, which has been considered the ideal excitation wavelength with the least heating impact on biological tissues. Generally, the ability of water absorption at 800 nm is about 20 times lower than that at 980 nm. In this regard, by co-doping with Nd^{3+} ions, Han and Shen creatively innovated a UCNP system (β-$NaYF_4$:Nd,Yb,Er(Tm)@$NaYF_4$) that is excitable with 800 nm CW laser (Shen et al. 2013). In this novel strategy, Nd^{3+} ions were doped and acted as 800 nm light absorber. As shown in Figure 6.2a, the absorbed energy is subsequently transferred to Er^{3+} or Tm^{3+} emitters through the bridge of Yb^{3+} ions. In regard to the bridging sensitizer of Yb^{3+} ions, they also investigated the Yb^{3+} doping concentration in order to optimize bridging energy transfer efficiencies. In their study, they found increasing Yb^{3+} concentration showed no improvement in the Er^{3+}(Tm^{3+}) emission efficiency. They suggested that 20%–30% of Yb^{3+} dopants are sufficient enough to deliver/transfer energy from Nd^{3+} ions with 1% doping concentration. Specifically, they showed in Figure 6.2b that 0.15 W of 800 nm CW laser for Nd/Yb/Er UCNPs was able to generate the same amount of upconversion emission as 0.049 W of 980 nm CW laser excited traditional Yb/Er UCNPs. Meanwhile, 0.15 W of 800 nm CW laser exhibits

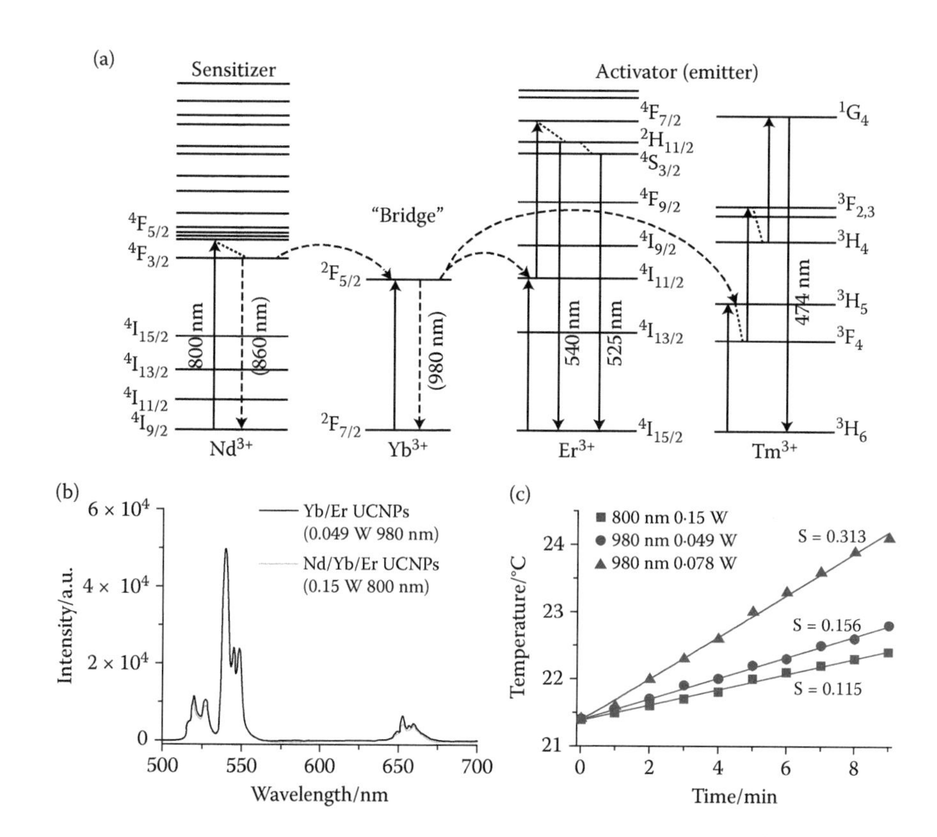

FIGURE 6.2
(a) Upconversion process of $Nd^{3+} \rightarrow Yb^{3+} \rightarrow Er^{3+}(Tm^{3+})$ tri-dopants system with 800 nm excitation. (b) 0.15 W of 800 nm CW laser for β-$NaYF_4$:Nd,Yb,Er/β-$NaYF_4$ UCNPs generates the same upconversion emission amount as 0.049 W of 980 nm CW laser for β-$NaYF_4$:Yb,Er/β-$NaYF_4$ UCNPs. (c) And the temperature increase of water under continuous irradiation of 980 nm and 800 nm laser with 0.15 W of 800 nm CW laser, 0.049 W of 980 CW laser and 0.078 W of 980 nm CW laser. (Modified from Shen, J., G. Y. Chen et al., *Adv. Optical Mater.*, 1, 644–650, 2013. With permission.)

lower calefaction slope of water heating, compared to 0.049 W of 980 nm CW laser, as can be seen from Figure 6.2c. Therefore, to achieve the same upconversion emission counts, their pioneered tri-doped cascade UCNPs with 800 nm excitation induced less heat effect in water than the traditional dual-doped UCNPs using 980 nm excitation source (Shen et al. 2013).

Despite the advantages of large absorption cross section of Nd^{3+} and the high $Nd^{3+} \rightarrow Yb^{3+}$ energy transfer efficiency, the introduction of Nd^{3+} at high concentration may cause quench of the upconversion emission, due to the deleterious energy transfer and relaxations between activators and Nd^{3+} (Wang et al. 2010). Spatially separating the activator and Nd^{3+} is proposed by Wang et al. (2013) to address this concern. They employed a core–shell structure, which could separate the activator and Nd^{3+} in the core and shell (β-$NaGdF_4$:Yb,Er@$NaYF_4$:Yb,Nd) thus largely avoiding such energy quenching effect (Wang et al. 2013). In their approach, the core is doped with Yb^{3+} ions along with activators, where the Yb^{3+} → activators UC process is supposed to occur. At the same time, the shell is doped with both Nd^{3+} and Yb^{3+}, where the excitation of Nd^{3+} and subsequent Nd^{3+} → Yb^{3+} energy transfer take place. As shown in Figure 6.3a and b, by adopting this core–shell structure, successive Nd^{3+} → Yb^{3+} → activator energy transfer would be dominant, and UC efficiency comparable with that of the Yb^{3+}-sensitized UC process was found. Figure 6.3c and d discussed the *in vivo* heating experiment monitored with an IR thermal imager, low temperature incensement was found under 800 nm laser irradiation compared with 980 nm laser irradiation. A notable

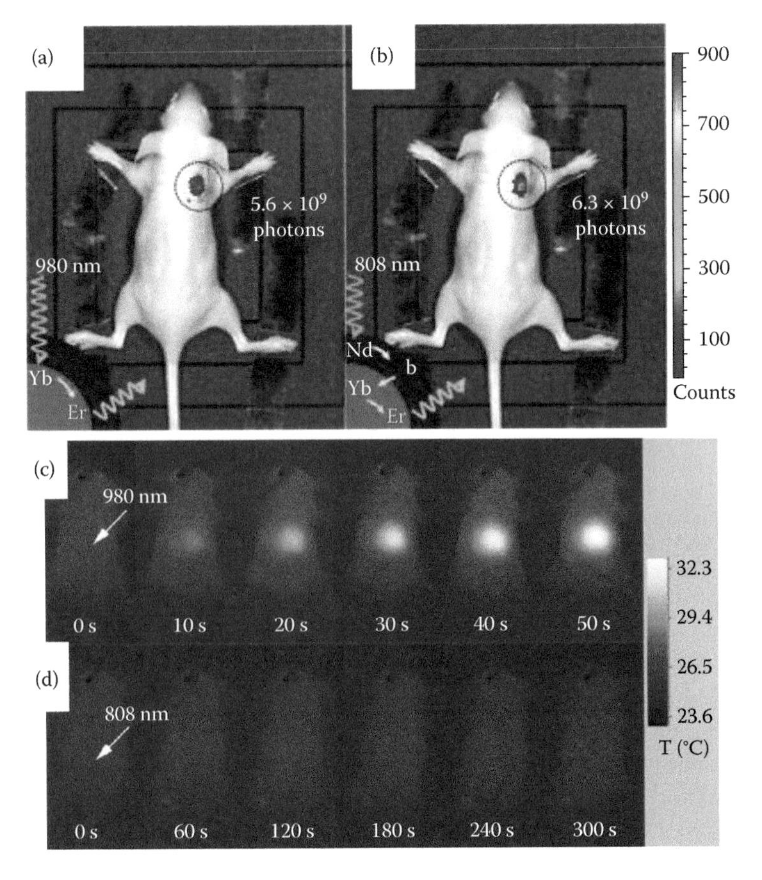

FIGURE 6.3
In vivo UC imaging of a nude mouse subcutaneously injected with Er@Nd NPs. The images were obtained with 980 nm laser (a) and 808 nm laser (b) irradiation, both with a power density of 200 mW/cm². And IR thermal image of a nude mouse during continuous (c) 980 nm laser irradiation for 50 S and (d) 808 nm laser irradiation for 300 s. Irradiation spots are denoted with the white arrows. (Modified from Wang, Y. F., G. Y. Liu et al., *ACS Nano*, 7, 7200–7206, 2013. With permission.)

local heating effect and significant rise in temperature were observed under irradiation of the 980 nm laser after 50 S, and a burn wound was discernible 1 day after irradiation. In contrast, for the 808 nm laser irradiation for as long as 5 min, only a slight rise in temperature for the whole mouse body was detected.

To further engineering the absorbability at 800 nm with confined cross-relaxation, Liu and coworkers developed a way of doping a high concentration of Nd^{3+} outside the UCNP's core as light harvesting reservoirs. Their strategy also utilized a core–shell structure, however, with spatially confined Nd^{3+} ions in the active shell layer, markedly enhanced upconversion emission was generated (Xie et al. 2013). Taking advantage of this active shell design, they were able to dope a higher amount of Nd^{3+} ions into the nanoparticles and achieved enhanced emission from Tm^{3+}, Er^{3+}, and Ho^{3+} activators. For example, the $NaYF_4$:Yb/Tm/Nd@$NaYF_4$:Nd core–shell nanoparticles with 20% of Nd^{3+} content in the shell layer exhibited an integrated emission intensity which is ~405 times higher than achievable by the nanoparticles without the shell coating. Importantly, the active shell coating strategy is capable of achieving an emission intensity that is about seven times higher than the one obtained with an inert $NaYF_4$ shell. Again, the overheating effect was greatly reduced in the cellular experiment, as most HeLa cells containing Nd^{3+}-sensitized UCNPs survived under 800 nm illumination. Under identical irradiation power densities, however, 980 nm excitation of Yb^{3+}-based UCNPs led to cell death within 5 min.

Even though, under 800 nm excitation the above mentioned structure can enhance UC luminescence by seven times compared with triply doped $NaYF_4$:Yb,Er,Nd@$NaYF_4$ UCNPs, there is still much space to further reduce the quenching effect between the Nd^{3+} ions and activators. In addition, the reported Nd^{3+} concentration in the shell was increased up to 10%–20%, although there is also room to increase the Nd^{3+} concentration to obtain the maximum absorbability of NIR light. To this end, Zhong and coworkers reported a design of a transition layer into the core–shell structure to fabricate a quenching-shield sandwich structure that enables the blocking of energy back-transfer from activators (Er^{3+}, Tm^{3+}, and Ho^{3+}) to the sensitizer Nd^{3+}. As shown in Figure 6.4a, the Nd^{3+} ions are confined in the outer shell and serve as the sensitizer to harvest 800 nm photons, which results in population of the $^4F_{5/2}$ state of Nd^{3+}. Figure 6.4b illustrated the energy transferring process, the Yb^{3+} ion is used to extract the excitation energy from the sensitizer Nd^{3+} through interionic cross-relaxation $[(^4F_{3/2})Nd,(^2F_{7/2})Yb] \rightarrow [(^4I_{9/2})Nd,(^2F_{5/2})Yb]$, followed by excitation-energy migration over the Yb sublattice and finally entrapment by the activator ions (Er^{3+}, Tm^{3+}, and Ho^{3+}) that are embedded in the inner core. As can be seen the transition layer between core and outer shell serves as a shield to prevent quenching interactions between the activator and sensitizer, while the efficient excitation energy transfer from Nd^{3+} to activators can still be achieved through the Yb-mediated core–shell–shell interface. Thus the unwanted quenching effects that generated

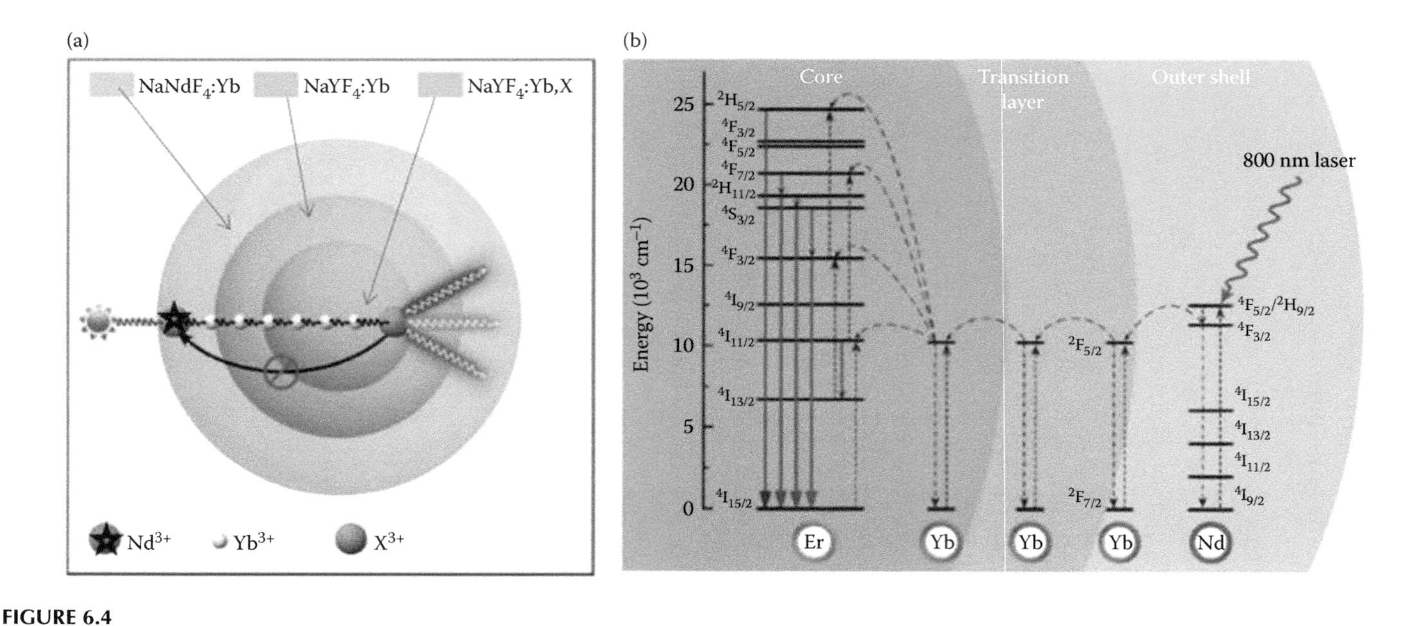

FIGURE 6.4
(a) Schematic illustration of the proposed energy-transfer mechanisms in the quenching-shield sandwich-structured UCNPs upon 800 nm excitation. (b) Proposed energy-transfer mechanisms in the quenching-shield sandwich nanoparticle upon 800 nm diode-laser excitation. (Modified from Xie, X. J. et al., *J. Am. Chem. Soc.*, 135, 12608–12611, 2013. With permission.)

from Nd^{3+} ions can be largely avoided even with a pure $NaNdF_4$ coating as the outer-most layer. By optimizing the thickness of the interlayer, the emission intensity of the quenching-shield sandwich structure upon 800 nm excitation reaches a maximum when the interlayer thickness is controlled at about 1.45 nm, and is even brighter than conventional 980 nm excited nanoparticles at low excitation power density (0.5 W/cm^2), as a result of the higher absorption cross section of the sensitizer Nd^{3+} ions (Zhong et al. 2014).

6.2.2 Nd^{3+}-Sensitized 745 nm LED Excitation

To date, UCNPs are mostly excited by laser source (including pulse laser and CW lasers) for upconversion luminescence. However, the use of lasers increases the risk of radiation accidents, especially those involving eye injuries (Sliney 1995). Moreover, the high price, limited lifetime, and complex design of laser systems also hinder the further application of the UCNPs in biological and clinical usage. In comparison, the light-emitting diode (LED) holds great promise as a radiation source for bio-application (Desmet et al. 2006; Li et al. 2015). The multiple absorption peaks and high extinction coefficient make Nd^{3+} a potential sensitizer for LED-excitable UCNPs. The commercial available 740 nm LED emits from 719 to 751 nm (half-peak width), which is perfectly matching the excitation peaks of Nd^{3+} UCNPs around 739 nm, referring to the $^4I_{9/2} \rightarrow {}^4F_{7/2}$ transition of the Nd^{3+} ions. UCNPs ($NaYF_4$:Er,Yb@$NaYF_4$:Yb@$NaNdF_4$:Yb) with varied concentration of Yb^{3+} ions content in the inner, middle, and outer layers were synthesized to further enhance the UCL by optimizing the Yb^{3+} ions distribution to achieve the most efficient energy migration over the Yb^{3+}-doped sublattice (Zhong et al. 2015). As a feasibility experiment, peptide-labeled UCNPs were developed for targeted imaging of human epidermal growth factor receptor 2 (Her2) positive cancer cells by using 740 nm LED as the excitation source. Human mammary gland carcinoma SKBR-3 cell line that overexpresses the HER2 gene product was chosen as an example for target-specific imaging. Besides, Her2 low-expressing human embryonic kidney 293A cell line, human lung carcinoma A549 cell line, and human cervix carcinoma HeLa cell line served as negative controls. As can be seen from Figure 6.5, the blue emission from the Hoechst was excited by 365 nm UV light, and the green emission from the UCNPs was excited by 740 nm NIR light.

6.2.3 Organic Dye-Sensitized NIR Excitation

The limitations of Yb^{3+} as sensitizer arise not only from the small absorption cross sections, but also from the narrow excitation range. Thus, it is still a significant challenge to overcome these restrictions. Zou et al. (2012) have proposed and demonstrated the viability of a new strategy to enhance the upconversion luminescence of lanthanide nanoparticles based on increased and spectrally broadened absorption, using organic IR dyes as sensitizers.

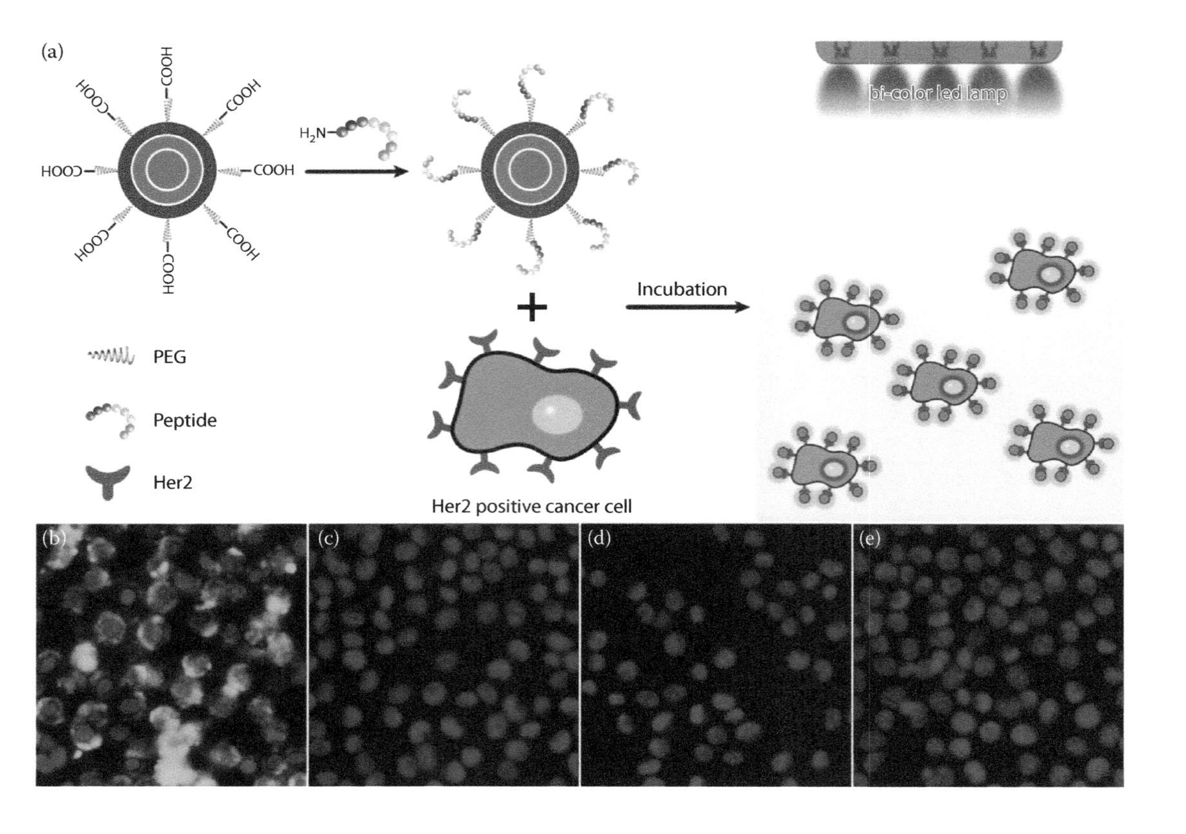

FIGURE 6.5
(a) Schematic representation of the surface chemistry of the peptide-labeled UCNPs, and UCNPs binding to membranes of SKBR-3 cells containing Her2. In vitro bioimaging (blue: Hoechst, green: UCNPs) of the peptide-labeled UCNPs incubated with (b)SKBR-3 cells, (c) 293A cells, (d) A549 cells, and (e) HeLa cells. (Adapted from Zhong, Y. T. et al., *Adv. Mater.*, 27, 6418–6422, 2015. With permission.)

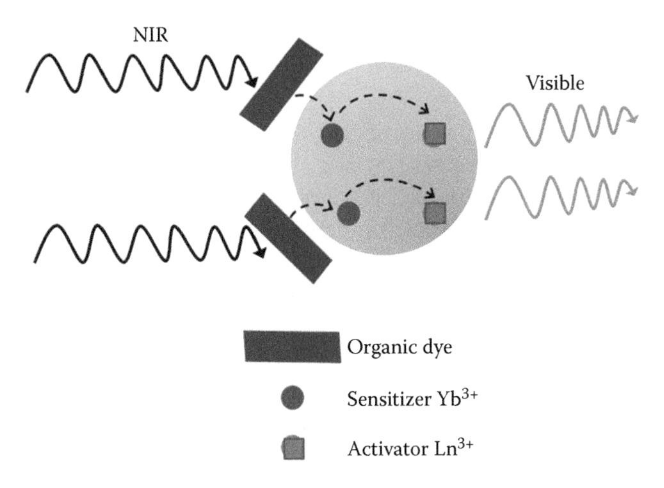

FIGURE 6.6
An illustration of dye-sensitized upconversion processes.

The main idea was inspired by the natural light-harvesting systems in which various absorber molecules surround and transfer the solar energy to the central reaction center, albeit for upconversion instead of charge transfer. In their approach, dye molecules function as antennas, absorbing incident light, and transferring their excitation energy to the upconverting metal ions encapsulated in a nanoparticle (Zou et al. 2012). Using cyanine dye with a carboxylic acid anchor as an antenna, they trapped 740–850 nm wavelength photons and transferred the resulting energy to Yb^{3+} ions. This was eventually transmitted to Er^{3+} ions embedded in the host matrices to generate strikingly improved upconversion, as shown in Figure 6.6. Their work introduced the idea of broadband dye-sensitized upconversion. The highly efficient energy transfer from organic dyes to UCNPs is explained by the classical Förster theory, which contends that the physical distance between the energy donor and acceptor must be close, typically no more than 10 nm, for resonance energy transfer to occur. In contrast to standard UCNPs that lack organic dye modifications, ones with dye antennae were reported to have up to 3300 times the emission enhancement in the excitation range of 720–1000 nm.

6.3 Tuning Emission Properties

For the application in biological imaging, on one side, diverse colors are required for more flexible bio-assays. Especially for multiplexed biological labeling, the ability to create multicolor upconversion outputs from UCNPs is particularly important (Jaiswal et al. 2003; Kumar et al. 2008). Ideal probes for multiplexed detection should absorb light at a given irradiation wavelength, give well-resolved emission peaks, and be photochemically stable (Li et al. 2005; Stoeva et al. 2006). One of the most effective ways of tuning a UCNP's emission color is by modifying the sensitizer/activator dopant species and their respective concentrations.

On the other hand, strong and pure red emissions are preferable for bioimaging applications with deep tissue penetration and low side effects in UV to visible region. However, it is difficult to achieve single-band upconversion with high chromatic purity, because lanthanide ions have multiple metastable excited states. In general, Yb/Er UCNPs only give weak red emissions, and for Yb/Tm UCNPs the accompanied strong blue emissions might lead to many unwanted side effects and lose recording sensitivity.

6.3.1 Multicolor Arrays

6.3.1.1 By Combination of Lanthanide Emitters

In order to generate multicolor emissions, scientists have studied the optical behaviors of Er^{3+} and Tm^{3+} in various doping matrix (Wang et al. 2009). Three combinations of dopants, Yb^{3+}/Er^{3+}, Yb^{3+}/Tm^{3+}, and $Yb^{3+}/Er^{3+}/Tm^{3+}$, can yield the three distinct green, magenta, and greenish white upconversion emission. For example, Yi and coworkers (2005) have carefully studied Yb^{3+} co-doped Er^{3+}, Tm^{3+} in LaF_3 matrices. The resulting nanocrystals are easily dispersed into organic solutions to form a transparent solution. Using NIR excitation at wavelength of 980 nm, blue, green, and red emission bands were observed, showing a potential application as bio-related probes and assays (Yi and Chow 2005).

In addition, Han et al. have developed multicolor emissions based on ultra-small YF_3 nanocrystals. The color output of these YF_3 nanocrystals, which were doped with Yb^{3+}/Er^{3+}, Yb^{3+}/Tm^{3+}, and $Yb^{3+}/Er^{3+}/Tm^{3+}$, was all finely modulated by adjusting the concentrations of Yb^{3+}, as can be seen from Figure 6.7 (Chen et al. 2012). In their YF_3:Yb^{3+}/Er^{3+} system, green upconversion emission at 521/543 nm was dominant when Yb^{3+} ion concentration was low (e.g., 10%), however, when Yb^{3+} concentration was high, the red emission at 650 nm became increasingly intense. This phenomenon was explained by the suppressed energy back transfer process from Er^{3+} to Yb^{3+} ions, thus increasing the population of the intermediate level $^4I_{13/2}$(Er) of red emission in the YF_3 host. And the colloidal nanocrystals YF_3 doped with 2% Tm^{3+} and 10% Yb^{3+} appear to be magenta to the naked eye when excited by 980 nm laser. This is explained by the fact that the negligible weak UC band at 700 nm in other host lattices becomes unignorable in the YF_3 host material due to its high efficiency for UC emission. As displayed in Figure 6.7, the UC color output in YF_3:$Yb^{3+}/Er^{3+}/Tm^{3+}$ nanocrystals was gradually modulated from greenish white to yellow, light red, and then to pure red, as the concentration of Yb^{3+} ions increased. Through their work, lanthanide-doped YF_3

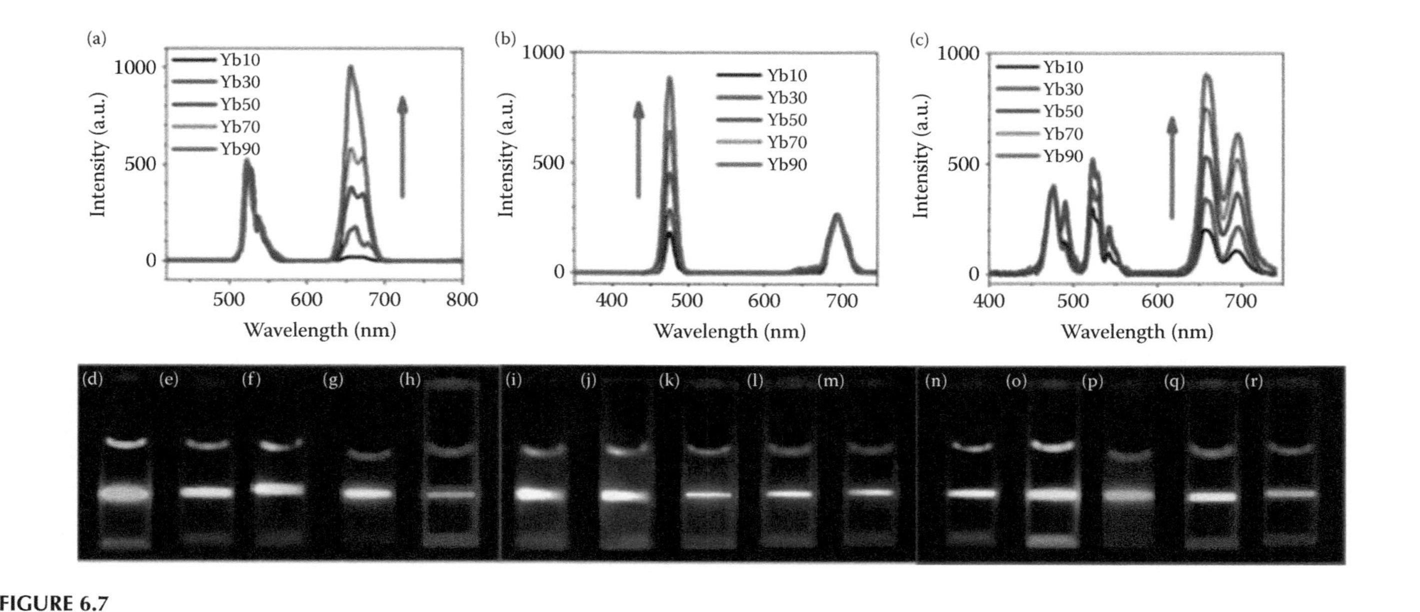

FIGURE 6.7
Upconversion emission spectra of the colloidal nanocrystals of YF_3 (dissolved in cyclohexane, 5 mg/mL) doped with (a) Er^{3+} ions of 0.5% and Yb^{3+} ions of 10%–90%. (b) Tm^{3+} ions of 2% and Yb^{3+} ions of 10%–90%. (c) Er^{3+} ions of 0.5%, Tm^{3+} ions of 2%, and Yb^{3+} ions of 10%–90%. The spectra in (a)–(c) in order were normalized to the emissions of Er^{3+} ions at 520 nm, Tm^{3+} ions at 475 nm, respectively. They complied upconversion photoluminescent images for the colloidal nanocrystals of YF_3 doped with (d)–(h) Er^{3+} ions of 0.5% and Yb^{3+} ions of 10%–90%. (i)–(m) Tm^{3+} ions of 2% and Yb^{3+} ions of 10%–90%. (n)–(r) Er^{3+} ions of 0.5%, Tm^{3+} ions of 2%, and Yb^{3+} ions of 10%–90%. The samples were excited at 980 nm with a power density of 300 W/cm^2. (Adapted from Chen, G. Y., H. L. Qiu et al., *J. Mater. Chem.*, 22, 20190–20196, 2012. With permission.)

nanocrystals with ultrasmall size, optimized UC PL intensity, and modulated color output show promise with respect to their use as effective imaging probes for bio-applications.

In general, the lanthanide doping concentration controls the average distance between neighboring ions, and has a strong influence on optical properties of UCNPs such as, the relative intensity of emission peaks (Zhang et al. 2011a). Mai and coworkers (2007) investigated this concept by adjusting the composition of $NaYF_4$:Yb,Er/Tm UCNPs to tune emission colors. After dissolving the arrays in cyclohexane and exciting the UCNPs with a 980 nm laser, multicolor upconversion emissions were observed by the naked eye (Mai et al. 2007). Wang and Liu (2008) further explored this phenomenon in-depth with Yb^{3+}/Er^{3+} and Tm^{3+} co-doped UCNP systems. They reported that tunable upconversion emission colors were obtained using a single-host source of $NaYF_4$ doped with different combinations and concentrations of lanthanide ions (Yb^{3+}, Tm^{3+}, and Er^{3+}).

Multicolor fine tuning of UCNPs was alternatively achieved with a lanthanide-doped system of $NaYF_4$:Yb/Ho/Tm in a multi-photon emission process, by Zhao et al. (Zhang et al. 2011a). Normally, the co-doped $NaYF_4$:20%Yb/0.2%Tm nanocrystals have a blue emission color, due to the transition from $^1D_2$ to $^3F_4$ and $^1G_4$ to $^3H_6$. The introduction of a second emitter lanthanide ion (Ho^{3+}) at different concentrations resulted in a precisely controllable relative intensity ratio of the dual emissions. For example, variations in the Ho^{3+} concentration from 0.2% to 2.0% led to prominent changes in the emission intensities in the green and red region, allowing one to explicitly change the color output from blue to green. Similar effects have been observed in other systems, like Y_2O_3:Yb,Ho/Tm (Zhang et al. 2011b).

Recently, a facile route for fabricating upconversion multiplexed probes by controlling the doping position has been reported by Gu and coworkers (2013). They synthesized a series of consistently sized nanoparticles containing equal amount of dopants at varying positions in the core and shell. As the doping position gradually moved from the inner core to the outer shell, the local relative concentration of Er^{3+} and Yb^{3+} dopants was increased (Li et al. 2013). As shown in Figure 6.8, the red/green emission ratio also intensified, due to the increasingly efficient cross-relaxation processes of Er^{3+} at high doping concentrations.

6.3.1.2 Through the Pulse Width of IR Laser Beams

An appealing feature of UCNPs is that the emission color can be tuned under single wavelength excitation by tailoring the intensity ratio of different emission peaks (Mahalingam et al. 2008). However, this color-tuning technique requires stringent experimentation involving repeated trials to obtain optimal dopant concentrations (Vetrone et al. 2004). Recently, Liu and coworkers (2015) demonstrated a laser pulse-width-modulated approach that offers exquisite control over the emission color of lanthanide-doped UCNPs.

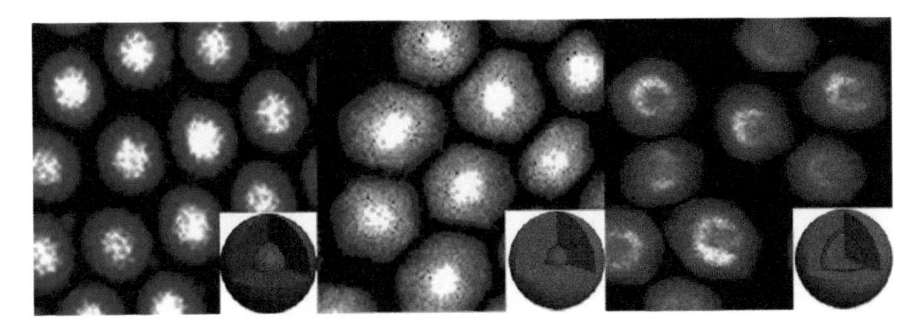

FIGURE 6.8
UCNPs multicolor output conducted via tunable doping positions (core doping and shell doping). From left to right, $NaGdF_4$:Yb,Er@$NaYF_4$, $NaYbF_4$:Er@$NaYF_4$, $NaYF_4$@$NaYbF_4$:Er@$NaYF_4$. (Adapted from Li, X. M. et al., *Chem. Mater.*, 25, 106–112, 2013. With permission.)

Material wise, β-$NaYF_4$ was chosen as the host containing layer-by-layer shells incorporated with five sets of lanthanide ions (Yb^{3+}, Nd^{3+}, Tm^{3+}, Ho^{3+}, and Ce^{3+}) at specific layers of precisely defined concentrations. On 808 nm CW laser excitation, the nanocrystals gave rise to a dominant blue emission band at 470 nm, which can be attributed to the $^1G_4 \rightarrow {}^3H_6$ transition of Tm^{3+}. However, when irradiated at 980 nm with a diode laser, these nanocrystals showed marked suppression in the blue emission, while exhibiting a weak green emission band at 541 nm and a strong red emission band at 646 nm. The green and red emission bands can be ascribed to the $^5F_4$, $^5S_2 \rightarrow {}^5I_8$ and $^5F_5 \rightarrow {}^5I_8$ transitions of Ho^{3+}, respectively. In addition, the green/red emission intensity ratio can be tuned by using a pulsed diode laser with different pulse widths. For instance, a relatively long pulse (6 ms, 100 Hz) yielded a red/green ratio of 7, providing a visible red emission from the colloidal nanocrystals. In contrast, on decreasing the pulse width (200 μs, 100 Hz), an intense green emission was provided as a result of a markedly reduced red/green intensity ratio to 0.5. As demonstrated by Figure 6.9, by adjusting the pulse width within the range of 200 μs to 6 ms, they could therefore achieve dynamic control of the emission color from green to red (Deng et al. 2015). As a proof of concept, they incorporated the nanocrystals into a polydimethylsiloxane monolith to construct a transparent display matrix. By supplementing the 808 nm CW light source with a pulse-modulated 980 nm beam, they generated a 3D imagery by moving the focal point of the beam within the volume of the display matrix. Using this approach, true multi-perspective color images with high spatial resolution and precision were achieved.

6.3.1.3 By Tuning Laser Power

The separation of white light into its components has been widely adopted in modern displays to obtain pure red-green-blue (RGB) colors for pixels (He et al. 2009; Jang et al. 2010; Shirasaki et al. 2012; Sivakumar et al. 2005).

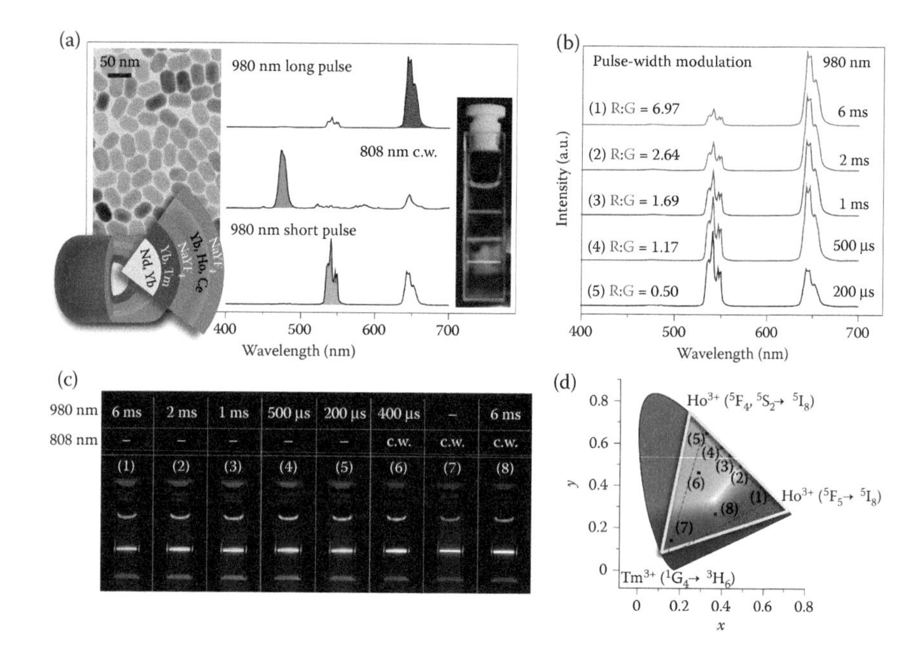

FIGURE 6.9
(a) Design of $NaYF_4$-based core–shell nanocrystals capable of emitting tunable colors when irradiated with NIR laser. (b) Upconversion emission spectra of the nanocrystals obtained through the use of different pulse durations, demonstrating the temporal tuning of the red and green emission intensities. (c) Luminescence photographs showing multicolor tuning for the samples through combined use of 980 and 808 nm laser. (d) Corresponding color gamut of the emission colors from the sample shown in (c), compared to the color spaces accessible by conventional high-definition televisions. (Adapted from Deng, R. R. et al., *Nat. Nanotech.*, 10, 237–242, 2015. With permission.)

Recently, Zhang et al. (2015) presented an integrated full-spectral upconversion nanostructure prepared by optimizing the pathway of photon transitions from a set of selected lanthanide ions. The RGB emitters exhibited different spectral sensitivities to the excitation laser power density and thus allowed the modulation of a wide range of emission colors. In their work, the $NaGdF_4$:Yb/Tm/Er@$NaGdF_4$:Eu@$NaYF_4$ nanostructure accommodates six kinds of lanthanide ions to obtain white light by upconversion. A series of upconversion spectra at different excitation power densities were measured, and the integrated intensity of each peak was calculated. The three color intensities increased as the laser power density increased, but exhibited very different increase rates, namely red blue green, leading to a change of proportions of the three colors in the total emission intensity (Zhang et al. 2015). In a proof-of-concept experiment, they made a flat-panel display on a nanocrystal-coated substrate surface. Pixels with different colors can be produced by fast scanning of a laser that is synchronously modulated by controlling the pump current. They subsequently fabricated a prototype of a flat-panel

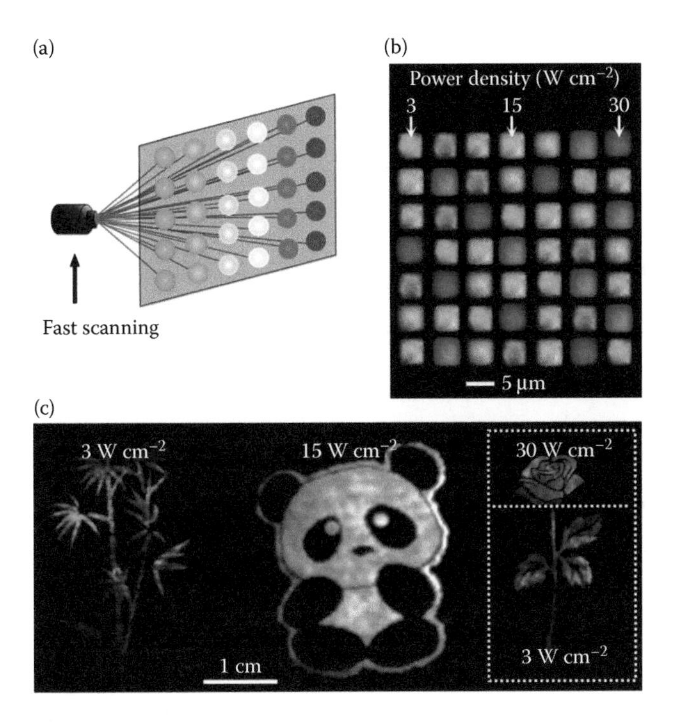

FIGURE 6.10
(a) Concept of flat-panel display by the fast scanning of a laser with modulated power. (b) Dot-array display on a piece of silicon wafer (the upconversion nanocrystals were loaded into 5×5 μm etched microwells on a silicon wafer). (c) Picture of bamboo, panda, and rose printed on a paper using colloidal nanostructure as colorless ink. (Adapted from Zhang, C. et al., *Angew. Chem. Int. Ed.*, 54, 11531–11535, 2015. With permission.)

display by using the color-tunable nanocrystals. Bamboo, panda, and rose motifs were printed onto standard A4 paper by an inkjet printer that used colloidal nanostructure as colorless ink (Mei and Zhang 2012). The green bamboo and white-black panda appeared on the paper when laser power densities of 3 and 15 W/cm^2 were used, respectively. Moreover, as shown in Figure 6.10, a rose with red petals and green stem and leaves was visible upon the synchronous use of two laser powers.

6.3.2 Single-Band Upconversion Emission

Upconversion luminescence imaging *in vivo* is expected to be the next generation photoluminescence imaging technique since it provides high sensitivity and spatial resolution (Zhou et al. 2010). More importantly, it could lead to predictive models for potential clinical applications (Xiong et al. 2010). However, due to the strong tissue absorption of short-wavelength light below 600 nm, *in vivo* imaging based on UCNPs as luminescent probes is still limited (Boyer et al. 2006; Zeng et al. 2005). It is generally believed that the NIR spectral range (700–1100 nm) and the red region (600–700 nm) are referred to as the "optical window" of the biological tissues, where the light scattering, absorbance, and autofluorescence of tissue are minimum in view of the lack of efficient endogenous absorbers (Yu et al. 2008). Thus, tuning both the excitation and emission peaks into the "optical window" is essential for the deep tissue imaging of fluorescent labels.

In 2012, Zhao and coworkers reported a strategy to achieve pure dark red emissions from $NaYF_4$:Yb/Er UCNPs via a manganese ion (Mn^{2+}) doping methodology. The presence of Mn^{2+} ions supposedly blocks the possible transitions between green and red emissions of Er^{3+}, while simultaneously aiding the emission of red light. In addition, the single-band red-emitting UCNPs exhibited exceptional photostability. The UCL intensity remained almost unchanged under 3 h of continuous 980 nm laser irradiation, which makes UCNPs suitable for long-term bioimaging. *In vivo* bioimaging studies using the UCNPs at different doping concentrations of Mn^{2+} subsequently revealed its potential applications for diagnosis and therapy (Tian et al. 2012). In another experiment, Yin et al. (2012) showed amplified red emission from GdF_3:Yb^{3+},Er^{3+} UCNPs with Li^+ doping. These improved UCNP materials were used further for both cell imaging and *in vivo* imaging studies (Yin and Zhao 2012).

Single-band upconversion emission has been reported by Liu and coworkers. They developed an oil-based procedure for the synthesis of lanthanide-doped $KMnF_3$ nanocrystals, in which a narrow band around 660 nm was observed for Er^{3+}-doped $KMnF_3$. This can be ascribed to a nonradiative energy transfer from $^1H_{9/2}$ and $^4S_{3/2}$ levels of Er^{3+} to $^4T_1$ level of Mn^{2+}, followed by a back energy transfer to the $^4F_{9/2}$ level of Er^{3+}. The complete disappearance of emissions in blue and green region shows a highly efficient energy exchange between Er^{3+} and Mn^{2+} ions. As a proof-of-concept experiment, they injected polymer-modified $KMnF_3$:Yb/Er nanocrystals into pork muscle tissue at varied depth (0–10 mm) and imaged them by a modified Maestro *in vivo* imaging system. They found the nanocrystals can be visualized even at a depth of 10 mm under an excitation power density of approximately 0.2 W/cm^2. Under identical experimental settings, however, $NaYF_4$ nanocrystals co-doped with Yb^{3+}/Er^{3+} at different ratios can only be detected at about 5 mm beneath the tissue surface. Notably, the emission color of the $KMnF_3$:Yb/Er nanocrystals did not change as a function of sample imaging depth, as confirmed by the recorded emission spectra. In contrast, the $NaYF_4$ nanocrystals injected at different depths showed significant changes in emission color, which was attributed to rapid attenuation of the green emission relative to red emission in tissues (Wang et al. 2011a,b).

6.3.3 Tuning Emission Lifetime

For imaging studies and data securities, emission lifetime is as important as emission wavelength (color) for simultaneous identification and quantification of multiple distinctive species (Han et al. 2001; Pregibon et al. 2007). In 2013, Jin and coworkers developed a strategy for controlling the luminescent lifetimes of upconversion nanocrystals. These tunable UCNPs, which researchers have named τ-dots, have distinct lifetimes that are independent of the particles' color and intensity. Lifetime tunability was accomplished by modifying the energy transfer from the sensitizer to the emitter ion at varying sensitizer–emitter physical distances. Lu et al. (2014) demonstrated this tunability by printing three overlapping patterns with three different $NaYF_4$:Yb,Tm UCNPs (Yb^{3+}:Tm^{3+} ratios of 20:4, 20:1, and 20:0.5). Normal luminescence color imaging gave a convoluted set of images, while time-resolved scanning cytometry easily separated each individual pattern. Figure 6.11 showed the resulted from the UCNPs' three distinct lifetimes of 52, 160, 455 μs, respectively (Lu et al. 2014). The tunability of lifetime emissions opens new opportunities for nanoparticles in bioimaging, data storage, and data security (Deng and Liu 2014).

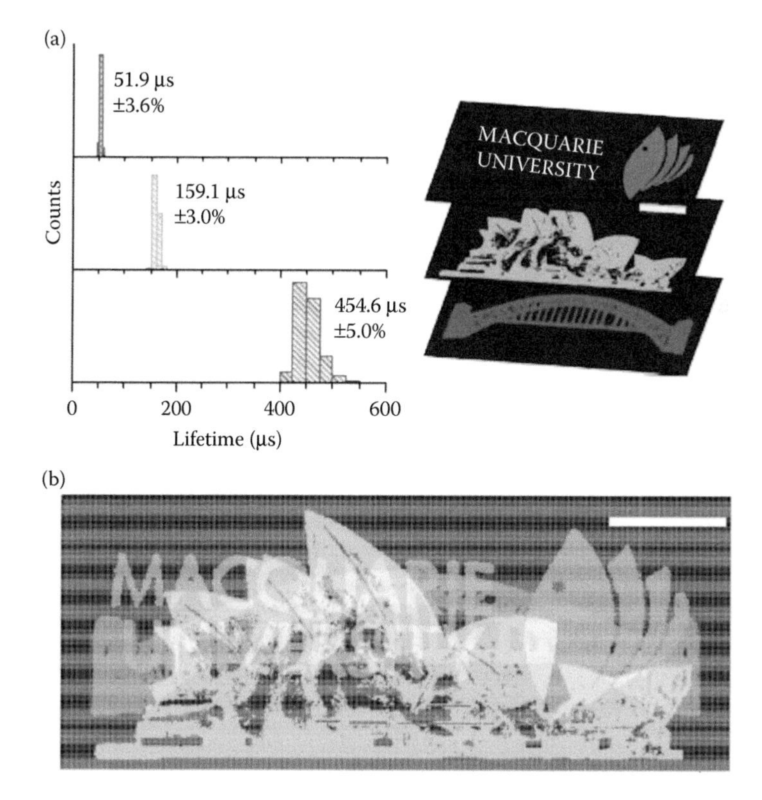

FIGURE 6.11
Example of lifetime encoded document for security and photonic data storage. (a) Lifetime of $NaYF_4$:Yb,Tm UCNPs with Yb^{3+}: Tm^{3+} ratios of 20:4, 20:1, and 20:0.5, respectively. (b) Pseudocolor exposing each individual pattern based on luminescence lifetime. (Adapted from Lu, Y. Q. et al. 2014. *Nat. Photo.* 8:33–37. With permission.)

6.4 Conclusion and Outlook

UCNPs have exhibited outstanding advantages as next generation advanced materials with uses in bioimaging, drug delivery, light-activated therapy, and many others. The evolution has seen the growth of this new tool from materials science to far-reaching biological applications that allow researchers to study and diagnose processes underneath deep tissue. However, further research is necessary to improve their functional applications. This includes enhancing upconversion efficiency as well as fabricating smaller nanoparticles (i.e., sub 10 nm) to increase mobility and reduce metabolic toxicity. Furthermore, manufacturing tunable UCNPs with controllable excitation/emission windows is crucial to generating more diverse bioassays. This chapter has focused primarily on recent engineering techniques that contribute more versatile operational wavelengths and bio-friendly UCNPs.

The route of using an efficient antenna as light absorber provides opportunities to enhancing UC luminescence by orders of magnitude as well as an approach to tuning the excitation wavelength. In addition, the development of more controllable photo-physical properties could be fulfilled by accurately and spatially controlling lanthanide ions doped in the core or shell. In this way, the energy transfer from activator to emitter could be regulated, while unwanted cross-relaxations between different lanthanide ions could be refined.

Challenges remain in regard to developing more advanced practical applications of lanthanide-doped nanoparticles. In this field more studies are needed in order to gain a better understanding of the principles and mechanisms of upconversion processes for use as a general guide. A better understanding of the effects of factors such as electronic structures and concentration of rare-earth ions, nanoparticle shape, size and morphology, and corresponding synthesis methods are essential for future controllable UCNP fabrication. Meanwhile, incorporating UCNPs into more complicated devices or structures is another key goal in the nano-science field.

References

Bloembergen, N. 1959. Solid state infrared quantum counters. *Phys. Rev. Lett.* 2:84.

Boyer, J. C., F. Vetrone, L. A. Cuccia, and J. A. Capoboianco. 2006. Synthesis of colloidal upconverting NaYF4 nanocrystals doped with Er3+, Yb3+ and Tm3+, Yb3+ via thermal decomposition of lanthanide trifluroacetate precursors. *J. Am. Chem. Soc.* 128:7444–7445.

Chen, G. Y., H. L. Qiu, R. Fan, S. Hao, S. Tan, C. Yang, and G. Han. 2012. Lanthanide-doped ultrasmall yttrium fluoride nanoparticles with enhanced multicolor upconversion photoluminescence. *J. Mater. Chem.* 22, 20190–20196.

Chen, G. Y., H. L. Qiu, P. N. Prasad, and X. Chem. 2014. Upconversion nanoparticles: Design, nanochemistry, and applications in theranostics. *Chem. Rev.* 114:5161–5214.

Deng, R. R. and X. G. Liu. 2014. Optical multiplexing tunable lifetime nanocrystals. *Nat. Photon.* 8:10–12.

Deng, R. R., F. Qin, R. Chen, W. Huang, M. Hong, and X. Liu. 2015. Temporal full-color tuning through non-steady-state upconversion. *Nat. Nanotech.* 10:237–242.

Desmet, K. D., D. A. Paz, J. J. Corry, J. T. Eells, M. T. T. Wong-Riley, M. M. Henry, E. V. Buchmann et al. 2006. Clinical and experimental applications of NIR-LED photobiomodulation. *Photomed. Laser Surg.* 24:121–128.

Dong, H., L. D. Sun, and C. H. Yan. 2015. Energy transfer in lanthanide upconversion studies for extended optical applications. *Chem. Soc. Rev.* 44:1608–1634.

Feng, W., X. J. Zhu, and F. Li. 2013. Recent advances in the optimization and functionalization of upconversion nanomaterials for *in vivo* bioapplications. *NPG Asia Mater.* 5:E57. doi: 10.1038/am.2013.63.

Fisher, L. H., G. S. Harms, and O. S. Wolfbeis. 2011. Upconverting nanoparticles for nanoscale thermometry. *Angew. Chem. Int. Ed.* 50:4546–4551.

Gamelin, D. R. and H. U. Gudel. 1998. Two-photon spectroscopy of d(3) transition metals: Near-IR-to-visible upconversion luminescence by Re4+ and Mo3+. *J. Am. Chem. Soc.* 120:12143–12144.

Gu, Z. J., L. Yan, G. Tian, S. Li, Z. Chai, and Y. Zhao. 2013. Recent advances in design and fabrication of upconversion nanoparticles and their safe theranostic application. *Adv. Mater.* 25:3758–3779.

Haase, M. and H. Schafer. 2011. Upconverting nanoparticles. *Angew. Chem. Int. Ed.* 50:5808–5829.

Han, M. Y., X. H. Gao, J. Z. Su, and S. Nie. 2001. Quantum-dot-tagged microbeads for multiplexed optical coding of biomolecules. *Nat. Biotech.* 19:631–535.

He, G. J., D. Guo, C. He, X. Zhang, X. Zhao, and C. Duan. 2009. A color-tunable europium complex emitting three primary colors and white light. *Angew. Chem. Int. Ed.* 48:6132–6135.

Heer, S., K. Kompe, H. Gudel, and M. Haase. 2004. Highly efficient multicolour upconversion emission in transparent colloids of lanthanide-doped NaYF4 nanocrystals. *Adv. Mater.* 16:2102–2105.

Heer, S., O. Lehmann, M. Haase, and H. Gudel. 2003. Blue, green and red upconversion emission from lanthanide-doped LuPO4 and YbPO4 nanocrystals in a transparent colloidal solution. *Angew. Chem. Int. Ed.* 42:3179–3182.

Idris, N. M., M. K. Gnanasammandhan, J. Zhang, P. C. Ho, R. Mahendran, and Y. Zhang. 2012. *In vivo* photodynamic therapy using upconversion nanoparticles as remote-controlled nanotransducers. *Nat. Med.* 18:1580–U1190.

Idris, N. M., M. K. G. Jayakumar, A. Bansal, and Y. Zhang. 2015. Upconversion nanoparticles as versatile light nanotransducers for photoactivation applications. *Chem. Soc. Rev.* 44:1449–1478.

Jaiswal, J. K., H. Mattoussi, J. M. Mauro, and S. M. Simon. 2003. Long-term multiple color imaging of live cells using quantum dot bioconjugates. *Nat. Biotech.* 21:47–51.

Jang, E., S. Jun, H. Jang, J. Lim, B. Kim, and Y. Kim. 2010. White-light-emitting diodes with quantum dot color converters for display backlights. *Adv. Mater.* 22:3076–3080.

Joubert, M. F., S. Guy, and B. Jacquier. 1993. Model of the photon-avalanche effect. *Phys. Rev. B* 48:10031–10037.

Khan, A. F., R. Yadav, P. K. Mukhopadhya, S. Singh, C. Dwivedi, V. Dutta, and S. Chawla. 2011. Core-shell nanophosphor with enhanced NIR-visible upconversion modifier for enhancement of solar cell efficiency. *J. Nanopart. Res.* 13:6837–6846.

Kik, P. G. and A. Polman. 2003. Cooperative upconversion as the gain-limiting factor in Er doped miniature Al_2O_3 optical waveguide amplifiers. *J. Appl. Phys.* 93:5008–5012.

Kou, L. H., D. Labrie, and P. Chylek. 1993. Refractive-indexes of water and ice in the 0.65-Mu-M to 2.5-Mu-M spectral range. *Appl. Optics* 32:3531–3540.

Kumar, S., J. Aaron, and K. Sokolov. 2008. Directional conjugation of antibodies to nanoparticles for synthesis of multiplexed optical contrast agents with both delivery and targeting moieties. *Nat. Protoc.* 3:314–320.

Lenth, W. and R. M. Macfarlane. 1990. Excitation mechanisms for upconversion lasers. *J. Lumin.* 45:346–350.

Li, X. M., Y. L. Liu, X. Song, H. Wang, H. Gu, and H. Zeng. 2015. Intercrossed carbon nanorings with pure surface states as low-cost and environment-friendly phosphors for white-light-emitting diodes. *Angew. Chem. Int. Ed.* 54:1759–1764.

Li, X. M., D. K. Shen, J. Yang, C. Yao, R. Che, F. Zhang, and D. Zhao. 2013. Successive layer-by-layer strategy for multi-shell epitaxial growth: Shell thickness and doping position dependence in upconverting optical properties. *Chem. Mater.* 25:106–112.

Li, Y. G., Y. T. H. Cu, and D. Luo. 2005. Multiplexed detection of pathogen DNA with DNA-based fluorescence nanobarcodes. *Nat. Biotech.* 23:885–889.

Li, Z. Q., Y. Zhang, and S. Jiang. 2008. Multicolor core/shell-structured upconversion fluorescent nanoparticles. *Adv. Mater.* 20:4765–4769.

Liu, Q. W., W. Feng, T. Yang, T. Yi, and F. Li. 2013. Upconversion luminescence imaging of cells and small animals. *Nat. Protoc.* 8:2033–2044.

Lu, Y. Q., J. B. Zhao, R. Zhang, Y. Liu, D. Liu, E. M. Goldys, X. Yang. et al. 2014. Tunable lifetime multiplexing using luminescent nanocrystals. *Nat. Photo.* 8:33–37.

Mader, H. S., P. Kele, S. M. Saleh, and O. S. Wolfbeis. 2010. Upconverting luminescent nanoparticles for use in bioconjugation and bioimaging. *Curr. Opin. Chem. Biol.* 14:582–596.

Mahalingam, V., F. Mangiarini, F. Vetrone, V. Venkatramu, M. Bettinelli, A. Speghini, and J. A. Capobianco. 2008. Bright white upconversoin emission from Tm3+/Tb3+/Er3+-doped Lu3Ga5O12 nanocrystals. *J. Phys. Chem. C* 112:17745–17749.

Mai, H. X., Y. W. Zhang, L. D. Sun, and C. H. Yan. 2007. Highly efficient multicolor up-conversion emissions and their mechanisms of monodisperse NaYF4:Yb,Er core and core/shell-structured nanocrystals. *J. Phys. Chem. C* 111:13721–13729.

McNichols, R. J., A. Gowda, M. Kangasniemi, J. A. Bankson, R. E. Price, and J. D. Hazle. 2004. MR thermometry-based feedback control of laser interstitial thermal therapy at 980 nm. *Lasers Surg. Med.* 34:48–55.

Mei, Q. S. and Z. P. Zhang. 2012. Photoluminescent graphene oxide ink to print sensors onto microporous membranes for versatile visualization bioassays. *Angew. Chem. Int. Ed.* 51:5602–5606.

Naczynski, D. J., M. C. Tan, M. Zevon, B. Wall, J. Kohl, A. Kulesa, S. Chen, C. M. Roth, R. E. Riman, and P. V. Moghe. 2013. Rare-earth-doped biological composites as *in vivo* shortwave infrared reporters. *Nat. Commun.* 4:2199.

Nam, S. H., Y. M. Bae, Y. II Park, J. H. Kim, H. M. Kim, J. S. Choi, K. T. Lee, T. Hyeon, and Y. D. Suh. 2011. Long-term real-time tracking of lanthanide ion doped upconverting nanoparticles in living cells. *Angew. Chem. Int. Ed.* 50:6093–6097.

Pregibon, D. C., M. Toner, and P. S. Doyle. 2007. Multifunctional encoded particles for high-throughput biomolecule analysis. *Science* 315:1393–1396.
Shalav, A., B. S. Richards, T. Trupke, K. W. Kramer, and H. U. Gudel. 2005. Application of NaYF4:Er3+ up-converting phosphors for enhanced near-infrared silicon solar cell response. *Appl. Phys. Lett.* 86:013505.
Shen, J., G. Y. Chen, A. M. Vu, W. Fan, O. S. Bilsel, C. C. Chang, and G. Han. 2013. Engineering the upconversion nanoparticle excitation wavelength: Cascade sensitization of tri-doped upconversion colloidal nanoparticles at 800 nm. *Adv. Optical Mater.* 1:644–650.
Shirasaki, Y., G. J. Supran, M. G. Bawendi, and V. Bulovic. 2012. Emergence of colloidal quantum-dot light-emitting technologies. *Nat. Photo.* 7:13–23.
Sivakumar, R., F. C. J. M. van Veggel, and M. Raudsepp. 2005. Bright white light through up-conversion of a single NIR source from sol–gel-derived thin film made with Ln(3+)-doped LaF3 nanoparticles. *J. Am. Chem. Soc.* 127:12464–12465.
Sliney, D. H. 1995. Laser safety. *Lasers Surg. Med.* 16:215–225.
Stoeva, S. I., J. S. Lee, J. E. Smith, S. T. Rosen, and C. A. Mirkin. 2006. Multiplexed detection of protein cancer markers with biobarcoded nanoparticle probes. *J. Am. Chem. Soc.* 128:8378–8379.
Strohhofer, C. and A. Polman. 2001. Relationship between gain and Yb3+ concentration in Er3+-Yb3+ doped waveguide amplifiers. *J. Appl. Phys.* 90:4314–4320.
Tian, G., Z. J. Gu, L. Zou, W. Yin, X. Liu, L. Yan, S. Jin et al. 2012. Mn2+ dopant-controlled synthesis of NaYF4:Yb/Er upconversion nanoparticles for *in vivo* imaging and drug delivery. *Adv. Mater.* 24:1226–1231.
Vetrone, F., J. C. Boyer, J. A. Capobianco, A. Speghini, and M. Bettinelli. 2004. Significance of Yb3+ concentration on the upconversion mechanisms in codoped Y2O3:Er3+, Yb3+ nanocrystals. *J. Appl. Phys.* 96:661–667.
Wang, F., Y. Han, C. S. Lim, Y. Lu, J. Wang, J. Xu, H. Chen, C. Zhang, M. Hong, and X. Liu. 2010. Simultaneous phase and size control of upconversion nanocrystals through lanthanide doping. *Nature* 463:1061–1065.
Wang, F. and X. G. Liu. 2008. Upconversion multicolor fine-tuning: Visible to near-infrared emission from lanthanide-doped NaYF4 nanoparticles. *J. Am. Chem. Soc.* 130:5642–5643.
Wang, G. F., Q. Peng, and Y. Li. 2009. Upconversion luminescence of monodisperse CaF2:Yb3+/Er3+ nanocrystals. *J. Am. Chem. Soc.* 131:14200–14201.
Wang, G. F., Q. Peng, and Y. Li. 2011a. Lanthanide-doped nanocrystals: Synthesis, optical-magnetic properties, and applications. *Acc. Chem. Res.* 44:322–332.
Wang, J., F. Wang, C. Wang, Z. Liu, and X. Liu. 2011b. Single-band upconversion emission in lanthanide-doped KMnF3 nanocrystals. *Angew. Chem. Int. Ed.* 50:10369–10372.
Wang, Y. F., G. Y. Liu, L. D. Sun, J. W. Xiao, J. C. Zhou, and C. H. Yan. 2013. Nd3+ sensitized upconversion nanophosphors: Efficient *in vivo* probes with minimized heating effect. *ACS Nano* 7:7200–7206.
Wermuth, M. and H. U. Gudel 1999. Photon avalanche in Cs2ZrBr6:Os4+. *J. Am. Chem. Soc.* 121:10102–10111.
Xie, X. J., N. Y. Gao, R. Deng, Q. Sun, Q. H. Xu, and X. Liu. 2013. Mechanistic investigation of photon upconversion in Nd3+ sensitized core-shell nanoparticles. *J. Am. Chem. Soc.* 135:12608–12611.
Xiong, L. Q., T. S. Yang, Y. Yang, C. Xu, and F. Li. 2010. Long-term *in vivo* biodistribution imaging and toxicity of polyacrylic acid-coated upconversion nanophosphors. *Biomaterials* 31:7078–7085.

Yi, G. S. and G. M. Chow. 2005. Colloidal LaF_3:Yb,Er, LaF_3:Yb,Ho and LaF_3:Yb,Tm nanocrystals with multicolor upconversion fluorescence. *J. Mater. Chem.* 15:4460–4464.

Yin, W. Y., L. N. Zhao, L. Zhou, Z. Gu, X. Liu, G. Tian, S. Jin. et al. 2012. Enhanced red emission from GdF3:Yb3+,Er3+ upconversion nanocrystals by Li+ doping and their application for bioimaging. *Chem. Eur. J.* 18:9239–9245.

Yu, X. F., L. D. Chen, M. Li, M. Xie, L. Zhou, Y. Li, and Q. Wang. 2008. Highly efficient fluorescence of NdF3/SiO2 core/shell nanoparticles and the applications for *in vivo* NIR detection. *Adv. Mater.* 20:4118–4123.

Zeng, J. H., J. Su, Z. Li, R. Yan, and Y. Li. 2005. Synthesis and upconversion luminescence of hexagonal-phase NaYF4:Yb,Er3+, phosphors of controlled size and morphology. *Adv. Mater.* 17:2119–2123.

Zhan, Q. Q., J. Qian, H. J. Liang, G. Somesfalean, D. Wang, S. L. He, Z. G. Zhang, and S. Andersson-Engels. 2011. Using 915 nm laser excited Tm3+/Er3+/Ho3+-doped NaYF4 upconversion nanoparticles for *in vitro* and deeper *in vivo* bioimaging without overheating irradiation. *ACS Nano* 5:3744–3757.

Zhang, C., L. Yang, J. Zhao, B. Liu, M. Y. Han, and Z. Zhang. 2015. White-light emission from an integrated upconversion nanostructure: Towards multicolor displays modulated by laser power. *Angew. Chem. Int. Ed.* 54:11531–11535.

Zhang, F., R. C. Haushalter, R. W. Haushalter, Y. Shi, Y. Zhang, K. Ding, D. Zhao, and G. D. Stucky. 2011a. Rare-earth upconverting nanobarcodes for multiplexed biological detection. *Small* 7:1972–1976.

Zhang, F., Q. H. Shi, Y. Zhang, Y. Shi, K. Ding, D. Zhao, and G. D. Stucky. 2011b. Fluorescence upconversion microbarcodes for multiplexed biological detection: Nucleic acid encoding. *Adv. Mater.* 23:3775–3779.

Zhang, Y. W., X. Sun, R. Si, L. P. You, and C. H. Yan. 2005. Single-crystalline and monodisperse LaF3 triangular nanoplates from a single-source precursor. *J. Am. Chem. Soc.* 127:3260–3261.

Zheng, W., P. Huang, D. Tu, E. Ma, H. Zhu, and X. Chen. 2015. Lanthanide-doped upconversion nano-bioprobes: Electronic structures, and biodetection. *Chem. Soc. Rev.* 44:1379–1415.

Zhong, Y. T., I. Rostami, Z. Wang, H. Dai, and Z. Hu. 2015. Energy migration engineering of bright rare-earth upconversion nanoparticles for excitation by light-emitting diodes. *Adv. Mater.* 27. 6418–6422.

Zhong, Y. T., G. Tian, Z. Gu, Y. Yang, L. Gu, Y. Zhao, Y. Ma, and J. Yao. 2014. Elimination of photon quenching by a transition layer to fabricate a quenching-shield sandwich structure for 800 nm excited upconversion luminescence of Nd3+ sensitized nanoparticles. *Adv. Mater.* 26:2831–2837.

Zhou, J., Y. Sun, X. Du, L. Xiong, H. Hu, and F. Li. 2010. Dual-modality *in vivo* imaging using rare-earth nanocrystals with near-infrared to near-infrared (NIR-to-NIR) upconversion luminescence and magnetic resonance properties. *Biomaterials* 31:3287–3295.

Zou, W. Q., C. Visser, J. A. Maduro, M. S. Pshenichikov, and J. C. Hummelen. 2012. Broadband dye-sensitized upconversion of near-infrared light. *Nat. Photo.* 6:560–564.

7

Upconversion Enhancement Using Epitaxial Core–Shell Nanostructures

Shuwei Hao, Jing Liu, Meiling Tan, and Guanying Chen

CONTENTS

7.1 Introduction 163
7.2 Nanochemistry for Well-Defined Core–Shell UCNPs 165
  7.2.1 Nanochemistry for Synthesis of Core UCNPs 165
    7.2.1.1 Thermal Decomposition 165
    7.2.1.2 Ostwald Ripening 166
    7.2.1.3 Hydro(solvo)thermal Method 167
  7.2.2 Seed-Mediated Epitaxial Growth of Core–Shell UCNPs 169
    7.2.2.1 Homogenous Epitaxial Core–Shell UCNPs 170
    7.2.2.2 Heterogeneous Core–Shell UCNPs 170
  7.2.3 Seed-Mediated Epitaxial Growth of Core–Multishell UCNPs 171
7.3 Enhancing UCL with an Active Core–Inert Shell Structure 172
  7.3.1 Homogeneous Core–Shell UCNPs with Enhanced UCL 172
  7.3.2 Heterogeneous Core–Shell UCNPs with Enhanced UCL 174
7.4 Enhancing UCL with an Active Core–Active Shell Structure 178
  7.4.1 Active Shell Containing Yb^{3+} Sensitizers 178
  7.4.2 Active Shell Containing Nd^{3+} Sensitizers 181
7.5 Core–Multishell Nanostructure for Enhanced UCL 183
7.6 Conclusion 186
References 187

7.1 Introduction

Trivalent rare earth (RE)-doped upconversion nanoparticles (UCNPs) can be considered as a dilute guest-host system, where trivalent lanthanide ions are dispersed as guests in an appropriate inorganic host lattice with a dimension of less than 100 nm. They are able to convert near-infrared (NIR) light into visible or ultraviolet luminescence through a set of coupled linear processes in a system of real energy levels of RE ions incorporated at the Bravais lattice points of the host material (Auzel 1990, 2004; Chen et al. 2014a; Haase and Schäfer 2011; Wang and Liu 2009). An important advantage offered by coupled linear excitation is the ability to generate nonlinear UC luminescence (UCL) with an excitation power density as low as ~10^{-1} W/cm^2, which can be easily provided by low cost continuous-wave laser diodes or incoherent light sources (light emitting diodes and incandescent light bulbs). Along with the low cytotoxicity entailed by the inorganic host lattice, the frequency converting capability of UCNPs imparts a number of advantages for them such as absence of autofluorescence, deep penetration of light in biological tissues, and minimum photodamage to living organisms (Achatz et al. 2011; Wang et al. 2010a). These merits engage them for a plethora of applications in bioimaging and therapy. As a rule of thumb, the luminescence efficiency of UCNPs is of particular importance for their embodiment in biophotonic applications, which is, however, severely limited by nanosize-induced surface-related quenching effects and detrimental interactions between the doped RE ions. The reported maximum upconversion quantum yield (UCQY) of UCNPs, which are devoid of any shell, is typically less than 1% even under a high laser irradiance of ~10^2 W/cm^2.

A hierarchical design of core–shell UCNPs, whereby each shell layer can be defined within a regulated scale to introduce a specific feature, provides a range of unique nanophotonic solutions to address the limitation of low efficiency. With the ability of nanochemistry to produce a shell of stoichiometric composition and a shell thickness with monolayer precision, the electronic structure as well as the physiochemical structure of the upconverting system can be purposely aligned for nanophotonic control of upconversion (UC). A direct result is that a core–shell structure enables suppression of surface-related quenching effect produced by an epitaxial growth of a shell layer. The shell layer is able to provide an effective passivation of lattice defects located on the surface of a core nanoparticle (NP), as well as to create a perfect spatial isolation of the core NP from the surrounding environment. As a consequence, a significant increase of UC efficiency can be reached. For example, the core–shell ($NaYF_4$:Yb^{3+} 30%, Tm^{3+} 0.5%)/$NaYF_4$ nanocrsytals can be about 100 times higher than the core-only $NaYF_4$:Yb^{3+} 30%, Tm^{3+} 0.5% nanocrystals, possessing an absolute UCQY as high as ~3.9% under an excitation density of 18 W/cm^2 (Xu et al. 2012). Moreover, nanoscopic control of ion–ion interactions can be achieved through spatial confinement of different types of lanthanide dopants in separated shell layers or the core domain. The interaction between lanthanide ions of the same or different types can be precisely engineered by appropriate arrangement of the electronic energy level hierarchy for lanthanides in different shell layers as well as by manipulating the shell thickness to control the interaction strength. This directed interaction provides powerful tools to manipulate the energy transfer processes which are the essence to achieve highly efficient photon UC.

This chapter intends to cover practical nanochemical approaches for the preparation of various types of core–shell UCNPs, as well as the recent progress of core–inert shell, core–active shell, and core–multishell nanostructures to produce enhanced UCL.

7.2 Nanochemistry for Well-Defined Core–Shell UCNPs

High-quality core–shell structure NPs can be prepared using a seed-mediated layer-by-layer coating process. It involves the deposition of single or multilayers on the surface of as-synthesized core NPs. In order to prepare an excellent core–shell structure with defined properties, several aspects need to be paid special attention. Firstly, the ability to synthesize high-quality core NPs is of fundamental and technological importance for growth of high-quality epitaxial core–shell UCNPs. This is because the core nanocrystals can produce significant effects on the resulting size and morphology of designated epitaxial core–shell NPs. The core NPs can generally be prepared by solution-based colloidal nanochemistry via three main types of synthetic approaches, including thermal decomposition, Ostwald-ripening, and hydro(solvo)thermal method. As a result, we will first illustrate the fundamental aspects of these three nanochemical approaches, and then apply them for seed-mediated epitaxial growth of core–shell or core–multishell UCNPs.

7.2.1 Nanochemistry for Synthesis of Core UCNPs

7.2.1.1 Thermal Decomposition

In this approach, alkaline and lanthanide trifluoroacetates were decomposed in the presence of octadecene (ODE) as well as oleic acid (OA) and/or oleylamine (OM) at ~270–330°C, which yields corresponding alkaline and lanthanide fluorides to create monomers. The noncoordinating ODE is used as the primary solvent due to its high boiling point (315°C), while OA and/or OM is chosen as a surface capping ligand, as it not only prevents the nanocrystals from agglomeration, but also controls the chemical reaction in nanoscale. A precise control over the nanocrystal phase, size, and morphologies can be achieved by varying reaction parameters such as, the composition and amount of precursors, the amount of surfactants, solvent (e.g., ODE) as well as reaction temperature and time. Moreover, the loading ways of precursors are also important for the final product, which mainly includes two types: "heating-up method" and "hot-injection method." In the "heating-up" process, the temperature of reaction solution is elevated from room temperature to a specific high temperature to produce NPs, while the "hot-injection" process injects pertinent precursors into a hot solution containing the high boiling solvent and surfactants. The heating-up method has been well-illustrated by Mai et al. 2006 who reported the preparation of high-quality $NaYF_4$ nanocrystals (Figure 7.1a) by heating appropriate metallic trifluoroacetates (RE trifluoroacetate, sodium trifluoroacetate) in the presence of OA and ODE (Du et al. 2008), while Capobianco et al. (Boyer et al. 2007a) prepared the uniform high-quality Yb/Er and Yb/Tm co-doped cubic $NaYF_4$ NPs by injecting sodium and RE trifluoroacetates into a mixture of OA and ODE at 310°C solution in a precisely defined manner (Figure 7.1c). Maintaining temperature at

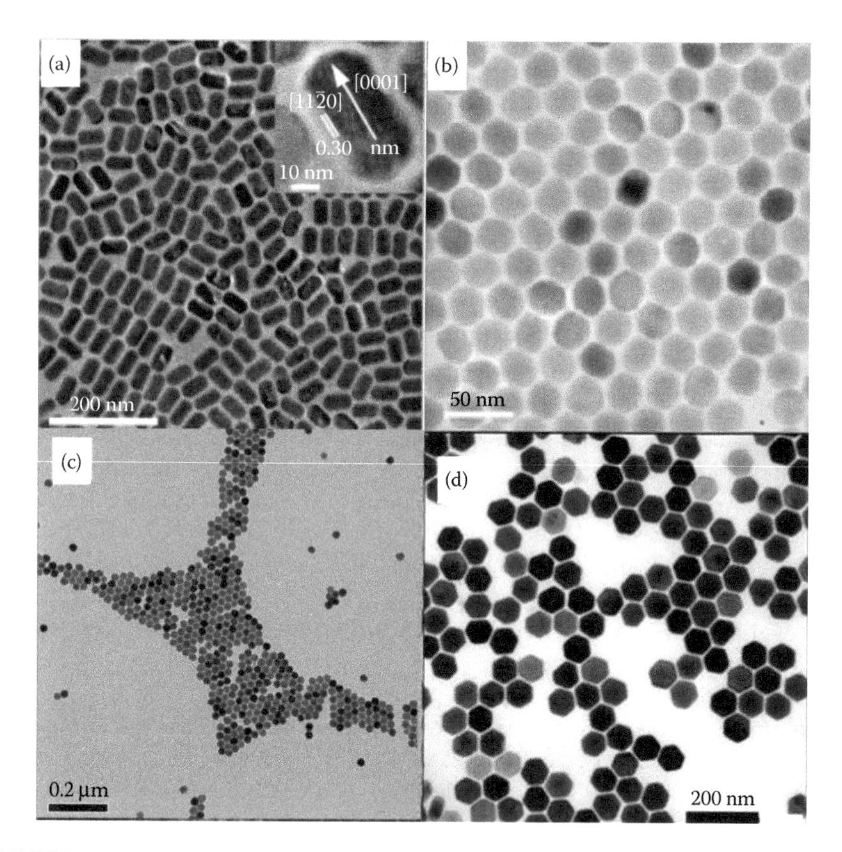

FIGURE 7.1
Representative shapes of various RE-doped fluoride nanomaterials synthesized by thermal decomposition and Ostwald-ripening processes. (Adapted from Mai, H.X., Y.W. Zhang, R. Si et al., *J. Am. Chem. Soc.*, 128, 6426–6436, 2006; Li, Z.Q. and Y. Zhang, *Nanotechnology*, 19, 345606, 2008; Boyer, J. C., L. A. Cuccia, and J. A. Capobianco. 2007a. *Nano Lett.* 7:847–852; Hao, S.W., W. Shao, H.L. Qiu et al., *RSC Adv.*, 4, 56302–56306, 2014. With permission.)

310°C is of vital importance, because all the metal trifluoroacetates are concurrently allowed to decompose at this temperature. In this way, a precise control of the nuclei and growth process can be implemented through a fine definition of the injecting rate of precursors.

7.2.1.2 Ostwald Ripening

Ostwald ripening refers to the process that larger particles with smaller surface-to-volume ratios are favored over energetically less stable smaller particles, resulting in the growth of larger particles at the expense of smaller ones. In the Ostwald-ripening approach, a mixture of OA and ODE is typically utilized as a high boiling reaction solution. This approach begins with heating inorganic metal salts in OA at elevated temperatures (~150°C) to produce metal oleate precursors, which complex the OA with the metal ions. After reducing to a temperature around (~150°C), subsequent addition of sodium hydroxide and fluorides (dissolved methanol) into the mixture will yield monodisperse nuclei and monomers. When heating up, these nuclei gradually grow into sacrificial small size NPs by consuming monomers that will eventually evolve into uniform nanocrystals through an Ostwald-ripening process (Figure 7.1b, Li and Zhang 2008). It is worth noting that the RE doping concentration can affect the size and shape of resulting NPs utilizing this method. We have observed that the size of $NaYF_4$:Yb^{3+}/Pr^{3+} NPs can vary along with the varied sensitizer ytterbium concentration (Figure 7.1d, Hao et al. 2014).

7.2.1.3 Hydro(solvo)thermal Method

Hydro(solvo)thermal methods perform chemical reactions in a solvent in a sealed environment under high-pressure and temperature, usually above the critical point of the solvent in order to increase the solubility and reactivity of the inorganic substances. In a hydro(solvo)thermal process for preparation of UCNPs, RE and fluoride precursors, solvents, and certain surfactants are mixed in aqueous or organic solvents and then heated in a specialized reaction vessel known as an autoclave. A classical hydrothermal approach to prepare small-sized UCNPs is the liquid–solid–solution (LSS) approach, whereby the reaction, phase transfer, and separation take place at the interfaces (Wang et al. 2005). Taking the synthesis of cubic $NaYF_4$ nanocrystals (Figure 7.2) as an example,

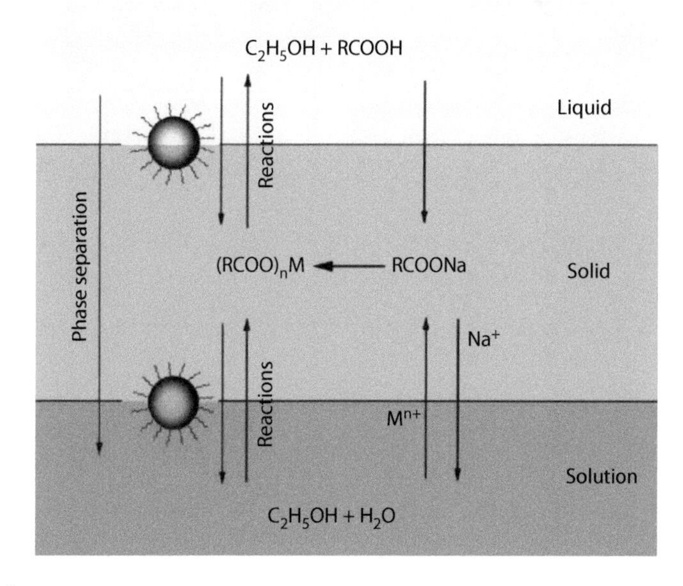

FIGURE 7.2
Schematic of LSS phase transfer and separation synthesis. (From Wang, X., J. Zhuang, and Q. Peng, *Nature*, 437, 121–124, 2005.)

the linoleic acid, NaOH, ethanol and deionized water, and $Y(NO_3)_3$ and NaF solution can be mixed and loaded into the autoclave. A solution system of three phases will be formed, which encompasses the solid phase (sodium linoleate), the liquid phase (ethanol and linoleic acid), and the solution phase containing RE ions (water/ethanol). The chemical reaction involves an ion exchange of RE ions across the interface of sodium linoleate (solid) and the water/ethanol solution (solution) to take place. When RE ions migrate from the aqueous to the solid phase, the cubic $NaYF_4$ nanocrystals can be formed through co-precipitation of RE ions and NaF. The hydro(solvo)thermal method has been proven powerful to fabricate RE-doped fluoride micro- or nanocrystals. Fluoride particles with a multitude of uniform morphologies have been achieved, such as $NaYF_4$ nanorods (Figure 7.3a, Wang et al. 2010b), flower-like $NaREF_4$ arrays (Figure 7.3b, Zhang et al. 2009), and hexagonal columns of $NaYF_4$ with long height (Figure 7.3c, Li et al. 2009) or short height (Figure 7.3d, Shang et al. 2015).

FIGURE 7.3
Different shapes of RE-doped fluoride nano-/microcrystals by hydro(solvo)thermal method. (Adapted from Wang, F., Y. Han, C. S. Lim et al., *Nature*, 463, 1061–1065, 2010b; Zhang, F., J. Li, J. Shan et al., *Chem. Eur. J.*, 15, 11010–11019, 2009; Li, C.X., C.M. Zhang, Z.Y. Hou et al., *J. Phys. Chem. C*, 113, 2332–2339, 2009; Shang, Y.F., S.W. Hao, J. Liu et al., *Nanomaterials*, 5, 218–232, 2015. With permission.)

7.2.2 Seed-Mediated Epitaxial Growth of Core–Shell UCNPs

The seed-mediated epitaxial growth is an effective approach to grow epitaxial core–shell UCNPs. A low lattice mismatch between the core host material and the shell host material is generally required to entail a well-defined structure with controllable morphology and size. This approach typically encompasses two steps. The first step is to synthesize the core UCNPs using solution-based nanochemical approaches (thermal decomposition, Ostwald-ripening, and hydrothermal methods) described above. The second step is to utilize the core nanocrystals as seeds to induce further epitaxial growth, which involves the use of precursor of the shell material. In this step, the monomer of the shell material, produced by either mixing or thermal decomposition of corresponding chemical precursors in the coordinating solvents, for example, of OA and ODE, is able to gradually and stably deposit on the seeding core NPs and finally form a uniform shell layer (Figure 7.4a). The shell precursors can be either directly injected into the growing solution of the core NPs to induce a further growth of the shell, or the core NPs can be centrifuged out, and the shell layer is grown in a separate step (Feng et al. 2010). Moreover, the growth of shell on top of a core nanocrystal is inclined to occur in organic solvents, as a homogeneous nucleation of shell is kinetically favored in an aqueous environment (Chen et al. 2015b,c). The concentration of shell precursor typically has to keep low in regard to the concentration of core NPs to avoid a formation of new nucleus by shell precursors that might then evolve into NPs of shell host material alone rather than the desired core–shell UCNPs. The thickness of the shell can be controlled via adjusting the shell precursor concentration, the amount of which can be determined according to a chemical stoichiometric calculation of the corresponding material mass increase.

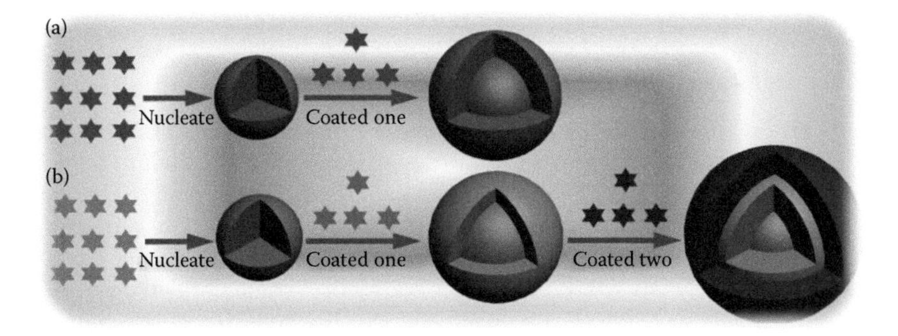

FIGURE 7.4
Strategies for constructing UC core–shell structure NPs. (a) Synthesis of an inert shell on the surface of NPs through one-step growth method. (b) Facile fabricating the core–multishell structure NPs via a layer-by-layer growth technique.

7.2.2.1 Homogenous Epitaxial Core–Shell UCNPs

Yi and Chow reported for the first time in 2007 an epitaxial shell growth strategy to prepare homogenous core–shell $NaYF_4$:Yb/Er@$NaYF_4$UCNPs, whereby the host lattice and the shell lattice are identical (Yi and Chow 2007). In their report, metal tri-fluoacetate (NaTFA, YTFA, YbTFA, and ErTFA) precursors with defined ratio were heated to 340°C in OM for 30 min to prepare the $NaYF_4$:Yb/Er core NPs first. The OM(oleyl amine) here not only functions as a high boiling solvent but also as a coordinating ligand. When the growth process of the core NPs was terminated, the shell precursors of NaTFA and YTFA dispersed in OM were subsequently injected into the reaction solution to induce shell growth. The null lattice mismatch between the core and the shell guarantees a direct deposition of undoped $NaYF_4$ shell layer on the surface of the $NaYF_4$:Yb/Er core NPs. This method has been adapted and applied later to synthesize RE-doped $NaGdF_4$@$NaGdF_4$ core–shell UCNPs with uniform morphology and narrow size distribution (Vetrone et al. 2009), and $LiLuF_4$@$LiLuF_4$ homogenous core–shell NPs with a tunable shell thickness (Huang et al. 2014). In the case of $LiLuF_4$@$LiLuF_4$ UCNPs, the shell thickness was controlled with a monolayer precision by controlling the added amount of precursors. A multiple injection of stoichiometric shell precursors was implemented to increase the inert $LiLuF_4$ shell layer thickness in a defined manner. An epitaxial growth of homogeneous core–shell UCNPs that base on other fluoride matrixes has also reported, such as $NaYF_4$:Yb/Er(Tm)@$NaYF_4$ (Wang et al. 2009), KYF_4:Yb/Er@KYF_4 (Schafer et al. 2008), and YF_3:Yb/Er/Tm@YF_3 (Chen et al. 2012b). However, it is hard to discern the core and the shell layer of the homogeneous core–shell NPs, as the core and the shell have zero phase contrast in the transmission electron microscopy. The formation of a homogenous core–shell structure has to be indirectly suggested such as by the size increase, by the prolonging of luminescence lifetime, as well as by the upconversion luminescence (UCL) enhancement.

7.2.2.2 Heterogeneous Core–Shell UCNPs

The method for epitaxial growth of homogenous core–shell structure can be utilized in straightforward manner to construct the heterogeneous core–shell NPs, such as $NaYbF_4$:Tm^{3+}/$NaGdF_4$ NPs, in which the shell layer incorporates trivalent Gd^{3+} ions for magnetic resonance imaging, while the core contains optically active Yb^{3+}/Tm^{3+} ion pairs for UCL bio-imaging (Chen et al. 2011). Heterogeneous core–shell nanostructures of ($NaYbF_4$:Tm^{3+})/CaF_2, $NaYF_4$:Yb/Er@CaF_2, $NaYF_4$/$NaGdF_4$ core–shell NP have been prepared using a similar strategy (Chen et al. 2012c; Wang et al. 2012; Zhang et al. 2012). In these reports, the synthetic chemistries to produce the core and to produce the shell are identical, which, however, are not mandatory. We showed that we were able to prepare heterogeneous CaF_2:Yb^{3+}/Ho^{3+}@$NaGdF_4$ core–shell UCNPs by combining two quite different approaches of LSS and thermal decomposition. In our approach, the 4 nm sized CaF_2:Yb^{3+}/Ho^{3+} core nanocrystals were produced by a LSS strategy (a typical hydrothermal process) first, and then transferred into an organic solution as seeds to induce the growth of uniform $NaGdF_4$ shell using a thermal decomposition method (Hao et al. 2015).

7.2.3 Seed-Mediated Epitaxial Growth of Core–Multishell UCNPs

The chemistry to produce core–multishell UCNPs relies on the deposition of shells in a layer-by-layer fashion (Figure 7.4b). The chemistry involved here is the same as that utilized for preparation of epitaxial core–shell NP with single inert shell. A straightforward approach is to use a multistep approach to prepare each shell using seed-mediated epitaxial growth described above. That is to say the core/1st shell UCNPs are utilized as the seeds for the core/1st shell/2nd shell UCNPs, which are in turn utilized as the seeds for the core/1st shell/2nd shell/3rd shell UCNPs, and so forth (Chen et al. 2015b). The stoichiometric nature of involved chemical reactions producing a shell causes its thickness to be precisely proportional to the shell precursor dosage, which can be calculated according to the increase in molar mass caused by deposition of the shell of a defined thickness. This essentiality provides an exciting opportunity to deposit each shell layer with a single monolayer precision. By controlling the type and amount of shell precursor in each shell growth step, the host lattice, thickness, crystal structure, as well as concentration and type of dopants in this specific shell can be well defined. A multi-step method has been employed to prepare sandwich-structured UCNPs with a $NaYbF_4$ matrix sandwiched between two $NaYF_4$ layers, namely the $NaYF_4$:Yb^{3+}/Er^{3+}@$NaYbF_4$:Tm^{3+}@$NaYF_4$:Yb^{3+}/Tm^{3+} core–shell–shell UCNPs which can achieve tunable UC emissions for multicolor cellular imaging (Dou et al. 2013). A hierarchical architecture of active core–active shell–inert shell ($NaYF_4$:Yb^{3+}30%/Tm^{3+}0.5%)/$NaYbF_4$/$NaYF_4$ has been prepared (Qiu et al. 2014). This multi-step approach has also been utilized to prepare $NaGdF_4$:Er^{3+}@$NaGdF_4$:Ho^{3+}@$NaGdF_4$ core–shell–shell structure in which different RE ions were spatially isolated into different shell layers to suppress detrimental interactions between the Er^{3+} and the Ho^{3+} ions (Chen et al. 2012a). Other core–multishell structures of $NaYF_4$:Yb^{3+}/Er^{3+}@$NaYF_4$:Yb^{3+}@$NaYF_4$:Nd^{3+} (Zhong et al. 2014), and $NaGdF_4$:Nd^{3+}/$NaYF_4$/$NaGdF_4$:Nd^{3+}, Yb^{3+}, Er^{3+}/$NaYF_4$ (Li et al. 2013b) have also been reported to realize efficient UCL at a new excitation wavelength of ~800 nm. The ability of nanochemistry to produce shells in a precisely defined manner provides numerous opportunities for nanophotonic control of the upconverting excitation dynamics to produce significantly enhanced UCL.

7.3 Enhancing UCL with an Active Core–Inert Shell Structure

Because of the high surface-to-volume ratio of NPs, a large number of RE dopants are located on the surface and exposed to the environment, resulting in surface-related quenching effects (caused by surface defects, lattice strains, organic ligands, and solvents). Surface deactivations generally occur in two ways: (i) direct luminescence quenching. The RE dopants located on or close to surface are directly deactivated by neighboring quenching centers through nonradiative relaxation involving phonons from the surrounding environment. This deactivation pathway applies to a range of intermediate excited states involved in producing UCL, resulting in more pronounced quenching than in the case of Stokes-shifted (downconversion) luminescence, which involves one exclusive emitting state. Direct luminescence quenching is a general surface-related quenching mechanism for all types of luminophores with nanoscale dimension. (ii) Energy migration-induced quenching. The electronic excitation energy of dopants located inside can randomly move inside UCNPs and travel to other dopant centers on/around the surface or directly to the surface quenching sites. This type of surface-related quenching mechanisms requires the existence of long-lived energy states favorable for excitation energy hopping. The latter mechanism is a premium quenching mechanism for the generally investigated UCNPs, which are typically doped with Yb^{3+}/X^{3+} (X = Er, Tm, and Ho) ion pairs. This is because the involved sensitizer of Yb^{3+} ion has an exclusive excited state, whose lifetime can be as long as milliseconds. The core–shell structure passivates the imperfects in the host lattice (often located at the core particle surface), and shields the core from the surrounding ligands (which are necessary for solvent dispersion) and solvents of high phonon energy. The distance between the UCNP core and the surrounding environment, created by the inert epitaxial shell, prevents quenching of the excited states of the surface RE ions and curbs the migration of excitation energy into the surrounding environment. As a result, the epitaxial core–shell structure is an effective strategy to achieve high efficient UCL by suppressing surface-related quenching mechanisms. For example, compared with the only core counterparts, the green, blue, and yellow UCL have been increased by more than one order of magnitude in $NaGdF_4$:Yb,Er@$NaGdF_4$:Yb, $NaYF_4$:Yb,Tm@$NaYF_4$, and KYF_4:Yb,Er@KYF_4:Yb core–shell NPs, respectively (Figure 7.5). The extent of suppression of surface-related quenching mechanisms is pertinent to the type, the quality, and the thickness of the shell material. The following describe some variations of this type of core–shell structures.

7.3.1 Homogeneous Core–Shell UCNPs with Enhanced UCL

The homogenous core–shell coating can reduce the number of lattice quenching sites on the surface of the core particle, and create a spatial isolation of the

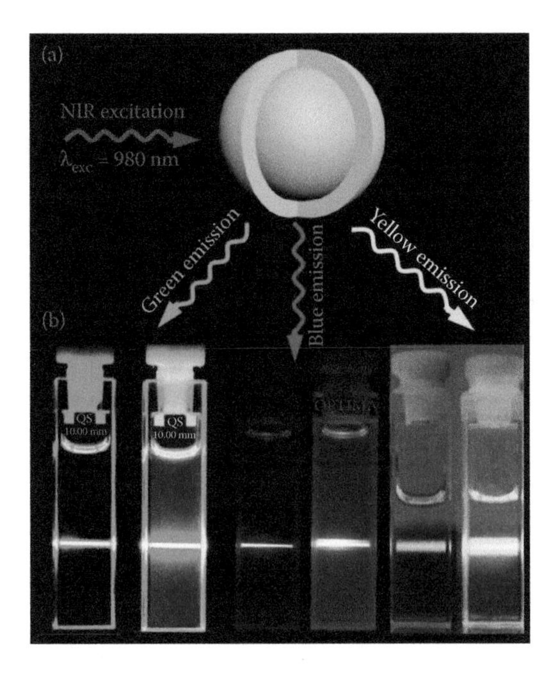

FIGURE 7.5
(a) General depiction of the core–shell lanthanide UCNPs architecture showing the absorption of NIR light by the Yb^{3+} and subsequent energy transfer to the Er^{3+} or Tm^{3+}, which leads to green, blue, and yellow upconverted emissions. (b) Photographs of colloidal solutions of the only core NPs solution and the corresponding $NaGdF_4$:Yb,Er@$NaGdF_4$:Yb core–shell NPs, $NaYF_4$:Yb,Tm@$NaYF_4$ core–shell NPs, and KYF_4:Yb,Er@KYF_4:Yb core–shell NPs, following excitation with 980 nm. (From Vetrone, F., R. Naccache, V. Mahalingam et al., *Adv. Funct. Mater.*, 19, 2924–2929, 2009; Schafer, H., P. Ptacek, O. Zerzouf et al., *Adv. Funct. Mater.*, 18, 2913–2918, 2008.)

core NPs from the quenchers in the surrounding environment. The identical host material between the core and the shell imparts a homogeneous interface between them, thus providing an effective passivation of lattice defects of the core surface. On the other hand, the spatial isolation prevents the interaction of the core NPs with the surrounding quenchers, thus preserving and confining the harvested excitation energy within the core NP with or without limited quenching dissipation. As a result, the homogeneous core–shell structure can enable a great increase of UCL efficiency. Yi and Chow fist reported on utilization of this strategy to suppress UCL quenching of $NaYF_4$:Yb,Er (Tm) cores through coating an undoped $NaYF_4$ shell (Yi and Chow 2007). The hexagonal phase $NaYF_4$: Yb^{3+}/Er^{3+} and $NaYF_4$: Yb^{3+}/Tm^{3+} cores with uniform morphology and size were successfully prepared using a thermolysis approach. They were then utilized as seeds to induce a thin layer growth of $NaYF_4$ (~2 nm). The formation and thickness of the inert $NaYF_4$ shell were confirmed by the x-ray diffraction (XRD) and the size distribution analysis. As compared with the only core NPs, the green band emissions of designed $NaYF_4$: Yb^{3+}/Er^{3+}@$NaYF_4$ core–shell NPs showed a 7.4 times increase, while the blue emission of $NaYF_4$: Yb^{3+}/Tm^{3+}@$NaYF_4$ core–shell NPs achieved a 29.6 times enhancement. Subsequently, Veggel et al. reported an epitaxial layer-by-layer growth method to construct the core–inert shell structure with precisely defined shell thickness. The Ostwald-ripening process was involved to grow the shell layer on large core NPs using small sacrificial NPs as the shell precursors (Johnson et al. 2012). Figure 7.6a–e shows the change of morphologies and sizes of coated core–shell NPs by successively injecting four 0.2 mmol sacrificial small NPs, demonstrating the feasibility of controlling the shell thickness. The inductively coupled plasma mass spectrometry (ICP-MS) elemental analysis indicates that the content of Y^{3+} was increased with the successive growth of shell layers, while the content of optically active ions ($Yb^{3+}$and Er^{3+}) remains unchanged (Figure 7.6f). This clearly confirms the formation of a core–shell structure with varied shell thickness. Moreover, the intensity of UCL is manifested to increase with an increase of the thickness of the shell layer (Figure 7.6g), demonstrating that the extent of suppression of surface-related quenching mechanisms is related to the shell thickness. Many studies have observed a similar UCL enhancement effect using the core–inert shell structure. $NaYF_4$: Yb^{3+}/Er^{3+}@$NaYF_4$ core–shell UCNPs were shown to yield a three times higher quantum yield than that of $NaYF_4$: Yb^{3+}/Er^{3+} core NPs (Boyer and van Veggel 2010). Helmut Schafer reported a deposition of a thin layer of undoped KYF_4 (~2.5 nm) on KYF_4:Yb^{3+}/Er^{3+}core NPs (Schafer et al. 2008), achieving a ~20 fold UCL enhancement. The advantage of using an inert shell layer to enhance UCL has also been demonstrated in other homogenous RE-doped core–shell structures, such as SrF_2 (Du et al. 2009), LaF_3 (Sivakumar et al. 2006), $NaGdF_4$ (Boyer et al. 2007b), YF_3 (Darbandi and Nann 2006), $NaYF_4$ (Abel et al. 2009), and Y_2O_3 (Zhang et al. 2011).

7.3.2 Heterogeneous Core–Shell UCNPs with Enhanced UCL

Heterogeneous core–inert shell NP refers to the one where the core and the shell materials have different compositions. This type of core–shell structure possesses a superior advantage that is inaccessible to the homogenous core–shell structure described in Section 7.3.1, as it allows a discernible trasmission electron microscopy (TEM) imaging contrast between the core and shell layer owing to the difference in their composition. This provides a solid foundation for mechanistic investigation of the UCL enchantment mechanism. Figure 7.7a and b shows TEM images of the exemplified $NaYbF_4$:0.5% Tm^{3+} core and the ($NaYbF_4$:0.5% Tm^{3+})/CaF_2 core–shell UCNPs (Chen et al. 2012c). High-angle annular dark-field (HAADF) scanning transmission electron microscopy (STEM) has been utilized to discriminate the epitaxial core–shell structure, as the contrast of HAADF is approximately proportional to the square of the atomic number of the element contained, producing a z-contrast image. Indeed, a clear contrast between the core which appears bright (ytterbium $Z = 70$) and the shell which appears dark (calcium $Z = 20$) can be clearly seen (Figure 7.7c). It was found that this hetero-shell of CaF_2 can

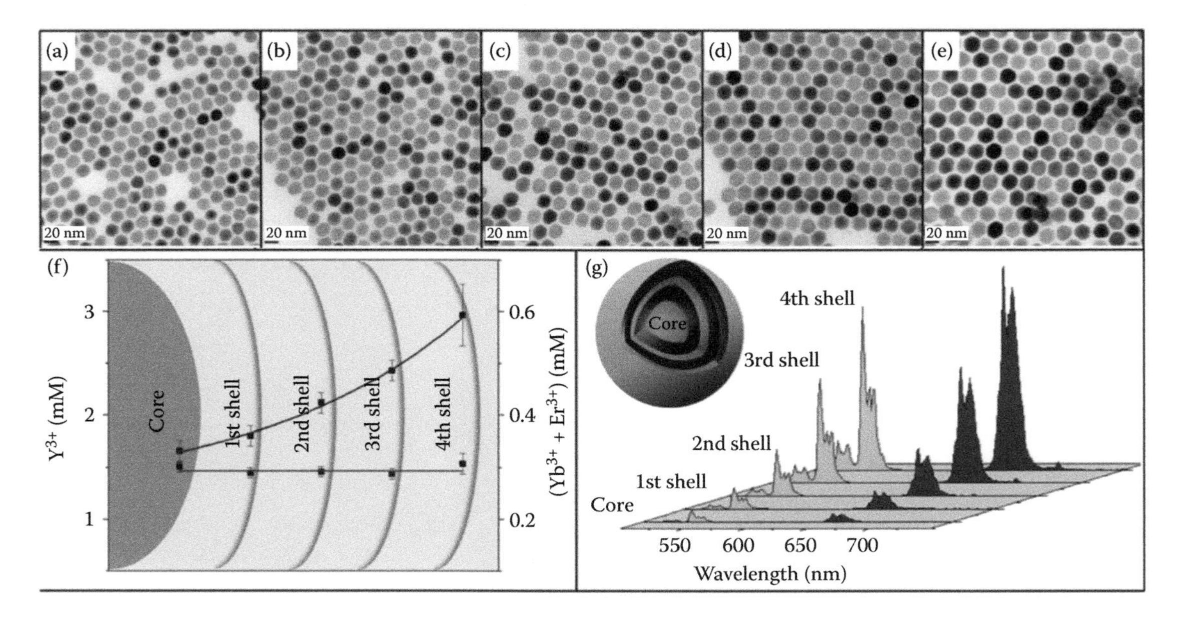

FIGURE 7.6
(a–e) TEM images of $NaYF_4$:Yb^{3+}/Er^{3+} (15/2%) core NCs ($t = 0$), $NaYF_4$:Yb^{3+}/Er^{3+} (15/2%) core/$NaYF_4$ shell NCs after successive layer-by-layer epitaxial growth at $t = 5$, 10, 15, and 20 min, respectively; (f) ICP-MS elemental analysis of the core and core–shell NCs with same number concentration of NCs; and (g) UC emission spectra of the hexane dispersions of core and core–shell NCs with same number concentration of NCs under 980 nm excitation. (From Johnson, N.J.J., A. Korinek, C.H. Dong et al., *J. Am. Chem. Soc.*, 134, 11068–11071, 2012.)

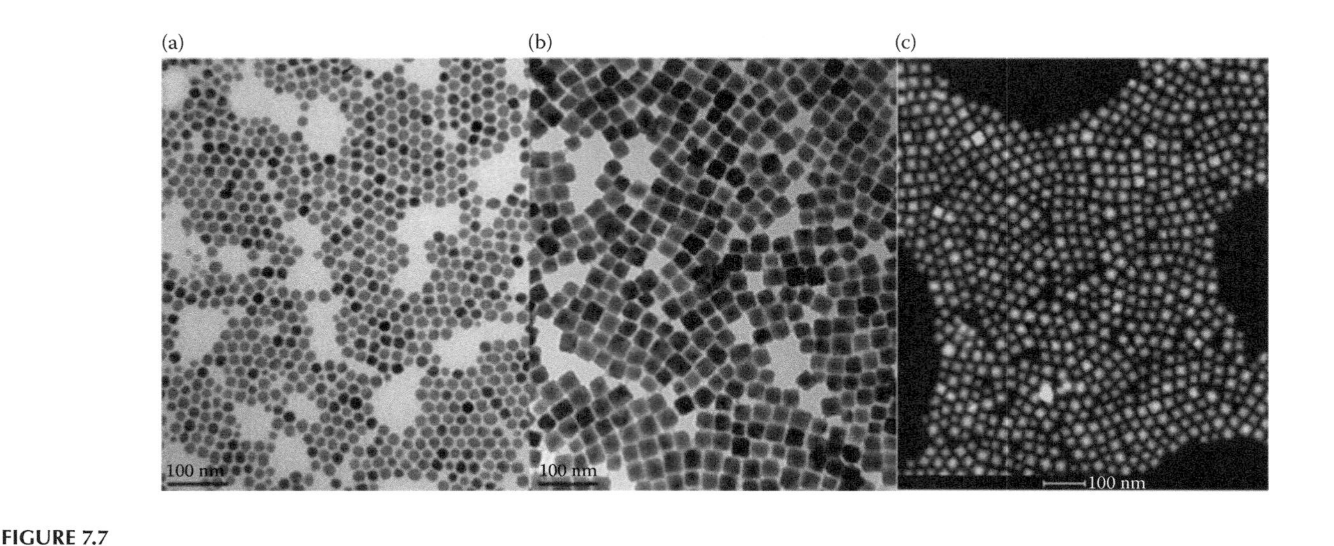

FIGURE 7.7
Successful epitaxial growth of CaF_2 shells on $NaYbF_4$:0.5% Tm^{3+} core NPs, resulting in uniform and monodispersed ($NaYbF_4$:0.5% Tm^{3+})/CaF_2 core–shell NPs. Transmission electron microscopy images of (a) $NaYbF_4$:0.5% Tm^{3+} core and (b) ($NaYbF_4$:0.5% Tm^{3+})/CaF_2 core–shell NPs. (c) HAADF STEM image of ($NaYbF_4$:0.5% Tm^{3+})/CaF_2 NPs with resolved core–shell structures; both the core (bright) and the shell (dark) are clearly visible. (From Chen, G. Y., J. Shen, T. Y. Ohulchanskyy et al. *ACS Nano* 6,8280–8287, 2012c.)

enhance NIR-to-NIR upconverting-photoluminescence (UCPL) from ~20 nm $NaYbF_4:Tm^{3+}$ by about 35-fold, yielding upconversion quantum yield (UCQY) to be as high as 0.6% under low excitation power density of 0.3 W/cm^2 (Figure 7.8). Wang et al. reported ~300-fold UCL enhancement of the 10–13 nm cubic $NaYF_4:Ln^{3+}$ when a $NaYF_4:Ln^{3+}@CaF_2$ heterogeneous core–shell structure is formed (Wang et al. 2012) (Figure 7.9). The UCL from heterogeneous cubic

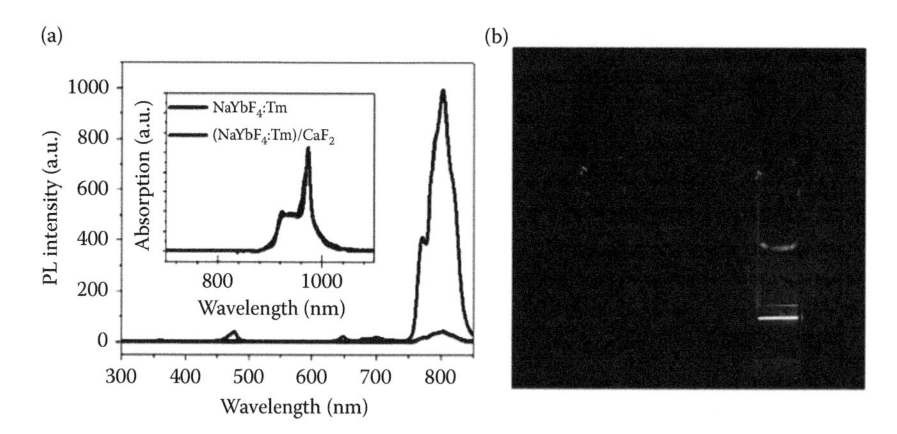

FIGURE 7.8
Optical characterizations of $(NaYbF_4:0.5\%\ Tm^{3+})/CaF_2$ core–shell UCNPs (hexane suspensions). (a) UC photoluminescence spectrum under laser excitation at 975 nm and (b) photographic images of cuvettes with suspensions of the core and the core–shell NPs under laser excitation at 975 nm. (From Chen, G. Y., J. Shen, T. Y. Ohulchanskyy et al. *ACS Nano*, 6,8280–8287, 2012c.)

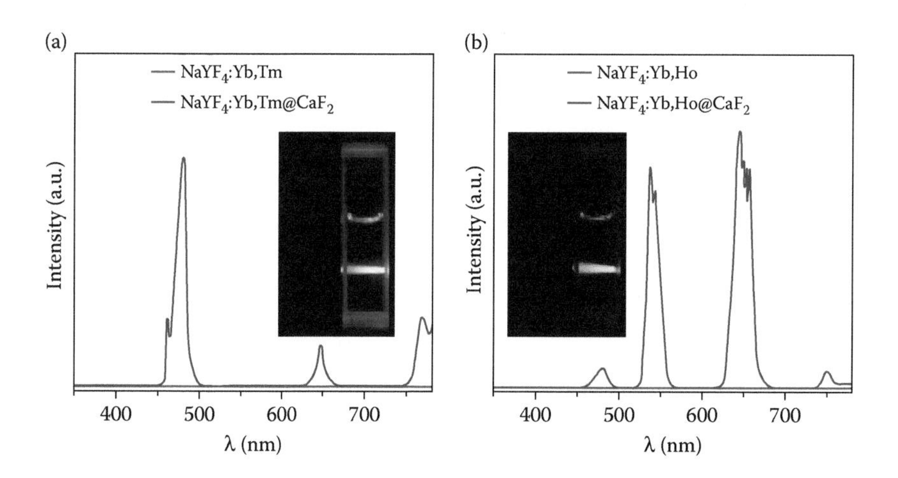

FIGURE 7.9
UC emission spectra and digital photographs (inset) of (a) $NaYF_4$:Yb,Tm and $NaYF_4$:Yb,Tm@CaF_2 NPs and (b) $NaYF_4$:Yb,Ho and $NaYF_4$:Yb,Ho@CaF_2 NPs. (From Wang, Y.F., L.D. Sun, J.W. Xiao et al., *Chem. Eur. J.*, 18, 5558–5564, 2012.)

$NaYF_4$:Yb^{3+}/Tm^{3+}@CaF_2 core–shell NPs can even be ~10 times higher than that of the corresponding hexagonal $NaYF_4$:Yb^{3+}/Tm^{3+}@$NaYF_4$ core–shell NPs (Chen et al. 2012c). Note that the $NaYF_4$ host lattice of hexagonal phase has ubiquitously been known to be more efficient than the one of cubic phase. This indicates the importance of selecting shell materials that have the different ability to minimize the surface-related quenching effects; and heterogeneous epitaxial core–shell structures definitely hold prime promise for such purpose. Indeed, many other reported observations support this conclusion. The $NaYF_4$:20%Yb^{3+}, 2%Er^{3+}@$NaYbF_4$ core–shell NPs produce enhanced UCL for the core NP at an optimized thickness of shell layer of 3 nm (Zhang and Lee 2013). Tremendous UCL enhancement has been observed in α-$NaYF_4$@$NaYbF_4$:Er@$NaYF_4$ core–shell–shell UCNPs (Li et al. 2013a), $NaGdF_4$:Yb/Tm@$NaGdF_4$:A@$NaYF_4$ core–shell–shell NPs (A = Dy, Sm, Tb, Eu) (Su et al. 2012), as well as in $NaYF_4$:Yb,Er@$NaGdF_4$ core–shell NPs (Zhang et al. 2012). It is worthwhile to mention that besides UCL enhancement, another unique advantage of the heterogeneous core–shell structure is that the UCNPs can be endowed with multifunctionalities for theranostic applications by incorporation of designated composition and doping ions into different shell layers (Dong et al. 2012; Guo et al. 2010; Lu et al. 2008).

7.4 Enhancing UCL with an Active Core–Active Shell Structure

7.4.1 Active Shell Containing Yb^{3+} Sensitizers

UCNPs have severely limited light absorbing ability due to their low extinction coefficient and narrow band absorption of RE ions. This problem mainly arises from the dipole–dipole forbidden nature of 4f–4f transitions as well as from well-shielded localized electronic wave function for RE ions in UCNPs, which severely limit the use of UCNPs in many applications. An engineering of the core–shell structure provides a new dimension to alleviate or resolve the absorption-related issues by introducing absorption from new type of REs or by coupling RE ions with electronic states from organic dyes. Indeed, instead of utilizing an inert epitaxial shell material as discussed in above section, a reasonable concentration of light absorbing dopants can be introduced in the shell layer or multiple shell layers to enhance light harvesting. An active shell not only suppresses surface-related deactivations, but also improves the interaction between RE ions in core and the sensitizers located in the shell layer, thus providing an elevated absorption for excitation and enhanced UCL.

Capobianco et al. first proposed to enhance the green and red emission of $NaGdF_4$:Yb^{3+},Er^{3+} UCNPs by introducing appropriate concentration of Yb^{3+} sensitizer into the $NaGdF_4$ shell layer (Vetrone et al. 2009). The core-only NPs were reported to have a mean diameter of 11.3 nm. After depositing an

Yb^{3+}-containing active layer onto the core NPs, the active core–active shell NPs grow into a size of ~15.9 nm (a shell thickness of ~2 nm). The active core–active shell NPs of $NaGdF_4$:Yb^{3+}, Er^{3+}@$NaGdF_4$:Yb^{3+} are about three times more efficient than the active core–inert shell NPs of $NaGdF_4$:Yb^{3+}, Er^{3+}@$NaGdF_4$ of almost identical size (Figure 7.10a). Since both structures have a shell layer to suppress the surface-related quenching effect, this enhancement can be ascribed to the increased absorption of the Yb^{3+} sensitizers in the active $NaGdF_4$:Yb^{3+} shell, which can then transfer the absorbed energy to the $NaGdF_4$:Yb^{3+}, Er^{3+} core (Figure 7.10b). Patra and coworkers have reached a similar conclusion in the designated $LaPO_4$:Er^{3+}@$LaPO_4$:Yb^{3+} active core–active shell nanorods (Ghosh et al. 2008), which showed similar UCL enhancement effect.

However, one should note that the cross relaxation process between Yb^{3+} ions enables energy migration within the Yb^{3+} sub-lattice. Opposite to the beneficial effect of increased absorption, this energy migration increases the possibility of deactivating the excitation energy at the quenching centers on the surface of the core–shell NPs. Higher Yb^{3+} ion concentration can increase more absorption ability, but concurrently favors energy migration due to a shorter distance between Yb^{3+} ions. As a consequence, it is of particular importance to optimize the concentration of Yb^{3+} ions in the active shell to reach an optimized UCL enhancement. Toward this end, Lin et al. investigated the Yb^{3+} content in the shell of the $BaGdF_5$:Yb^{3+}, Er^{3+}@$BaGdF_5$:Yb^{3+} active core–active shell UCNPs (Yang et al. 2011). Figure 7.11a and b indicates that after coating the $BaGdF_5$:Yb^{3+} active shell, the morphology and monodispersity of the $BaGdF_5$:Yb^{3+}, Er^{3+} core were retained. The UCL enhancement fold has been investigated versus the Yb^{3+} concentration in the shell, manifesting an optimized Yb^{3+} ion concentration of 5%. The optimized UCL enhancement can be clearly seen from photographic images displayed in Figure 7.11d. The optimized Yb^{3+} concentration in the shell of ~10 nm sized $BaGdF_5$:Yb^{3+}, Er^{3+}@$BaGdF_5$:Yb^{3+} active core–active shell UCNPs is different from that of the 20 nm sized $NaYF_4$:Yb^{3+}, Er^{3+}@$NaYF_4$:Yb^{3+} active core–active shell UCNPs (Vetrone et al. 2009), where an optimized concentration of 20% was manifested. This indicates that the optimized dopant concentration in the active shell varies with the host material and the size of UCNPs, and should be specifically investigated on a given case.

There are also some variations of this active core–active shell strategy to enhance UCL in UCNPs. Liu and Zhang exercise a spatial isolation of RE dopants in the core as well as various shell domains to control the doping area of the emitter; it is observed that the conventional concentration quenching effect of the emitter was alleviated (Liu et al. 2011). Huang and Lin introduced the Nd^{3+}/Yb^{3+} ion pair into the shell of $NaYF_4$: Nd^{3+}/Yb^{3+}/X^{3+} (X = Er and Tm), yielding a 522-fold UCL enhancement when excited by an 808 nm laser Huang and Lin 2015. The active shell strategy has also been applied to the $NaYF_4$:Yb^{3+}/Er^{3+}@$NaYF_4$:Yb^{3+} (Chen et al. 2014b; Zeng et al. 2013), the (CaF_2:Yb^{3+}/Tm^{3+}/Ho^{3+})@$NaYF_4$:Yb^{3+} (Zhou et al. 2013),

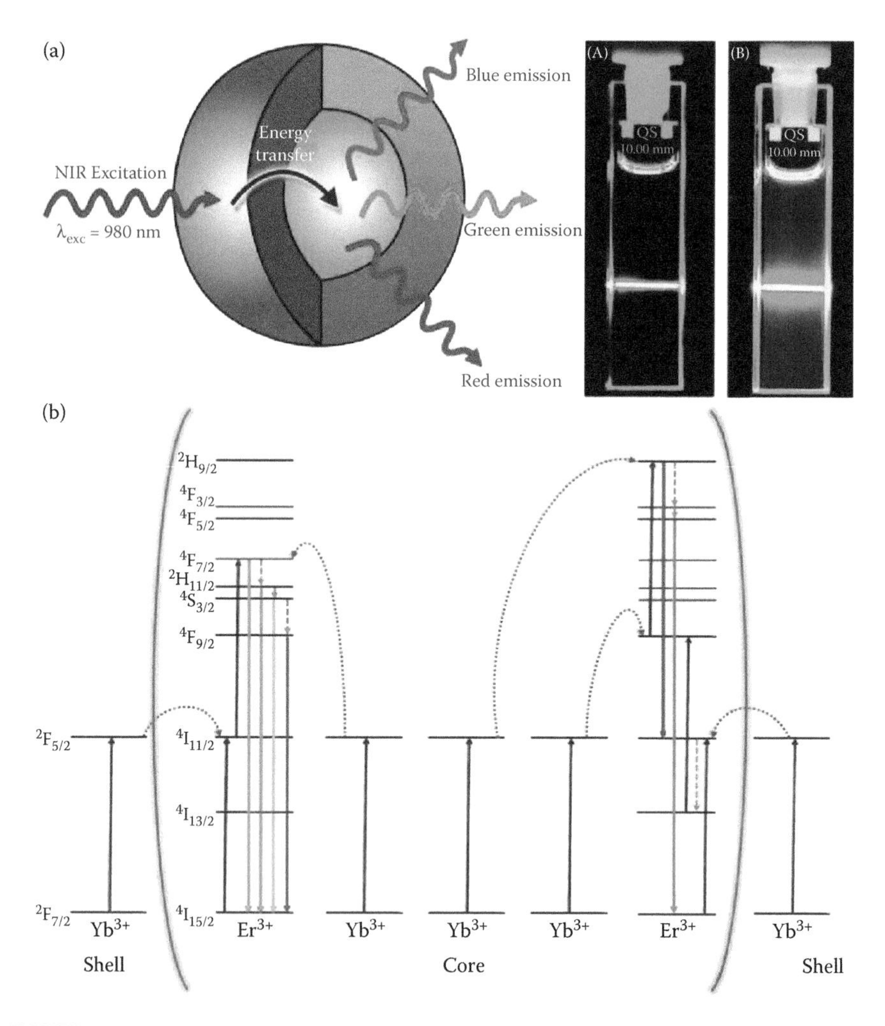

FIGURE 7.10
(a) Schematic illustration of the active core–active shell NP architecture showing the absorption of NIR light by the Yb^{3+}-rich shell (represented in red) and subsequent energy transfer to Er^{3+} ions in core (represented in green), which leads to upconverted blue, green, and red emissions. (b) The possible energy level diagrams of the Yb^{3+} in outer shell and Er^{3+}/Yb^{3+} dopant ions in core under 980 nm excitation. The full arrows pointing upwards represent energy absorption, full arrows pointing downwards represent visible emission, dotted arrows represent multiphonon relaxation (nonradiative decay), and the curly arrows represent energy transfer (note: only relevant energy levels of Er^{3+} are shown for simplicity). (From Vetrone, F., R. Naccache, V. Mahalingam et al., *Adv. Funct. Mater.*, 19, 2924–2929, 2009.)

the ($NaLuF_4$:$Gd^{3+}/Yb^{3+}/Er^{3+}$)@$NaLuF_4$:Yb^{3+} active core–active shell UCNPs (Juan et al. 2014), and the ($NaYF_4$:Yb^{3+}/Tm^{3+})/$NaYbF_4$/$NaYF_4$ active core–active shell–inert shell UNCPs (Qiu et al. 2014). One to two orders of UCL enhancement has been observed, indicating the effectiveness of this strategy for UCNPs of various host lattices.

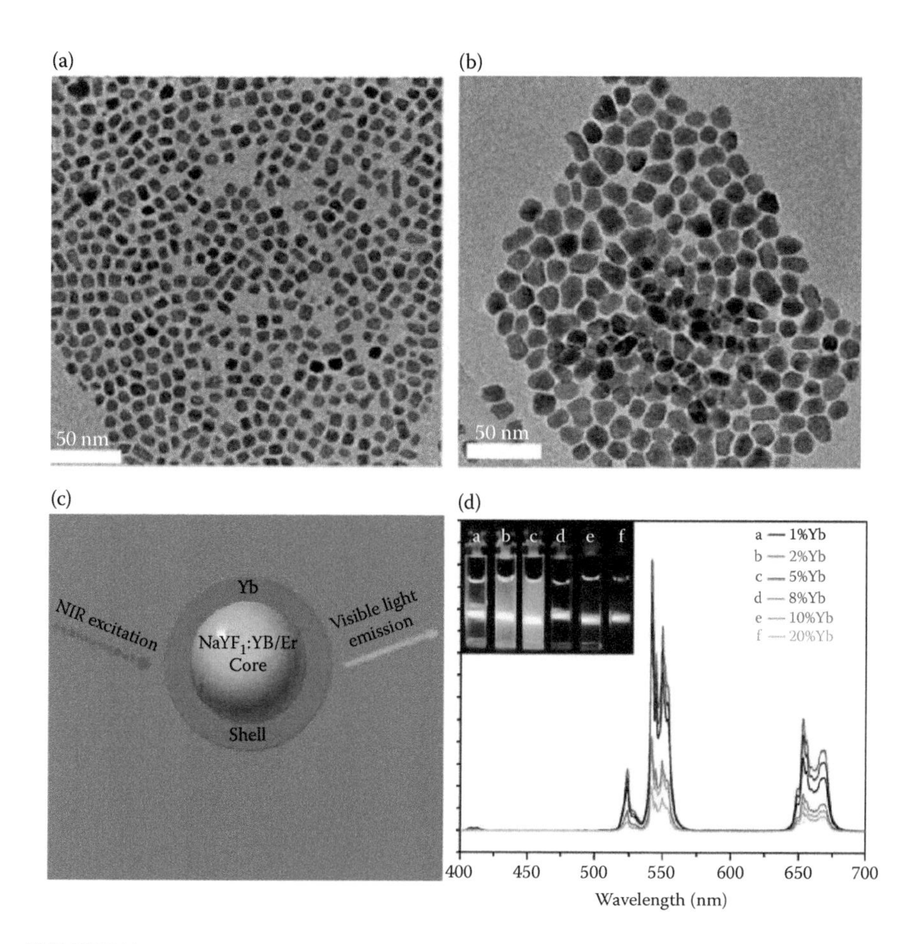

FIGURE 7.11
(a) TEM images of the $BaGdF_4$:Yb,Er core and (b) core–shell architectured $BaGdF_4$:Yb,Er@ $BaGdF_4$:Yb UCNPs. (c) Schematic illustration of the active core–active shell structure showing the absorption of NIR excitation by the Yb^{3+}-rich shell and subsequent energy transfer to Er^{3+} ions in core (represented in green), which leads to visible emissions. (d) The corresponding UCL spectra and photographs. (From Yang, D.M., C.X. Li, G.G. Li et al., *J. Mater. Chem.*, 21, 5923–5927, 2011.)

7.4.2 Active Shell Containing Nd^{3+} Sensitizers

The absorption peak of Yb^{3+} sensitizers limited the excitation of most current UCNPs to be performed at ~980 nm, which is located on the outside of biological optical window (Figure 7.12). This elicits the heating problem which can cause possible damage to cells and tissues due to strong water absorption at the excitation wavelength (Kobayashi et al. 2010; McNichols et al. 2004; Xie and Liu 2012; Zhan et al. 2011). This issue becomes particularly prominent when a high-power density and a long-time excitation are both required. Most recently, extensive efforts have been devoted to the development of UC materials that can be excited at better excitation wavelengths but with

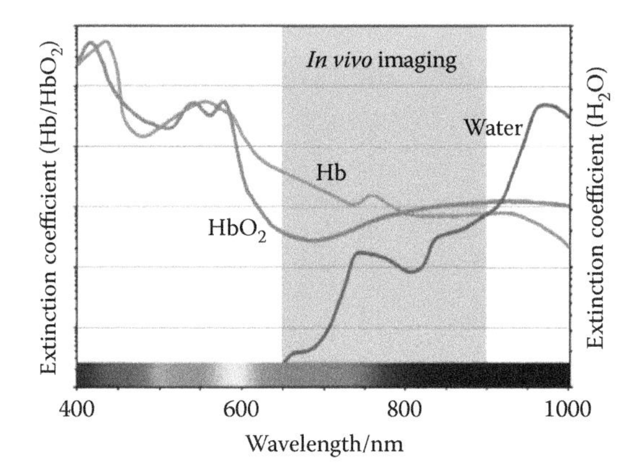

FIGURE 7.12
Spectral profiles of tissue optical window. The extinction coefficient of water at 800 nm is about 20 times lower than that at 980 nm (Hb: hemoglobin; HbO_2: oxyhemoglobin). (From Shen, J., G.Y. Chen, A.M. Vu et al., *Adv. Opt. Mater.*, 1, 644–650, 2013.)

minimized adverse effects (Zhan et al. 2013; Zou et al. 2012). Doping of Nd^{3+} ion as a sensitizer into the current UC systems would be appealing toward this purpose, as the Nd^{3+} ions can be excited by the commercial diode laser sources at ~808 nm. The absorption coefficient of water at ~808 nm is around 0.02 cm^{-1}, more than 24 times lower than the value of 0.482 cm^{-1} at ~980 nm (Weber 1971). The Nd^{3+} sensitizer displays high absorption cross section (~1.2×10^{-19} cm^2) at ~808 nm (Kushida et al. 1968), which is one order of magnitude higher than that of the Yb^{3+} sensitizer (~1.2×10^{-20} cm^2) (Xie and Liu 2012). Moreover, efficient energy transfers from Nd^{3+} to Yb^{3+} and then from Yb^{3+} to other activators, have been established in bulk materials (Strohhöfer 2001). This provides opportunities to produce UCL being excited at ~808 nm. As a consequence, a lot of attention has been placed on the development of Nd^{3+}-sensitized UC materials to prevent the overheating effect associated with Yb^{3+}-sensitized UCNPs by 980 nm laser excitation. Toward this end, the $NaGdF_4$:Yb,Er@$NaGdF_4$:Nd,Yb UCNPs were prepared; a set of experiments were designed to confirm the energy transfer process of $Nd^{3+} \rightarrow Yb^{3+} \rightarrow Er^{3+}$ in the active core–active shell NPs after absorbing excitation photons (Wang et al. 2013). In the shell, the Nd^{3+} ion can effectively absorb the energy from the 800 nm laser and be excited to its $^4F_{5/2}$ state; it can subsequently transfer its absorbed energy to the Yb^{3+} ions in the shell, and then to the Yb^{3+} in the core (Figure 7.13). Replacing the Nd^{3+} in the shell with other types of RE ions (Dy, Tm, Pr, Ho) has been attempted, but no UC emissions were observed from the active core–active shell structure. This result validates the rational utilization of Nd^{3+} as the sensitizer. A spatial isolation of the sensitizer Nd^{3+} from the activator Er^{3+} in a core–shell structure is rational, as there exist deleterious cross-relaxation processes between them that can severely restrict the effect of the Nd^{3+} sensitization. Indeed, it has been shown that only a

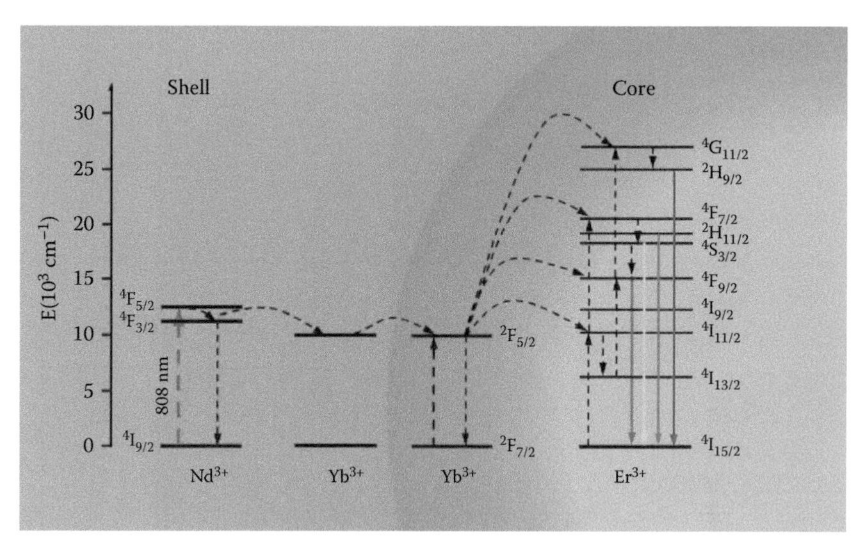

FIGURE 7.13
Energy transfer pathway from Nd^{3+} to Yb^{3+}-activated Er^{3+} UC emission in core–shell structured UCNPs under 808 nm excitation.

small amount of Nd^{3+} ions (about 1%) is permitted in the ($NaYF_4$:Nd,Yb,Er)/$NaYF_4$ NPs to produce a discernible UCL; while higher doping concentration of Nd^{3+} sensitizer leads the UCL to be quenched (Shen et al. 2013). In a marked contrast, a Nd^{3+} concentration as high as ~20% can be incorporated in the shell layer of the active core–active shell structure, which not only increases the absorption ability of the excitation light, but also can improve the efficiency of the cascade energy process of $Nd^{3+} \rightarrow Yb^{3+} \rightarrow Er^{3+}$ (Wang et al. 2013; Xie et al. 2013). The intensity of UCL from the active core–active shell $NaYF_4$:Yb/A/Nd@$NaYF_4$:Nd (A = Tm, Er, Ho) UCNPs can be about 17 times higher than that from the active core–inert shell of $NaYF_4$:Yb/A/Nd@$NaYF_4$ (A = Tm, Er, Ho) UCNPs (Figure 7.14, Xie et al. 2013). We would like to mention that this active core–active shell design has been adapted to produce enhanced UCL from $NaYF_4$:Yb,Er@$NaYF_4$:Yb@$NaNdF_4$:Yb@$NaYF_4$:Yb employing triple active shell layers (Liu et al. 2015) and ($NaYF_4$:Nd^{3+}/Yb^{3+}/Ho^{3+})/ ($NaYF_4$:Nd^{3+}/Yb^{3+}) core–shell NPs (Huang and Lin 2015), or single-band red UCL from the Yb/Ho/Ce:$NaGdF_4$$NaYF_4$ active core@active shell UCNPs (Chen et al. 2015a).

7.5 Core–Multishell Nanostructure for Enhanced UCL

The use of core–shell architecture to suppress the luminescence quenching caused by the surfaces-related effect is a well-established strategy to

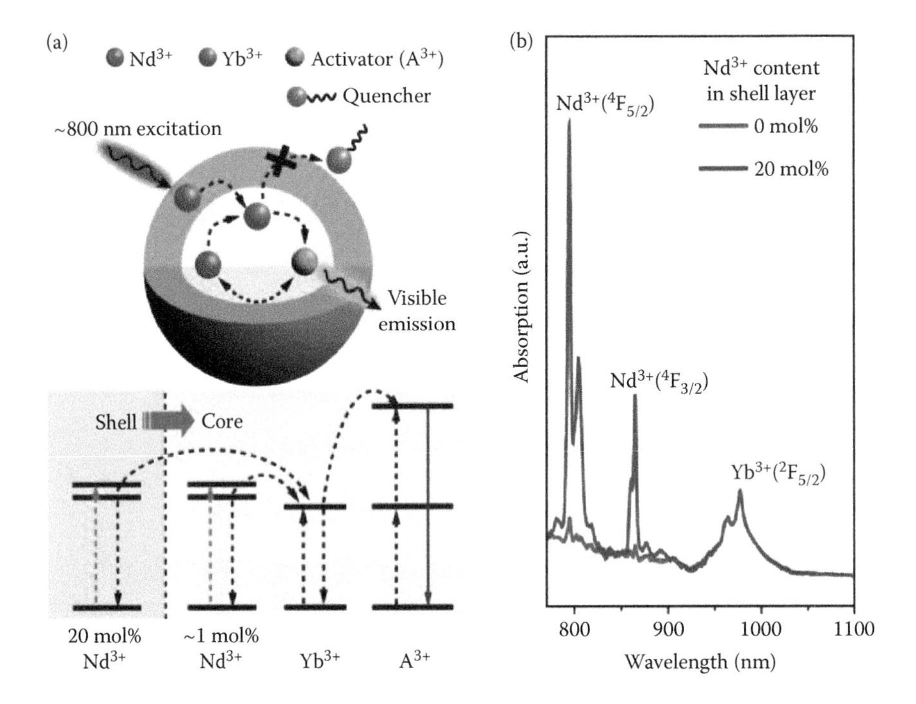

FIGURE 7.14
(a) Schematic design (top) and simplified energy level diagram (bottom) of a core–shell NP for photon UC under 800 nm excitation. Nd^{3+} ions doped in the core and shell layers serve as sensitizers to absorb the excitation energy and subsequently transfer it to Yb^{3+} ions. After energy migration from the Yb^{3+} ions to activator ions, activator emission is achieved via the Nd^{3+}-sensitization process. (b) NIR absorption spectra of $NaYF_4$:Yb/Nd(30/1%) NPs coated with an inert $NaYF_4$ shell or an active $NaYF_4$:Nd(20%) shell. The absorption spectra were normalized at 976 nm for comparison. (From Xie, X.J., N.Y. Gao, R.R. Deng et al., *J. Am. Chem. Soc.*, 135, 12608–12611, 2013.)

improve the UCL efficiency. Moreover, high-quality core–shell NPs with RE ions incorporated into the core or/and different shells are also designated to control the UC color output. However, the UCL efficiency is still limited by the deleterious interactions (e.g., the cross relaxation) between RE activators. To overcome this adverse effect, a core–multishell design for spatial isolation of RE activators in different core and shell layers has been developed. Chen reported a design to incorporate Er^{3+} and Ho^{3+} ions into the $NaGdF_4$ core and the $NaGdF_4$ shell, respectively, realizing a spatial isolation of different activators (Chen et al. 2012a). Figure 7.15a shows the UCL spectra of $NaGdF_4$: Er^{3+}/Ho^{3+}, $NaGdF_4$: Er^{3+} core NPs, and $NaGdF_4$: Er^{3+}@$NaGdF_4$, $NaGdF_4$: Er^{3+}@$NaGdF_4$: Ho^{3+}@$NaGdF_4$ core–shell UCNPs. As one can be seen, the $NaGdF_4$: Er^{3+}@$NaGdF_4$: Ho^{3+}@$NaGdF_4$ core–shell UCNPs exhibit the higher emission intensity compared to the $NaGdF_4$: Er^{3+}/Ho^{3+} and the $NaGdF_4$: Er^{3+} core NPs. It is noted that the UCL intensity of Er^{3+}/Ho^{3+} co-doped $NaGdF_4$ UCNPs is about 10 times lower than single Er^{3+}-doped or single Ho^{3+}-doped $NaGdF_4$

UCNPs, revealing the existence of cross relaxation-induced quenching effects between the Er^{3+} and the Ho^{3+} ions. When the Er^{3+} and the Ho^{3+} ions were separately incorporated into the core and shell, the lifetimes of the $^2H_{11/2}$ and $^4S_{3/2}$ states of the Er^{3+} ion in $NaGdF_4$:Er^{3+}@$NaGdF_4$: Ho^{3+}@$NaGdF_4$ core–shell UCNPs are longer than that of Er^{3+} ion in the $NaGdF_4$:Er^{3+}/Ho^{3+}nanoparicles, and approximated to the same as the lifetime of Er^{3+} ion in $NaGdF_4$:Er^{3+}@ $NaGdF_4$ core–shell NPs (Figure 7.15c,d). Therefore, considering the remarkable change of UCL intensity and lifetime, spatial isolation of activators in core–multishell structure is an essential way to suppress deleterious cross relaxations between the RE activators. However, owing to the high content of Er^{3+} and Ho^{3+} in the core–shell layers, adverse relaxations still induce quenching beyond the interfaces.

To reduce the quenching effect beyond the interfaces of core–shell, Chen and his coworkers reported a design of hexagonal core–multishell $NaYF_4$:10%Er^{3+}@$NaYF_4$@$NaYF_4$:10%Ho^{3+}@$NaYF_4$@$NaYF_4$:1%Tm^{3+}@$NaYF_4$

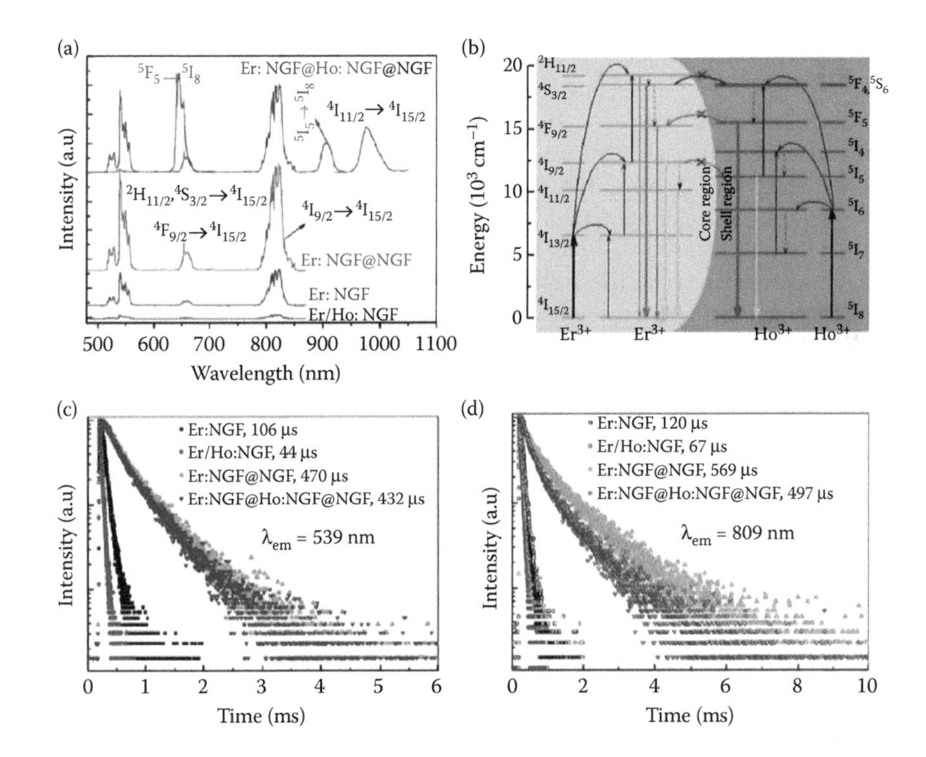

FIGURE 7.15
(a) UC emission spectra of $NaGdF_4$: Er^{3+}/Ho^{3+}, $NaGdF_4$: Er^{3+} core NPs and $NaGdF_4$: Er^{3+}@ $NaGdF_4$, $NaGdF_4$: Er^{3+}@$NaGdF_4$: Ho^{3+}@$NaGdF_4$ core–shell UCNPs; (b) possible energy processes of Er^{3+} and Ho^{3+} ions, showing possible energy transfer UC processes in the $NaGdF_4$: Er^{3+}@$NaGdF_4$: Ho^{3+}@$NaGdF_4$ core–shell UCNPs. Corresponding UC decay curves of (c) $^2H_{11/2}$, $^4S_{3/2}$ states of Er^{3+} and (d) $^4H_{9/2}$ states in core and core–shell NCs. (From Chen, G.Y., J. Shen, T.Y. Ohulchanskyy et al., *ACS Nano*, 6, 8280–8287, 2012.)

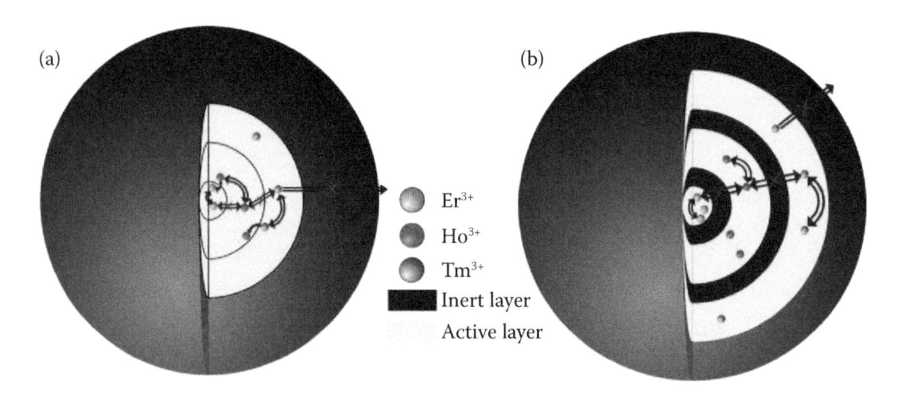

FIGURE 7.16
Schematic illustration of (a) core–shell UCNPs with an inert shell layer and (b) core–multishell UCNPs is exemplified using five layers, in which layers 1 and 3 (the gray striped ones) can be either inert layers to avoid undesirable interactions between the active layers (the colored ones), or active layers to enhance the interactions between different active domains. (From Shao, W., G.Y. Chen, T.Y. Ohulchanskyy et al., *Adv. Opt. Mater.*, 3, 575–582, 2015.)

UCNPs (Shao et al. 2015). In this design, the Er^{3+} ions are doped into the core, while the third and the fourth layers were doped with Ho^{3+} and Tm^{3+} ions, respectively. Two inert $NaYF_4$ shells were inserted into the middle of active layers to eliminate the detrimental cross relaxation and/or nonraditaive energy dissipation across the boundaries (see Figure 7.16a,b). Indeed, the UCL intensity from the designed core–multishell UCNPs is about two times higher than that from the counterpart without the inert $NaYF_4$ layers $NaYF_4$:10%Er^{3+}@$NaYF_4$:10%Ho^{3+}@$NaYF_4$:1%Tm^{3+}@$NaYF_4$, and about 10 times higher than that of the core–shell $NaYF_4$:10%Er^{3+}, 10%Ho^{3+}, 1%Tm^{3+}@$NaYF_4$ NPs without spatial isolation of the RE activators. These core–multishell NPs can be excited at ~1120–1190 nm (due to Ho^{3+}), ~1190–1260 nm (due to Tm^{3+}), and ~1450–1580 nm (due to Er^{3+}), collectively covering a broad spectral range of ~270 nm in the infrared range that can be very useful for infrared photosensitization of solar cells. The ability to spatially isolate RE activators and to manipulate the distance between them provides a versatile tool for nanophotonic control of UC processes for enhanced UCL.

7.6 Conclusion

In this chapter, we have discussed nanochemical approaches to prepare core–shell nanostructures such as epitaxial core–shell and core–multishell UCNPs. These methods allow the building of nanostructures in a precisely defined way, which enable controlling the excitation dynamics of UC. Indeed, the low efficiency of UCNPs, usually orders of magnitude lower than that of the corresponding bulk materials, can be alleviated using an epitaxial core–inert shell nanostructure that passivates the imperfects in the host lattice (often located at the core particle surface), and shields the core from the surrounding ligands (which are necessary for solvent dispersion) and solvents of high phonon energy. Besides this, the absorption cross section of UCNPs can be enhanced by utilizing the active core–active shell structure whereby sensitizers ions (Yb^{3+} or Nd^{3+}) can be doped into the shell layer on purpose, providing a new way to enhance UCL. Moreover, the core–multishell nanostructure provides a way for spatial isolation of intended REs at nanoscale, which avoids unwanted interactions between them and thus, enables intense UC output. Indeed, core–shell nanostructures are able to produce significantly enhanced UCL in RE-doped UCNPs to meet their needs for emerging photonic and biophotonic applications.

References

Abel, K. A., J. C. Boyer, and F. C. J. M. van Veggel. 2009. Hard proof of the $NaYF_4$/$NaGdF_4$ nanocrystal core/shell structure. *J. Am. Chem. Soc.* 131:14644.

Achatz, D. E., R. Ali, and O. S. Wolfbeis. 2011. Luminescent chemical sensing, biosensing, and screening using upconverting nanoparticles. *Top. Curr. Chem.* 300:29–50.

Auzel, F. 1990. Upconversion processes in coupled ion systems. *J. Lumin.* 45:341–345.

Auzel, F. 2004. Upconversion and anti-Stokes processes with f and d ions in solids. *Chem. Rev.* 104:139–173.

Boyer, J. C., L. A. Cuccia, and J. A. Capobianco. 2007a. Synthesis of colloidal upconverting $NaYF_4$: Er^{3+}/Yb^{3+} and Tm^{3+}/Yb^{3+} monodisperse nanocrystals. *Nano Lett.* 7:847–852.

Boyer, J. C., J. Gagnon, L. A. Guccia et al. 2007b. Synthesis, characterization, and spectroscopy of $NaGdF_4$: Ce^{3+}, Tb^{3+}/$NaYF_4$ core/shell nanoparticles. *Chem. Mater.* 19:3358–3360.

Boyer, J. C. and F. C. J. M. van Veggel. 2010. Absolute quantum yield measurements of colloidal $NaYF_4$: Er^{3+}, Yb^{3+} upconverting nanoparticles. *Nanoscale* 2:1417–1419.

Chen, D. Q., L. Lei, A. P. Yang et al. 2012a. Ultra-broadband near-infrared excitable upconversion core/shell nanocrystals. *Chem. Commun.* 48:5898–5900.

Chen, D. Q., L. Liu, P. Huang et al. 2015a. Nd^{3+}-sensitized Ho^{3+} single-band red upconversion luminescence in core–shell nanoarchitecture. *J. Phys. Chem. Lett.* 6:2833–2840.

Chen, G. Y., H. Ågren, T. Y. Ohulchanskyya et al. 2015b. Light upconverting core–shell nanostructures: Nanophotonic control for emerging applications. *Chem. Soc. Rev.* 44:1680–1773.

Chen, G. Y., T. Y. Ohulchanskyy, W. C. Law et al. 2011. Monodisperse $NaYbF_4$:Tm^{3+}/$NaGdF_4$ core/shell nanocrystals with near-infrared to near-infrared upconversion photoluminescence and magnetic resonance properties. *Nanoscale* 3:2003–2008.

Chen, G. Y., H. L. Qiu, R. W. Fan et al. 2012b. Lanthanide-doped ultrasmall yttrium fluoride nanoparticles with enhanced multicolor upconversion photoluminescence. *J. Mater. Chem.* 22:20190–20196.

Chen, G. Y., H. L. Qiu, P. N. Prasad et al. 2014a. Upconversion nanoparticles: Design, nanochemistry, and applications in theranostics. *Chem. Rev.* 114(10):5161–5214.

Chen, G. Y., J. Shen, T. Y. Ohulchanskyy et al. 2012c. (alpha-$NaYbF_4$:Tm^{3+})/CaF_2 core/shell nanoparticles with efficient near-infrared to near-infrared upconversion for high-contrast deep tissue bioimaging. *ACS Nano* 6:8280–8287.

Chen, M. L., Y. Ma, and M. Y. Li. 2014b. Facile one-pot synthesis of hydrophilic $NaYF_4$:Yb,Er@$NaYF_4$:Yb active-core/active-shell nanoparticles with enhanced upconversion luminescence. *Mater. Lett.* 114:80–83.

Chen, X., D. F. Peng, J. Qiang et al. 2015c. Photon upconversion in core–shell nanoparticles. *Chem. Soc. Rev.* 44:1318–1330.

Darbandi, M. and T. Nann. 2006. One-pot synthesis of YF_3@silica core/shell nanoparticles. *Chem. Commun.* 7:776–778.

Dong, C. H., A. Korinek, B. Blasiak et al. 2012. Cation exchange: A facile method to make $NaYF_4$:Yb,Tm-$NaGdF_4$ core–shell nanoparticles with a thin, tunable, and uniform shell. *Chem. Mater.* 24:1297–1305.

Dou, Q. Q., N. M. Idris, and Y. Zhang. 2013. Sandwich-structured upconversion nanoparticles with tunable color for multiplexed cell labeling. *Biomaterials* 34:1722–1731.

Du, Y. P., X. Sun, Y. W. Zhang et al. 2009. Uniform alkaline earth fluoride nanocrystals with diverse shapes grown from thermolysis of metal trifluoroacetates in hot surfactant solutions. *Cryst. Growth Des.* 9:2013–2019.

Du, Y. P., Y. W. Zhang, L. D. Sun et al. 2008. Luminescent monodisperse nanocrystals of lanthanide oxyfluorides synthesized from trifluoroacetate precursors in high-boiling solvents. *J. Phys. Chem. C.* 112:405–415.

Feng, W., L. D. Sun, Y. W. Zhang et al. 2010. Synthesis and assembly of rare earth nanostructures directed by the principle of coordination chemistry in solution-based process. *Coordin. Chem. Rev.* 254:1038–1053.

Ghosh, P., J. Oliva, E. De la Rosa et al. 2008. Enhancement of upconversion emission of $LaPO_4$:Er@Yb core–shell nanoparticles/nanorods. *J. Phys. Chem. C.* 112:9650–9658.

Guo, H., Z. Q. Li, H. S. Qian et al. 2010. Seed-mediated synthesis of $NaYF_4$:Yb,Er/$NaGdF_4$ nanocrystals with improved upconversion fluorescence and MR relaxivity. *Nanotechnology* 21:125602.

Haase, M. and H. Schäfer. 2011. Upconverting nanoparticles. *Angew. Chem. Int. Ed.* 50:5808–5829.

Hao, S. W., W. Shao, H. L. Qiu et al. 2014. Tuning the size and upconversion emission of $NaYF_4$:Yb^{3+}/Pr^{3+} nanoparticles through Yb^{3+} doping. *RSC Adv.* 4:56302–56306.

Hao, S. W., L. M. Yang, H. L. Qiu et al. 2015. Heterogeneous core/shell fluoride nanocrystals with enhanced upconversion photoluminescence for *in vivo* bioimaging. *Nanoscale* 7:10775–10780.

Huang, P., W. Zheng, S. Zhou et al. 2014. Lanthanide-doped $LiLuF_4$ upconversion nanoprobes for the detection of disease biomarkers. *Angew. Chem. Int. Ed.* 53:1252–1257.

Huang, X. Y. 2015. Giant enhancement of upconversion emission in ($NaYF_4$:Nd^{3+}/Yb^{3+}/Ho^{3+})/($NaYF_4$:Nd^{3+}/Yb^{3+}) core/shell nanoparticles excited at 808 nm. *Opt. Lett.* 40:3599–3602.

Huang, X. Y. and J. Lin. 2015. Active-core/active-shell nanostructured design: An effective strategy to enhance Nd^{3+}/Yb^{3+} cascade sensitized upconversion luminescence in lanthanide-doped nanoparticles. *J. Mater. Chem. C.* 3:7652–7657.

Johnson, N. J. J., A. Korinek, C. H. Dong et al. 2012. Self-focusing by Ostwald ripening: A strategy for layer-by-layer epitaxial growth on upconverting nanocrystals. *J. Am. Chem. Soc.* 134:11068–11071.

Juan, O. Y., D. G. Yin, X. Cao et al. 2014. Synthesis of $NaLuF_4$-based nanocrystals and large enhancement of upconversion luminescence of $NaLuF_4$:Gd, Yb, Er by coating an active shell for bioimaging. *Dalton Trans.* 43:14001–14008.

Kobayashi, H., M. Ogawa, R. Alford et al. 2010. New strategies for fluorescent probe design in medical diagnostic imaging. *Chem. Rev.* 110:2620–2640.

Kushida, T., H. M. Marcos, and J. E. Geusic. 1968. Laser transition cross section and fluorescence branching ratio for Nd^{3+} in yttrium aluminum garnet. *Phys. Rev.* 167:289.

Li, C. X., C. M. Zhang, Z. Y. Hou et al. 2009. Beta-$NaYF_4$ and beta-$NaYF_4$:Eu^{3+} Microstructures: Morphology control and tunable luminescence properties. *J. Phys. Chem. C.* 113:2332–2339.

Li, X. M., D. K. Shen, J. P. Yang et al. 2013a. Successive layer-by-layer strategy for multi-shell epitaxial growth: Shell thickness and doping position dependence in upconverting optical properties. *Chem. Mater.* 25:106–112.

Li, X. M., R. Wang, F. Zhang et al. 2013b. Nd^{3+} sensitized up/down converting dual-mode nanomaterials for efficient *in-vitro* and *in-vivo* bioimaging excited at 800 nm. *Sci. Rep.* 3:3536.

Li, Z. Q. and Y. Zhang. 2008. An efficient and user-friendly method for the synthesis of hexagonal-phase $NaYF_4$:Yb, Er/Tm nanocrystals with controllable shape and upconversion fluorescence. *Nanotechnology* 19:345606.

Liu, B., Y. Y. Chen, C. X. Li et al. 2015. Poly(acrylic acid) modification of Nd^{3+}-sensitized upconversion nanophosphors for highly efficient UCL imaging and pH-responsive drug delivery. *Adv. Funct. Mater.* 25:4717–4729.

Liu, X. M., X. G. Kong, Y. L. Zhang et al. 2011. Breakthrough in concentration quenching threshold of upconversion luminescence via spatial separation of the emitter doping area for bio-applications. *Chem. Commun.* 47:11957–11959.

Lu, Q., A. H. Li, F. Y. Guo et al. 2008. Experimental study on the surface modification of Y_2O_3:Tm^{3+}/Yb^{3+} nanoparticles to enhance upconversion fluorescence and weaken aggregation. *Nanotechnology* 19:145701.

Mai, H. X., Y. W. Zhang, R. Si et al. 2006. High-quality sodium rare-earth fluoride nanocrystals: Controlled synthesis and optical properties. *J. Am. Chem. Soc.* 128:6426–6436.

McNichols, R. J., A. Gowda, M. Kangasniemi et al. 2004. MR thermometry-based feedback control of laser interstitial thermal therapy at 980 nm. *Laser Surg. Med.* 34:48–55.

Qiu, H. L., C. H. Yan, W. Shao et al. 2014. Enhanced upconversion luminescence in Yb^{3+}/Tm^{3+}-codoped fluoride active core/active shell/inert shell nanoparticles through directed energy migration. *Nanomaterials* 4:55–68.

Schafer, H., P. Ptacek, O. Zerzouf et al. 2008. Synthesis and optical properties of KYF_4/Yb, Er nanocrystals, and their surface modification with undoped KYF_4. *Adv. Funct. Mater.* 18:2913–2918.

Shang, Y. F., S. W. Hao, J. Liu et al. 2015. Synthesis of upconversion beta-$NaYF_4$:$Nd^{3+}/Yb^{3+}/Er^{3+}$ particles with enhanced luminescent intensity through control of morphology and phase. *Nanomaterials* 5:218–232.

Shao, W., G. Y. Chen, T. Y. Ohulchanskyy et al. 2015. Lanthanide-doped fluoride core/multishell nanoparticles for broadband upconversion of infrared light. *Adv. Opt. Mater.* 3:575–582.

Shen, J., G. Y. Chen, A. M. Vu et al. 2013. Engineering the upconversion nanoparticle excitation wavelength: Cascade sensitization of tri-doped upconversion colloidal nanoparticles at 800 nm. *Adv. Opt. Mater.* 1:644–650.

Sivakumar, S., P. R. Diamente, and F. C. J. M. van Veggel. 2006. Silica-coated Ln^{3+}-doped LaF_3 nanoparticles as robust down- and upconverting biolabels. *Chem. Eur. J.* 12:5878–5884.

Strohhöfer, C. 2001. Optical properties of ion beam modified waveguide materials doped with erbium and silver. PhD dissertation, p. 50, Universiteit Utrecht, Utrecht, the Netherlands.

Su, Q. Q., S. Y. Han, X. J. Xie et al. 2012. The effect of surface coating on energy migration-mediated upconversion. *J. Am. Chem. Soc.* 134:20849–20857.

Vetrone, F., R. Naccache, V. Mahalingam et al. 2009. The active-core/active-shell approach: A strategy to enhance the upconversion luminescence in lanthanide-doped nanoparticles. *Adv. Funct. Mater.* 19:2924–2929.

Wang, F., D. Banerjee, Y. S. Liu et al. 2010a. Upconversion nanoparticles in biological labeling, imaging, and therapy. *Analyst* 135:1839–1854.

Wang, F., Y. Han, C. S. Lim et al. 2010b. Simultaneous phase and size control of upconversion nanocrystals through lanthanide doping. *Nature* 463:1061–1065.

Wang, F. and X. G. Liu. 2009. Recent advances in the chemistry of lanthanide-doped upconversion nanocrystals. *Chem. Soc. Rev.* 38:976–989.

Wang, X., J. Zhuang, and Q. Peng. 2005. A general strategy for nanocrystal synthesis. *Nature.* 437:121–124.

Wang, Y., L. P. Tu, J. W. Zhao et al. 2009. Upconversion luminescence of beta-$NaYF_4$:Yb^{3+},Er^{3+}@beta-$NaYF_4$ core/shell nanoparticles: Excitation power, density and surface dependence. *J. Phys. Chem. C.* 113:7164–7169.

Wang, Y. F., G. Y. Liu, L. D. Sun et al. 2013. Nd^{3+}-sensitized upconversion nanophosphors: Efficient *in vivo* bioimaging probes with minimized heating effect. *ACS Nano* 7:7200–7206.

Wang, Y. F., L. D. Sun, J. W. Xiao et al. 2012. Rare-earth nanoparticles with enhanced upconversion emission and suppressed rare-earth-ion leakage. *Chem. Eur. J.* 18:5558–5564.

Weber, M. J. 1971. Optical properties of Yb^{3+} and Nd^{3+}–Yb^{3+} energy transfer in $YAlO_3$. *Phys. Rev. B.* 4:3153–3159.

Xie, X. and X. Liu. 2012. Photonics upconversion goes broadband. *Nat. Mater.* 11:842–843.

Xie, X. J., N. Y. Gao, R. R. Deng et al. 2013. Mechanistic investigation of photon upconversion in Nd^{3+}-sensitized core–shell nanoparticles. *J. Am. Chem. Soc.* 135:12608–12611.

Xu, C.T., P. Svenmarker, H. Liu et al. 2012. High-resolution fluorescence diffuse optical tomography developed with nonlinear upconverting nanoparticles. *ACS Nano* 6(6):4788–4795.

Yang, D. M., C. X. Li, G. G. Li et al. 2011. Colloidal synthesis and remarkable enhancement of the upconversion luminescence of $BaGdF_5$: Yb^{3+}/Er^{3+} nanoparticles by active-shell modification. *J. Mater. Chem.* 21:5923–5927.

Yi, G. S. and G. M. Chow. 2007. Water-soluble $NaYF_4$:Yb,Er(Tm)/$NaYF_4$/polymer core/shell/shell nanoparticles with significant enhancement of upconversion fluorescence. *Chem. Mater.* 19:341–343.

Zeng, Q. H., B. Xue, Y. L. Zhang et al. 2013. Facile synthesis of $NaYF_4$:Yb, Ln/$NaYF_4$:Yb core/shell upconversion nanoparticles via successive ion layer adsorption and one-pot reaction technique. *CrystEngComm* 15:4765–4772.

Zhan, Q., S. He, J. Qian et al. 2013. Optimization of optical excitation of upconversion nanoparticles for rapid microscopy and deeper tissue imaging with higher quantum yield. *Theranostics* 3:306–316.

Zhan, Q., J. Qian, H. Liang et al. 2011. Using 915 nm laser excited Tm^{3+}/Er^{3+}/Ho^{3+}-doped $NaYbF_4$ upconversion nanoparticles for *in vitro* and deeper *in vivo* bioimaging without overheating irradiation. *ACS Nano* 5:3744–3757.

Zhang, C. and J. Y. Lee. 2013. Prevalence of anisotropic shell growth in rare earth core–shell upconversion nanocrystals. *ACS Nano* 7:4393–4402.

Zhang, F., R. C. Che, X. M. Li et al. 2012. Direct imaging the upconversion nanocrystal core/shell structure at the subnanometer level: Shell thickness dependence in upconverting optical properties. *Nano Lett.* 12:2852–2858.

Zhang, F., R. C. Haushalter, R. W. Haushalter et al. 2011. Rare-earth upconverting nanobarcodes for multiplexed biological detection. *Small* 7:1972–1976.

Zhang, F., J. Li, J. Shan et al. 2009. Shape, size, and phase-controlled rare-earth fluoride nanocrystals with optical up-conversion properties. *Chem. Eur. J.* 15:11010–11019.

Zhong, Y. T., G. Tian, Z. J. Gu et al. 2014. Elimination of photon quenching by a transition layer to fabricate a quenching-shield sandwich structure for 800 nm excited upconversion luminescence of Nd^{3+} sensitized nanoparticles. *Adv. Mater.* 26:2831–2837.

Zhou, B., L. L. Tao, Y. H. Tsang et al. 2013. Core–shell nanoarchitecture: A strategy to significantly enhance white-light upconversion of lanthanide-doped nanoparticles. *J. Mater. Chem. C.* 1:4313–4318.

Zou, W., C. Visser, J. A. Maduro, M. S. Pshenichnikov, and J. C. Hummelen. 2012. Broadband dye-sensitized upconversion of near-infrared light. *Nat. Photonics.* 6:560–564.

Section II

Applications

8

Active-Core–Active-Shell Upconverting Nanoparticles: Novel Mechanisms, Features, and Perspectives for Biolabeling

K. Prorok, D. Wawrzyńczyk, M. Misiak, and A. Bednarkiewicz

CONTENTS

8.1 Introduction: Background and Driving Forces 196
8.2 Lanthanide-Doped NPs in Biomedical Sciences 198
8.3 Biomedical Applications of UCNPs 203
8.4 Lanthanide-Doped NPs: The Issues and Solutions 206
 8.4.1 Crystallographic Structure Tuning 207
 8.4.2 Surface Passivation with (Un) Doped Crystalline or Amorphous Shells 208
 8.4.3 Plasmonic and Photonic Effects 210
 8.4.4 Increasing Active Ions Concentration 211
8.5 Core–Shell UCNPs: Novel Possibilities 212
 8.5.1 Shell Increases Functionality and Biocompatibility 212
 8.5.2 Shell Protects NPs from Surface Quenching 214
 8.5.3 Shells Allow the Optimization of Activator Distribution 215
 8.5.4 Shell Improves Excitation Schemes 218
8.6 Core–Shell UCNPs: Synthesis, Properties, and Issues 221
 8.6.1 Host Materials for Cores 224
  8.6.1.1 Sodium Fluoride 224
  8.6.1.2 Calcium Fluoride 225
  8.6.1.3 RE Fluoride 225
  8.6.1.4 Metal Oxide 226
 8.6.2 Shell Synthesis Methods and Compositions 227
 8.6.3 Silica Amorphous Layer 229
 8.6.4 Crystalline Homo- and Hetero-Structural Shells 232
 8.6.5 Methods to Confirm Shell Formation Upon Core NPs 233
 8.6.6 Nonhomogenous Shell Composition 236
8.7 Conclusions 236
Acknowledgments 239
References 240

8.1 Introduction: Background and Driving Forces

The optical properties of lanthanide (Ln^{3+}) ions were recognized many years ago in several important fields of science and technology. Lanthanide-doped crystal, glasses, and fibers have been used as optically active materials for compact solid-state laser crystals (Geusic et al. 1964; Moncorgé et al. 1999; Quarles et al. 1990), fiber lasers (Digonnet 2001), TV tubes or lamp phosphors (Lakshmanan 2008), as well as IR quantum cutters (Esterowitz et al. 1968). The unique optical features of lanthanide-doped nanomaterials arise from the electronic configuration of optically active ions doped into the crystal matrix. The 4f–4f intra-configurational transitions are Laporte forbidden, and simultaneously the electrons at 4f energy levels are shielded by the filled higher orbitals. The consequences of forbidden character of electronic transitions observed in lanthanides are among others: low transition probability, and further long luminescence lifetimes, which vary between micro to milliseconds. These long decays facilitate ultrahigh sensitivity detection, since in time-resolved mode, the background free detection becomes possible. In addition, due to the weak interaction of electrons at 4f orbitals with the crystal field of the host matrix, the absorption and emission lines are very narrow and occur in strictly defined spectral ranges, which on the other hand, enable the engineered design of multicolor emission and further multiplex detection.

Novel possibilities of lanthanide-doped materials were discovered in the 1960s by *Francois Auzel*, who noticed the occurrence of anti-Stokes, short-wavelength emission under long-wavelength excitation due to the energy transfer between two ions (Auzel 1966). Previously, many anti-Stokes mechanisms have been recognized, such as second (or higher) harmonics generation or two photon absorption in nonlinear materials, however, the real electronic levels of Ln^{3+} ions with low decay rates, which participate in the energy upconversion in lanthanides, make the anti-Stokes emission by far (i.e., orders of magnitude) more efficient in lanthanides as compared to conventional nonlinear optical materials. In general, low energy photons are absorbed by one type of ion, and through either ladder-like mechanisms, or sequential absorption, can populate higher excited states of another lanthanide ions, what is followed by the emission of photons with larger energy than the energy of pumping photons.

In these early years, the interest in the efficient anti-Stokes emission of lanthanides initiated new studies and brought new discoveries—many host materials have been verified to optimize the effectiveness of the observed visible emission, and novel upconversion mechanisms have been found in lanthanide ions, such as energy transfer upconversion (ETU), excited state absorption (ESA), cooperative energy transfer, and photon avalanche (PA) (Auzel 2004). Those processes, together with downconversion (DC) and nonlinear optical phenomena, are summarized in Figure 8.1. The relatively

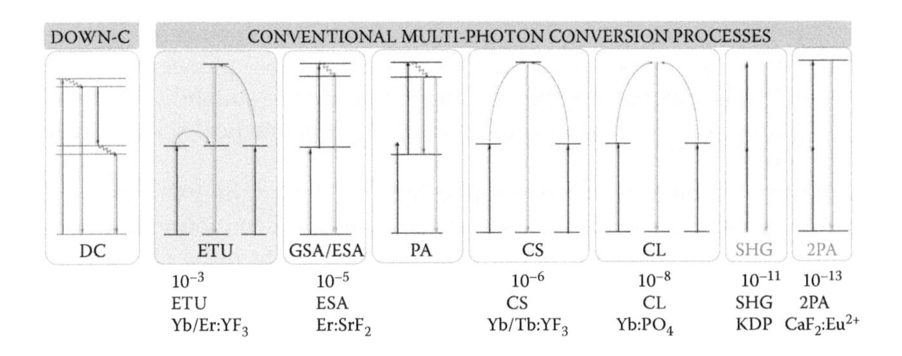

FIGURE 8.1
Schematic illustration of Stokes and anti-Stokes photophysical process, A—absorption, DC—downconversion, ETU—energy transfer upconversion, GSA/ESA—ground/excited state absorption, PA—photon avalanche, CS—cooperative sensitization, CL—cooperative luminescence, SHG—second harmonic generation, 2PA—two photon absorption, and emission cross sections of various anti-Stokes processes are compared (Based on Auzel, F. 2004. *Chem. Rev.* 1:139–173).

high quantum efficiency of such anti-Stokes processes (Figure 8.1) made numerous new applications, such as volumetric displaying (Downing et al. 1996), remote temperature fiber sensing (Berthou and Jorgensen 1990), or upconversion-based lasing (Lenth et al. 1987) possible. However, yet another opportunity for the lanthanide co-doped materials has been discovered in biomedical sciences. The pioneering work and first biomedical demonstration of fluorescence bioimaging with upconverting (UC) nanoparticles (NPs) as optical markers was performed by Zijlmans et al. with 200–400 nm sized Y_2O_2S:Yb:Tm particles (Zijlmans et al. 1999). The major conclusions were unprecedentedly low background signal and thus high imaging contrast, as well as the photostability of the phosphors. Although successful as proof-of-concept, real biological applications required upconverting phosphors of much smaller sizes and higher brightness.

A significant step forward has been done by the ground-breaking work of Haase et al. in 2003, who demonstrated the possibility of synthesizing optically clear, chloroform colloidal solutions of upconverting $LuPO_4$:Yb^{3+},Tm^{3+} and $YbPO_4$:Er^{3+} NPs (Heer et al. 2003). Through proper, post-synthetic surface functionalization, such nanophosphors could be transferred to water solutions, and with further attachment of biomolecules could act as transducers in biosensing, bioimaging, or biolabeling (Gnach and Bednarkiewicz 2012). The most important property of these materials, which makes them so unique, is efficient energy upconversion leading to anti-Stokes emission (Auzel 2004). Such ability allows the shifting of excitation lines to the far-red or near-infrared (NIR) spectral region, and thus prevents numerous deteriorative biophysical processes such as the presence of tissue autofluorescence under short-wavelength excitation, nonselective excitation of Stokes type dyes/labels, and may further increase excitation light penetration depth into highly scattering and heterogeneous tissues. Since those pioneering reports, enormous work has been devoted to demonstrate the suitability of various types of lanthanide-doped upconverting NPs in biomedical science including nanomedicine.

Although upconversion has been originally found in Er^{3+} ions singly doped materials, the addition of sensitizing ions has been found to enhance the upconversion emission quantum yield (QY) significantly. The sensitizer ions (e.g., Yb^{3+} and Nd^{3+}) absorb the excitation light and relatively efficiently transfer this energy to activator ions responsible for visible photon emission (Auzel and Pecile 1973). Recently, owing to the new capabilities in designing and reproducible synthesis of active-core@active-shell (ACAS) upconverting NPs, novel routes of ETU have been designed, that is, energy migration-mediated upconversion (EMU) (Wang et al. 2011b, 2013; Zhong et al. 2014), which are of special importance for the biomedical field.

8.2 Lanthanide-Doped NPs in Biomedical Sciences

The possible energy transfer and nonradiative/radiative processes in multiple lanthanide ions doped nanomaterials have been already reviewed (Chen et al. 2015b; Liu 2015; Tsang et al. 2015; Tu et al. 2015), with a simultaneous growing number of excellent review articles regarding their biomedical properties and applications (Chen et al. 2014; Dong et al. 2014; Gnach et al. 2015; Gnach and Bednarkiewicz 2012; Gu et al. 2013; Idris et al. 2014; Li et al. 2014b; Sun et al. 2014; Xu et al. 2012; Yang et al. 2014; Zheng et al. 2015; Zhou et al. 2014). In particular, there are three types of transitions in lanthanides, which are of some importance for biomedical applications. Most obvious one is *Stokes VIS emission under UV* excitation. Typically, under short-wavelength excitation (<400 nm), visible narrowband emission from Eu^{3+} (~620 nm), Tb^{3+} (540 nm), Sm^{3+} (650 nm), or Dy^{3+} (570 nm)-doped complexes or crystals is observed. The photoexcitation originates from either *f–f* electronic absorption, charge transfer (CT) (Ln–O or Ln-ligand), or from *f–d* electronic absorption. Due to the forbidden character of *f–f* transitions, the CT and *f–d* processes are usually much more efficient. The organic complexes of Eu^{3+} and Tb^{3+} ions have been often used in bioassays or for bioimaging, but unfortunately those materials require short-wavelength excitation. The application of downconverting colloidal lanthanide-doped NPs in the biomedical field is, however, much less frequent. The spectral overlap of excitation or emission bands of these compounds with respective absorption or emission of endogenous chromophores (Figure 8.2) can significantly decrease the sensitivity of bioassays or diminish the contrast of *in vivo* imaging. However, owing to very long luminescence lifetimes of Tb^{3+} and Eu^{3+} (millisecond range), time-gated

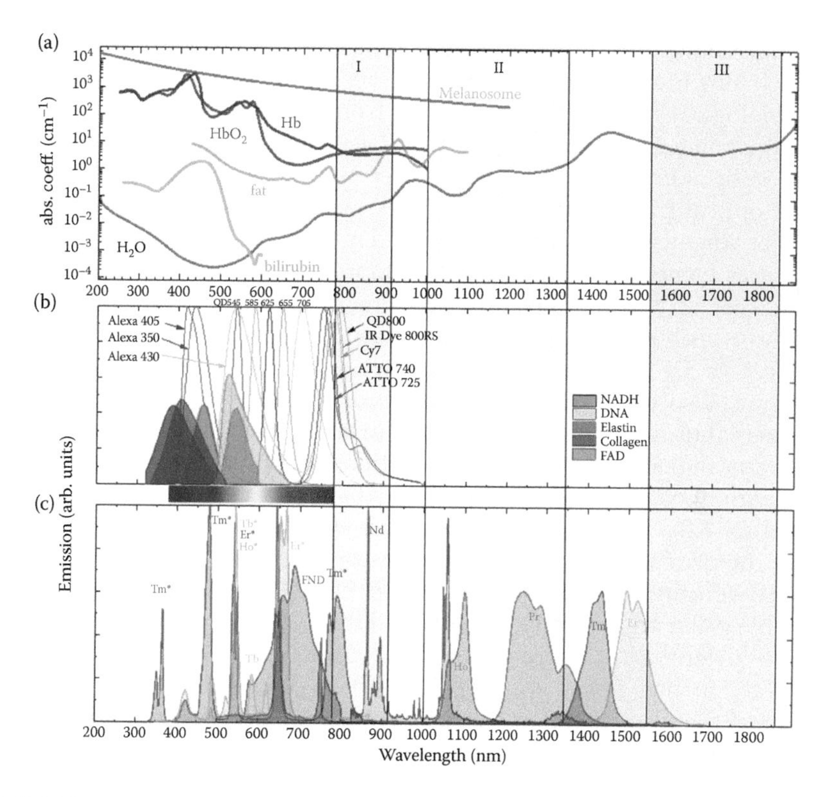

FIGURE 8.2
Absorption coefficients (a) and fluorescence spectra (b) of major tissue components (nicotinamide adenine dinucleotide (NADH), DNA, elastin, collagen, and flavin adenine dinucleotide (FAD)) combined with representative fluorescence spectra of commercial QDs and organic fluorophores (b), and luminescence spectra (stokes and anti-Stokes) of lanthanide-doped nanoparticles (NPs) (c). The spectra on B, C are just for spectral range demonstration and not for intensity comparison. The spectra are a combination of own measurements (lanthanides), spectra available at www.semrock.com (dyes, QDs), fat (R. L. P. van Veen, H. J. C. M. Sterenborg, A. Pifferi, A. Torricelli, and R. Cubeddu, *OSA Annual BIOMED Topical Meeting,* 2004.), absorption coeff. (http://omlc.ogi.edu/spectra), endogenous fluorophores (Fotodynamicz name to darozpoznawania I leczenianowotworów, ed. A. Graczykowa, Domywydawniczy Bellona 1999, Poland.), sensitivity of detectors (*Photomultipliers Low-Light, High-Speed Specialists* Ken Kaufmann, Hamamatsu Corporation, http://www.photonics.com/EDU/Handbook.aspx?AID=25145.), Ag2Se (Bohua Dong, Chunyan Li, Guangcun Chen, Yejun Zhang, Yan Zhang, Manjiao Deng and Qiangbin Wang. Facile synthesis of highly photoluminescent Ag_2Se quantum dots as a new fluorescent probe in the second near-infrared window for in vivo imaging. *Chem. Mater.* 2013, 25, 2503–2509.), Ag_2S (Yejun Zhang, Yongsheng Liu, Chunyan Li, Xueyuan Chen, and Qiangbin Wang. Controlled synthesis of Ag_2S quantum dots and experimental determination of the exciton Bohr radius. *J. Phys. Chem. C.* 2014, 118, 4918–4923), fluorescent nanodiamonds (*Physical Sciences—Physics,* Chi-Cheng Fu et al. Characterization and application of single fluorescent nanodiamonds as cellular biomarkers. *PNAS,* 2007, 104(3), 727–732; published ahead of print January 9, 2007, doi:10.1073/pnas.0605409104.), QDs (S. Shen and Q. Wang. Rational tuning the optical properties of metal sulfide nanocrystals and their applications. *Chem. Mater.* 2013, DOI:10.1021/cm302482d.), and new generation of magnetic and luminescent NPs for *in vivo* real-time imaging (L.-M. Lacroix, F. Delpech, C. Nayral, S. Lachaize, Bruno Chaudret Published April 23, 2013. DOI: 10.1098/rsfs.2012.0103.).

techniques have been efficient in removing the unwanted background signal. Such long lifetimes make also Förster resonance energy transfer (FRET)-based biological studies less challenging from the technology (laser sources, detection, and acquisition) perspective (Rajendran et al. 2014).

Anti-Stokes (upconversion) emission under NIR photoexcitation is the second type of optical transitions in lanthanides, which was recognized as one of the major advantages of lanthanide-doped materials in biodetection and bioimaging as compared to conventional fluorescent labels. Typically, 980 nm excitation of Yb^{3+} ions or 808 nm excitation of Nd^{3+} ions sensitizers is used to achieve visible and multicolor emission from activators such as Tm^{3+}, Er^{3+}, Ho^{3+}, or Tb^{3+} ions. Except for water, no organic chromophores absorb at 980 nm (absorption of Yb^{3+} sensitizer) or at 808 nm (absorption of Nd^{3+} sensitizer), thus the ratio of useful signal to background is usually very high. Only water molecules exhibit an absorption band at around 980 nm, which under high photoexcitation densities may bring some risk of local overheating (Figure 8.2). These effects can be, however, diminished by using Nd^{3+} ions as the primary sensitizer, whose absorption cross-section at 808 nm is around five times larger than that of Yb^{3+} at 980 nm, and simultaneously negligible water absorption is observed at this wavelength, which is around 20 times lower than at 980 nm (Figure 8.2; Zhong et al. 2014).

Recently, further new possibilities have been discovered by the nanoengineering of the host materials and development of core@multi-shell formation with independent lanthanide ions doping of every single shell. These possibilities have opened new avenues in terms of increased light penetration depth, limited local overheating, and multicolor emission capability (Xie et al. 2013). Moreover, different upconverting NPs hosts have been studied and comprehensively compared by Haase and Schäfer (2011). However, due to optimal phonon frequency which maximizes upconversion emission intensity and QY, the most up-to-date studied matrix are lanthanide-doped tetrafluorides MBF_4 (M = K,Na,Li B = Y,Gd,Lu). Since nonradiative relaxation mechanisms in fluoride $NaYF_4$ matrices are suppressed in comparison to other hosts, and multiphonon assisted energy upconversion is reasonably high, these type of hosts has been shown to be the most efficient nanocrystalline material for upconversion (Guanying et al. 2009). Moreover, the $NaYF_4$ materials found an extraordinary interest since low temperature synthesis protocols have been developed (Boyer et al. 2007a; Chen and Zhao 2012; Lin et al. 2012, in press; Wang et al. 2011c), which allow easy tuning of material size and morphology (Wang et al. 2010; Ye et al. 2010), facile and reproducible design of core@shell NPs, with simultaneous facile dispersion and stable colloidal solutions formation (Boyer et al. 2007a). The detailed description of the synthesis procedures used for upconverting $NaYF_4$ core@shell NPs have been included in Chapter 6. Furthermore, ligand exchange or other functionalization protocols have been mastered for these materials (Wang and Liu 2009).

Finally, the third type of optical transitions in lanthanide ions, that is, *Stokes emission in NIR* ($\lambda > 1$ μm) is particularly important in the biomedical field due to the negligible spectral overlap with endogenous or exogenous biochromophores. Most of the lanthanides demonstrate Stokes emission with obviously higher quantum efficiency than the upconversion processes. Some of these ions can generate emission in the first (NIR I) or second (NIR II) optical windows of biological cells and tissues. The most prominent rare earth (RE) ions for *in vivo* NIR I and NIR II imaging are Nd^{3+} (emission at 860, 1060, and 1330 nm), Er^{3+} (at 1530 nm), or Ho^{3+} (at 1450 nm) (Wang and Zhang 2014). Most of the studies of bioapplications of NIR emitting lanthanide-doped NPs concerned the crystal matrices of LaF_3 (Rocha et al. 2013, 2014), NdF_3 (Yu et al. 2008), and Y_2O_3:Yb/Er (Kamimura et al. 2011), but most promises are devoted again to $NaYF_4$ materials, for example, $NaGdF_4$:Nd^{3+}@$NaGdF_4$ and $NaYF_4$:Yb/Er/Ho/Tm/Pr@$NaYF_4$core@shell downconverting NPs (Chen et al. 2012a; Naczynski 2013). Unlike LaF_3, the $NaYF_4$ type of fluorides are synthesized in a more predictable manner, that is, the mono-distribution of size, biofunctionalization protocols, ability of making core@shell designs are much better controlled and reproducible.

Inorganic, lanthanide-doped NPs also present exceptional chemical- and photostability, with neither photoblinking nor photobleaching behavior, which is characteristic for organic dyes or quantum dots (QDs). This feature gives the possibility for long observation times under focused laser beam in microscopy setups, which are required for studying time-dependent processes or for tracing biofunctionalized labels circulating within living organisms.

The next interesting possible applications of lanthanide-doped NPs arise also from their unique photophysical properties. Due to the shielding of 4f electrons by filled higher lying energy levels, lanthanide ions show narrow- and multiband absorption and emission lines, with large Stokes shifts. This feature allows for efficient separation of lanthanide emission from the much stronger excitation laser line, and also for multiplexing. Changing the doping ions compositions and relative concentration allows the designing of numerous, easily distinguishable, spectral codes to label multiple biological targets in the same sample, for example, to detect multiple organelles in a single cell or different disease markers in human samples. However, there are some challenges to be overcome when facing multiplex imaging with lanthanide-doped NPs. Since the emission form Er^{3+}, Tm^{3+}, Ho^{3+}, Sm^{3+}, Tb^{3+}, and Eu^{3+} ions overlap partially, the post measurements spectra decomposition usually allows distinguishing between the spectral codes, which is hindered or impossible with broadly emitting QDs or organic dyes. In addition, the emission color of Ln^{3+} ions can be further optimized by engineering the energy transfer pathways by either passive way (varying size (Lim et al. 2010; Schietinger et al. 2009), shape, composition, morphology (Damasco et al. 2014; Zhang et al. 2009), host matrix (Haase and Schäfer

2011; Renero-Lecuna et al. 2011), surface ligands (Niu et al. 2010), admixing optically inactive ions such as K^+, Li^+ (Cheng et al. 2012b), Fe^{3+} (Ramasamy et al. 2013), or replacing Gd^{3+} for Y^{3+} in $NaYF_4$ (Li et al. 2012)). Also, active modification is possible by varying the relative concentration of Ln^{3+} dopants (Dou et al. 2013; Qian and Zhang 2008; Wang and Liu 2008), adding spectrally active ions such as Mn^{2+} (Tian et al. 2012; Wang et al. 2015) or Ce^{3+} (Chen et al. 2009; Gao et al. 2014).

The design of core@shell layer-by-layer structures, with spatial separation of optically active ions, has also been demonstrated to simply cascade multicolor emission (Peng et al. 2015). Beside engineering the size, shape, or composition at the single NP level, the multiplexing can be achieved either by simple homogenous mixing of lanthanide NPs doped with different ions inside bar codes, for example, within single SiO_2 or polyethylene glycol (PEG) beads (Gerver et al. 2012), or by doping lanthanides independently into core and shells parts in ACAS NPs (Zhong et al. 2014). The *optical coding* with lanthanide-doped NPs *in the time domain* have also been predicted (Bednarkiewicz and Strek 2005) and have demonstrated (Lu et al. 2014; Zhao et al. 2013), what opens, up to now, a neglected area of intensity independent sensing applications.

The well-separated lanthanide emission bands give further the opportunity for another type of intensity independent sensing technique—*ratiometric detection*. The multiple upconversion emission spectra spanning over visible and NIR spectral regions allows for feasible selection of the emission bands which will serve as a reference, with the other being modulated proportionally to the concentration of the analytes. The biosensors based on lanthanide luminescence resonance energy transfer (LRET) have been demonstrated for sensing DNA hybridization, enzyme activity, pH, or [Hg] ions concentration. For those sensors construction, one of Ln^{3+} ions upconverted emission bands (e.g., the one at 470 nm of Tm^{3+} emission) overlapped with the absorption of acceptor or sensitive chromophores and could be quenched by either LRET or inner filter effect (IFE). Simultaneously, the other emission bands (the ones at 650 or 800 nm of Tm^{3+} emission) remained unaffected. In other words, only one emission band of lanthanide ions is being indirectly modulated by the changes in the local environment (e.g., by the change of the acceptor absorption spectra upon increase Hg ions concentration), while the other bands serve as a reference for quantitative measurements. The numerous examples of such ratiometric biosensors can be recalled, including pH (Sun et al. 2009), carbon dioxide (Ali et al. 2010), ammonia (Mader and Wolfbeis 2010), mercury (Liu et al. 2011c), glucose (Wang and Li 2007), cyanide anions (Liu et al. 2011b), hydroxyl radicals (Mei et al. 2015), and oxygen (Achatz et al. 2010) sensing. A similar idea was employed by Kang to study ibuprofen (IBU) drug release, where the upconversion emission QY was proportional to the amount of IBU in the system (Kang et al. 2011), forming a platform for drug delivery and drug release monitoring.

8.3 Biomedical Applications of UCNPs

The physical, optical, and chemical properties of lanthanide-doped upconverting nanoparticles (UCNPs), which have been discussed above, gave the advantages for their application as: (i) passive, (ii) modulating, and (iii) active biomedical optical agents (Gnach and Bednarkiewicz 2012). *Passive applications* have regard to the use of lanthanide-doped NPs as active markers in fluorescence microscopy, magnetic resonance imaging (MRI) imaging (owing to Gd^{3+} ions building, e.g., $NaGdF_4$ matrix (Ju et al. 2011; Kumar et al. 2009; Petoral et al. 2009; Xing et al. 2012)), and X-ray imaging or positron emission tomography (PET) imaging (owing to F^{18} isotopes attached to the NPs surface (Liu et al. 2011d; Zhou et al. 2011)). The core@shell $NaLuF_4$ NPs with radioactive, magnetic, X-ray attenuation, and upconversion luminescence properties were synthesized and applied as nanoprobes to image tumor angiogenesis. The $NaLuF_4$ NPs doped with lanthanide active ions (Tm^{3+}, Yb^{3+}) allowed obtaining both intense upconversion luminescence emission and X-ray CT imaging. This luminescent core was coated with $NaGdF_4$ shell doped with radioactive $153Sm^{3+}$ ions, while the Gd^{3+} ions have been used in MRI. In addition, as the $153Sm^{3+}$ ions emit gamma radiation, designed multimodal NPs could be used in radiotherapy and as a single-photon emission computed tomography (SPECT) imaging probe. In addition, multimodal imaging of pairs (Kumar et al. 2009; Xia et al. 2011, 2014; Yin et al. 2012; Zhou et al. 2010) or three different methods have been proposed (Xia et al. 2012), with the cumulating hexamodal imaging with porphyrin–phospholipid (PoP)-coated upconverting NPs, which combined computer and positron emission tomography (PET), upconversion and fluorescence, Cerenkov luminescence, and finally photoacoustic (PA) imaging (Rieffel et al. 2015). The fluorescence contrasts of upconverting NPs labels were so sensitive, that as few as 10 stem cells could be detected *in vivo* (Wang et al. 2012) for at least 1 week after delivery (Idris et al. 2009).

The *modulation* type of applications, requires predictable change of the luminescent properties of upconverting NPs by external factors (e.g., bioresponsive molecules, physical or chemical environment) with either direct, indirect, and hybrid sensing methodology (Gnach and Bednarkiewicz 2012b). Direct sensing includes the straightforward impact of the chemical environment on spectral properties of lanthanide-doped NPs. Examples include H_2O_2 sensing (changing the valence state of $Eu^{3+} \rightarrow Eu^{2+}$) or temperature sensing, which modifies the emission branching ratio of Stark components in Er^{3+} or Nd^{3+} ions (Jaque and Vetrone 2012; Vetrone et al. 2010; Wawrzynczyk et al. 2012). Additional functionalities in concentration independent ratiometric temperature sensing have been proposed in the core@shell structures of NPs, where both core and shell parts of NPs showed optical response to local temperature changes (Marciniak et al. 2016).

Indirect sensing rely on the other hand on modulating, for example, through the LRET process, luminescent properties of lanthanides by the presence of acceptor chromophores (Saleh et al. 2010; Wang et al. 2005, 2009a) or another type of quenching NPs (Rantanen et al. 2008) in close vicinity to the surface of the lanthanide-doped upconverting NPs. The examples include immuno-reactions (e.g., *E. coli* [Niedbala et al. 2001] and human chorionic gonadotropin [Hampl et al. 2001] detection), DNA hybridization (van de Rijke et al. 2001; Ylihärsilä et al. 2011), or enzyme cutting proteins (Rantanen et al. 2008). Finally, hybrid biosensing rely on the so-called inner filter effect (IFE), where light from one of the Ln^{3+} ions emission bands (usually the short-wavelength one) is selectively absorbed by organic dyes being responsive to pH (Sun et al. 2009), carbon dioxide (Ali et al. 2010), ammonia (Mader and Wolfbeis 2010), mercury (Liu Peng et al. 2011c), glucose (Wang and Li 2007), cyanide anions (Liu et al. 2011b), or oxygen (Achatz et al. 2010). A good and simple example of this IFE is phenol red as pH probe. The absorption band of ammonia sensitive phenol red rises at 475–600 nm for growing pH value. The phenol red is thus suitable to absorb green ($^2H_{11/2} + {}^4S_{3/2} \rightarrow {}^4I_{15/2}$) upconversion emission from Er^{3+} ions in relation to the red one ($^4F_{9/2} \rightarrow {}^4I_{15/2}$), which remains unmodified. These types of sensors expand the well-known behavior of some organic dyes with the additional advantage, from the point of view of biomedical applications, of the ability to excite them in NIR wavelength range.

In *active applications*, the emission or nonradiative transitions in lanthanide ions affect the cells or biological tissues directly. The examples include the thermotherapy (local overheating of cancerous tissues) with upconverting NPs directly (Bednarkiewicz et al. 2011a,b) or with UCNPs bound to plasmonic or magnetic NPs (Cheng et al. 2011, 2012a; Qian et al. 2010), drugs delivery (Wang et al. 2011a), or upconversion triggered photodynamic therapy (Chatterjee and Yong 2008; Cui et al. 2012; Qian et al. 2009; Shan et al. 2011; Zhao et al. 2012; Zhou et al. 2012). Due to the cross-relaxation (CR) processes between energy levels in one or more types of lanthanide ions inside highly doped NPs, the excitation energy can be effectively turned into heat generation. However, due to the low absorption of light by lanthanides, this effect is often increased by the decoration of lanthanide-doped NPs with gold or iron nanostructures. As for the drug delivery applications, there is no possibility to include the organic drugs within the crystal matrix itself, however, the surface of NPs can be covered with mesoporous SiO_2 or polymeric shells, which may carry and release drugs. It have been already shown, that the mesoporous silica shell covering the $LaF_3:Yb^{3+},Er^{3+}$ (Yang et al. 2010) or $NaYF_4:Yb^{3+},Er^{3+}$ (Kang et al. 2011) NPs can be effectively loaded with the IBU and further used in drug delivery and release applications.

Going one step further with possible applications of lanthanide-doped NPs is the combination of their passive and active features in designing "smart" theranostics agents. The NIR initiated photodynamic therapy was presented in hybrid NPs, which include upconverting $NaYF_4$ NPs as luminescent bioprobes in the core and simultaneously photosensitizing molecules (e.g., Chlorin e6) either covalently attached to the surface of NPs, or embedded into mesoporous SiO_2/PEG shell. The theranostic applications of lanthanide-doped NPs are so promising, because of the capability to combine several functional features within single NP. In most of these applications, NPs act as light converter from NIR to visible or NIR spectral region, which basically decreases the signal to noise ratio, increases penetration depth of excitation light, and thus facilitates photo-biosensing and phototherapies into heterogeneous tissues. The briefly reviewed types of upconverting lanthanide-doped NPs applications above have been summarized in Figure 8.3, with the distinction between active, passive, and modulation applications.

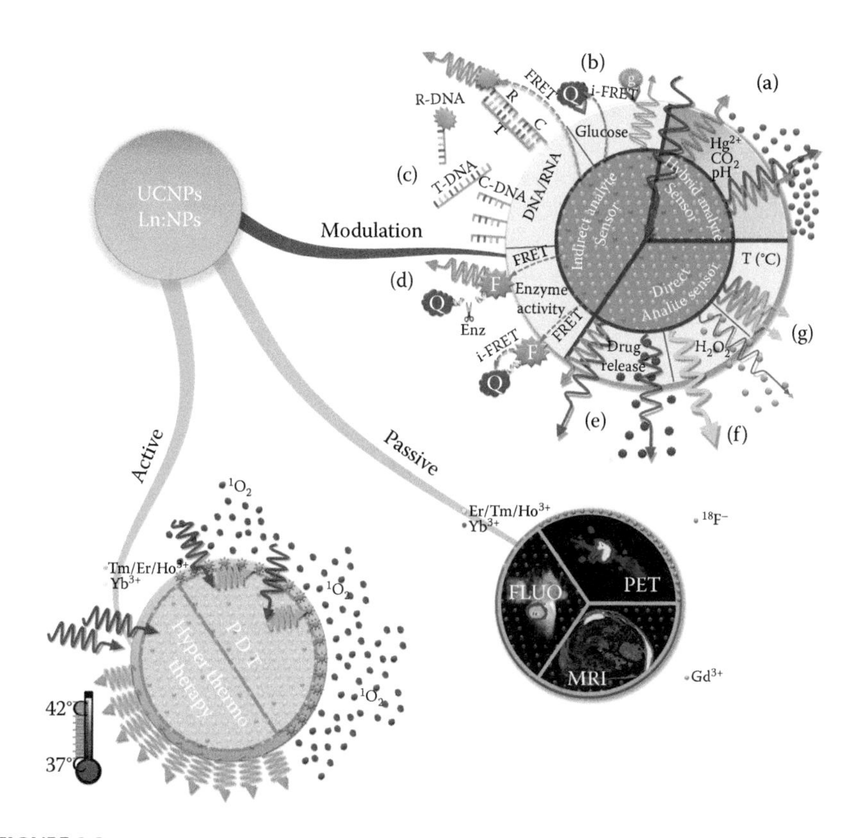

FIGURE 8.3
Illustration of reported applications to date of lanthanide-doped (upconverting) NPs in bio-related sciences. Active, passive, and modulation applications of upconverting NPs relate to the direct influence of upconverted emission on surrounding, often cancerous tissue, active markers type contrast agents (PET, MRI, fluorescence, etc.), and changing the spectral features, including the emission color or luminescence kinetics, of upconverting NPs by different analytes (direct analyte sensors, e.g., temperature (g), drug (e), H_2O_2 (f)), environment (e.g., CO_2, pH, etc. (a)) responsive biomolecules (hybrid sensors) or FRET-based (enzymatic (d), nucleic acids hybridization (c), quenching (b), etc.) sensors, respectively. (Reprinted from Lanthanide-doped up-converting nanoparticles: Merits and challenges, 7, Gnach, A. and A. Bednarkiewicz, *Nano Today*, 532–563, Copyright 2012, with permission from Elsevier.)

8.4 Lanthanide-Doped NPs: The Issues and Solutions

Despite many enthusiastically reported advantages, lanthanide-doped materials display some disadvantages and ongoing research in this field is continuously required. First of all, lanthanide-doped upconverting NPs can act, in the majority of cases, as lantern only, which means, they do not exhibit any type of biological specificity. Moreover, owing to their physical properties, that is, the intra-configurational character of 4f–4f electronic transitions, they display only limited self-functionality, which means they are hardly responsive to the local environment, and beside susceptibility to temperature changes and acceptor/quencher presence at close proximity, lanthanide-doped NPs cannot themselves act as biosensors. For these two reasons, an additional functional (bio)layer is usually deposited at the surface of NPs, which will be discussed in further chapters. Another drawback of lanthanide-doped upconverting NPs is their relatively low QY owing to low absorption and emission cross sections, which result from forbidden f–f optical transition. The QY, although much lower than in organic dyes or QDs, is not prohibitive for bioapplications and ultrasensitive immunoassays, and a few cancer cells detection have been demonstrated *in vivo* with these luminescent labels (Zhou et al. 2012). This low efficiency is also the outcome of large ratio of surface to volume which is found in nanomaterials. Significant amount of doping ions are located close to surface of the NP and their excited states are thus susceptible to quenching by crystal structure defects, local chemical environment, ligands, and the solvent. As a consequence, the upconversion QY drops from 3% in bulk materials, down to 0.005%–0.3% for 10–100 nm β-$NaYF_4$:2%Er^{3+}, 20%Yb^{3+} NPs (Boyer and van Veggel 2010; Suyver et al. 2006). There are also some indications for luminescent ions leakage from the volume of the NPs over time, which ultimately, in a very long-time scale, may lead to further change of luminescent properties, including gradual decrease of QYs. These drawbacks may be, however, diminished by using core@shell lanthanide-doped NPs architectures. It has been shown, that c.a. 4 nm thick passive (undoped) shell is suitable to passivate the surface and diminish the unfavorable effect of surface defects, solvent, and ligands quenching (Prorok et al. 2014b). Since the upconversion QY is the most closely related to the suitability of these materials for bioapplications, a number of methods are sought to improve the upconversion emission intensity in NPs. The currently explored "rescue" methods include (i) surface passivation by homo- (e.g., $NaYF_4$ shell over $NaYF_4$ core) or hetero- (e.g., CaF_2 shell over $NaYF_4$ core) shells, (ii) co-doping the NPs with passive mono- or tetravalent (e.g., K^+, Li^+, Hf^{4+}, or Ti^{4+}) ions to disrupt Ln^{3+} cluster formation and change the crystal field strength, (iii) co-doping the NPs with additional active ions (e.g., Nd^{3+} and Mn^{2+}) suitable to increase the absorption cross section and to change the excitation or emission schemes, (iv) interaction of lanthanide ions with plasmonic features of metal NPs, toward absorption and emission enhancement, (v) increasing concentration of sensitizer, that is, Yb^{3+}/Nd^{3+}, or activator ions, (vi) onion-like core-multiple shell structure (to diminish more equal activators distribution and limit concentration quenching), and finally (vii) proper host selection. The evidence of overall upconverted emission enhancement on the strategies mentioned above is discussed here in more detail.

8.4.1 Crystallographic Structure Tuning

The selection of host crystal matrix is essential for optimization of optical properties of upconverting NPs, such as high QY and controllable emission profile. For efficient upconversion generation, it is very important to prevent bridging of the lanthanide ions excited states by phonon modes of the matrix, which can assist in nonradiative depopulation of emitting levels. On the other hand, the upconversion process requires phonon assistance in some cases, thus the proper balance between the phonon modes available in the crystal host and the excitation intensity should be provided. Several types of host materials have been studied as matrices for upconverting lanthanide ions pairs (e.g., Er^{3+}/Yb^{3+}, Tm^{3+}/Yb^{3+}, and Ho^{3+}/Yb^{3+}), including different types of glasses, phosphides, garnets, oxides, and fluorides (Güdel and Pollnau 2000; Haase and Schäfer 2011), however, up to date, the hexagonal β-$NaYF_4$ crystal matrix is regarded as the most efficient host for upconverting lanthanide ions. This is due to the very low energy of phonons characteristic for this crystal lattice (~350 cm^{-1}), when compared to, for example, YVO_4 material (~600 cm^{-1}) or phosphate glass (~1200 cm^{-1}) (Gorris and Wolfbeis 2013). In addition, the β-$NaYF_4$ matrix has three independent crystal sites (Mai et al. 2007a), energy donor (Yb^{3+}) and acceptor (Er^{3+}, Tm^{3+}, or Ho^{3+}) ions are thus located in separate symmetry sites, which multiplies the number of possible energy transfer processes and increases the efficiency of the upconversion process. Unfortunately, the $NaYF_4$ matrix can exist in two crystallographic structures, cubic α-$NaYF_4$ and hexagonal β-$NaYF_4$, and the upconversion efficiency is much lower for α-$NaYF_4$ one. Numerous papers have reported the increase of upconversion emission intensity and luminescence lifetimes' elongation upon $\alpha \rightarrow \beta$ transformation of $NaYF_4$ matrix, and readers are referred to the latest review on this topic by Kar et al. (2015) and the references therein. This is why the particular interest should be limited to the β-$NaYF_4$ structure. The transformation between cubic and hexagonal crystallographic phases of $NaYF_4$ NPs can be achieved either through increasing the reaction time and temperature (Mai et al. 2007b), or by incorporating doping ions (Chen and Wang 2013). Simply by replacing the Y^{3+} with Gd^{3+} ions, with the same reaction conditions, it is possible to obtain cubic to hexagonal phase conversion (Wang et al. 2010). This fact was explained based on density functional theory calculations, which revealed the increase in formation energy per atom when the Y^{3+}ions were replaced by Gd^{3+} ions. Further experiments showed that addition of other lanthanide ions with ionic radius larger then Yb^{3+} has similar effect as that of Gd^{3+} ions. Recently, cubic to hexagonal crystal $NaYF_4$ structure transformation was achieved with the assistance of phosphate salts, in which additional F^- ions acted as an accelerator of the crystal growth process (Yin et al. 2014). Another approach to tune the crystal phase of $NaYF_4$ matrix is by doping with passive ions, either monovalent (e.g., Li^+ or K^+) or tetravalent (e.g., Hf^{4+}) (Dou and Zhang 2011). When small amount of Li^+ ions is doped into the crystal matrix, they may locate in either Na^+ lattice sites or in interstitial positions, further induce the breakdown of crystallographic site symmetry and phase transformation (Lin et al. 2015). However, by increasing the amount of Li^+ ions up to 50%, the NPs crystal structure is changed back from hexagonal to cubic. The Li^+ ions can enhance the upconversion luminescence, nevertheless the proper amount of this type of impurity doping should be always carefully optimized. Similar effect was observed for $NaYF_4$ NPs intentionally doped with Hf^{4+} (Huang et al. 2015) or Ti^{4+} (Chen et al. 2011a) ions. Interestingly, the coordination environment of lanthanide ions in cubic $NaYF_4$ matrix were varied stronger by Hf^{4+} doping, than in the hexagonal phase, which resulted in much larger upconversion emission intensity enhancement (Huang et al. 2015).

8.4.2 Surface Passivation with (Un) Doped Crystalline or Amorphous Shells

The selection of crystal matrix with low phonon energy and suitable coordination symmetry for lanthanide ions is not the only requirement for intense upconversion emission generation, since for any nanosized particles surface-related defects and vibrational modes of ligands and solvent molecules can effectively quench the luminescence. Current synthetic approaches allow for the preparation of lanthanide-doped β-$NaYF_4$ NPs with different sizes and shapes, and the colloidal stability of NPs is ensured by the surface attached ligands, which also stabilize the crystal growth during the synthesis process. However, as the size of obtained NPs is reduced down to sub-10 nm range, the emission is strongly decreased. There are three main reasons of observed upconversion diminishing, which can arise from: (i) vibrational modes of ligands and solvents molecules, (ii) optical traps, and (iii) surface defects, which cause the so-called surface quenching effect. The available vibrational modes depend on the type of ligand molecules used as capping agents and also on the type of solvent in which the NPs are dispersed. From the view of any bio-related applications, NPs should be dispersed in a nontoxic environment, which in most cases is water. Unfortunately, it has been shown that water molecules can effectively depopulate the excited $^7F_{5/2}$ energy state in Yb^{3+} ions, and thus quench the upconverted luminescence. From the crystallographic analysis, most defects in the crystal structure are located near the surface of NPs, and as the surface to volume ratio increases for smaller NPs, the defects and optical surface traps created during synthesis start to play a key role in decreasing the observed upconverted luminescence. For the applications of lanthanide-doped NPs as labels in standard imaging devices, it is necessary to compress the decrease in luminescence intensity caused by the high surface to volume ratio, and this can be achieved by surface passivation with either crystalline (e.g., $NaYF_4$ or CaF) or amorphous (e.g., SiO_2 or PEG) shells. The synthesis of core@shell nano crystal structures makes lanthanide ions doped core NPs covered with crystalline shell composed from either isostructural $NaYF_4$ or other type of crystal structure, for example, CaF. One of the first demonstrations of upconverted luminescence enhancement by undoped shell growth on $NaYF_4$ NPs was reported by Mai et al. (2007a). The authors have observed increased green Er^{3+} upconverted emission intensity, with simultaneous reduced nonradiative decay pathways upon covering the $NaYF_4$:Er,Yb NPs with 2.5 nm thick $NaYF_4$ shell. The 300% increase in total QY after the growth of an undoped $NaYF_4$ shell on β-$NaYF_4$:Yb^{3+}, Er^{3+} nanocrystals was also reported by Boyer and van Veggel (2010). They showed that the core@shell nanocrystals of size c.a. 30 nm had the same QY as that of 100 nm bare nanocrystals, which clearly demonstrated the advantages of the core@shell structures (Boyer and van Veggel 2010). On the other hand, we have shown that similar protective properties, and upconversion emission enhancement up to 40 times, could be achieved with CaF_2 shell (Prorok et al. 2014a). Up to date, the obvious enhancement of Er^{3+}, Tm^{3+}, and Ho^{3+} ions upconverted emission, upon covering the NPs with undoped shell have been confirmed for NPs with different crystal structures, morphologies, and sizes proving the destructive role of surface states on upconverted emission. Chen et al. (2015b) have recently published a tutorial review regarding the enhancement of upconversion emission intensity in core@shell structures. Besides using the crystalline shell formation, the surface located optically active ions could be separated from solvent and ligands molecules by covering the NPs with amorphous silica or polymeric shell. In addition, this kind of surface covering assures the NPs hydrophilization, and further allows for functionalization with biologically important molecules. The upconverting $NaYF_4$ NPs covered with SiO_2 and PEG shells could be, for example, functionalized with photodegradable ortho-nitrobenzyl alcohol derivate and doxorubicin for controlled drug release upon 976 nm laser irradiation (Alonso-Cristobal et al. 2015). The mesoporous silica shell was also used as container for nitroxide radicals, which in combination with upconverted emission from $NaYF_4$ core NPs resulted in dual (i) luminescence and (ii) magnetic resonance modality imaging (Chen et al. 2015a). The formation of noncrystalline, polymeric shells on $NaYF_4$ NPs was shown to diminish the water quenching effect and to enhance the possibility for bioconjugation of synthesized NPs. Unfortunately, the presence of OH- groups on PEG coating can promote the nonradiative depopulation of lantanide ions exited levels.

8.4.3 Plasmonic and Photonic Effects

The increase of upconverted emission intensity can be further enhanced by the engineered coupling of lanthanide ions emission with surface plasmon resonance modes of metallic NPs. In the proximity of noble metal NPs, the electric field of both incident light and luminescence from the NPs can be greatly modified. In general, a fluorophore in its excited state has the properties of an oscillating dipole, and can induce oscillation of the electrons in metal NPs. This interaction is bilateral, and induced oscillation in the metal NP can alter the fluorophore luminescence (Lakowicz 2006). The metal NPs can cause not only increased rates of excitation due to the more concentrated electric field around such a NP, but also increased rates of radiative decay. This effect can allow for decreased incident light intensities, increased photostability, increased sensitivity, and decreased background. Chen et al. (2015c) recently observed huge, up to 180 times, enhancement of Tm^{3+} ions upconverted emission intensity after coupling the $NaYF_4$:Tm^{3+},Yb^{3+} NPs with thin film of Au–Ag alloy composite "islands." Beside the increase in overall upconverted emission intensity, the plasmonic structures can also modify the energy transfer rates between optically active ions and provide prominent nonlinear gain (Wang et al. 2015). It has been further shown theoretically, that the QYs of upconverted emission can be substantially increased in the neighborhood of metallic nanostructures (Elhalawany et al. 2014). The experimental approaches have also proven the fact that changing the distance between the metallic and the lanthanide-doped NPs can cause enhancement or quenching of the luminescence, concentrate the incident light field, and increase the radiative decay rate. Often, amorphous silica shell is used as a spacer, and in such a manner both the protection of NPs surface and luminescence enhancement are provided (Shen et al. 2013). An extensive review on upconversion enhancement by metallic NPs have been recently published by Park et al. (2015), and interested readers are referred to this particular paper. Further increase in upconverted emission intensities could be obtained by combining metal NPs plasmons and photonic crystal effect. Liao et al. (2014) showed synergistic effect of opal/Ag hybrid substrates, which enhanced the Er^{3+} ions upconverted emission intensity up to 10 times. Combination of the unique optical features of lanthanide-doped NPs, with the opportunity to modify output luminescence rendered by the metal NPs, enabled a broad range of possible applications. For example, Zhang et al. (2009) reported pH-induced reversible luminescence switching with controlled assembly of gold NPs. The presented system enabled construction of an optical pH sensor. The core@shell structures of $NaYF_4$:Yb^{3+}, Er^{3+}@Ag NPs, showed theranostic features enabling simultaneous imaging and photothermal therapy (Dong et al. 2011a). Very recently, lanthanide-doped NPs decorated with gold nanorods (Au NRs), whose surface was modified with 2-thiouracil, were used for nonenzymatic detection of uric acid, showing a detection limit as low as 1 pM (Kannan et al. 2013). Chen et al. (2013) showed sensitive and selective detection of thrombin based on mutual interaction between $NaYF_4$:Yb^{3+}, Er^{3+} NPs and Au NRs. A similar system was used by Wang et al. (2009a) to develop a sandwich-type immunoassay for the detection of goat antihuman immunoglobulin G. In general, by construction of hybrid NPs systems, composed of lanthanide-doped NPs + amorphous spacer + metallic NPs, it is possible to enhance the lanthanide luminescence, and shorten the luminescence lifetime, and thus provide longer and better optical stability of studied systems. However, it is necessary to carefully optimize the structure properties by selection of proper distance between optically active species, type, size, and shape of metallic NPs. This is why such a strategy to enhance lanthanide luminescence can be troublesome in a complex biological environment.

8.4.4 Increasing Active Ions Concentration

Besides engineering the crystal structure, surface, or close surrounding of lanthanide-doped NPs, the increase in upconverted emission intensity can be achieved by simple doping NPs with a higher concentration of optically active ions to raise the absorption cross sections. Prasad et al. (Chen et al. 2010) demonstrated such an effective increase of absorption cross section by high Yb^{3+} doping level at NIR through the use of stoichiometric $NaYbF_4$ nanocrystals. Lei et al. (2014) have performed detailed studies for sandwich-like core@shell@shell structures, where different amounts of Yb^{3+} ions were doped in separate shell layers. The authors found, that appropriate Yb^{3+} ions concentration should be introduced into the shell or transition layer to prevent back energy transfer from Er^{3+} ions in the core to Yb^{3+} in the shell to achieve high upconversion efficiency. Similar effect was observed by Qiu et al. (2014) who reported approximately 240 times increase in Tm^{3+} ions upconversion emission intensity in $NaYF_4$:30%Yb^{3+},0.5% Tm^{3+}@$NaYbF_4$@$NaYF_4$ NPs when compared to simple core ones. While it is possible to increase the concentration of harvesting energy Yb^{3+} ions up to 100%, the emitting activator ions concentrations usually remain at low <1 at % level to maintain optimal balance between improved emission intensity and deteriorative luminescence quenching. However, writing in Nature Nanotechnology Letters D. Jin et al., (Zhao et al. 2013) as well as Gargas et al. (2014) have demonstrated that concentration quenching does not necessarily need to be the limiting factor. Through careful and eye-opened analysis, it has been shown that under high excitation intensity, new doping rules have to be developed since heavy doping, otherwise prone to parasitic CRs, under intense excitation, brings significant improvements in the brightness of upconverting NPs emission. Due to the altering of population densities of discreet and long living excited energy levels of lanthanides, single upconverting NPs sensitivity have been experimentally demonstrated and theoretically predicted. These new findings became possible thanks to approximately 70-fold enhancement of upconversion intensity owing to lanthanide doping at the unprecedentedly high 7% of Tm^{3+} or 20% of Er^{3+} levels, respectively, under high 10^6 W/cm^2 photo excitation density in confocal raster scanned imaging configuration. What has been shown by the two groups change the perspective and expel the canonical approach to lanthanide-doped upconverting NPs designs.

8.5 Core–Shell UCNPs: Novel Possibilities

Not all of the approaches mentioned in the previous chapter have been successful in significant upconversion emission intensity and QYs enhancement. For example, the traditional synthesis of NPs does not allow for controlling active ions segregation and clustering, moreover, concentration quenching, owing to parasitic CR between respective energy levels (Figure 8.1 CR) prevents doping the upconverting NPs with activators above approximately 2%. Moreover, although upconversion emission sensitization by Yb^{3+} has been considered a great advantage over conventional fluorescent labels, the significant absorption coefficient of water at the respective excitation wavelength (980 nm) is prohibitive for prolonged *in vivo* experiments due to local overheating occurring at higher excitation intensities. Fortunately, completely new possibilities have been opened with recent developments in ACAS NPs designs. The stepwise covering of the NPs core with crystalline layers, what is the most important, allows for predictable spatial distribution of optically active ions. In addition, the protection form quenching effect arising from interaction between emissive ions and ligands and solvent molecules becomes possible. The main advantages of using ACAS engineered NPs include:

1. Increased functionality and biocompatibility
2. Upconversion enhancement through the increase of absorption cross section
3. Capability to improve (homogenize) activators distribution
4. Capability to shift the excitation wavelength

The mentioned features, discovered by the synthesis procedure optimization for core@shell structures, opened new perspectives for upconversion emission enhancement, and will be discussed below in greater detail. It should be stressed once again, that depending on the desired functionality, it is possible to select the type, crystal structure, or the thickness of the outer layer.

8.5.1 Shell Increases Functionality and Biocompatibility

In order to ensure the compatibility of the synthesized NPs with the biological environment, the shell should be composed of nontoxic, and most preferably, organic molecules. Many polymeric shells, such as PEG, poly-acrylic acid (PAA), or $mSiO_2$ (mesoporous silica), have shown the increase in functionality and biocompatibility of coated NPs (Liu et al. 2012). First of all, such surface coatings form a biocompatible layer, which allows to intentionally control circulation time within bloodstream and clearance *in vivo* (Kumar et al. 2010; Xiong et al. 2010), with simultaneous protein corona formation (Budijono et al. 2010; Zhang et al. 2015a). The biological studies performed for upconverting NPs functionalized with polymeric shells showed their potential as optical labels for long-term targeted imaging and therapy studies *in vivo.* This was, among others, due to the increased surface charge of functionalized NPs. Moreover, such biocompatible layers, allow for further easier biofunctionalization, since polyelectrolyte molecules are reached in functional groups (e.g., –COOH, $-NH_2$, and –SH), which form the basis to attach bioresponsive molecules, such as antibodies or DNA, which in the next step enable active targeting. Recently, Raj et al. (2015) and Alonso-Cristobal et al. (2015) have independently shown the application of $NaYF_4$ upconverting NPs with amorphous polymeric shells surface covers for detecting the target DNA at the picomolar concentrations. Another interesting functionality, which could be provided by polymeric shells is the possibility to carry an active cargo (e.g., hydrophobic cytostatic drugs or toxins), which can be slowly released proportionally to the local pH of the environment, temperature, or the presence of the magnetic field (Yang et al. 2015). The other method reported for controlled drug release is optical activation with light of proper wavelength and intensity (Li et al. 2013; Zhang et al. 2015b). Zhang et al. (2015b) showed theranostic applications of Tm^{3+}/Yb^{3+} co-doped $NaYF_4$ NPs covered with polymeric shell loaded with anticancer drug. Upon NIR excitation, the slow release of cisplatin and the death of HeLa cells was observed, while the upconverted emission from Tm^{3+} ions could be used for bioimaging. Another good example of upconverting ACAS NPs application is NIR-activated photodynamic therapy (Wang et al. 2013). In this approach, the photosensitizers are again loaded inside the polymeric shell, the NIR-excited upconversion emission is then reabsorbed or nonradiatively transferred to drugs, which when excited can effectively generate the reactive oxygen species for anticancer treatment. These hybrid structures, that is, upconverting NPs + photosensitizers, become activated with UV–Vis upconverted emission light upon NIR photoexcitation, which enable an increase in the effective penetration depth of activation light and non-invasive deep tumor treatment. Ai et al. (2015) have recently shown how the 808 nm activated $NaYbF_4$:Nd@$NaGdF_4$:Yb/Er@$NaGdF_4$core@shell@shell NPs covered with Chlorin e6-loaded PEG shell can be used for simultaneous photodynamic cancer treatment and fluorescence imaging. The polymeric shells may also contain organic molecules, which respond to changes in local environment (concentration of ions, pH, etc.) by changing their absorption spectra. Since lanthanide-doped NPs usually emit light at a few discreet and well-separated emission lines, only part of this emission is absorbed by the molecules proportional to external stimulus, thus ratiometric (self-calibration) biosensing could be easily implemented.

8.5.2 Shell Protects NPs from Surface Quenching

As already mentioned in this chapter, the undoped shell has been proven to significantly enhance the upconverted emission by the decrease of the quenching effect arising from solvent and ligands molecules. The first demonstration of luminescence enhancement by shell growth on lanthanide-doped NPs was observed by Haase et al. in cerium phosphate ($CePO_4$) NPs singly doped with terbium ions and coated with a lanthanum phosphate ($LaPO_4$) shell. The authors studied core@shell NPs (8–10 nm in size) with QY of Tb^{3+} emission reaching 70% as compared to 86% for bulk materials. To date, the most widely investigated upconversion core@shell structure is the $NaREF_4$ core coated with $NaREF_4$ shell, where RE stands for lanthanide and yttrium ions. In case of $NaYF_4$ NPs, the first demonstration of upconversion emission intensity enhancement was shown by Yi and Chow (2006). In hexagonal $NaYF_4$:Yb^{3+}, Er^{3+} and $NaYF_4$:Yb^{3+}, Tm^{3+} NPs, they reported a 7- and 29-times enhancement of the emission intensity after coating with undoped $NaYF_4$ shell, respectively (Yi and Chow 2006). The 300% increase in total QY after growth of an undoped $NaYF_4$ shell on β-$NaYF_4$:Yb^{3+}, Er^{3+} NPs was also reported by Boyer and van Veggel (2010). They showed that the core@shell NPs of size c.a. 30 nm had the same QY as that of 100 nm bare nanocrystals, which clearly demonstrates the advantages of the core@shell structure (Boyer and van Veggel 2010). In addition, improvement of upconverting NPs performance could be obtained by intentional increase of sensitizer ions concentration. By doping the outer part of NP with additional donor ions, that is, Yb^{3+} or Nd^{3+}, the effective absorption cross section of excitation wavelength can be increased. This ACAS approach has been already investigated in Er^{3+}/Yb^{3+} co-doped upconverting NPs, as well as in Ce^{3+}/Tb^{3+} downconverting ones. Huang et al. (2015) have recently reported up to 990-fold increase in Ho^{3+} ions upconversion emission enhancement with the ACAS NPs design. The additional Nd^{3+} and Yb^{3+} sensitizer ions in the shell part of NPs provided the increased NIR absorption and efficient energy transfer from Nd^{3+} primary sensitizers to Ho^{3+} activators via Yb^{3+} bridging sensitizers. A similar effect was observed for core@shell NPs with only additional Yb^{3+} donor ions present in the shell (Lei et al. 2014; Qiu et al. 2014). However, to prevent the quenching effect produced by random excitation energy migration between sensitizer Yb^{3+} ions within the active shell to the deactivation sites on the outer surface, and thus decrease upconverted emission excitation efficiency, careful optimization of Yb^{3+} ions concentration inside the core and shell parts of NPs should be performed. Qiu et al. (2014) have also emphasized the advantages of sandwich-like ACAS@passive-shell structures, where the intermediate layer could be doped with 100% of light harvesting Yb^{3+} ions. The spatial separation of Yb^{3+} ions from both emissive

Tm^{3+} ions present in the core and surface defects and ligand molecules, resulted in approximately 240 or approximately 11 times enhancement of upconversion emission when compared to core NPs or to active-core@passive-shell NPs, respectively. Despite the fact that doping upconverting NPs with high concentration of sensitizers Yb^{3+} or Nd^{3+} ions can significantly increase NIR light absorption cross section and thus overall emission intensity, it can also influence the crystal structure and morphology of the NPs (Wang et al. 2010). This effect was more pronounced for Nd^{3+} ions with larger ionic radius and dipole polarizability, thus for any multiple shell NPs design, the careful optimization of active ions doping levels should be performed, in order to balance the emission intensity, absorption cross sections and crystal structure, and surface-related quenching. The example structure of the sandwich-like core@shell NPs is presented in Figure 8.4. In simple core NPs (Figure 8.4a), the upconverted emission is quenched by surface defects, ligands and solvents molecules. By the addition of either Yb^{3+}-doped or undoped shell (Figure 8.4b and c), the emission quenching can be diminished, however, the excited state of donor ions is still susceptible to vibrational modes of surface covering molecules. The highest upconversion emission intensity and QYs enhancement was obtained to date by covering the ACAS NPs with another undoped passive shell (Figure 8.4d).

Depending on the design of chemical architecture of upconverting NPs, different scenarios may be expected. For the passive shell over the active core (Figure 8.4c; Wang et al. 2012b; Zhang et al. 2012), the surface quenching states are passivated with undoped shell, but further upconversion enhancement may be expected by doping the shell with sensitizing ions (Vetrone et al. 2009), which shall increase the absorption cross section (Figure 8.4b). Unfortunately, Yb^{3+} ions can effectively remove the excitation energy from the shell and the core parts of NPs through the migration of energy over the Yb^{3+}–Yb^{3+} network, leading to quenching, especially for growing concentrations of Yb^{3+} ions sensitizers. The ultimate solution could be the passivation of such ACAS NPs with one another passive shell (Figure 8.4d). Such design assures an increased absorption cross section and diminishes surface quenching (Qiu et al. 2014). Unfortunately, while such a solution is good option for the application of upconverting NPs as a luminescent label, the use of such NPs may hinder their applicability for FRET bioassays since the relative distance between luminescent lanthanide ions and acceptor molecules is increased.

8.5.3 Shells Allow the Optimization of Activator Distribution

A recent approach for emission enhancement, which became available with the reproducible synthesis of multi-shell NPs, is the spatial separation of different lanthanide ions. Due to the ladder-like energy levels for lanthanide ions, these levels are located near each other and multiple CR processes between one or several types of active ions can occur (Wang

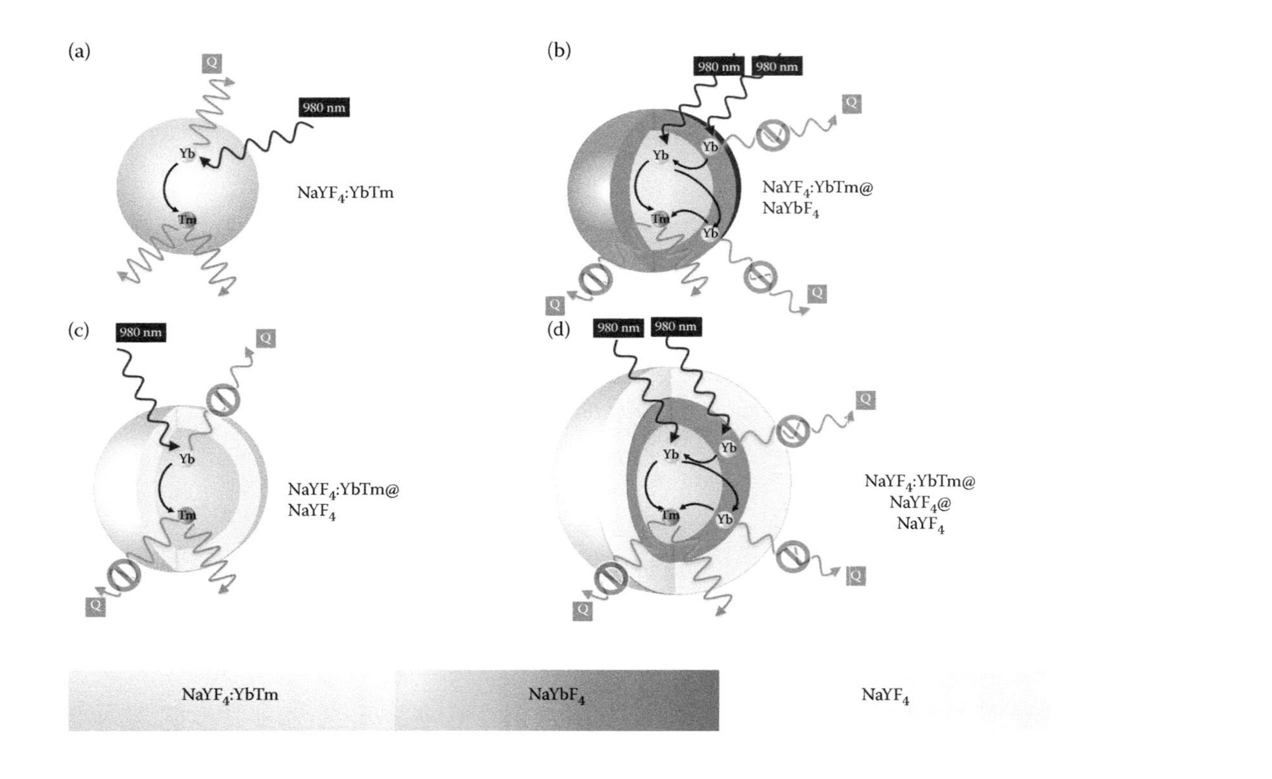

FIGURE 8.4
Illustration of the luminescence quenching mechanism in (a) the core NPs and (b) the active-core/active-shell architecture. The reader can find a similar example in the work of F. Vetrone et al. (2009). (c) The luminescence enhancement mechanism by diminishing surface quenching in the active-core/inert shell NPs as, for example, in the Wang et al. article (Wang, Y.-F. et al. 2012a). (d) The most advanced structure active-core/active-shell/inert shell NPs studied in work of Qiu et al. (2014). The last architecture combines an advantage of controlled spatial separation of ions, obtained by active-core/active-shell structure, with the surface protection against quenching via the passive shell preventing excited ions interaction with environment.

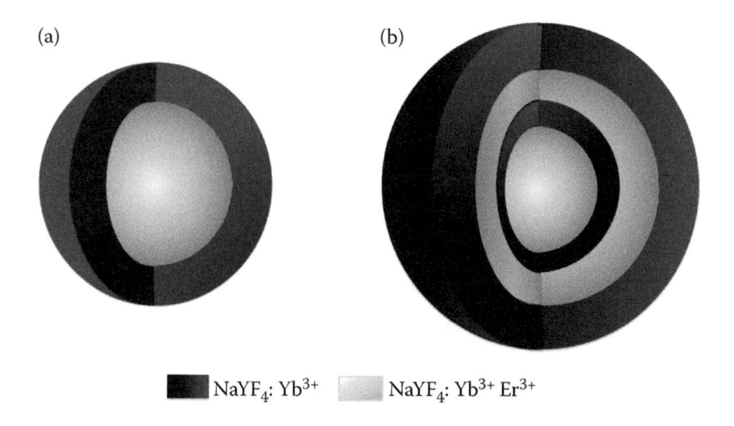

FIGURE 8.5
(a) The classical core/active-shell structure. (b) The designed strategy with emitters doping area spatially separated. It contains four parts: the core ($NaYF_4$:Yb^{3+}, Er^{3+}), the first separating active shell ($NaYF_4$:Yb^{3+}), the second illuminating shell ($NaYF_4$:Yb^{3+}, Er^{3+}), and the final active shell ($NaYF_4$:Yb^{3+}). (Based on *Chem. Commun.*, 2011, 47, 11957–11959.)

and Liu 2009b). Those processes become one of the main reasons of emission intensity decrease when the concentration of Er^{3+}, Tm^{3+}, or Ho^{3+} ions go beyond approximately 2%. We have recently shown (Misiak et al. 2013), that for Tm^{3+}/Yb^{3+} co-doped $NaYF_4$ NPs, the blue and red emission lines are progressively quenched with increasing Tm^{3+} concentration in the course of CR processes, and were the most intense for 0.1% of Tm^{3+} ions. The possible solution, and decrease in those unwanted processes, could be obtained by the spatial separation of optically active ions (Figure 8.5) within the separate crystal layers of NPs (Wang et al. 2011a). Wang et al. (2011a) reported the sublattice-mediated energy migration and excitation energy trapping by the activator ions in core@shell structures, what resulted in overall upconverted emission enhancement. Even more interesting results could be achieved by the additional separation of doped parts of NPs with passive, undoped shells. Shao et al. (2015) introduced a novel core@multilayer NPs design, the inert $NaYF_4$ layers in between upconverting domains in $NaYF_4$:10%Er^{3+}@$NaYF_4$@ $NaYF_4$:10%Ho^{3+}@$NaYF_4$@$NaYF_4$:1%Tm^{3+}@$NaYF_4$ NPs were shown to efficiently suppress the detrimental CR processes between different types of lanthanide ions, thus yielding about twice more efficient upconversion emission than the counterpart $NaYF_4$:10%Er^{3+}@$NaYF_4$:10%Ho^{3+}@$NaYF_4$:1%Tm^{3+}@$NaYF_4$ NPs without the inert $NaYF_4$ layers, and about one order of magnitude higher than the core@shell $NaYF_4$:10%Er^{3+},10%Ho^{3+},1%Tm^{3+}@$NaYF_4$ NPs without spatially isolating the Er^{3+}, Ho^{3+}, and Tm^{3+} ions. In addition, high doping levels of emissive ions allowed also for Stokes excitation in the NIR region between 1100 and 1500 nm.

The CR processes in highly doped NPs are usually perceived as deleterious, and are found to be responsible for upconverted emission quenching.

Nevertheless, Wei et al. (2014) have shown that CRs between activator ions can be successfully used in achieving a pure red upconversion emission color, and also act as an alternative approach for precise upconversion emission color tuning and provide further insight into the upconversion processes mechanism. The authors have synthesized a series of sensitizer-reach $NaYbF_4$ NPs doped with different amounts of Tm^{3+} (1%–20%) and Er^{3+} (1%–100%) ions and based on spectroscopic investigation have shown that with an increase in emissive ions concentration, and thus CR processes, pure red emission for both types of investigated systems could be achieved.

8.5.4 Shell Improves Excitation Schemes

Despite significant advantages of using upconversion luminescence for biomedical applications, the use of approximately 980 nm photoexcitation (adjusted to the maximum of Yb^{3+} sensitizer absorption in $NaYF_4$) is problematic, because water—a major component of living tissues, has its absorption band at this wavelength. The 980 nm excitation energy is not only attenuated by tissue components, and its penetration depth is limited, but also contributes to the considerable elevation of temperature at the illuminated volume. It is important to add, that due to low QY of the up-conversion (UC) in NPs, power density of 980 excitation reaches its maximum allowed levels in order to enable imaging in the upconversion mode. In the context of bioimaging, overheating is an undesired side effect that can affect not only cell viability but also induce tissue injury. As alternative to 980 nm, absorption side bands of Yb^{3+} at around 915, 940, or 1550 nm absorption of Er^{3+} were engaged to improve the penetration depth and reduce the overheating of tissues in the course of prolonged bioimaging. Unfortunately, such an approach leads to a reduced absorption cross section and weakens upconversion intensity.

Recently, a new way to bypass the necessity of using approximately 980 nm photoexcitation was proposed. Neodymium ions are known to possess strong absorption bands in the visible and NIR spectral range. The 780–920 nm spectral region (called the first optical window of the skin) seems to be the most appropriate for imaging tissues and organisms *in vivo*, mostly because of low scattering of light, and lowest absorption coefficient of tissue as compared to other wavelengths. Interestingly, one of the most intense absorption bands of Nd^{3+} is located in this range. This band can be easily excited with powerful and easily commercially available 808 nm laser diodes, which were technologically advanced for pumping Nd^{3+} yttrium aluminum garnet (YAG) lasers.

Although some efforts have been devoted to sensitize upconversion by replacing Yb^{3+} sensitizer with Nd^{3+} ions, the homogenous mixing of activators led to parasitic quenching of activating ions through numerous energy transfer (ET) and cross-relaxation (CR) processes and efficient activator → Nd^{3+} (e.g., $Tm^{3+}/Er^{3+} \rightarrow Nd^{3+}$) back energy transfer, which all led to decreased UC efficiency. The solution of those problems was found as soon as ACAS synthesis were developed and optimized, which allowed the doping of individual core and shell with activator ions (Er^{3+} or Tm^{3+}) and Nd^{3+} primary sensitizer, respectively, with Yb^{3+} ions (present in both core and shell) as an energy migration network. Although Yb^{3+} ions do not absorb excitation energy directly, they remain critically important and responsible for transferring the absorbed energy from Nd^{3+} ions to the activator. Owing to the favorable properties of neodymium ions, such as larger absorption cross section at approximately 800 nm as compared to the one of Yb^{3+} at approximately 980 nm, up to 70% effciency of the $Nd^{3+} \rightarrow Yb^{3+}$ energy transfer with insignificant back energy transfer, and water absorption at 800 nm being around 20-fold lower than at approximately 980 nm excitation, the Nd^{3+} ions were proposed as primary sensitizer to novel $Nd^{3+} \rightarrow Yb^{3+} \rightarrow$ activator EMU (Figure 8.6). It is very important to mention that the suitability of Nd^{3+} sensitization for upconversion had been recognized much earlier, but only the developments of ACAS nanostructures and separating primary sensitizing Nd^{3+} ions from activator (e.g., Tm^{3+}, Er^{3+}) ions enabled the improvement of the Nd^{3+}-sensitized ETU to an acceptable level.

The EMU process is an extremely interesting way to engineer the spectral properties of upconverting NPs. Recently, another version of EMU, exploiting Gd^{3+} ions as migrating ions was proposed, which allowed the achieving of the upconverting emission from Eu^{3+} or Tb^{3+}, which nevertheless is not very efficient for practical applications (Wang et al. 2011b).

The two described EMU mechanisms display many similarities (Figure 8.7). In the harvesting layer (HL), primary sensitizer ions (Yb^{3+} in EMU1 and Nd^{3+} or Yb^{3+} in EMU2) absorb excitation, and transfer that thought migration layer (ML) to the emission layer (EL). Both Gd^{3+}or Yb^{3+} mediating ions display a significant energy gap between ground and excited states, which basically prevents nonradiative de-excitation. The ML splits the HL from the EL (i.e., sensitizing ions from activating ions), which enables either more efficient excitation of ions such as Tb^{3+} or Eu^{3+} in Stokes mode (EMU1) or prevents parasitic activator $\rightarrow$ sensitizer back energy transfer (EMU2). Both processes contain an energy upconversion step, however, in EMU1, this occurs in ML ($Tm^{3+} \rightarrow Gd^{3+}$) before exciting the activators, while in EMU2, the ETU occurs between migrating Yb^{3+} network and conventional activators (Tm^{3+}, Er^{3+}). It is therefore obvious, that due to high ($N \geq 4$) number of NIR photons required to reach high-energy levels of Tm^{3+} in order to populate Gd^{3+} ions, the currently known EMU1 is less efficient than the EMU2 process.

Many novel capabilities have been opened with the rational design of such active-core–active-shell NPs. For example, a broadened temperature sensitivity range (from typically $\Delta T = 150$ K up to 300 K (150–450 K) has been demonstrated by employing a Yb/Er thermometer in the core and Nd/Yb thermometer in the shell {Marciniak, 2016}. Under 808 nm, the relative intensity of NIR emission of Nd and Yb as well as EMU2 originated Er emission, were combined in a single core–shell NP. When homogenously doped, Nd^{3+} ions would be responsible for $Er^{3+} \rightarrow Nd^{3+}$ energy transfer, but spatial splitting of both ions enabled increased functionality and enhanced properties.

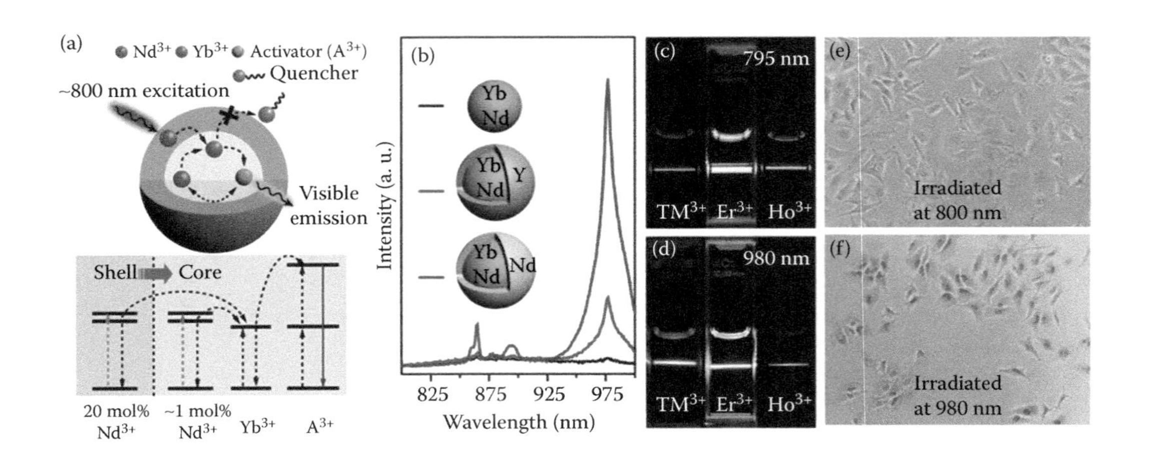

FIGURE 8.6
(a) The proposed energy transfer mechanisms of EMU in lanthanide-doped $Er^{3+}/Yb^{3+}:NaYF_4$ @ $Yb^{3+}/Nd^{3+}:NaYF_4$ core–shell nanocrystals (NCs); (b) demonstration of capability to design different color spectral codes in core–shell–shell EMU-based upconverting NPs; comparison of UC and EMU efficiency (c and d) and impact of excessive overheating on the survival of living cells irradiated at 800 and 980 nm (e and f). (Reprinted with permission from Xie et al. 2013. Mechanistic investigation of photon upconversion in Nd^{3+}-sensitized core–shell nanoparticles. *J. Am. Chem. Soc.* 135, 12608–12611. Copyright 2013 American Chemical Society.)

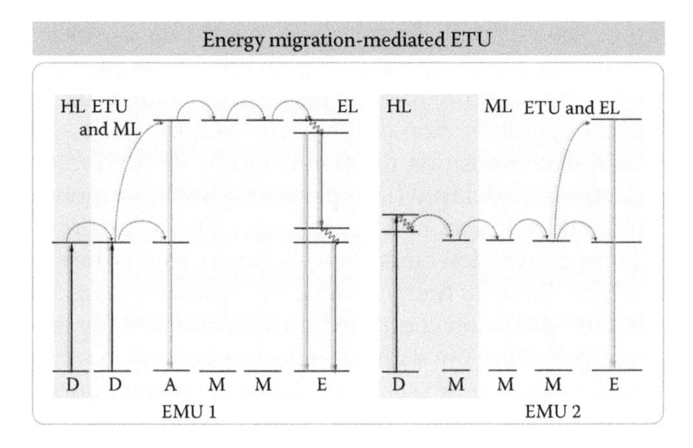

FIGURE 8.7
Schematic illustration of two types of energy migration-mediated energy transfer upconversion, HL, ETU, ML, EL describe harvesting layer, energy transfer upconversion layer, migration layer, and emission layer, respectively, and D, A, M, and E denote donor, activator, migrating, and emissive ions.

8.6 Core–Shell UCNPs: Synthesis, Properties, and Issues

Besides engineering the optical properties of upconverting NPs by the dopant concentration and spatial separation, intensive research has been devoted to the synthesis procedures optimization. Several reports have described efficient synthesis approaches to produce the shape-controlled, stable, biocompatible, and monodispersed NPs. The most common chemical synthesis strategies including co-precipitation, thermal decomposition, hydrothermal and solvothermal synthesis, sol–gel procedure, combustion synthesis, and cation exchange have been demonstrated to synthesize lanthanide-doped luminescence NPs. In particular, the NPs used as bioprobes should be characterized by narrow size distribution, suitable shape, and high aqueous solubility, and thus require a special method of synthesis. One of the most successful and popular methods for the synthesis of good quality upconverting NPs are hydro(solvo) thermal and thermal decomposition synthesis. Both methods allow obtaining good quality NPs with precise and narrow size control. Thermal decomposition reaction was developed for the first time by Yan's group to synthesize monodisperse triangular LaF_3nanoplates (Zhang et al. 2005), and next refined and extended as a common route to the synthesis of $NaYF_4$ nanocrystals (Mai et al. 2006). This method usually involves dissolving organic metal salts (commonly used trifluoroacetate or acetate compounds) in high-boiling point solvents, typically with polar capping groups and long hydrocarbon chains, such as 1-octadecene (ODE, noncoordinating solvent with high-boiling point: 315°C), oleic acid (OA), and oleylamine (OM), which perform the function of both solvents for the reaction mixture and coordinating agents which stabilize the NPs surface as well as prevent agglomeration of the particles during the synthesis process. The thermal decomposition reaction could be divided into four stages including nucleation in a delayed time, particles growth by monomer supply, size shrinkage by dissolution, and aggregation. The formation of the most suitable host for upconversion emission generation, that is, hexagonal phase $NaYF_4$, proceeds as follows, first a kinetic precipitation stage for the cubic phase, and then the diffusion controlled growth and size focusing stage for the formation of β-$NaYF_4$. Yan's group reported general synthesis of high-quality cubic and hexagonal group of lanthanide sodium fluoride NPs (Mai et al. 2006). By tuning the ratio of Na/RE, solvent composition, reaction temperature, and time, they could manipulate the phase, shape, and size of the nanocrystals (Mai et al. 2006). Through controlling the nanocrystals growth kinetics, the nanopolyhedra, NRs, nanoplates, and nanospheres were successfully obtained (Figure 8.8). Shan et al. reported the thermal decomposition method using single solvent—trioctylphosphine oxide (TOPO, used as a boiling solvent and capping reagent) rather than the previous mentioned

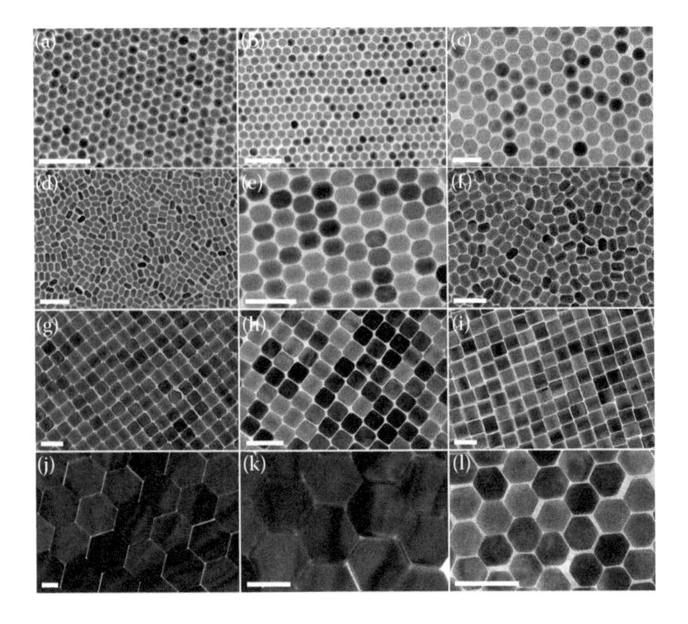

FIGURE 8.8
Transmission electron microscopy (TEM) images of β-NaYF4-based UCNPs. (a, d, g, j) NaYF4: Yb/Er (20/mol%) UCNPs. (b, e, h, k) NaYF4: Yb/Tm (22/mol%) UCNPs. (f, i) NaYF4: Yb/Ho (20/mol%) UCNPs. (c, l) NaYF4: Yb/Ce/Ho (20/11/mol%) UCNPs. All scale bars represent 100 nm. (Reprinted with permission from Ye et al., *Proc. Natl. Acad. Sci. USA*, 107, 22430–22435, 2010.)

component solvent (OA/ODE or OA/OM/ODE) (Jingning et al. 2007). The obtained NPs had a controllable size in range 5–20 nm with narrow size distribution, and presented efficient upconversion (Jingning et al. 2007). They showed that the energy barrier of the cubic to hexagonal phase transition was reduced using TOPO, allowing more efficient formation of hexagonal phase NPs (Jingning et al. 2007). By a suitable choice of the synthesis parameters, the shape and uniformity of the NPs could be controlled. Comparing the size, shape, aggregation, and luminescent properties of NPs synthesized in different conditions and with different reagents, it was further possible to define optimal parameters of the synthesis process for the desired application of upconverting NPs.

The thermal decomposition method was also adapted to produce metal oxides (Paik et al. 2013; Wawrzynczyk et al. 2014) and magnetic NPs (Brollo et al. 2014; Xiao et al. 2015). Although thermal decomposition is demonstrated to be the most effective in controlling the shape and size of NPs, there remain some disadvantages such as required rigorous synthesis conditions (high temperature, anhydrous, and oxygen free). In addition, resulting NPs are generally only dissolved in nonpolar solvent (e.g., chloroform and hexane), and due to the hydrophobic ligands stabilizing NPs surfaces, post-synthetic functionalization is necessary to transfer the as-synthesized NPs to aqueous solutions and to finally achieve biocompatibility (Gnach and Bednarkiewicz 2012). These hydrophobic NPs may be transformed into hydrophilic via ligand exchange with hydrophilic polymers or through silica coating. Another disadvantage of the thermal decomposition reaction procedure is associated with some safety concerns (Liu et al. 2011a). The decomposition of the metal trifluoroacetates produces various fluorinated and oxyfluorinated carbon species (such as trifluoroacetic anhydride, trifluoroacetyl fluoride, carbonyl difluoride, and tetrafluoroethylene) which are considered heavily toxic (Kannan et al. 2013). One another commonly used synthesis type, that is, the hydrothermal/solvothermal method, refers to a synthesis procedure within a sealed environment under high-pressure and temperature above the critical point of solvent to increase the solubility of the solid. The reaction is carried out in a reaction vessel (autoclave) to provide sealed a reaction condition. The advantage of adopting this method for synthesis of NPs includes the ability to create a highly crystalline phase at a lower temperature and the easy control of size, structure, and morphology of the NPs. In addition, the hydrothermal/solvothermal method allows for obtaining water-soluble NPs by using series polymers such as polyvinylpyrrolidone (PVP), PEG, PAA, and polyethylenimine (PEI) (Wang et al. 2009c). These polymers which were capped on the surface of the NPs made them hydrophilic and acted as capping ligands to stabilize the NPs against aggregation. However, in most cases, the hydrophilicity of the NPs was not sufficient due to the presence of hydrophobic ligands on the surface of the NPs. To improve the water solubility, surface modification was also necessary (Wang et al. 2011a).

8.6.1 Host Materials for Cores

The properties of host lattice and its interaction with dopant ions have a strong influence on the upconversion process in two ways: (i) by the phonon dynamics and (ii) by the local crystal field. The host lattice strongly affects the energy exchange interactions between dopant ions, promoting the f–f electronic transitions within lanthanide ions through perturbing the 4f wave function by crystal fields. This fact is of great importance for energy transfer between lanthanide ions, which are typically characterized by narrow emission and absorption bands that are unfavorable for the spectral overlap. The crystal structure of the lattice of host materials determines also the distance between the dopant ions, their relative spatial position, coordination numbers, and the type of anions surrounding the dopant ions. In many lanthanide-doped host materials, the upconversion luminescence have been observed, but highly efficient emission requires low phonon energy and small lattice mismatch to the dopant ions.

An ideal host matrix should have low lattice phonon energy in order to minimize nonradiative losses and maximize radiative emission (Cheng et al. 2013). Among the available types of host materials for upconverting lanthanide ions pairs, fluorides $AREF_4$ (A = alkali metal, RE = rare earth element), such as $NaREF_4$, $KLnF_4$ and LnF_3, CaF_2, $KMnF_3$, have proved to be ideal host candidates, due to their low photon energies and high chemical stability, and thus are often used as host materials for upconversion. Other heavy halides such as bromides and iodides exhibit low phonon energy of less than 300 cm^{-1}, however, they are hygroscopic and show low chemical stability. Metal oxides present good chemical stability, but often exhibit large phonon energies above 500 cm^{-1}. In addition, lattice impurities may increase the multiphonon relaxation thereby reducing the overall upconversion emission. Host lattice based on cations like Na^+, Ca^{2+}, and Y^{3+} with ionic radii close to lanthanide dopant ions prevent the formation of crystals defects and lattice stress. For this reason, generally Na^+ and Ca^{2+} fluorides are preferable host materials for the upconversion process. The properties of upconverting NPs are determined by their phase, structure, size, morphology, synthesis procedure, and composition. In addition, surface to volume ratio, type of surface ligands, and defects in the crystal structure of NPs can have great influence on their optical properties.

8.6.1.1 Sodium Fluoride

In cubic sodium RE fluoride systems ($NaREF_4$) containing one type of high-symmetry cation site, fluorite structures (CaF_2) are formed, with the Ca^{2+} sites randomly occupied by Na^+ and RE^{3+} ions. In contrast, the crystal structure of hexagonal phase $NaREF_4$ consists of an ordered array of F^- ions with two types of relatively low-symmetry cation sites selectively occupied by Na^+ and RE^{3+} ions, resulting in significant electron cloud distortion of the cations to accommodate the structural change. Importantly, light lanthanides with large ionic radii exhibit a high tendency toward electron cloud distortion owing to increased dipole polarizability, and thus favor the hexagonal structures. Doping $NaYF_4$ host lattices with lanthanide ions larger in size than Y^{3+} ($r = 1.159$ Å) should enhance the formation of pure hexagonal phase $NaYF_4$ nanocrystals. Sodium RE fluoride may be prepared in two polymorphic forms, cubic and hexagonal phases. In the cubic structure (space group $Fm\text{-}3m$, $Z = 4$) containing one type of cation site, the RE^{3+} and Na^+ ions are equally and randomly distributed in the cationic sublattice. In contrast, the crystal structure of hexagonal phase $NaREF_4$ consists of an ordered array of F^- ions with two types of relatively low-symmetry cation sites selectively occupied by Na^+ and RE^{3+} ions. Numerous investigations have shown that the hexagonal $NaYF_4$ is a much better host lattice for the luminescence of RE ions than the cubic $NaYF_4$. Hexagonal β-phase is controversial over the cation sites and distribution. Usually $NaNdF_4$ structure with space group $P\text{-}6$ and $Z = 1.5$ is used as a model for the hexagonal form. It has three different cation sites, a 9-fold coordinated Y^{3+} position, 9-fold coordinated mixed Na^+/Y^{3+} position, and 6-fold coordinated site that may be occupied by Na^+ exclusively. Recently, it was shown, however, that β-phase is better described in $P6_3/m$ symmetry where Na^+ ions may locate in the centers of trigonally compressed octahedrons whereas Y^{3+} ions may occupy only one mixed Y^{3+}/Na^+ site with 9-fold coordination. The Y^{3+} sites are substituted randomly with other lanthanides ions (Tu et al. 2013).

8.6.1.2 Calcium Fluoride

CaF_2 material shows high chemical and thermal stability, wide transmission range, and low phonon energy (328 cm^{-1}). CaF_2 is an ionic crystal with the fluorite structure. The lattice is a face centered cubic structure with three sublattices. The unit cell of the material is described as a simple cubic lattice formed by the F^- ions where a Ca^{2+} cation is contained in every second cube. The remaining empty cubes are important for accommodation of lanthanide ions or dopants. When Ca^{2+} ions are substituted by trivalent lanthanide ions, charge balance compensating F^- ions should enter the fluorite structure (Masahiko et al. 2004). Lanthanide-doped CaF_2 upconverting NPs with different size and shape and have been investigated and proved their bioapplicability (Dong et al. 2011b; Pedroni et al. 2011).

8.6.1.3 RE Fluoride

Lanthanum fluoride has low phonon energy (~350 cm^{-1}), and due to high ionicity of the La^{3+} to fluorine bond nonradiative loss of energy is suppressed (He et al. 2008; Xiaoting et al. 2015). Chow's group synthesized hexagonal LaF_3: Er^{3+}, Yb^{3+} nanocrystals with an average size of 5 nm and narrow size distribution. Compared with cubic $NaYF_4$ nanocrystals, for LaF_3 NPs, five times higher visible emission was observed under the same experimental conditions (Yi and Chow 2005). The highest upconversion efficiency of bulk YF_3 materials doped with Yb^{3+}/Er^{3+} or Yb^{3+}/Tm^{3+} has been reported to be higher than that of hexagonal $NaYF_4$ materials doped with Yb^{3+}/Er^{3+} or Yb^{3+}/Tm^3 (Auzel et al. 2012). In addition, the emission intensity of 3.7 nm sized YF_3 NPs doped with Yb^{3+}, Er^{3+}/Tm^{3+} was found to be about one order of magnitude higher than that of 4.5 nm sized CaF_2 NPs.

8.6.1.4 Metal Oxide

Yttrium oxide (Y_2O_3) is one of the most studied ceramics as a host matrix, especially when doped with europium (Eu^{3+}), resulting in a bright red emission. In case of Y_2O_3, the luminescence of monoclinic Y_2O_3 doped with Eu^{3+} is lower that of cubic Y_2O_3. On the contrary, in case of terbium ions, the monoclinic form demonstrates higher emission intensity. In cubic Y_2O_3 structure, there are two sites where lanthanide ions can substitute Y^{3+}: one with C_2 symmetry and without any inversion center (75% of the all sites), and one with S6 symmetry with one inversion center. In monoclinic Y_2O_3, all sites have Cs symmetry without any inversion center. The theoretical studies shown that for green emission, the Tb^{3+} ions should be embedded in the lattice the symmetry of which does not have inversion sites, such as the monoclinic Y_2O_3 (Sotiriou et al. 2012). This phenomenon can be explained by a better distribution of Tb^{3+} ions in the monoclinic structure because it is of smaller size than the cubic one, which facilitates the Tb^{3+} radiative transitions. In addition, the probability of the $^5D_4 \rightarrow {}^7F_5$ electric-dipole transition of Tb^{3+} contains contributions from linear and third-order terms of crystal lattice. Some of the crystal point groups in the cubic Y_2O_3 do not have linear terms and therefore, a large transition probability may be realized by embedding Tb^{3+} in lattice sites with crystals having linear terms such as monoclinic Y_2O_3 (Sotiriou et al. 2012). It is therefore important to select the optimal structure of the host lattice, which will correspond to the properties of the doped ions. $Y_2O_3:Eu^{3+}$ material is one of the most useful red emitting phosphor with almost 100% quantum efficiency (Ronda 1997). In $Y_2O_3:Eu^{3+}$, the electronic transition from 2p orbital of O^{2-} to the 4f orbital of Eu^{3+} forms excited CT band, which excited to luminescence with great efficiency and is mostly used as the excitation level for commercial phosphor. The CT band in cubic nanocrystalline $Y_2O_3:Eu^{3+}$ depends on local structure surrounding Eu^{3+} (Wang et al. 2005). For the preparation metal oxide NPs, a number of chemical techniques such as spray drying, freeze drying, sol–gel, co-precipitation, and self-sustaining combustion syntheses have been developed (Dhanaraj et al. 2001). Murray's group demonstrated synthesis of Gd_2O_3 doped with Er^{3+} and Yb^{3+}nanoplates with morphology tunable from tripodal to triangular through addition lithium hydroxide, which is used as a shape-directing

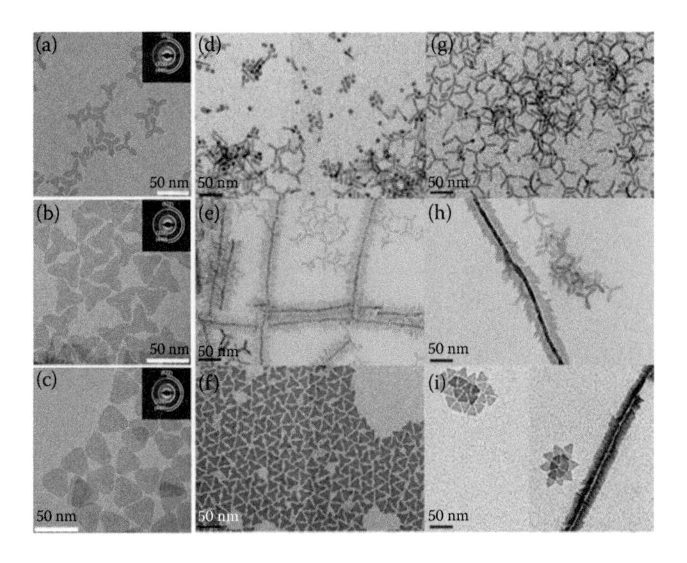

FIGURE 8.9
TEM images of Gd_2O_3:Eu^{3+} obtained after 45 min (a), 1.5 h (b), and 3.5 h (c) time of synthesis. (Reprinted from Wawrzynczyk, D., Nyk, M., Bednarkiewicz, A., Strek, W., and Samoc, M., *J. Nanoparticle Res.*, 16, 1–13, 2014.) Temperature-dependent shape transformation. TEM images of nanoplates synthesized at (d) 280°C for 30 min, (g) 280°C, (e) 290°C, (h) 300°C, (f) 310°C, and (i) 320°C for 1 h. (Reprinted with permission from Paik, T et al. 2013. Designing tripodal and triangular gadolinium oxide nanoplates and self-assembled nanofibrils as potential multimodal bioimaging probes. *ACS Nano* 7, 2850–2859. Copyright 2013 American Chemical Society.)

agent (Paik et al. 2013). Wawrzynczyk et al. (2014) investigated the correlation between the size and shape of colloidal Gd_2O_3 nanoplates and the Eu^{3+} ions luminescence (Figure 8.9).

8.6.2 Shell Synthesis Methods and Compositions

Currently, NPs are widely coated with a variety of materials such as silica, polymers, metal NPs, or other inorganic materials. Coating the surface of the NPs, as reviewed in the previous subchapters, allows for improvement of the luminescent properties, functionalization for biomedical application, and creation of multifunctional structures. The most effective strategy to improve the luminescent properties of NPs is creation of core@shell structures, where a shell layers are grown around the luminescent core. Having coated the core with the protective shell, the distance between lanthanide ions and surface quenchers is increased. Thanks to this, the reduction of the nonradiative pathways and energy quenching in the upconversion energy transfer process occurs. By coating a shell, the total intensity improved from several times to several hundred times in different upconversion NPs structures (Chen 2012c; Wang 2012b).

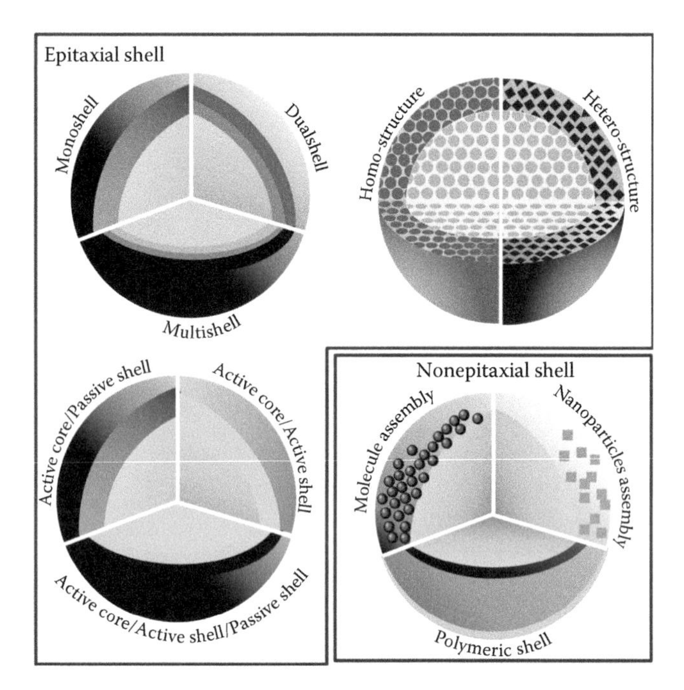

FIGURE 8.10
Schematic illustration of the typical architectures of UC core–shell NPs.

Based on the host matrix, the core@shell NPs can be classified as homo- and hetero-structured (Figure 8.10). Homo-structured core@shell NPs (e.g., $NaYF_4$@$NaYF_4$, $NaGdF_4$@$NaGdF_4$) have the same crystalline host in both core and shell, except for the distribution of dopant ions. In hetero-structures (e.g., $NaYF_4$@$NaGdF_4$, $NaGdF_4$@$NaYF_4$), the core and shell materials are of different composition and structures. The first type of core@shell structures allow for luminescence enhancement, while the hetero-structure allows for both luminescence enhancement and integration of different useful properties. For example, the $NaYF_4$:Yb,Er@$NaGdF_4$core@shell structure has an upconverting core coated with $NaGdF_4$ shell, where the paramagnetic Gd^{3+} ions in the shell allows for integrating MRI properties (Chen et al. 2011b). The $NaLuF_4$:Yb,Tm@$NaGdF_4$ (^{153}Sm) core@shell NPs with radioactive, magnetic, X-ray attenuation, and luminescence properties were applied as nanoprobe to image tumor angiogenesis (Sun et al. 2013). The core NPs allowed obtaining both intense upconversion luminescence emission and X-ray CT imaging. The Gd^{3+} ions have been used in MRI. The radioactive ^{153}Sm ions emits a gamma photon that can be used in radiotherapy and also as a SPECT imaging probe (Sun et al. 2013). Recently, Wang et al. (2012b) demonstrated $NaYF_4$:Er^{3+}, Yb^{3+} NPs coated with CaF_2 layer shell NPs. They observed approximately 300-fold enhancement of upconversion emission yield in the 10–13 nm large α-$NaYF_4$:Yb^{3+}, Er^{3+}@CaF_2core@shell NPs in comparison to pristine α-$NaYF_4$:Yb^{3+}, Er^{3+} NPs. The enhancement was higher than typically found in $NaYF_4$@$NaYF_4$core@shell approaches, which was explained by the heterogeneous nature of CaF_2 shell. In addition, CaF_2 shell protects the lanthanide ions from leaking into the local environment. The dialysis experiments and inductively coupled plasma (ICP) analysis show the advantage of CaF_2-coated NPs over the $NaYF_4$-shelled ones, especially in biological experiments, thus lowering the risk of possible ionic toxicity (Wang et al. 2012b). It has been also shown, that core@shell structures of completely different chemical compositions can be constructed. The hetero-shell of CaF_2 was chosen by Prasad et al. to coat α-$NaYbF_4$:Tm^{3+} NPs. They observed a 35-fold enhancement of NIR–NIR upconversion intensity (Chen et al. 2012c). A similar enhancement for cubic $NaYF_4$ NPs doped with terbium and ytterbium ions and coated with a CaF_2 shell was observed by our group (Prorok et al. 2014a). We have demonstrated the spectral properties of cubic $NaYF_4$:Tb^{3+},Yb^{3+} colloidal NPs coated with $NaYF_4$ or CaF_2 shell. The upconversion emission intensity obtained from the core@shell NPs increased up to 40 fold in comparison with the bare core NPs. Our results combined with relative upconversion intensity enhancement suggest that both types of shells counteract nonradiative decay to the same extent. Moreover, both shell materials, that is, $NaYF_4$ and CaF_2 showed similar protective properties (Prorok et al. 2014a).

8.6.3 Silica Amorphous Layer

The nanocomposites based on upconverting NPs have attracted much interest due to their unique multifunctionality that can lead to a wider range of applications. The silica coating of NPs is one of the most commonly used methods to render them water soluble, biocompatible, and photostable. The surface chemistry of silica spheres is well established and silica is known to have mild effects in biological systems. Moreover, the SiO_2 shell can shield the NPs from direct interaction with the environment or the influence of physiological factors (Qian et al. 2009). The fluorescence intensity of SiO_2-modified Y_2O_3 doped with Tm^{3+} and Yb^{3+} ions was enhanced by 9.4 times in comparison to the upconversion intensity of bare NPs (Lü et al. 2008b). In case of NPs synthesized with using the annealing process, SiO_2 shell prevented the unwanted particles growth and phase transformation (Sotiriou et al. 2012). Silica shell protects $NaYF_4$ cores from aggregation and growing up, but also prevent the phase change of nanocrystals from cubic to hexagonal phase during annealing (Liu et al. 2014). One of the greatest advantages in the use of silica shell is the facility to modify the surface with various functional groups introduced by employing suitable organosilane derivatives, which further offer the possibility to combine them (e.g., via $-NH_2$ or –SH group) with biomolecules or other organic compounds (Bharali et al. 2005; Wang et al. 2009b). There are couple of different techniques to synthesize silica shell. They can be divided based on the polar nature of the capping ligand on the NP surface, or on the type of silica shell, that is, mesoporous or dense SiO_2. Around many different variations of synthesis method, the sonogel synthesis is one that deals with unstable powder NPs dispersion (Chen et al. 2008). The Stöber method, on the other hand, is one of the simplest processes where an amorphous silica may be deposited on the surface of NPs (Cong et al. 2010; Sivakumar et al. 2006; Stober et al. 1968). Nevertheless, this process is usually limited to hydrophilic NPs. In case of hydrophobic NPs, the method of reverse microemulsion (water in oil) is often used to synthesize amorphous silica shell (Bagwe et al. 2006; Hlaváček et al. 2014; Jana et al. 2007; Li et al. 1999). This method is based on the hydrolysis of tetraethoxysilane in nanoreactors made by a homogeneous mixture of cyclohexane, surfactant (e.g., Igepal CO-520, TritonX-100), ammonia, and tetraethoxysilane. The modified silica, that is, mesoporous, with a higher surface area, large pore volume, and tunable pore sizes which enabled the construction of drug-carrier composites was intensively studied in the past few years. The formation of mesoporous silica demands the presence of the structure-directing agent, for example, cetyltrimethylammonium bromide (CTAB) or octadecyltrimethoxysilane (C18TMS) (Cichos and Karbowiak 2014; Qian et al. 2009; Yi et al. 2006). The recently developed facile method for coating of hydrophobic nanocrystals with a thin layer of mesoporous silica based on the biphasic system allows coating NCs of various sizes with control over the thickness of the silica coating layer as shown in Figure 8.11. For more details, the reader is referred to Cichos and Karbowiak (2014). Titanium oxide (TiO_2) is, similarly to silica, one of the suitable candidates for the shell material, however, it has

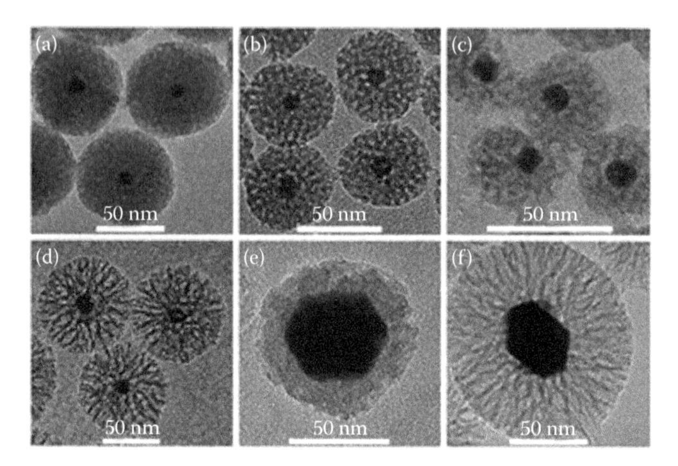

FIGURE 8.11
TEM images of different NPs with silica shell synthesized under numerous conditions such as type of organic layer solvent, time of reaction, size of core NPs, and amount of tetraethoxysilane. (Cichos, J. and M. Karbowiak. 2014. A general and versatile procedure for coating of hydrophobic nanocrystals with a thin silica layer enabling facile biofunctionalization and dye incorporation. *J. Mater. Chem. B* 2:556–568. Reproduced by permission of The Royal Society of Chemistry.)

not yet been extensively studied in the literature as an upconverting NPs shell (Lü et al. 2008a).

The features and conformity of surface coating ligands influence the stability of NPs under complex and biologically substantial conditions such as ionic strengths or various pH values. The covering with various type and differently charged polymers greatly influence the surface charge of upconverting NPs determining their cellular uptake efficiency (Jin et al. 2011). The leading approach to encapsulate NPs into polymers is based on employing hydrophobic interactions between the polymer chains and NPs surface ligands. The polymer encapsulation can be also obtained via the ligand exchange process. The commonly used polymers are PAA, PVP, PEG, and PEI. Herein, the reader can find some examples of using polymer encapsulation in the extension of upconverting NPs usage. One interesting example is the work of Budijono et al., who studied two different routes for polymer surface modification of the upconverting NPs, that is, direct ligand exchange using (PAA) and amphiphilc copolymer encapsulation via flash nanoprecipitation. The authors obtained colloidally stable NPs dispersions in deionized water, and what is even more essential in buffers and serum media (Budijono et al. 2009). Lu et al. designed the approach where upconverting NPs were first conjugated with PEG to gain physiological stability. Subsequently nanocrystals were coated with one or two layers of PEI polymers to acquire gene loading ability. As a result, NPs with reduced cytotoxicity and enhanced gene transfection ability were obtained (He et al. 2013). The application of amphiphilic polymer (C18PMH-PEG) modified NPs with cyclometallated ruthenium complex in the selective detection of Hg^{2+} ions in aqueous solution was demonstrated in the work of Li et al. (2014a). In their work, Guller et al. (2015) systematically investigated the cytotoxicity of upconverting NPs coated with different polymers, including PEI, poly(D,L-lactide) (PLa), poly(lactideco-glycolide) (PLG), and poly(maleicanhydride-alt-1-octadecene) (PMAO), or NPs embedded into polystyrene beads. Jin et al. (2011) studied a series of polymer-coated NPs with different surface charges (positive, neutral, and negative) and found that positively charged NPs have significantly enhanced cellular uptake in several human cell types. The polymer coating method and the possibility of integrating different functional groups onto the particles' surface lead to multifunctional upconverting NPs and a broader range of potential applications.

As it was stated in the previous chapters, the overall upconversion emission intensity and QY can be increased by the careful design of hybrid upconverting NPs + plasmonic NPs structures. In general, there are two main strategies for the synthesis of such nano-architectures: (i) Au or Ag shell can be directly deposited on the surface of upconverting NPs or (ii) the SiO_2 or TiO_2 spacer intermediate layer is first deposited between nanophosphor and plasmonic NP. The former approach requires to transfer the upconverting NPs to water solutions, since the synthesis of Ag and Au NPs or nanoshells is usually performed in aqueous environment. For that purpose, the surface of upconverting NPs is first covered with polymers and the Au or Ag NPs are then precipitated at the surface (Dong et al. 2011a; Zhang et al. 2010). Depending on the amount of the metal salts and the reaction time, the thickness of the metallic shell could be controlled. However, much better control over the mutual interaction between upconverting and metallic NPs was obtained, when the surface of the luminescent core was first covered with silica or titanium oxide shell of desired thickness (Xu et al. 2015; Zhao et al. 2014). In that way, the proper selection of the distance between fluorophore and plasmonic structures allowed for the designing of upconversion emission enhancement due to the metal enhanced fluorescence effect. To date, several papers have shown that the synthesis of SiO_2 or TiO_2 covered upconverting NPs, and further depositing additional metallic shell allows increased overall upconverted emission intensity. The strategy to obtain the spacer shell layer has been described in the previous chapter, and for the synthesis of another plasmonic shell, the seed-mediated reduction of metal salts (e.g., $HAuCl_4$ or $AgNO_3$) is used. The complete opposite approach was shown by Song et al. (2014), who used noble metal NPs (Au and Ag) as cores and Tb^{3+} ions doped $NaYF_4$ NPs as shell layer. They used a simple, low temperature co-precipitation method to obtain highly crystalline and monodispersive in size core@shell NPs with the potential application as photothermal agents.

8.6.4 Crystalline Homo- and Hetero-Structural Shells

The most effective strategy to improve the luminescent properties of NPs is the creation of a core@shell structure, where shell layers are grown around the luminescent core. Having coated the core with the protective shell, the distance between lanthanide ions and surface quenchers was increased. Thanks to that, the reduction of the nonradiative pathways and the energy quenching in the upconversion energy transfer process occurs. There are several different ways for synthesis of core@shell upconverting NPs with the use of different F^- and Na^+ sources and lanthanide precursors. In general, however, the synthesis strategies, can be classified into one-step (one-pot or shell synthesis on the purified core particles) and multi-step synthesis (Figure 8.12). The multi-step synthesis strategy can be further categorized into two, one-pot and multi-cycle methods (Ghosh Chaudhuri and Paria 2012). In the one-pot method, the core particles are synthesized *in situ*, that is, after the core NPs are formed, the reagents for shell formation are added. This method has one very important disadvantage. In *in situ* method, some impurities from reaction media may be trapped between the core and shell. The multi-cycle approach is a more complex approach, since the core particles are separately synthesized, centrifuged, and purified, and then the ready core NPs are added to the reaction mixture with the precursors for shell formation. The core NPs act as nucleation centers and the shell is preferentially crystallized at the core NPs' surface. It is, however, very important to assure proper concentration of both core NPs and shell precursor in the reaction, since the NPs with broad

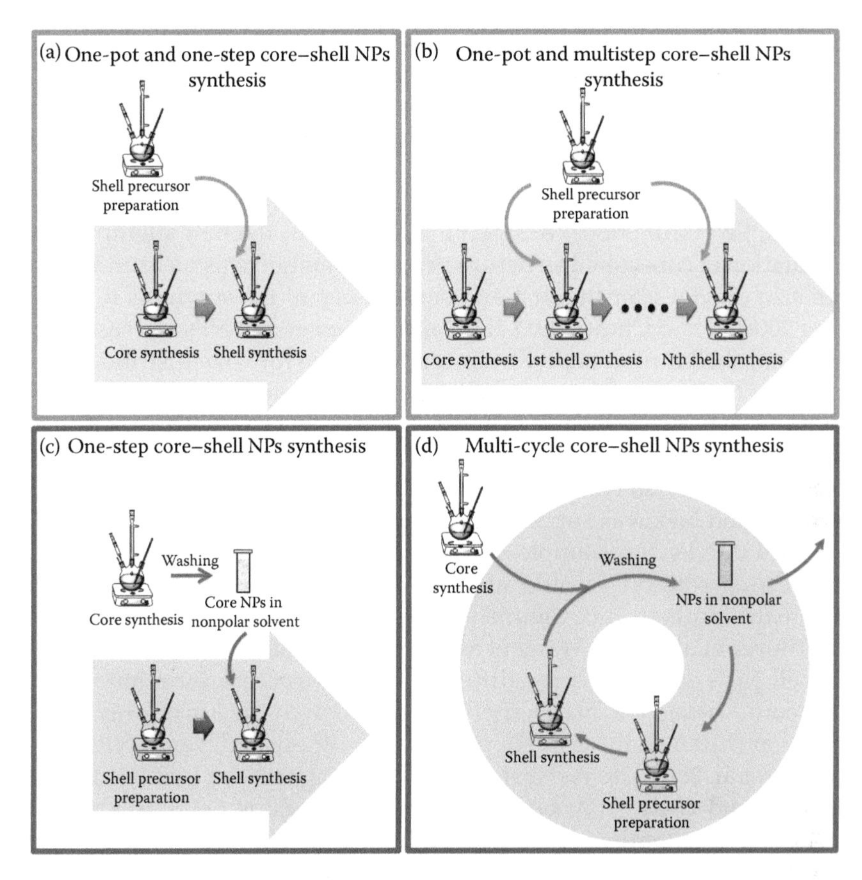

FIGURE 8.12
Illustration of various shell synthesis methods. (a) One-step and one-pot synthesis is the most facile way but not very beneficial for upconversion improvement and quality of the materials. (b) Multi-step and one-pot method that conquers an issue of uneven shell deposition. (c) One-step route with additional step, that is, washing the cores before addition to the shell precursor. (d) Multi-step (multi-cycle) process that employs nanocrystals' purification in every shell layer synthesis cycle.

size and shape distribution can be formed. The most unfavorable outcome of such reaction may be the utilization of shell precursors for the formation of new nanocrystals in the solution instead of the passivation of core NPs (i.e., nucleation centers). The most important benefit of this method is the fact that the core particles are available in pure form, so there is less possibility of any impurities from the reaction solution on the core surface.

8.6.5 Methods to Confirm Shell Formation Upon Core NPs

However, the synthesis procedures for core@multi-shell nanostructures are well established in the literature, the unequivocal confirmation of the efficient covering of core NPs with homogenous and crystalline shell remain challenging. As long as the shell was composed of amorphous polymers or had different composition from the core part of the NPs, the confirmation of high-resolution TEM images were entirely sufficient (Prorok et al. 2014a). The problem was with the NPs having core and shell parts composed of the same material, for example, widely discussed in this chapter upconverting $NaYF_4$:Er^{3+} (Tm^{3+},Ho^{3+})Yb^{3+}@$NaYF_4$ NPs. One of the first attempts of the confirmation of core@shell structure proper formation was measurements of mean size of NPs after the first and second step of the synthesis (Qian and Zhang 2008; Yi and Chow 2007). If the mean diameter of NPs increased upon intentional shell formation, with simultaneous narrow size distribution preserved, the formation of core@shell NPs could be indirectly confirmed. In other words, if after a two-step synthesis process, NPs with only one size were observed, with no additional smaller ones, the authors concluded a successful core@shell NPs formation. Often, the analysis of the morphology of synthesized NPs was supported by the investigation of the spectroscopic properties of NPs, for example, the formation of undoped sell covering the upconverting core NPs would increase the overall emission intensity and elongate the luminescence lifetimes, due to the decreased surface quenching effect (Boyer et al. 2007b; Wang et al. 2009). The intentional synthesis of core and shell parts of NPs having a different crystal structure, for example, cubic $NaYF_4$ core + hexagonal $NaYF_4$ shell, further allowed the use of x-ray powder diffraction (XRD) analysis to support the hypothesis of layered NPs formation (Mai et al. 2007a). K. A. Abel et al. (2009) designed $NaYF_4$@$NaGdF_4$core@shell NPs and used X-ray photoelectron spectroscopy (XPS) to show that Gd^{3+} ions are predominantly located at the surface of the nanocrystals, proving the same core@shell structure. The combination of those four methods, that is, TEM, XRD, XPS, and luminescence spectra and kinetics, allowed stating the fact of successful synthesis procedure modification and of obtaining the core@shell NPs. However, those measurements were done for the colloidal solutions, and gave to some extent the statistical picture of all NPs present in the stock. In addition, the improvement in upconverting properties of NPs after assumed shell formation could also result from: (i) dispersing the emissive ions inside the bigger NPs due to the Ostwald ripening or cation exchange, that would reduce the concentration quenching and (ii) incomplete core@shell formation and partial decrease of surface quenching effects (van Veggel et al. 2012). The advanced techniques, which nowadays enable the undoubted confirmation of the core@shell NPs structure formation are usually based on high-resolution electron microscopies, and include high-angle annular dark field imaging (HAADF), energy dispersive spectroscopy or energy dispersive X-ray spectroscopy (EDS or EDX), and electron energy-loss spectroscopy (EELS). These measurements can be additionally performed in line scan mode or in two-dimensional (2D) elemental mapping mode, that gives information about the ions' distribution at the single NPs level. EDS and EDX line scans of $NaYF_4$@$NaGdF_4$ NPs have shown symmetrical double-peak profile for Gd^{3+} ions, whereas the core of NPs was reached in Yb^{3+} ions (Abel et al. 2011; Wang et al. 2011a). Similar results were obtained for the 2D electron energy loss spectroscopy (EELS) mapping mode (Dong et al. 2012; Lee et al. 2011), Dong et al. (2012) have reported elemental maps of Gd^{3+} ions in $NaYF_4$:Yb^{3+},Tm^{3+}@$NaGdF_4$ NPs and showed NPs with bright other circles due to the relatively higher Gd^{3+} concentration at the edges of NP. The HAADF technique was also proven to be sensitive to the atomic number of NPs doping elements, and could be successfully used to distinguish the presence of Gd^{3+} ions in the shell of NPs. The reference examples of application of high-resolution electron microscopies for the confirmation of the core@shell NPs formation are presented in Figure 8.13.

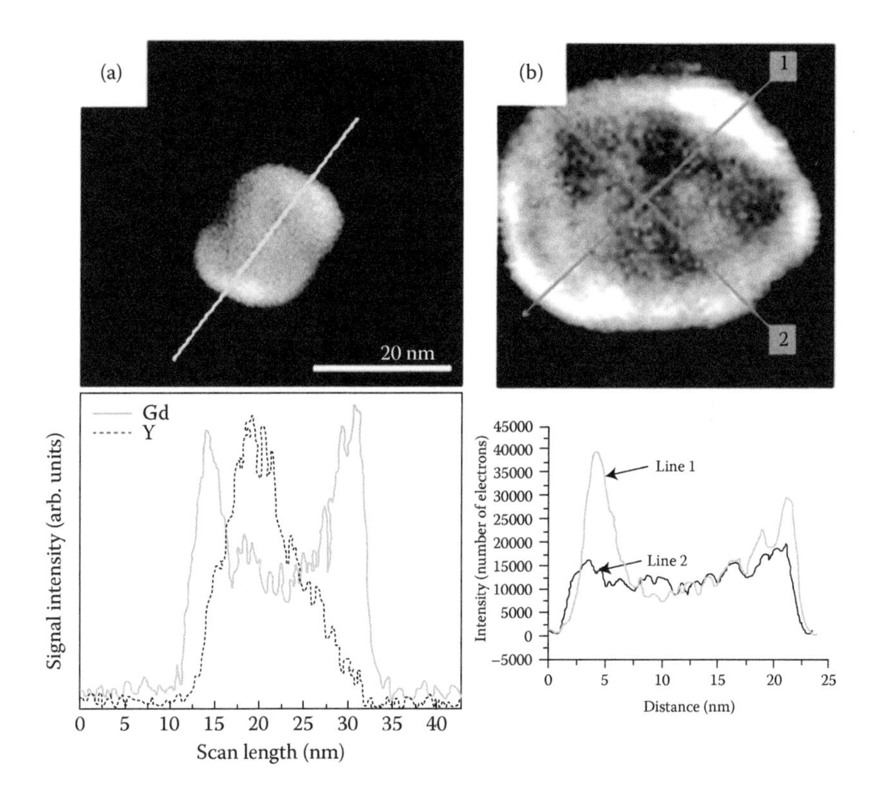

FIGURE 8.13
EDS line scan across a single $NaYF_4$@$NaGdF_4$core@shell NP showing Y^{3+} ions in the core of the particle and Gd^{3+} located in the shell (a, left side panel) (Reprinted with permission from Abel, 2011. Analysis of the shell thickness distribution on $NaYF_4$/$NaGdF_4$ core/shell nanocrystals by EELS and EDS. *J. Phys. Chem. Lett.* 3, 185–189. Copyright 2011 American Chemical Society.) and 2D EELS elemental maps and line profiles of Gd^{3+} in $NaYF_4$:Yb,Tm@$NaGdF_4$core A facile method to make NaYF4:Yb,Tm-$NaGdF_4$ core–shell nanoparticles with a thin, tunable, and uniform shell. *Chem. Mater.* 24, 1297–1305. Copyright 2012 American Chemical Society.)

8.6.6 Nonhomogenous Shell Composition

Another common disadvantage of the one-step synthesis of core@shell upconverting NPs is uneven shell distribution often found in this method. The noncentrosymmetric heterogeneous shell growth can be responsible for inefficient core passivation and some residual interaction of active ions with the environment, and luminescence quenching. One of the interesting examples in the literature is the work of Zhang et al. (2012), who compared the one-step shell synthesis method with the successive layer-by-layer approach. Besides higher luminescent intensity and lifetimes, the layer-by-layer synthesized core@shell NPs exhibited fewer intensity losses in the polar solvent under increasing water content than the core@shell NPs obtained by the one-step protocol. The illustrations of various kinds of shell deposition, that is, centrosymmetric and noncentrosymmetric are shown in Figure 8.14.

Another issue to consider is crystallization of NPs from the precursors dedicated to form shell material. The synthesis of thicker shell without any crystallization of small new nanocrystals can be attained through the use of the layer-by-layer approach (Prorok et al. 2014a). The reader can find the comparison of shell deposition results in one-step and layer-by-layer synthesis in the left panel of Figure 8.15.

8.7 Conclusions

Upconversion technology has experienced different periods of advancement and stagnation. While promising for biomedical applications, the upconverting phosphors were not studied as broadly as other luminescent NPs (e.g., QDs), mostly because joint efforts in many fields of science and technology were under development in the same time. The most severe drawback, which hindered the broad adoption of upconverting NPs as compared to QDs, was, and still is, the lack of widely available instrumentation suitable to detect or image biomolecules, cells, or tissues labeled with those materials in upconversion mode. Although some devices become available commercially (e.g., Hidex well plate reader or CytoViva upconversion microscope), most current studies were carried with home-made or home-modified conventional well plate readers or fluorescence microscopes. Despite spectroscopic equipment which allows measuring the upconversion spectra and kinetics, the quantitative comparison of quantum efficiency of upconverting materials is still a serious challenge and currently only a few groups worldwide can measure upconversion emission QY with self-made instruments. Lack of such technological backup with the simultaneous highly limited progress in developing such instruments slows down the further developments in the field of upconversion luminescence application.

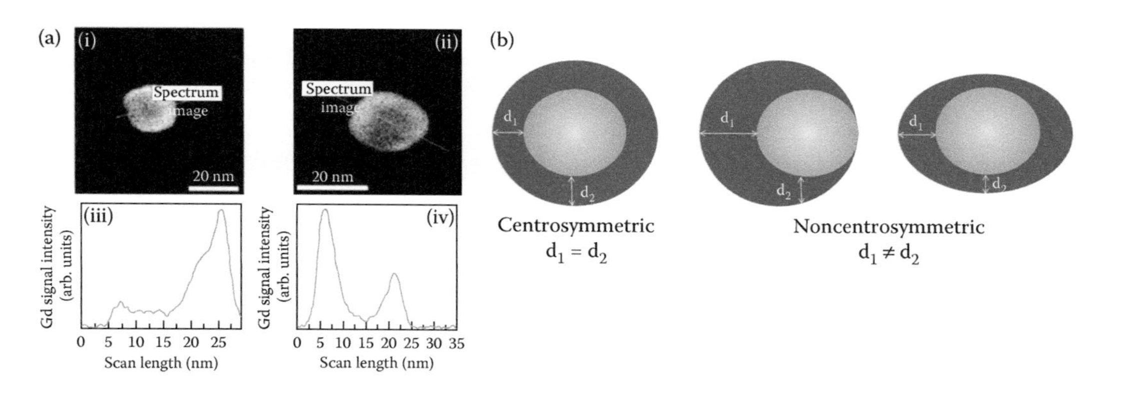

FIGURE 8.14
(a) TEM images of two representative single $NaYF_4/NaGdF_4$ core/shell nanocrystals (i and ii) and the corresponding electron energy loss spectroscopy (EELS) line scans (iii and iv), respectively. (Reprinted with permission from Abel et al. 2011. Analysis of the shell thickness distribution on $NaYF_4/NaGdF_4$ core/shell nanocrystals by EELS and EDS. *J. Phys. Chem. Lett.* 3, 185–189. Copyright 2011 American Chemical Society.) (b) The schematic illustration of structural composition of core–shell nanocrystals. The perfect core–shell structure with the same shell thickness around the core in all directions (centrosymmetric), core–shell structure with uneven distribution of the shell around the core (noncentrosymmetric), with not complete core passivation or complete coverage of the core.

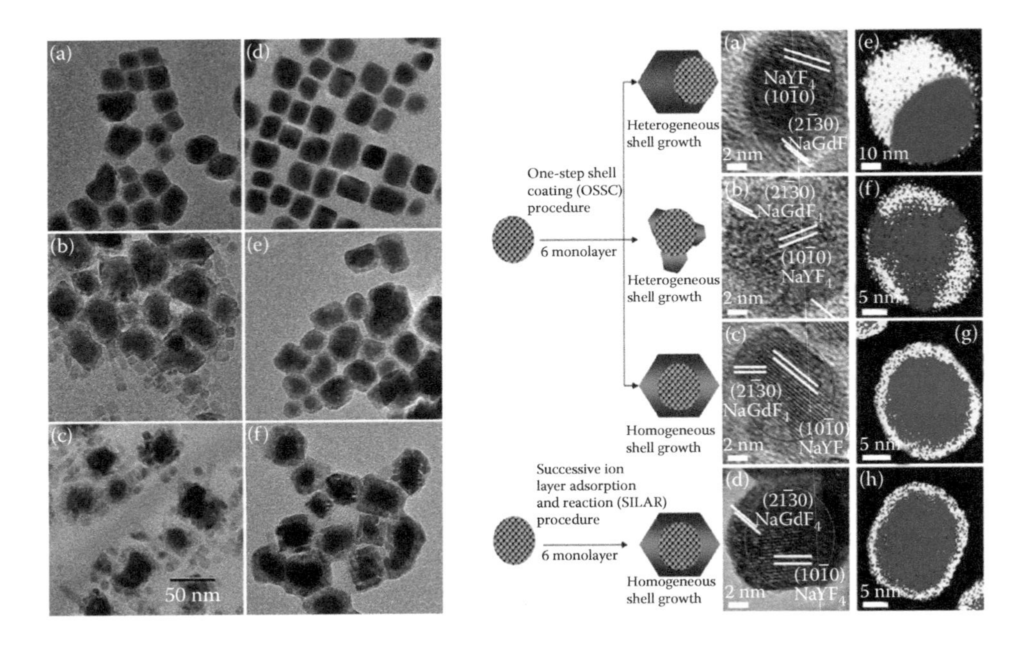

FIGURE 8.15
TEM images of $NaYF_4$:YbTb@CaF_2 core–shell NPs obtained by one-step (a–c) and one-pot layer-by-layer method (d–f). The shell:core molar ratio equal to 1 (a and d), 2 (b and e), and 3 (c and f). (Prorok, K., A. Bednarkiewicz, B. Cichy, A. Gnach, M. Misiak, M. Sobczyk, and W. Strek. 2014a. The impact of shell host $NaYF_4/CaF_2$ and shell deposition methods on the upconversion enhancement in Tb^{3+}, Yb^{3+} codoped colloidal α-$NaYF_4$ core–shell nanoparticles. *Nanoscale* 6:1855–1864. Reproduced by permission of The Royal Society of Chemistry). Right panel: High-resolution TEM (HRTEM) pictures of β-$NaYF_4$:YbEr@ $NaGdF_4$ core/shell nanocrystals prepared by one-step shell coating procedure (a–c) and layer-by-layer (here named successive ion layer adsorption and reaction) procedure (d), respectively. The composite color maps (e – h) show the location of different elements. Blue represents yttrium and yellow gadolinium atoms. (Reprinted with permission from Zhang et al. 2012. Direct imaging the upconversion nanocrystal core/shell structure at the subnanometer level: Shell thickness dependence in upconverting optical properties. *Nano Lett.* 12, 2852–2858. Copyright 2012 American Chemical Society.)

Independently from the above-mentioned issues, much advancement and major progress have been made in the field of materials science and physics of upconverting NPs since the first demonstrations of their biomedical applications. The most important milestones include (i) detailed control over NPs size, morphology, and phase, (ii) development of numerous upconversion enhancement methodologies such as plasmonics, passive (Li^+,K^+,Na^+), and active (Mn^{2+}) co-doping, and most importantly surface passivation with either homo- or heterogeneous shells, as well as (iii) the development of migration-mediated upconversion mechanism. The latter allowed to intentionally engineer the chemical architecture of upconverting NPs, and thus to design the spectral properties of these materials. Such approach enabled Nd^{3+} sensitizers become a new "star" in the field, since they allowed shifting the excitation wavelength from 980 to 808 nm. This is of great importance for biomedical applications since it allows the diminishing of local overheating effects, improves light penetration depth, enhances upconversion efficiency, and increases the functionality of such UCNPs. One may expect further developments in the chemistry, materials science, and applications of such Nd^{3+}-sensitized ACAS upconverting NPs. The very first examples, which took advantage of these innovations, include spectral color design or enhanced nanothermometer working range. In the first case, lanthanide ions (such as Tb^{3+}, Eu^{3+}, Sm^{3+}, or Dy^{3+}) were exploited, which due to low upconversion emission intensity were not used so far. In the latter case, two well-known temperature-dependent processes (i.e., Er^{3+} emission and $Nd^{3+} \rightarrow Yb^{3+}$ ET efficiency) were intentionally combined in a single ACAS NPs. Such designs and achievements were unavailable to homogenously co-doped upconverting NPs, and this is why the capability to synthesize ACAS NPs is, and will continue to be, of such great importance for novel developments, either in terms of materials science research or real applications in biotechnology and other fields of optical sensing.

Acknowledgments

A.B. acknowledges support from Polish National Science Centre Grant No. 2012/05/E/ST5/03901. A.B., K.P., and M.M. acknowledge the support from Wrocław Research Centre EIT + within the project "The Application of Nanotechnology in Advanced Materials"—NanoMat (POIG.01.01.02-02-002/08) financed by the European Regional Development Fund (Operational Programme Innovative Economy, 1.1.2). D.W. acknowledges support from the Ministry of Science and Higher Education under the grant "Iuventus Plus" in years 2015–2017, project no. IP2014 050273 and by a statutory activity, subsidy from the Polish Ministry of Science and Higher Education for the Faculty of Chemistry of Wroclaw University of Technology. M.M.

acknowledges support from Polish National Science Centre Grant No. DEC-2013/11/N/ST5/02716. K.P. acknowledges financial support from the National Science Center Poland (NCN) under the ETIUDA doctoral scholarship (DEC-2014/12/T/ST5/00646).

References

Abel, K. A., J. C. Boyer, C. M. Andrei, and F. C. J. M. van Veggel. 2011. Analysis of the shell thickness distribution on $NaYF_4$/$NaGdF_4$ core/shell nanocrystals by EELS and EDS. *J. Phys. Chem. Lett.* 3:185–189.

Abel, K. A., J. C. Boyer, and F. C. J. M. van Veggel. 2009. Hard proof of the $NaYF_4$/$NaGdF_4$ nanocrystal core/shell structure. *J. Am. Chem. Soc.* 41:14644–14645.

Achatz, D. E., R. Ali, and O. S. Wolfbeis. 2010. Luminescent chemical sensing, biosensing, and screening using upconverting nanoparticles. In *Luminescence Applied in Sensor Science*, edited by L. Prodi, M. Montalti, and N. Zaccheroni, 29–50. Berlin, Heidelberg: Springer.

Ai, F., Q. Ju, X. Zhang, X. Chen, F. Wang, and G. Zhu. 2015. A core–shell–shell nanoplatform upconverting near-infrared light at 808 nm for luminescence imaging and photodynamic therapy of cancer. *Sci. Rep.* 5: 10785.

Ali, R., S. M. Saleh, R. J. Meier, H. A. Azab, I. I. Abdelgawad, and O. S. Wolfbeis. 2010. Upconverting nanoparticle based optical sensor for carbon dioxide. *Sens. Actuators B* 1:126–131.

Alonso-Cristobal, P., O. Oton-Fernandez, D. Mendez-Gonzalez, J. F. Díaz, E. Lopez-Cabarcos, I. Barasoain, and J. Rubio-Retama. 2015. Synthesis, characterization, and application in HeLa cells of an NIR light responsive doxorubicin delivery system based on NaYF4:Yb,Tm@SiO_2–PEG nanoparticles. *ACS Appl. Mater. Inter.* 27:14992–14999.

Alonso-Cristobal, P., P. Vilea, A. El-Sagheer, E. Lopez-Cabarcos, T. Brown, O. L. Muskens, J. Rubio-Retama, and A. G. Kanaras. 2015. Highly sensitive DNA sensor based on upconversion nanoparticles and graphene oxide. *ACS Appl. Mater. Inter.* 23:12422–12429.

Auzel, F. 1966. Compteur quantique par transfert denergie entre deux ions de terres rares dans un tungstate mixte Et dans un verre (Quantum counter obtained by using energy transfer between two rare earth ions in a mixed tungstate and in a glass). *C. R. Hebd. Seances Acad. Sci. B (Comptes Rendus Hebdomadaires des Séances de l'Académie des Sciences. Séries A et B, Sciences Mathématiques et Sciences Physiques)* 15:1016–1019.

Auzel, F. 2004. Upconversion and anti-stokes processes with f and d ions in solids. *Chem. Rev.* 1:139–173.

Auzel, F. and D. Pecile. 1973. Radiation transfer between Yb^{3+} in mechanism for anti-Stokes fluorescence of matrices doped with Yb^{3+}–Er^{3+}. *C. R. Hebd. Seances Acad. Sci. B* 7:155–157.

Bagwe, R. P., L. R. Hilliard, and W. Tan. 2006. Surface modification of silica nanoparticles to reduce aggregation and nonspecific binding. *Langmuir ACS J. Surf. Colloids* 9:4357–4362.

Bednarkiewicz, A. and W. Strek. 2005. Interstitial single fiber multi-decay-probe for light dosimetry in photodynamic therapy: Modelling. *Proc. SPIE* 5862:586210.

Bednarkiewicz, A., D. Wawrzynczyk, M. Nyk, and W. Strek. 2011a. Optically stimulated heating using Nd^{3+} doped $NaYF_4$ colloidal near infrared nanophosphors. *Appl. Phys. B Lasers Opt.* 4:847–852.

Bednarkiewicz, A., D. Wawrzynczyk, M. Nyk, and W. Strek. 2011b. Synthesis and spectral properties of colloidal Nd^{3+} doped $NaYF_4$ nanocrystals. *Opt. Mater.* 10:1481–1486.

Berthou, H. and C. K. Jorgensen. 1990. Optical-fiber temperature sensor based on upconversion-excited fluorescence. *Opt. Lett.* 19:1100–1102.

Bharali, D. J., I. Klejbor, E. K. Stachowiak, P. Dutta, I. Roy, N. Kaur, E. J. Bergey, P. N. Prasad, and M. K. Stachowiak. 2005. Organically modified silica nanoparticles: A nonviral vector for *in vivo* gene delivery and expression in the brain. *Proc. Natl. Acad. Sci. USA* 32:11539–11544.

Boyer, J. C., L. A. Cuccia, and J. Capobianco. 2007a. Synthesis of colloidal upconverting $NaYF_4$: Er^{3+}/Yb^{3+} and Tm^{3+}/Yb^{3+} monodisperse nanocrystals. *Nano Lett.* 3:847–852.

Boyer, J. C., J. Gagnon, L. A. Cuccia, and J. A. Capobianco. 2007b. Synthesis, characterization, and spectroscopy of $NaGdF_4$: Ce^{3+}, $Tb^{3+}/NaYF_4$ core/shell nanoparticles. *Chem. Mater.* 14:3358–3360.

Boyer, J. C. and F. C. J. M. van Veggel. 2010. Absolute quantum yield measurements of colloidal $NaYF_4$: Er^{3+}, Yb^{3+} upconverting nanoparticles. *Nanoscale* 8:1417.

Brollo, M. E. F., R. López-Ruiz, D. Muraca, S. J. A. Figueroa, K. R. Pirota, and M. Knobel. 2014. Compact $Ag@Fe_3O_4$ core–shell nanoparticles by means of single-step thermal decomposition reaction. *Sci. Rep.* 4:6839.

Budijono, S. J., J. Shan, N. Yao, Y. Miura, T. Hoye, R. H. Austin, Y. Ju, and R. K. Prudhomme. 2009. Synthesis of stable block-copolymer-protected $NaYF_4$:Yb^{3+}, Er^{3+} up-converting phosphor nanoparticles. *Chem. Mater.* 2:311–318.

Budijono, S. J., J. Shan, N. Yao, Y. Miura, T. Hoye, R. H. Austin, Y. Ju, and R. K. Prudhomme. 2010. Synthesis of stable block-copolymer-protected $NaYF_4$:Yb^{3+}, Er^{3+} up-converting phosphor nanoparticles. *Chem. Mater.* 2:311–318.

Chatterjee, D. K. and Z. Yong. 2008. Upconverting nanoparticles as nanotransducers for photodynamic therapy in cancer cells. *Nanomedicine (Philadelphia, PA, USA)* 1:73–82.

Chen, C., N. Kang, T. Xu, D. Wang, L. Ren, and X. Guo. 2015a. Core–shell hybrid upconversion nanoparticles carrying stable nitroxide radicals as potential multifunctional nanoprobes for upconversion luminescence and magnetic resonance dual-modality imaging. *Nanoscale* 12:5249–5261.

Chen, D., P. Huang, Y. Yu, F. Huang, A. Yang, and Y. Wang. 2011b. Dopant-induced phase transition: A new strategy of synthesizing hexagonal upconversion $NaYF_4$ at low temperature. *Chem. Commun.* 47:5801–5803.

Chen, D. and Y. Wang. 2013. Impurity doping: A novel strategy for controllable synthesis of functional lanthanide nanomaterials. *Nanoscale* 5:4621–4637.

Chen, F., W. Bu, S. Zhang et al. 2011b. Positive and negative lattice shielding effects co-existing in Gd III ion doped bifunctional upconversion nanoprobes. *Adv. Funct. Mater.* 21:4285–4294. doi: 10.1002/adfm.201101663.

Chen, G., H. Liu, G. Somesfalean, H. Liang, and Z. Zhang. 2009. Upconversion emission tuning from green to red in Yb^{3+}/Ho^{3+}-codoped $NaYF_4$ nanocrystals by tridoping with Ce^{3+} ions. *Nanotechnology* 20:385704.

Chen, G., T. Y. Ohulchanskyy, R. Kumar, H. Ågren, and P. N. Prasad. 2010. Ultrasmall monodisperse $NaYF_4$:Yb^{3+}/Tm^{3+} nanocrystals with enhanced near-infrared to near-infrared upconversion photoluminescence. *ACS Nano* 4:3163–3168.

Chen, G., H. Qiu, R. Fan, S. Hao, S. Tan, C. Yang, and G. Han. 2012. Lanthanide-doped ultrasmall yttrium fluoride nanoparticles with enhanced multicolor upconversion photoluminescence. *J. Mater. Chem.* 22:20190–20196.

Chen, G., J. Shen, T. Y. Ohulchanskyy et al. 2012a. α-$NaYbF_4$:Tm^{3+}/CaF_2 core/shell nanoparticles with efficient near-infrared to near-infrared upconversion for high-contrast deep tissue bioimaging. *ACS Nano* 6:8280–8287.

Chen, G. Y., T. Y. Ohulchanskyy, S. Liu, W. C. Law, F. Wu, M. T. Swihart, H. Agren, and P. N. Prasad. 2012b. Core/shell $NaGdF_4$:Nd^{3+}/$NaGdF_4$ nanocrystals with efficient near-infrared to near-infrared downconversion photoluminescence for bioimaging applications. *ACS Nano* 6:2969–2977.

Chen, H., F. Yuan, S. Wang, J. Xu, Y. Zhang, and L. Wang. 2013. Aptamer-based sensing for thrombin in red region via fluorescence resonant energy transfer between $NaYF_4$:Yb,Er upconversion nanoparticles and gold nanorods. *Biosens. Bioelectron.* 48:19–25.

Chen, J. and J. X. Zhao. 2012. Upconversion nanomaterials: Synthesis, mechanism, and applications in sensing. *Sensors* 12:2414–2435.

Chen, Q., C. Boothroyd, G. H. Tan, N. Sutanto, and A. M. Soutar. 2008. Silica coating of nanoparticles by the sonogel process. *Langmuir* 24:650–653.

Chen, X., D. Peng, Q. Ju, and F. Wang. 2015b. Photon upconversion in core–shell nanoparticles. *Chem. Soc. Rev.* 44:1318–1330.

Chen, X., W. Xu, L. Zhang, X. Bai, S. Cui, D. Zhou, Z. Yin, H. Song, and D.-H. Kim. 2015c. Large upconversion enhancement in the islands Au–Ag alloy/$NaYF_4$: Yb^{3+}, Tm^{3+}/Er^{3+} composite films, and fingerprint identification. *Adv. Funct. Mater.* 25:5462–5471.

Cheng, L., C. Wang, and Z. Liu. 2013. Upconversion nanoparticles and their composite nanostructures for biomedical imaging and cancer therapy. *Nanoscale* 5:23–37.

Cheng, L., K. Yang, Y. Li, J. Chen, C. Wang, M. Shao, S. T. Lee, and Z. Liu. 2011. Facile preparation of multifunctional upconversion nanoprobes for multimodal imaging and dual-targeted photothermal therapy. *Angew. Chem. Int. Ed.* 50:7385–7390.

Cheng, L., K. Yang, Y. Li, X. Zeng, M. Shao, S.-T. Lee, and Z. Liu. 2012a. Multifunctional nanoparticles for upconversion luminescence/MR multimodal imaging and magnetically targeted photothermal therapy. *Biomaterials* 33:2215–2222.

Cheng, Q., J. Sui, and W. Cai. 2012b. Enhanced upconversion emission in Yb^{3+} and Er^{3+} codoped $NaGdF_4$ nanocrystals by introducing Li+ ions. *Nanoscale* 4:779–784.

Cichos, J. and M. Karbowiak. 2014. A general and versatile procedure for coating of hydrophobic nanocrystals with a thin silica layer enabling facile biofunctionalization and dye incorporation. *J. Mater. Chem. B* 2:556–568.

Cong, L., M. Takeda, Y. Hamanaka, K. Gonda, M. Watanabe, M. Kumasaka, Y. Kobayashi, M. Kobayashi, and N. Ohuchi. 2010. Uniform silica coated fluorescent nanoparticles: Synthetic method, improved light stability and application to visualize lymph network tracer. *PLoS One* 5:13167.

Cui, S., H. Chen, H. Zhu, J. Tian, X. Chi, Z. Qian, S. Achilefu, and Y. Gu. 2012. Amphiphilic chitosan modified upconversion nanoparticles for *in vivo* photodynamic therapy induced by near-infrared light. *J. Mater. Chem.* 22:4861–4873.

Damasco, J. A., G. Chen, W. Shao, H. Agren, H. Huang, W. Song, J. F. Lovell, and P. N. Prasad. (throughout) 2014. Size-tunable and monodisperse Tm^{3+}/Gd^{3+}-doped hexagonal $NaYbF_4$ nanoparticles with engineered efficient near infrared-to-near infrared upconversion for *in vivo* imaging. *ACS Appl. Mater. Inter.* 6:13884–13893.

Dhanaraj, J., R. Jagannathan, T. R. N. Kutty, and C.-H. Lu, 2001. Photoluminescence characteristics of $Y_2O_3:Eu^{3+}$ nanophosphors prepared using sol–gel thermolysis. *J. Phys. Chem. B* 105:11098–11105.

Digonnet, M. J. F. 2001. Rare-earth-doped fiber lasers and smplifiers. In *Electronic Materials: Science and Technology,* edited by M. Dekker, Vol. 2, 303–365. New York: CRC Press.

Dong, B., S. Xu, J. Sun, S. Bi, D. Li, X. Bai, Y. Wang, L. Wang, and H. Song. 2011a Multifunctional $NaYF_4:Yb^{3+}$, Er^{3+} @Ag core/shell nanocomposites: Integration of upconversion imaging and photothermal therapy. *J. Mater. Chem.* 21:6193–6200.

Dong, C., A. Korinek, B. Blasiak, B. Tomanek, and F. C. J. M. van Veggel. 2012. Cation exchange: A facile method to make $NaYF_4$:Yb, Tm-$NaGdF_4$ core–shell nanoparticles with a thin, tunable, and uniform shell. *Chem. Mater.* 24:1297–1305.

Dong, H., L. D. Sun, and C. H. Yan. 2014. Energy transfer in lanthanide upconversion studies for extended optical applications. *Chem. Soc. Rev.* 6:1608–1634

Dong, N. N., M. Pedroni, F. Piccinelli et al. 2011b. NIR-to-NIR two-photon excited $CaF_2:Tm^{3+},Yb^{3+}$ nanoparticles: Multifunctional nanoprobes for highly penetrating fluorescence bio-imaging. *ACS Nano* 5:8665–8671.

Dou, Q., N. M. Idris, and Y. Zhang. 2013. Sandwich-structured upconversion nanoparticles with tunable color for multiplexed cell labeling. *Biomaterials* 34: 1722–1731.

Dou, Q. and Y. Zhang. 2011. Tuning of the structure and emission spectra of upconversion nanocrystals by alkali ion doping. *Langmuir* 27 (21):13236–13241. doi: 10.1021/la201910t.

Downing, E., L. Hesselink, J. Ralston, and R. Macfarlane. 1996. A three-color, solid-state, three-dimensional display. *Science* 273:1185–1189.

Elhalawany, A., W. E. Hayenga, S. He, C. Lantigua, N. J. J. Johnson, A. Almutairi, and M. Khajavikhan. 2014. Increased upconversion quantum yield in plasmonic structures. In *Plasmonics: Metallic Nanostructures and Their Optical Properties XII,* 916331, edited by A. D. Boardman. San Diego: SPIE Press.

Esterowitz, L., A. Schnitzler, J. Noonan, and J. Bahler. 1968. Rare earth infrared quantum counter. *Appl. Opt.* 7:2053–2070.

Gao, W., H. Zheng, Q. Han, E. He, F. Gao, and R. Wanga. 2014. Enhanced red upconversion luminescence by codoping Ce^{3+} in β-$NaY(Gd_{0.4})F_4:Yb^{3+}/Ho^{3+}$ nanocrystals. *J. Mater. Chem. C* 2:5327–5334.

Gargas, D. J., E. M. Chan, A. D. Ostrowski et al. 2014. Unexpectedly bright sub-10-nm nanocrystals for single-molecule Imaging. *Nat. Nanotechnol.* 4:300–305.

Gerver, R. E., R. Gomez-Sjoberg, B. C. Baxter, K. S. Thorn, P. M. Fordyce, C. A. Diaz-Botia, B. A. Helms, and J. L. DeRisi. 2012. Programmable microfluidic synthesis of spectrally encoded microspheres. *Lab Chip* 12:4716–4723.

Geusic, J. E., H. M. Marcos, and L. G. VanUitert. 1964. Laser oscillations in nd-doped yttrium aluminum, yttrium gallium and gadolinium garnets. *Appl. Phys. Lett.* 4:182–184.

Ghosh Chaudhuri, R. and S. Paria. 2012. Core/shell nanoparticles: Classes, properties, synthesis mechanisms, characterization, and applications. *Chem. Rev.* 112:2373–2433.

Gnach, A. and A. Bednarkiewicz. 2012. Lanthanide-doped up-converting nanoparticles: Merits and challenges. *Nano Today* 7:532–563.

Gnach, A., T. Lipinski, A. Bednarkiewicz, J. Rybka, and J. Capobianco. 2015. Upconverting nanoparticles: Assessing the toxicity. *Chem. Soc. Rev.* 44:1561–1584

Gorris, H. H. and O. S. Wolfbeis. 2013. Photon-upconverting nanoparticles for optical encoding and multiplexing of cells, biomolecules, and microspheres. *Angew. Chem. Int. Ed.* 52:3584–3600.

Gu, Z., L. Yan, G. Tian, S. Li, Z. Chai, and Y. Zhao. 2013. Recent advances in design and fabrication of upconversion nanoparticles and their safe theranostic applications. *Adv. Mater.* 25:3758–3779.

Guanying, C., L. Haichun, S. Gabriel, L. Huijuan, and Z. Zhiguo. 2009. Upconversion emission tuning from green to red in Yb^{3+}/Ho^{3+}-codoped $NaYF_4$ nanocrystals by tridoping with Ce^{3+} ions. *Nanotechnology* 20:385704.

Güdel, H. U. and M. Pollnau. 2000. Near-infrared to visible photon upconversion processes in lanthanide doped chloride, bromide and iodide lattices. *J. Alloys Compd.* 303–304:307–315.

Guller, A., A. Generalova, E. Petersen et al. 2015. Cytotoxicity and non-specific cellular uptake of bare and surface-modified upconversion nanoparticles in human skin cells. *Nano Res.* 8:1546–1562.

Haase, M. and H. Schäfer. 2011. Upconverting nanoparticles. *Angew. Chem. Int. Ed.* 50:5808–5829.

Hampl, J., M. Hall, N. A. Mufti, Y. M. Yao, D. B. MacQueen, W. H. Wright, and D. E. Cooper. 2001. Upconverting phosphor reporters in immunochromatographic assays. *Anal. Biochem.* 288:176–187.

He, H., C. Zhigang, C. Tianye, Z. Qiang, Y. Mengxiao, L. Fuyou, Y. Tao, and H. Chunhui. 2008. Hydrothermal synthesis of hexagonal lanthanide-doped LaF_3 nanoplates with bright upconversion luminescence. *Nanotechnology* 19:375702.

He, L., L. Feng, L. Cheng, Y. Liu, Z. Li, R. Peng, Y. Li, L. Guo, and Z. Liu. 2013. Multilayer dual-polymer-coated upconversion nanoparticles for multimodal imaging and serum-enhanced gene delivery. *ACS Appl. Mater. Inter.* 5:10381–10388.

Heer, S., O. Lehmann, M. Haase, and H. U. Gudel. 2003. Blue, green, and red upconversion emission from lanthanide-doped $LuPO_4$ and $YbPO_4$ nanocrystals in a transparent colloidal solution. *Angew. Chem.* 42:3179–3182.

Hlaváček, A., A. Sedlmeier, P. Skládal, and H. H. Gorris. 2014. Electrophoretic characterization and purification of silica-coated photon-upconverting nanoparticles and their bioconjugates. *ACS Appl. Mater. Inter.* 6:6930–6935.

Huang, Q., H. Yu, E. Ma, X. Zhang, W. Cao, C. Yang, and J. Yu. 2015. Upconversion effective enhancement by producing various coordination surroundings of rare-earth ions. *Inorg. Chem.* 54:2643–2651.

Huang, X. 2015. Giant enhancement of upconversion emission in $NaYF_4$:$Nd^{3+}/Yb^{3+}/Ho^{3+}/NaYF_4$:$Nd^{3+}/Yb^{3+}$ core/shell nanoparticles excited at 808 nm. *Opt. Lett.* 40:3599–3602.

Idris, N. M., M. K. Jayakumar, A. Bansal, and Y. Zhang. 2014. Upconversion nanoparticles as versatile light nanotransducers for photoactivation applications. *Chem. Soc. Rev.* 44:1449–1478

Idris, N. M., Z. Li, L. Ye, E. K. W. Sim, R. Mahendran, P. C.-L. Ho, and Y. Zhang. 2009. Tracking transplanted cells in live animal using upconversion fluorescent nanoparticles. *Biomaterials* 30:5104–5113.

Jana, N. R., C. Earhart, and J. Y. Ying. 2007. Synthesis of water-soluble and functionalized nanoparticles by silica coating. *Chem. Mater.* 19:5074–5082.

Jaque, D. and F. Vetrone. 2012. Luminescence nanothermometry. *Nanoscale* 4 (15):4301–4326.

Jin, J., Y.-J. Gu, C. W. Y. Man, J. Cheng, Z. Xu, Y. Zhang, H. Wang, V. H.-Y. Lee, S. H. Cheng, and W. T. Wong. 2011. Polymer-coated $NaYF_4:Yb^{3+}$, Er^{3+} upconversion nanoparticles for charge-dependent cellular imaging. *ACS Nano* 5 (10):7838–7847.

Jingning, S., Q. Xiao, Y. Nan, and J. Yiguang. 2007. Synthesis of monodisperse hexagonal $NaYF_4$:Yb, Ln Ln = Er, Ho and Tm upconversion nanocrystals in TOPO. *Nanotechnology* 18:445607.

Ju, Q., D. Tu, Y. Liu, R. Li, H. Zhu, J. Chen, Z. Chen, M. Huang, and X. Chen. 2011. Amine-functionalized lanthanide-doped $KGdF_4$ nanocrystals as potential optical/magnetic multimodal bioprobes. *J. Am. Chem. Soc.* 134:1323–1330.

Kamimura, M., N. Kanayama, K. Tokuzen, K. Soga, and Y. Nagasaki. 2011. Near-infrared 1550 nm *in vivo* bioimaging based on rare-earth doped ceramic nanophosphors modified with PEG-b-poly4-vinylbenzylphosphonate. *Nanoscale* 3:3705–3713.

Kang, X., Z. Cheng, C. Li, D. Yang, M. Shang, P. Ma, G. Li, N. Liu, and J. Lin. 2011. Core–shell structured up-conversion luminescent and mesoporous $NaYF_4:Yb^{3+}/Er^{3+}@nSiO_2@mSiO_2$ nanospheres as carriers for drug delivery. *J. Phys. Chem. C* 115:15801–15811.

Kannan, P., F. Abdul Rahim, R. Chen, X. Teng, L. Huang, H. Sun, and D.-H. Kim. 2013. Au nanorod decoration on $NaYF_4$:Yb/Tm nanoparticles for enhanced emission and wavelength-dependent biomolecular sensing. *ACS Appl. Mater. Inter.* 5:3508–3513.

Kar, A., S. Kundu, and A. Patra. 2015. Lanthanide-doped nanocrystals: Strategies for improving the efficiency of upconversion emission and their physical understanding. *ChemPhysChem* 16:505–521.

Kumar, R., M. Nyk, T. Y. Ohulchanskyy, C. A. Flask, and P. N. Prasad. 2009. Combined optical and MR bioimaging using rare earth ion doped $NaYF_4$ nanocrystals. *Adv. Funct. Mater.* 19:853–859.

Kumar, R., I. Roy, T. Y. Ohulchanskky, L. A. Vathy, E. J. Bergey, M. Sajjad, and P. N. Prasad. 2010. *In vivo* biodistribution and clearance studies using multimodal organically modified silica nanoparticles. *ACS Nano* 4:699–708.

Lakowicz, J. R. 2006. *Principles of Fluorescence Spectroscopy*, New York: Springer. Available from http://site.ebrary.com/id/10229235.

Lakshmanan, A. 2008. *Luminescence and Display Phosphors: Phenomena and Applications*. Nova Science Publishers, New York; Gazelle Drake Academic, Lancaster [distributor].

Lee, J. H., J. T. Jang, J. S. Choi, S. H. Moon, S.-H. Noh, J. W. Kim, J. G. Kim, I. S. Kim, K. I. Park, and J. Cheon. 2011. Exchange-coupled magnetic nanoparticles for efficient heat induction. *Nat. Nanotechnol.* 6:418–422.

Lei, L., D. Chen, W. Zhu, J. Xu, and Y. Wang. 2014. Impact of high ytterbium(III) concentration in the shell on upconversion luminescence of core–shell nanocrystals. *Chem. Asian J.* 9:2765–2770.

Lenth, W., R. M. Macfarlane, and A. J. Silversmith. 1987. Green infrared-pumped upconversion lasers. *Opt. News* 13:26–27.

Li, C., D. Yang, P. A., Ma, Y. Chen, Y. Wu, Z. Hou, Y. Dai, J. Zhao, C. Sui, and J. Lin. 2013. Multifunctional upconversion mesoporous silica nanostructures for dual modal imaging and *in vivo* drug delivery. *Small* 9:4150–4159.

Li, T., J. Moon, A. A. Morrone, J. J. Mecholsky, D. R. Talham, and J. H. Adair. 1999. Preparation of Ag/SiO_2 nanosize composites by a reverse micelle and sol–gel technique. *Langmuir* 15:4328–4334.

Li, X., Y. Wu, Y. Liu, X. Zou, L. Yao, F. Li, and W. Feng. 2014a. Cyclometallated ruthenium complex-modified upconversion nanophosphors for selective detection of Hg^{2+} ions in water. *Nanoscale* 6:1020–1028.

Li, X., F. Zhang, and D. Zhao. 2014b. Lab on upconversion nanoparticles: Optical properties and applications engineering via designed nanostructure. *Chem. Soc. Rev.* 44:1346–1378

Li, Z., B. Wang, L. Xing, S. Liu, N. Tan, S. Xiao, and J. Ding. 2012. Enhancement of upconversion luminescence of $YAlO_3$: Er^{3+} by Gd^{3+} doping. *Chin. Opt. Lett.* 10:081602.

Liao, J., Z. Yang, S. Lai, B. Shao, J. Li, J. Qiu, Z. Song, and Y. Yang. 2014. Upconversion emission enhancement of $NaYF_4$:Yb,Er nanoparticles by coupling silver nanoparticle plasmons and photonic crystal effects. *J. Phys. Chem. C* 118:17992–17999.

Lim, S. F., W. S. Ryu, and R. H. Austin. 2010. Particle size dependence of the dynamic photophysical properties of $NaYF_4$:Yb, Er nanocrystals. *Opt. Exp.* 18:2309–2316.

Lin, H., D. Xu, A. Li, D. Teng, S. Yang, and Y. Zhang. 2015. Tuning of structure and enhancement of upconversion luminescence in $NaLuF_4$:Yb^{3+},Ho^{3+} crystals. *Phys. Chem. Chem. Phys.* 17:19515–19526.

Lin, M., Y. Zhao, S. Wang, M. Liu, Z. Duan, Y. Chen, F. Li, F. Xu, and T. Lu. 2012. Recent advances in synthesis and surface modification of lanthanide-doped upconversion nanoparticles for biomedical applications. *Biotechnol. Adv.* 6:1551–1561.

Liu, C., L. Zhang, Q. Zheng, F. Luo, Y. Xu, and W. Weng. 2012. Advances in the surface engineering of upconversion nanocrystals. *Sci. Adv. Mater.* 4:1–22.

Liu, D., D. Zhao, D. Zhang, K. Zheng, and W. Qin. 2011a. Synthesis and characterization of upconverting $NaYF_4$:Er^{3+}, Yb^{3+} nanocrystals via thermal decomposition of stearate precursor. *J. Nanosci. Nanotechnol.* 11:9770–9773.

Liu, G. 2014. Advances in the theoretical understanding of photon upconversion in rare-earth activated nanophosphors. *Chem. Soc. Rev.* 44:1635–1652.

Liu, J., Y. Liu, Q. Liu, C. Li, L. Sun, and F. Li. 2011b. Iridium(III) complex-coated nanosystem for ratiometric upconversion luminescence bioimaging of cyanide anions. *J. Am. Chem. Soc.* 133:15276–15279.

Liu, L., C. Tang, Y. Zhang, C. Zang, D. Zhang, H. Xiao, R. Qin, and Z. Bao. 2014. Influence of SiO_2 coating on morphology, phase and upconversion luminescence properties of $NaYF_4$:Yb^{3+},Er^{3+} submicrocubes during annealing. *J. Alloys Compd.* 591:320–325.

Liu, Q., J. J. Peng, L. N. Sun, and F. Y. Li. 2011c. High-efficiency upconversion luminescent sensing and bioimaging of Hg(II) by chromophoric ruthenium complex-assembled nanophosphors. *ACS Nano* 5:8040–8048.

Liu, Q., Y. Sun, C. Li, J. Zhou, C. Li, T. Yang, X. Zhang, T. Yi, D. Wu, and F. Li. 2011. ^{18}F-labeled magnetic-upconversion nanophosphors via rare-earth cation-assisted ligand assembly. *ACS Nano* 5:3146–3157.

Lü, Q., F. Guo, L. Sun, A. Li, and L. Zhao. 2008a. Silica-/titania-coated Y_2O_3:Tm^{3+}, Yb^{3+} nanoparticles with improvement in upconversion luminescence induced by different thickness shells. *J. Appl. Phys.* 103:123533.

Lü, Q., A. Li, F. Guo, L. Sun, and L. Zhao. 2008b. Experimental study on the surface modification of Y_2O_3:Tm^{3+}/Yb^{3+} nanoparticles to enhance upconversion fluorescence and weaken aggregation. *Nanotechnology* 19:145701.

Lu, Y. Q., J. B. Zhao, R. Zhang et al. 2014. Tunable lifetime multiplexing using luminescent nanocrystals. *Nat. Photonics* 8:33–37.

Mader, H. S. and O. S. Wolfbeis. 2010. Optical ammonia sensor based on upconverting luminescent nanoparticles. *Anal. Chem.* 82:5002–5004.

Mai, H. X., Y. W. Zhang, R. Si, Z. G. Yan, L. D. Sun, L. P. You, and C. H. Yan. 2006. High-quality sodium rare-earth fluoride nanocrystals: Controlled synthesis and optical properties. *J. Am. Chem. Soc.* 128:6426–6436.

Mai, H. X., Y. W. Zhang, L.-D. Sun, and C.-H. Yan. 2007a. Highly efficient multicolor up-conversion emissions and their mechanisms of monodisperse $NaYF_4$:Yb,Er core and core/shell-structured nanocrystals. *J. Phys. Chem. C* 111:13721–13729.

Mai, H. X., Y. W. Zhang, L. D. Sun, and C. R. Yan. 2007b. Size- and phase-controlled synthesis of monodisperse $NaYF_4$:Yb,Er nanocrystals from a unique delayed nucleation pathway monitored with upconversion spectroscopy. *J. Phys. Chem. C* 111:13730–13739.

Marciniak, L., K. Prorok, L. Francés-Soriano, J. Pérez-Prieto, and A. Bednarkiewicz. 2016. A broadening temperature sensitivity range with a core–shell YbEr@YbNd double ratiometric optical nanothermometer. *Nanoscale* 8: 5037–5042. doi: 10.1039/C5NR08223D

Masahiko, I., G. Christelle, G. Yannick, L. Kheirreddine, F. Tsuguo, and B. Georges. 2004. Crystal growth, Yb^{3+} spectroscopy, concentration quenching analysis and potentiality of laser emission in $Ca_{1-X}Yb_XF_{2+X}$. *J. Phys. Condens. Matter* 16:1501.

Mei, Q. S., Y. Li, B. N. Li, and Y. Zhang. 2015. Oxidative cleavage-based upconversional nanosensor for visual evaluation of antioxidant activity of drugs. *Biosens. Bioelectron.* 64:88–93.

Misiak, M., K. Prorok, B. Cichy, A. Bednarkiewicz, and W. Strek. 2013. Thulium concentration quenching in the up-converting alpha-Tm^{3+}/Yb^{3+} $NaYF_4$ colloidal nanocrystals. *Opt. Mater.* 35:1124–1128.

Moncorgé, R., L. D. Merkle, and B. Zandi. 1999. UV–visible lasers based on rare-earth ions. *MRS Bull.* 24:21–26.

Naczynski, D. J., M. C. Tan, M. Zevon, B. Wall, J. Kohl, A. Kulesa, S. Chen, C. M. Roth, R. E. Riman, and P. V. Moghe. 2013. Rare-earth-doped biological composites as *in vivo* shortwave infrared reporters. *Nat. Commun.* 4:2199.

Niedbala, R. S., H. Feindt, K. Kardos, T. Vail, J. Burton, B. Bielska, S. Li, D. Milunic, P. Bourdelle, and R. Vallejo. 2001. Detection of analytes by immunoassay using up-converting phosphor technology. *Anal. Biochem.* 293:22–30.

Niu, W., S. Wu, and S. Zhanga. 2010. A facile and general approach for the multicolor tuning of lanthanide-ion doped $NaYF_4$ upconversion nanoparticles within a fixed composition *J. Mater. Chem.* 20:9113–9117.

Paik, T., T. R. Gordon, A. M. Prantner, H. Yun, and C. B. Murray. 2013. Designing tripodal and triangular gadolinium oxide nanoplates and self-assembled nanofibrils as potential multimodal bioimaging probes. *ACS Nano* 7:2850–2859.

Park, W., D. Lu, and S. Ahn. 2015. Plasmon enhancement of luminescence upconversion. *Chem. Soc. Rev.* 44:2940–2962.

Pedroni, M., F. Piccinelli, T. Passuello, M. Giarola, G. Mariotto, S. Polizzi, M. Bettinelli, and A. Speghini. 2011. Lanthanide doped upconverting colloidal CaF_2 nanoparticles prepared by a single-step hydrothermal method: Toward efficient materials with near infrared-to-near infrared upconversion emission. *Nanoscale* 3:1456–1460.

Peng, D. F., Q. Ju, X. Chen, R. H. Ma, B. Chen, G. X. Bai, J. H. Hao, X. S. Qiao, X. P. Fan, and F. Wang. 2015. Lanthanide-doped energy cascade nanoparticles: Full spectrum emission by single wavelength excitation. *Chem. Mater.* 27:3115–3120.

Petoral, R. M., F. Söderlind, A. Klasson, A. Suska, M. A. Fortin, N. Abrikossova, L. Selegård, P. O. Käll, M. Engström, and K. Uvdal. 2009. Synthesis and characterization of Tb^{3+}-doped Gd_2O_3 nanocrystals: A bifunctional material with combined fluorescent labeling and MRI contrast agent properties. *J. Phys. Chem. C* 113:6913–6920.

Prorok, K., A. Bednarkiewicz, B. Cichy, A. Gnach, M. Misiak, M. Sobczyk, and W. Strek. 2014a. The impact of shell host $NaYF_4/CaF_2$ and shell deposition methods on the up-conversion enhancement in Tb^{3+}, Yb^{3+} codoped colloidal α-$NaYF_4$ core–shell nanoparticles. *Nanoscale* 6:1855–1864.

Qian, H. S., H. C. Guo, P. C. L. Ho, R. Mahendran, and Y. Zhang. 2009. Mesoporous-silica-coated up-conversion fluorescent nanoparticles for photodynamic therapy. *Small* 5:2285–2290.

Qian, H. S. and Y. Zhang. 2008. Synthesis of hexagonal-phase core–shell $NaYF_4$ nanocrystals with tunable upconversion fluorescence. *Langmuir ACS J. Surf. Colloids* 24:12123–12125.

Qian, L. P., L. H. Zhou, H.-P. Too, and G.-M. Chow. 2010. Gold decorated $NaYF_4$:Yb,Er/$NaYF_4$/silica core/shell/shell upconversion nanoparticles for photothermal destruction of BE(2)-C neuroblastoma cells. *J. Nanoparticle Res.* 13:499–510.

Qiu, H., C. Yang, W. Shao, J. Damasco, X. Wang, H. Ågren, P. Prasad, and G. Chen. 2014. Enhanced upconversion luminescence in Yb^{3+}/Tm^{3+}-codoped fluoride active core/active shell/inert shell nanoparticles through directed energy migration. *Nanomaterials* 4:55–68.

Quarles, G. J., A. Rosenbaum, C. L. Marquardt, and L. Esterowitz. 1990. Efficient room-temperature operation of a flash-lamp-pumped, Cr,Tm:YAG laser at 2.01 microm. *Opt. Lett.* 15:42–44.

Raj, J. G. J., M. Quintanilla, K. A. Mahmoud, A. Ng, F. Vetrone, and M. Zourob. 2015. Sensitive detection of ssDNA using an LRET-based upconverting nanohybrid material. *ACS Appl. Mater. Inter.* 7:18257–18265.

Rajendran, M., E. Yapici, and L. W. Miller. 2014. Lanthanide-based imaging of protein–protein interactions in live cells. *Inorg. Chem.* 53:1839–1853.

Ramasamy, P., P. Chandra, S. W. Rhee, and J. Kim. 2013. Enhanced upconversion luminescence in $NaGdF_4$:Yb,Er nanocrystals by Fe^{3+} doping and their application in bioimaging. *Nanoscale* 5:8711–8717.

Rantanen, T., M. L. Järvenpää, J. Vuojola, K. Kuningas, and T. Soukka. 2008. Fluorescence-quenching-based enzyme-activity assay by using photon upconversion. *Angew. Chem. Int. Ed.* 47:3811–3813.

Renero-Lecuna, C., R. Martín-Rodríguez, R. Valiente, J. González, F. Rodríguez, K. W. Krämer, and H. U. Güdel. 2011. Origin of the high upconversion green luminescence efficiency in β-$NaYF_4$:2%Er^{3+},20%Yb^{3+}. *Chem. Mater.* 23:3442–3448.

Rieffel, J., F. Chen, J. Kim et al. 2015. Hexamodal imaging with porphyrin–phospholipid-coated upconversion nanoparticles. *Adv. Mater.* 27:1785–1790.

Rocha, U., C. J. da Silva, W. F. Silva et al. 2013. Subtissue thermal sensing based on neodymium-doped LaF_3 nanoparticles. *ACS Nano* 7:1188–1199.

Rocha, U., K. U. Kumar, C. Jacinto et al. 2014. Neodymium-doped LaF_3 nanoparticles for fluorescence bioimaging in the second biological window. *Small* 10: 1141–1154.

Ronda, C. R. 1997. Recent achievements in research on phosphors for lamps and displays. *J. Lumin.* 72–74:49–54.

Saleh, S. M., R. Ali, T. Hirsch, and S. W. Otto. 2010. Detection of biotin–avidin affinity binding by exploiting a self-referenced system composed of upconverting nanoparticles and gold nanoparticles. *Biosens. Bioelectron.* 13:4603–4611.

Schietinger, S., L. d. S., Menezes, B. Lauritzen, and O. Benson. 2009. Observation of size dependence in multicolor upconversion in single Yb^{3+}, Er^{3+} codoped $NaYF_4$ nanocrystals. *Nano Lett.* 9:2477–2481.

Shan, J., S. J. Budijono, G. Hu, N. Yao, Y. Kang, Y. Ju, and R. K. Prud'homme. 2011. Pegylated composite nanoparticles containing upconverting phosphors and meso-tetraphenyl porphine TPP for photodynamic therapy. *Adv. Funct. Mater.* 21:2488–2495.

Shao, W., G. Chen, T. Y. Ohulchanskyy, A. Kuzmin, J. Damasco, H. Qiu, C. Yang, H. Ågren, and P. N. Prasad. 2015. Lanthanide-doped fluoride core/multishell nanoparticles for broadband upconversion of infrared light. *Adv. Opt. Mater.* 3:575–582.

Shen, J., Z. Q. Li, Y. R. Chen, X. H. Chen, Y. W. Chen, Z. Sun, and S. M. Huang. 2013. Influence of SiO_2 layer thickness on plasmon enhanced upconversion in hybrid $Ag/SiO_2/NaYF_4$:Yb, Er, Gd structures. *Appl. Surf. Sci.* 270:712–717.

Sivakumar, S., P. R. Diamente, and F. C. J. M. van Veggel. 2006. Silica-coated Ln^{3+}-doped LaF_3 nanoparticles as robust down- and upconverting biolabels. *Chem. Eur. J.* 12:5878–5884.

Song, Y., G. Liu, J. Wang, X. Dong, and W. Yu. 2014. Synthesis and luminescence resonance energy transfer based on noble metal nanoparticles and the $NaYF_4:Tb^{3+}$ shell. *Phys. Chem. Chem. Phys.* 16:15139–15145.

Sotiriou, G. A., M. Schneider, and S. E. Pratsinis. 2012. Green, silica-coated monoclinic $Y_2O_3:Tb^{3+}$ nanophosphors: Flame synthesis and characterization. *J. Phys. Chem. C* 116:4493–4499.

Stober, W., A. Fink, and E. Bohn. 1968. Controlled growth of monodisperse silica spheres in the micron size range. *J. Colloid Interface Sci.* 26:62–69.

Sun, L. N., H. Peng, M. I. J. Stich, D. Achatz, and O. S. Wolfbeis. 2009. pH sensor based on upconverting luminescent lanthanide nanorods. *Chem. Commun.* 33:5000–5002.

Sun, Y., W. Feng, P. Yang, C. Huang, and F. Li. 2014. The biosafety of lanthanide upconversion nanomaterials. *Chem. Soc. Rev.* 44:1509–1525.

Sun, Y., X. Zhu, J. Peng, and F. Li. 2013. Core–shell lanthanide upconversion nanophosphors as four-modal probes for tumor angiogenesis imaging. *ACS Nano* 7:11290–11300.

Suyver, J. F., J. Grimm, M. K. van Veen, D. Biner, K. W. Kramer, and H. U. Gudel. 2006. Upconversion spectroscopy and properties of $NaYF_4$ doped with Er^{3+}, Tm^{3+} and/or Yb^{3+}. *J. Lumin.* 117:1–12.

Tian, G., Z. Gu, L. Zhou et al. 2012. Mn^{2+} dopant-controlled synthesis of $NaYF_4$:Yb/Er upconversion nanoparticles for *in vivo* imaging and drug delivery. *Adv. Mater.* 24:1226–1231.

Tsang, M. K., G. Bai, and J. Hao. 2015. Stimuli responsive upconversion luminescence nanomaterials and films for various applications. *Chem. Soc. Rev.* 44:1585–1607

Tu, D., Y. Liu, H. Zhu, R. Li, L. Liu, and X. Chen. 2013. Breakdown of crystallographic site symmetry in lanthanide-doped $NaYF_4$ crystals. *Angew. Chem. Int. Ed.* 52:1128–1133.

Tu, L., X. Liu, F. Wu, and H. Zhang. 2014. Excitation energy migration dynamics in upconversion nanomaterials. *Chem. Soc. Rev.* 44:1331–1345

van de Rijke, F., H. Zijlmans, S. Li, T. Vail, A. K. Raap, R. S. Niedbala, and H. J. Tanke. 2001. Up-converting phosphor reporters for nucleic acid microarrays. *Nat. Biotechnol.* 19:273–276.

van Veggel, F. C. J. M., C. Dong, N. J. J. Johnson, and J. Pichaandi. 2012. Ln^{3+}-doped nanoparticles for upconversion and magnetic resonance imaging: Some critical notes on recent progress and some aspects to be considered. *Nanoscale* 4:7309–7321.

Vetrone, F., R. Naccache, V. Mahalingam, C. G. Morgan, and J. A. Capobianco. 2009. The active-core/active-shell approach: A strategy to enhance the upconversion luminescence in lanthanide-doped nanoparticles. *Adv. Funct. Mater.* 19:2924–2929.

Vetrone, F., R. Naccache, A. Zamarrón, A. Juarranz de la Fuente, F. Sanz-Rodríguez, L. Martinez Maestro, E. Martín Rodriguez, D. Jaque, J. García Solé, and J. A. Capobianco. 2010. Temperature sensing using fluorescent nanothermometers. *ACS Nano* 4:3254–3258.

Wang, C., L. Cheng, and Z. Liu. 2011a. Drug delivery with upconversion nanoparticles for multi-functional targeted cancer cell imaging and therapy. *Biomaterials* 32:1110–1120.

Wang, C., L. Cheng, and Z. Liu. 2013. Upconversion nanoparticles for photodynamic therapy and other cancer therapeutics. *Theranostics* 3:317–330.

Wang, C., L. Cheng, H. Xu, and Z. Liu. 2012. Towards whole-body imaging at the single cell level using ultra-sensitive stem cell labeling with oligo-arginine modified upconversion nanoparticles. *Biomaterials* 33:4872–4881.

Wang, F., R. Deng, J. Wang, Q. Wang, Y. Han, H. Zhu, X. Chen, and X. Liu. 2011b. Tuning upconversion through energy migration in core–shell nanoparticles. *Nat. Mater.* 10:968–973.

Wang, F., Y. Han, C. S. Lim, Y. Lu, J. Wang, J. Xu, H. Chen, C. Zhang, M. Hong, and X. Liu. 2010. Simultaneous phase and size control of upconversion nanocrystals through lanthanide doping. *Nature* 463:1061–1065.

Wang, F. and X. Liu. 2008. Upconversion multicolor fine-tuning: Visible to near-infrared emission from lanthanide-doped $NaYF_4$ nanoparticles. *J. Am. Chem. Soc.* 130:5642–5643.

Wang, F. and X. Liu. 2009. Recent advances in the chemistry of lanthanide-doped upconversion nanocrystals. *Chem. Soc. Rev.* 38:976–989.

Wang, H., W. Lu, Z. Yi, L. Rao, S. Zeng, and Z. Li. 2015. Enhanced upconversion luminescence and single-band red emission of $NaErF_4$ nanocrystals via Mn^{2+} doping. *J. Alloys Compd.* 618:776–780.

Wang, J.-W., Y.-M. Chang, H.-C. Chang, S.-H. Lin, L. C. L. Huang, X.-L. Kong, and M.-W. Kang. 2005. Local structure dependence of the charge transfer band in nanocrystalline Y_2O_3:Eu^{3+}. *Chem. Phys. Lett.* 405:314–317.

Wang, L. and Y. Li. 2007. Luminescent nanocrystals for nonenzymatic glucose concentration determination. *Chem. Eur. J.* 13:4203–4207.

Wang, L., R. Yan, Z. Huo, L. Wang, J. Zeng, J. Bao, X. Wang, Q. Peng, and Y. Li. 2005. Fluorescence resonant energy transfer biosensor based on upconversion-luminescent nanoparticles. *Angew. Chem. Int. Ed.* 44:6054–6057.

Wang, M., G. Abbineni, A. Clevenger, C. Mao, and S. Xu. 2011c. Upconversion nanoparticles: Synthesis, surface modification and biological applications. *Nanomed. Nanotechnol. Biol. Med.* 7:710–729.

Wang, M., W. Hou, C.-C. Mi, W.-X. Wang, Z.-R. Xu, H.-H. Teng, C.-B. Mao, and S.-K. Xu. 2009a. Immunoassay of goat antihuman immunoglobulin G antibody based on luminescence resonance energy transfer between near-infrared responsive $NaYF_4$:Yb, Er upconversion fluorescent nanoparticles and gold nanoparticles. *Anal. Chem.* 81:8783–8789.

Wang, M., C.-C. Mi, J.-L. Liu, X.-L. Wu, Y.-X. Zhang, W. Hou, F. Li, and S.-K. Xu. 2009b. One-step synthesis and characterization of water-soluble $NaYF_4$:Yb,Er/polymer nanoparticles with efficient up-conversion fluorescence. *J. Alloys Compd.* 485:L24–L27.

Wang, M., C. Mi, Y. Zhang, J. Liu, F. Li, C. Mao, and S. Xu. 2009c. NIR-responsive silica-coated $NaYbF_4$:Er/Tm/Ho upconversion fluorescent nanoparticles with tunable emission colors and their applications in immunolabeling and fluorescent imaging of cancer cells. *J. Phys. Chem. C* 113:19021–19027.

Wang, R. and F. Zhang. 2014. NIR luminescent nanomaterials for biomedical imaging. *J. Mater. Chem. B* 2:2422–2443.

Wang, Y., F. Nan, Z. Cheng, J. Han, Z. Hao, H. Xu, and Q. Wang. 2015. Strong tunability of cooperative energy transfer in Mn^{2+}-doped Yb^{3+}, $Er^{3+}/NaYF_4$ nanocrystals by coupling with silver nanorod array. *Nano Res.* 8:2970–2977.

Wang, Y., L. Tu, J. Zhao, Y. Sun, X. Kong, and H. Zhang. 2009. Upconversion luminescence of beta-$NaYF_4$: Yb^{3+}, Er^{3+}@beta-$NaYF_4$ core/shell nanoparticles: Excitation power, density and surface dependence. *J. Phys. Chem. C* 113:7164–7169.

Wang, Y. F., G. Y. Liu, L. D. Sun, J. W. Xiao, J. C. Zhou, and C. H. Yan. 2013. Nd^{3+}-sensitized upconversion nanophosphors: Efficient *in vivo* bioimaging probes with minimized heating effect. *ACS Nano* 7:7200–7206.

Wang, Y. F., L. D. Sun, J. W. Xiao, W. Feng, J. C. Zhou, J. Shen, and C. H. Yan. 2012. Rare-earth nanoparticles with enhanced upconversion emission and suppressed rare-earth-ion leakage. *Chem. Eur. J.* 18:5558–5564.

Wawrzynczyk, D., A. Bednarkiewicz, M. Nyk, W. Strek, and M. Samoc. 2012. Neodymium(III) doped fluoride nanoparticles as non-contact optical temperature sensors. *Nanoscale* 4:6959–6961.

Wawrzynczyk, D., M. Nyk, A. Bednarkiewicz, W. Strek, and M. Samoc. 2014. Morphology- and size-dependent spectroscopic properties of Eu^{3+}-doped Gd_2O_3 colloidal nanocrystals. *J. Nanoparticle Res.* 16:1–13.

Wei, W., Y. Zhang, R. Chen, J. Goggi, N. Ren, L. Huang, K. K. Bhakoo, H. Sun, and T. T. Y. Tan. 2014. Cross relaxation induced pure red upconversion in activator- and sensitizer-rich lanthanide nanoparticles. *Chem. Mater.* 26:5183–5186.

Xia, A., M. Chen, Y. Gao, D. Wu, W. Feng, and F. Li. 2012. Gd^{3+} complex-modified $NaLuF_4$-based upconversion nanophosphors for trimodality imaging of NIR-to-NIR upconversion luminescence, X-ray computed tomography and magnetic resonance. *Biomaterials* 33:5394–5405.

Xia, A., Y. Gao, J. Zhou, C. Y. Li, T. S. Yang, D. M. Wu, L. M. Wu, and F. Y. Li. 2011. Core–shell $NaYF_4$:Yb^{3+},Tm^{3+}@FexOy nanocrystals for dual-modality T_2-enhanced magnetic resonance and NIR-to-NIR upconversion luminescent imaging of small-animal lymphatic node. *Biomaterials* 32:7200–7208.

Xia, A., X. Zhang, J. Zhang, Y. Deng, Q. Chen, S. Wu, X. Huang, and J. Shen. 2014. Enhanced dual contrast agent, Co^{2+}-doped $NaYF_4$:Yb^{3+},Tm^{3+} nanorods, for near infrared-to-near infrared upconversion luminescence and magnetic resonance imaging. *Biomaterials* 35:9167–9176.

Xiao, W., X. Liu, X. Hong, Y. Yang, Y. Lv, J. Fang, and J. Ding. 2015. Magnetic-field-assisted synthesis of magnetite nanoparticles via thermal decomposition and their hyperthermia properties. *CrystEngComm* 17:3652–3658.

Xiaoting, Z., T. Hayakawa, Y. Ishikawa, Y. Liushuan, and M. Nogami. 2015. Structural investigation and Eu^{3+} luminescence properties of LaF_3:Eu^{3+} nanophosphors. *J. Alloys Compd.* 644:77–81.

Xie, X., N. Gao, R. Deng, Q. Sun, Q.-H. Xu, and X. Liu. 2013. Mechanistic investigation of photon upconversion in Nd^{3+}-sensitized core–shell nanoparticles. *J. Am. Chem. Soc.* 135:12608–12611.

Xing, H., W. Bu, S. Zhang et al. 2012. Multifunctional nanoprobes for upconversion fluorescence, MR and CT trimodal imaging. *Biomaterials* 33:1079–1089.

Xiong, L., T. Yang, Y. Yang, C. Xu, and F. Li. 2010. Long-term *in vivo* biodistribution imaging and toxicity of polyacrylic acid-coated upconversion nanophosphors. *Biomaterials* 31:7078–7085.

Xu, C. T., P. Svenmarker, H. Liu, X. Wu, M. E. Messing, L. R. Wallenberg, and S. Andersson-Engels. 2012. High-resolution fluorescence diffuse optical tomography developed with nonlinear upconverting nanoparticles. *ACS Nano* 6:4788–4795.

Xu, Z., M. Quintanilla, F. Vetrone, A. O. Govorov, M. Chaker, and D. Ma. 2015. Harvesting lost photons: Plasmon and upconversion enhanced broadband photocatalytic activity in cre@shell microspheres based on lanthanide-doped $NaYF_4$, TiO_2, and Au. *Adv. Funct. Mater.* 25:2950–2960.

Yang, D., P. Ma, Z. Hou, Z. Cheng, C. Li, and J. Lin. 2014. Current advances in lanthanide ion Ln-based upconversion nanomaterials for drug delivery. *Chem. Soc. Rev.* 44:1416–1448.

Yang, D., P. A. Ma, Z. Hou, Z. Cheng, C. Li, and J. Lin. 2015. Current advances in lanthanide ion Ln^{3+}-based upconversion nanomaterials for drug delivery. *Chem. Soc. Rev.* 44:1416–1448.

Yang, Y., Y. Qu, J. Zhao, Q. Zeng, Y. Ran, Q. Zhang, X. Kong, and H. Zhang. 2010. Fabrication of and drug delivery by an upconversion emission nanocomposite with monodisperse LaF_3:Yb,Er core/mesoporous silica shell structure. *Eur. J. Inorg. Chem.* 33:5195–5199.

Ye, X., J. E. Collins, Y. Kang, J. Chen, D. T. N. Chen, A. G. Yodh, and C. B. Murray. 2010. Morphologically controlled synthesis of colloidal upconversion nanophosphors and their shape-directed self-assembly. *Proc. Natl. Acad. Sci. USA* 107:22430–22435.

Yi, D. K., S. S. Lee, G. C. Papaefthymiou, and J. Y. Ying. 2006. Nanoparticle architectures templated by SiO_2/Fe_2O_3 nanocomposites. *Chem. Mater.* 18:614–619.

Yi, G. S. and G. M. Chow. 2005. Colloidal LaF_3:Yb,Er, LaF_3:Yb,Ho and LaF_3:Yb,Tm nanocrystals with multicolor upconversion fluorescence. *J. Mater. Chem.* 15:4460–4464.

Yi, G. S. and G. M. Chow. 2006. Water-soluble $NaYF_4$:Yb,ErTm/$NaYF_4$/polymer core/shell/shell nanoparticles with significant enhancement of upconversion fluorescence. *Chem. Mater.* 19:341–343.

Yi, G. S. and G. M. Chow. 2007. Water-soluble $NaYF_4$:Yb,ErTm/$NaYF_4$/polymer core/shell/shell nanoparticles with significant enhancement of upconversion fluorescence. *Chem. Mater.* 19:341–343.

Yin, B., W. Zhou, Q. Long, C. Li, Y. Zhang, and S. Yao. 2014. Salt-assisted rapid transformation of $NaYF_4:Yb^{3+},Er^{3+}$ nanocrystals from cubic to hexagonal. *CrystEngComm* 16:8348–8355.

Yin, W., L. Zhou, Z. Gu et al. 2012. Lanthanide-doped $GdVO_4$ upconversion nanophosphors with tunable emissions and their applications for biomedical imaging. *J. Mater. Chem.* 22:6974–6981.

Ylihärsilä, M., T. Valta, M. Karp, L. Hattara, E. Harju, J. Hölsä, P. Saviranta, M. Waris, and T. Soukka. 2011. Oligonucleotide array-in-well platform for detection and genotyping human adenoviruses by utilizing upconverting phosphor label technology. *Anal. Chem.* 83:1456–1461.

Yu, X. F., L. D. Chen, M. Li, M. Y. Xie, L. Zhou, T. Li, and Q. Q. Wang. 2008. Highly efficient fluorescence of NdF_3 SiO_2 core/shell nanoparticles and the applications for *in vivo* NIR detection. *Adv. Mater.* 10:1–6.

Zhang, F., R. Che, X. Li, C. Yao, J. Yang, D. Shen, P. Hu, W. Li, and D. Zhao. 2012. Direct imaging the upconversion nanocrystal core/shell structure at the subnanometer level: Shell thickness dependence in upconverting optical properties. *Nano Lett.* 12:2852–2858.

Zhang, F., J. Li, J. Shan, L. Xu, and D. Y. Zhao. 2009. Shape, size, and phase-controlled rare-earth fluoride nanocrystals with optical up-conversion properties. *Chem. Eur. J.* 15:11010–11019.

Zhang, H., Y. Li, I. A. Ivanov, Y. Qu, Y. Huang, and X. Duan. 2010. Plasmonic modulation of the upconversion fluorescence in $NaYF_4$:Yb/Tm hexaplate nanocrystals using gold nanoparticles or nanoshells. *Angew. Chem. Int. Ed.* 49:2865–2868.

Zhang, J., F. Liu, T. Li, X. He, and Z. Wang. 2015a. Surface charge effect on the cellular interaction and cytotoxicity of $NaYF_4:Yb^{3+}$, $Er^{3+}@SiO_2$ nanoparticles. *RSC Adv.* 5:7773–7780.

Zhang, S. Z., L. D. Sun, H. Tian, Y. Liu, J. F. Wang, and C. H. Yan. 2009. Reversible luminescence switching of $NaYF_4$:Yb,Er nanoparticles with controlled assembly of gold nanoparticles. *Chem. Commun.* 18:2547–2549.

Zhang, Y.-W., X. Sun, R. Si, L.-P. You, and C.-H. Yan. 2005. Single-crystalline and monodisperse LaF_3 triangular nanoplates from a single-source precursor. *J. Am. Chem. Soc.* 127:3260–3261.

Zhang, Z., X. Ma, Z. Geng, K. Wang, and Z. Wang. 2015b. One-step synthesis of carboxyl-functionalized rare-earth fluoride nanoparticles for cell imaging and drug delivery. *RSC Adv.* 5:33999–34007.

Zhao, J., D. Jin, E. P. Schartner et al. 2013. Single-nanocrystal sensitivity achieved by enhanced upconversion luminescence. *Nat. Nanotechnol.* 8:729–734.

Zhao, J. B., Z. D. Lu, Y. D. Yin, C. Mcrae, J. A. Piper, J. M. Dawes, D. Y. Jin, and E. M. Goldys. 2013. Upconversion luminescence with tunable lifetime in $NaYF_4$:Yb,Er nanocrystals: Role of nanocrystal size. *Nanoscale* 5:944–952.

Zhao, P., Y. Zhu, X. Yang, X. Jiang, J. Shen, and C. Li. 2014. Plasmon-enhanced efficient dye-sensitized solar cells using core–shell-structured beta-$NaYF_4$:Yb,Er@SiO_2@Au nanocomposites. *J. Mater. Chem. A* 2:16523–16530.

Zhao, Z., Y. Han, C. Lin, D. Hu, F. Wang, X. Chen, Z. Chen, and N. Zheng. 2012. Multifunctional core–shell upconverting nanoparticles for imaging and photodynamic therapy of liver cancer cells. *Chem. Asia J.* 7:830–837.

Zheng, W., P. Huang, D. Tu, E. Ma, H. Zhu, and X. Chen. 2015. Lanthanide-doped upconversion nano-bioprobes: Electronic structures, optical properties, and biodetection. *Chem. Soc. Rev.* 44:1379–1415

Zhong, Y., G. Tian, Z. Gu, Y. Yang, L. Gu, Y. Zhao, Y. Ma, and J. Yao. 2014. Elimination of photon quenching by a transition layer to fabricate a quenching-shield sandwich structure for 800 nm excited upconversion luminescence of Nd-sensitized nanoparticles. *Adv. Mater.* 26:2831–2837.

Zhou, A., Y. Wei, B. Wu, Q. Chen, and D. Xing. 2012. Pyropheophorbide A and cRGDyK comodified chitosan-wrapped upconversion nanoparticle for targeted near-infrared photodynamic therapy. *Mol. Pharmaceut.* 9:1580–1589.

Zhou, J., Z. Liu, and F. Li. 2012. Upconversion nanophosphors for small-animal imaging. *Chem. Soc. Rev.* 41:1323–1349.

Zhou, J., Y. Sun, X. Du, L. Xiong, H. Hu, and F. Li. 2010. Dual-modality *in vivo* imaging using rare-earth nanocrystals with near-infrared to near-infrared NIR-to-NIR upconversion luminescence and magnetic resonance properties. *Biomaterials* 31:3287–3295.

Zhou, J., M. Yu, Y. Sun, X. Zhang, X. Zhu, Z. Wu, D. Wu, and F. Li. 2011. Fluorine-18-labeled $Gd^{3+}/Yb^{3+}/Er^{3+}$ co-doped $NaYF_4$ nanophosphors for multimodality PET/MR/UCL imaging. *Biomaterials* 32:1148–1156.

Zhou, L., X. Zheng, Z. Gu, W. Yin, X. Zhang, L. Ruan, Y. Yang, Z. Hu, and Y. Zhao. 2014. Mesoporous $NaYbF_4$@$NaGdF_4$ core–shell up-conversion nanoparticles for targeted drug delivery and multimodal imaging. *Biomaterials* 35:7666–7678.

Zijlmans, H. J., J. Bonnet, J. Burton, K. Kardos, T. Vail, R. S. Niedbala, and H. J. Tanke. 1999. Detection of cell and tissue surface antigens using up-converting phosphors: A new reporter technology. *Anal. Biochem.* 267:30–36.

9

Upconversion Nanoparticles for Phototherapy

Akshaya Bansal and Zhang Yong

CONTENTS

9.1 Introduction: Need for NIR-Based Phototherapy ... 255
9.2 Photoinduced ROS Production ... 257
9.2.1 Photodynamic Therapy ... 257
9.2.1.1 PSs for PDT ... 259
9.2.1.2 UCNPs for PDT: Host Matrix, Dopants, and Surface Coatings ... 261
9.2.1.3 PS Loading Strategies ... 264
9.2.1.4 *In Vitro* and *In Vivo* PDT Using UCNPs ... 266
9.2.2 Photochemical Internalization ... 268
9.2.2.1 Photosensitive Chemicals for PCI ... 270
9.3 Photocontrolled Release of Molecules ... 272
9.3.1 Photolabile Groups in Photocontrolled Delivery ... 272
9.3.2 UCNP-Based Phototriggered Release *In Vitro* and *In Vivo* ... 274
9.4 Photothermal Therapy ... 277
9.4.1 UCNP-Based PTT ... 278
9.5 UCNPs in Combination Therapy ... 279
9.6 Limitations of UCNPs ... 281
9.7 Conclusion and Outlook ... 283
References ... 284

9.1 Introduction: Need for NIR-Based Phototherapy

Photoactivation has garnered tremendous interest in the last decade, with widespread applications ranging from medicine to energy harvesting and even wastewater treatment. This increasing interest stems from the fact that photoactivation allows the use of light—a noninvasive, spatially, and temporally controllable stimulus to achieve specific outcomes. This is particularly useful for therapeutic applications where specificity and safety are a major concern. With the use of light, therapy can be targeted, reducing side effects and improving efficacy of treatment. In addition, the ability to control parameters such as wavelength, intensity, and duration of exposure, provide an added degree of control. To this end, phototherapies such as photodynamic therapy (PDT), photothermal therapy (PTT), photocontrolled release of drugs/nucleic acids, and combination therapies involving synergistic use of two or more of the previous techniques have come to the fore.

PDT involves the use of photoresponsive chemicals called photosensitizers (PSs), which produce reactive oxygen species (ROS) upon irradiation with light of a particular wavelength (usually in the visible range), subsequently killing cells in the vicinity. This type of targeted cell kill is being explored as a cancer-treatment modality. Besides PDT, localized generation of ROS at lower concentrations is also used to improve cytoplasmic delivery of biomolecules into cells, a technique called photochemical internalization (PCI). In PCI, the amount of ROS generated is not enough to kill the cell but can rupture endosomes into which the cells take up biomolecules of interest through endocytosis. This causes the cargo to be released into the cytoplasm. Besides generation of ROS, there are phototherapies working on different principles such as light to heat transduction, as is the case with PTT. Localized increase in temperature can also be used to kill cells in a targeted manner and is being developed as a cancer therapy. In addition to the phototherapies mentioned above, light can also be used for controlled delivery of biomolecules. This is usually done by either "caging" the molecule of interest or sequestering it through the use of photolabile moieties. Upon irradiation with light of a suitable wavelength (usually ultraviolet [UV] light), the caging or sequestering photolabile group is cleaved, thereby "uncaging" or releasing the biomolecule of interest at the desired site. This strategy has been used for site-specific activation/delivery of chemotherapeutic drugs, nucleic acids like siRNA, and other small molecules, with greater temporal control and reduced off target effects.

Although the techniques mentioned above have great therapeutic potential, their practical applications are hindered by certain constraints. Most photoresponsive moieties used in the techniques mentioned earlier, respond either to visible or UV light. These wavelengths have low tissue penetration and UV in particular is known to be toxic. In the near-infrared (NIR) range, light absorption and tissue scattering is minimal, allowing for greater tissue penetration. Thus, phototherapies that allow use of NIR light as opposed to visible or UV light would be advantageous. Even though efforts have been made to develop NIR light sensitive moieties for use in phototherapy, the field still in its infancy. Since most photoresponsive moieties cannot use NIR light directly, a means of transducing this light to the usable visible or UV regions is needed. Upconversion nanoparticles (UCNPs) are such transducers. They are excited by light in the NIR range (most commonly 980 nm) and can be tuned to emit in the UV, visible, and infrared ranges, in accordance with the absorption requirements of the photoresponsive moiety being used. The details on how UCNPs work and their synthesis can be found in the previous chapters. This chapter will delve into the applications of UCNPs for phototherapy, the limitations of using UCNPs, and efforts being made to overcome these limitations.

9.2 Photoinduced ROS Production

Chemicals called PSs produce ROS when irradiated with light of a suitable wavelength. This ROS production can be used therapeutically in different ways. It can be used to cause targeted cell kill in a technique called PDT or when generated in a localized manner at a lower concentration, be used to aid endosomal escape—a major bottleneck in delivery of biomolecules especially when using nanoparticles as carriers. This is called PCI (Berg et al. 2006). This section will describe both PDT and PCI in detail and explain how UCNPs are useful as both carriers and transducers, allowing the use of NIR instead of UV or visible light for these techniques, thereby improving their clinical potential.

9.2.1 Photodynamic Therapy

PDT involves dye-sensitized photooxidation of biological matter. When a photosensitizing agent is irradiated with light of a specific wavelength, oxygenated products (ROS) harmful to cell function arise, which eventually result in tissue destruction through a variety of mechanisms (Henderson and Dougherty 1992). There are three components essential for implementing PDT—a PS, light, and the presence of oxygen. Light of a suitable wavelength that matches the absorption maximum of the PS excites the electrons of this PS to a higher energy level, converting it to a short-lived singlet state and then through intersystem crossing to a relatively long-lived triplet state (Juzeniene et al. 2007). This excited state can transfer energy to a neighboring substrate, producing free radical species, which then react with the surrounding oxygen to generate superoxide anion radicals, hydroxyl radicals, and hydrogen peroxide. This type of ROS generation is classified as type I. Another mechanism, involves direct transfer of energy from a triplet-state PS to surrounding molecular oxygen resulting in the formation of singlet oxygen ($^1O_2$). This type II reaction is the prevailing mechanism of ROS generation for most PSs (Bonnett 1995; Idris et al. 2015) (Figure 9.1).

Successful application of PDT involves consideration of parameters such as PS type, its administration and distribution in the body, type of light source used, mode of light delivery, and finally the immediate and late effects of PDT. The tumoricidal effect of PDT is multifactorial and results from a variety of direct and indirect responses of the cells, vasculature, and immune system. The acute or lethal tumor cell killing effect arises from apoptosis or necrosis that results from direct damage to proteins, lipids, or nucleic acids in the

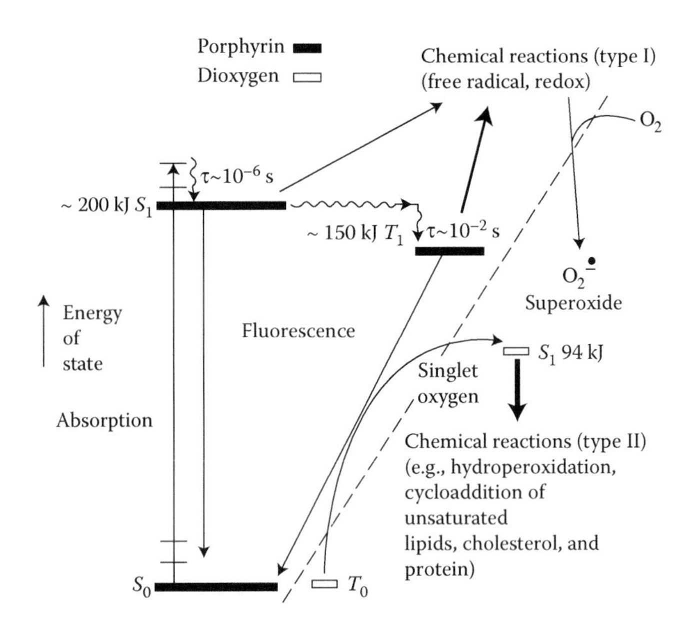

FIGURE 9.1
Schematic showing types I and II reactions for ROS generation. (Reproduced with permission from Bonnett, R., *Chem. Soc. Rev.*, 24 (1), 19–33, 1995.)

cells due to the ROS produced by the PS taken up by these cells or through oxygen and nutrient deprivation resulting from destruction of tumor vasculature. Secondary tumor killing responses on the other hand are caused due to activation of the immune system. Nonlethal photooxidative lesions of membrane lipids activate membranous phospholipidases leading to accelerated phospholipid degradation, release of lipid fragments, and metabolites of arachidonic acid that are powerful inflammatory mediators. In addition, even minor photodamage to the vasculature can attract circulating immune cells such as neutrophils and platelets, resulting in a progressive impairment of vascular function along with massive release of inflammatory cytokines. Thus, nonlethal photooxidation can set of a cascade of events leading to a strong inflammatory response further amplifying PDT-mediated tumoricidal response (Castano et al. 2006; Dougherty et al. 1998; Plaetzer et al. 2009).

The irreversible damage caused to the tissue during PDT has been used successfully for treating a variety of solid tumors such as those involving the lung, bladder, prostrate, skin, etc. Besides cancer therapy, PDT is also being explored as an antimicrobial tool with potential applications in treating infections, cardiovascular, dermatological, and ophthalmic diseases. Although promising, the widespread acceptance of PDT as a primary therapy faces certain hurdles. The most prominent of these, as mentioned earlier is the low penetration depth of light sources used. Most PSs used for PDT have an absorption maximum in the visible region of light, where tissue absorption and scattering is high and as such light can only penetrate to a few millimeters (1–3 mm). This limits the application of PDT to superficial tumors or those accessible by an endoscope. In the 700–1000 nm range, maximal tissue transparency is attained wherein scattering and absorption of light by biomolecules is minimal. Thus, to target deeper tumors, a longer wavelength of light such as NIR is desirable. Although efforts have been made to develop PSs that respond to this range of light, the field is still in its infancy and such PSs are rare. UCNPs provide an elegant solution to this problem, wherein the older tried and tested PSs can be used in conjunction with NIR excitation. Here, these UCNPs serve as nanotransducers that convert NIR light into the visible light needed for exciting the PSs. Besides their role as transducers, these nanoparticles can also serve as carriers of these PSs and can be targeted to specific tissues through various surface modifications. The following section will delve into the various PSs used for PDT, types of UCNPs that can be used in this technique, various surface modifications that allow loading of different types of PSs on to UCNPs, targeting moieties for selective accumulation in tumor regions, and finally how such systems can be implemented *in vitro* and *in vivo*.

9.2.1.1 PSs for PDT

PSs are the central component of the light–PS–oxygen system needed for PDT. Ideally, a PS used for cancer therapy should localize preferentially to the tumor region and not accumulate in off target organs such as the lungs or the skin (unless these are desired targets), be amphiphilic so as to allow blood transport (hydrophilicity) as well as uptake by cells (lipophilicity), have low dark toxicity such that it only causes cell kill when irradiated with light, good ROS production efficiency, a high absorption coefficient at longer wavelength of light so deeper tissues can be targeted, and lastly, be easy to synthesize. In short, an ideal PS should localize to the site of interest and kill cells only when excited with light of a suitable wavelength for a targeted therapy while mitigating unwanted side effects. With these considerations in mind, several PSs have been designed for PDT. Depending on their chemical composition, they can be classified as porphyrins, chlorins, 5-aminolevulinic acid (ALA), and naphthalocyanines. Of these PSs, porphyrins are the oldest and most widely studied, with many of this family approved for clinical use. Besides porphyrins, some chlorins and phthalocyanines are also available clinically. The PSs mentioned here differ from each other in terms of circulation times, localization in specific subcellular compartments, and the wavelength of light they absorb. Some PSs like Photofrin® (a hematoporphyrin derivative) have long-circulation times while others accumulate rapidly in tissues. Rapid accumulation and clearance may be advantageous for a single application of PS followed by treatment the same day. On the other hand, long-circulation times and low clearance may allow for multiple PDT sessions with a single infusion of the PS (Allison and Sibata 2010). In addition, those PSs with high circulation times can also be used to induce vascular shutdown following illumination and cut down of oxygen supply to the tumor, another means of bringing about tumor destruction.

A useful way to classify PSs can also be on the basis of the subcellular compartment to which they localize once the cell takes them up since the mechanism of cell lethality via PDT depends to some extent on the location of the PS in the cell. Sensitizers that tend to localize in the mitochondria like Photofrin or ALA (which produces the active compound protoporphyrin IX (PPIX) in the mitochondria), kill the cells usually through apoptosis while those that accumulate in the plasma membrane do so through necrosis in most cases (Dougherty et al. 1998). An association between mitochondrial damage due to PDT and apoptosis has been established with the release of cytochrome c and other mitochondrial factors following photodamage triggering an apoptotic reaction (Liu et al. 1996). However, use of carriers for delivering PSs can be used to significantly alter the circulation times and localization site. Details of UCNPs as carriers will be discussed in the next section.

In terms of the excitation wavelength for these PSs, the absorption maxima is usually in the visible region of light ranging from blue light around 400 nm to the red region around 650 nm. Tissue absorption and scattering increases with decreasing light wavelength, thus, the achievable tissue penetration across the visible range varies from 1 to 5 mm (approximately). Longer wavelength excitation is preferred, since it allows treatment of deeper regions in the body. Photofrin, one of the earliest developed PSs for PDT is an inefficient producer of singlet oxygen at 630 nm. In addition to 630 nm, it can also absorb at 408 and 510 nm with better ROS production efficiency (Bernstein et al. 1999). Some of its drawbacks include long-circulation times, low clearance, and accumulation in the skin, which can result in prolonged undesirable photosensitivity. However, with dose adjustments, these side effects can be minimized (Allison and Sibata 2008). ALA, a prodrug is enzymatically converted to the active form PPIX *in vivo*, which is subsequently converted to heme. This PS can also be excited using blue (410 nm), green (510 nm), or red (635 nm) wavelengths, with ROS production efficiency being highest at the shorter wavelengths (Peng et al. 1997). The phthalocyanine dye family along with its relative, the naphthalocyanines, are potent PSs for PDT-based applications. They have a porphyrin-based structure with a central atom usually of aluminum, zinc, or silicon to increase the production of singlet oxygen. Unlike the previous PSs, they have a strong absorption band in the 670 nm range and thus potential for treating relatively deeper seated tissues with lower irradiation power densities (Allen et al. 2001; Ben-Hur and Rosenthal 1985).

Although efforts have been made toward the development of PSs for PDT applications keeping in mind the ideal requirements of selectivity, amphiphilicity, long-wavelength excitation, and high ROS production efficiency, many of them only partially meet these considerations. Several of these PSs are fraught with one or more of the following undesirable characteristics ranging from aggregation in aqueous solutions due to hydrophobic nature, poor selectivity, low extinction coefficients, absorption at relatively short wavelengths, to high accumulation rates in the skin leading to normal tissue toxicity from unintentional sunlight exposure (Idris et al. 2015). This is particularly deleterious for PSs that can persist in the system for weeks. Prolonged photosensitivity especially in the skin will render treatment painful and mar acceptance of treatment by the end users, the patients (Allison and Sibata 2010). Development of nanocarriers for delivering these PSs has come a long way in overcoming a majority of these issues. UCNPs are particularly useful in this regard, as they not only serve as carriers, but also nanotransducers allowing the use of deeper tissue penetrating NIR light for excitation of PSs. Furthermore, by coating these particles with different moieties, targeted delivery to tumor specific regions can be achieved, providing a one stop solution to the multiple problems that plague PDT at present. The following section will describe in detail how different UCNPs can be used in the delivery of different types of PSs, the various surface coatings that allow loading of PSs and preparation of aqueous formulations as well as strategies for targeting these UCNP-PS formulations to desired sites in the body.

9.2.1.2 UCNPs for PDT: Host Matrix, Dopants, and Surface Coatings

A UCNP-based PDT system consists of two main components: the nanoparticles themselves, serving as both carriers and transducers and a conduit for generating ROS, the PS. To enable the UCNPs to carry these PSs, different types of surface coatings are required depending on the nature of the PS. In addition to allowing loading of cargo, these coatings also serve to impart biofunctionality and aqueous solubility needed for formulations meant for biological use. Besides the PSs described in the previous section, some surface coatings of the UCNPs (such as titanium oxide) themselves can produce ROS via upconverted light produced by the UCNPs. In short, when using UCNPs for PDT (Hou et al. 2015; Idris et al. 2014), either conventional PSs can be loaded onto UCNPs with appropriate surface coatings or the surface coating of the UCNPs themselves could be such that it produces the required ROS required for PDT.

When using UCNPs for PDT, the first step is to choose the right type of nanoparticle, in terms of the host matrix and dopant types used. This will determine the luminescence efficiency as well as emission wavelength, both factors essential for successful implementation of PDT. The emission wavelength the nanoparticle selects should match as closely as possible to the PS being used. This will ensure that the ROS production and thus the therapeutic effect is optimal. Of the various UCNP host matrix materials, hexagonal phase $NaYF_4$ is the most popular for use in PDT. This is not surprising since it is reported to be one of the most efficient upconversion host materials (Aebischer et al. 2006). Other than $NaYF_4$, materials like $NaGdF_4$ (Qiao et al. 2012; Zhao et al. 2012), $NaLuF_4$ (Ouyang et al. 2014), and $NaYbF_4$ (S. Jin et al.

2013), have also been reported but to a lesser extent. In addition to the choice of the host material, the geometric shape of the synthesized nanoparticles also affects the luminescence efficiency, with hexagonal phase nanocrystals displaying a remarkably higher upconversion luminescence than cubic-phase ones (Liu et al. 2011).

These host matrices are doped with sensitizers and activator ions that determine the emission wavelength upon excitation with NIR light (usually 980 nm for the host materials mentioned above). The most commonly used sensitizer is Yb^{3+} (absorbs 980 nm), while the activator ions used are usually Er^{3+} and Tm^{3+}. Yb^{3+}/Er^{3+} doped particles emit mostly red and green light and are thus the UCNPs of choice for most PSs (Krämer et al. 2004). For PSs such as titanium dioxide (TiO_2) and hypocrellin A that do not absorb green or red light, and have maximal absorption in the UV and/or blue light range, UCNPs doped with Yb^{3+}/Tm^{3+} can be recruited instead (Idris et al. 2014). Core–shell UCNPs consisting of a $NaYF_4$ Yb^{3+}/Er^{3+} core coated with a $NaYF_4$ Yb^{3+}/Tm^{3+} shell or vice versa can also be synthesized (X. Liu et al. 2013). These nanoparticles would have multiple emission peaks that can be used for PS activation in conjunction with other functions such as bioimaging. Aside from the activator ions mentioned above, ions such as Gd^{3+} and Mn^{2+} can also be doped in to the host matrix. Such ions improve the luminescence efficiency while imparting paramagnetic properties to the UCNPs, allowing them to double up as a T_1- or T_2-weighted magnetic resonance imaging (MRI) contrast agents (Ni et al. 2014; C. Wang et al. 2014). From the above discussion on choice of host matrix and dopants, it can be seen that these factors are paramount toward the eventual functionality of the UCNPs and must be chosen in accordance with the end requirements of the UCNP–PDT system.

After having selected the right UCNPs matched to the PS, the next step in the UCNP-mediated PDT system is to coat these particles with a suitable material that allows the PS to be loaded on to them and confers properties amenable for biological use such as aqueous solubility, low immunogenicity, and biofunctionality (for conjugation of targeting moieties, co-therapeutic molecules, etc.). There are numerous materials used to coat UCNPs ranging from polymers such as polyethylene glycol (PEG) to ceramics such as silica (Chatterjee et al. 2010). Of these materials, polymers are by far the most popular when it comes to UCNP-based PDT. One of their most attractive features is their ability to confer stability to colloidal UCNPs in aqueous solutions, a property that is crucial to allow for deliverability in biological systems (Idris et al. 2012). Although polymers such as polyethyleneimine, PEG, polyacrylic acid (PAA), and various block copolymers of PEG, polylactic acid, etc. have been reported as surface coatings for UCNPs, PEG stands out as an attractive candidate. Besides making UCNPs water dispersible, it is enormously useful as a linker for conjugating UCNPs to targeting moieties, other drugs, etc. In addition, it has low toxicity and immunogenicity, making it suitable for *in vivo* use. Polymers like polyethyleneimine (PEI) though imparting aqueous solubility are cationic and have the disadvantage of being toxic, thus limiting their practical use. Besides the synthetic polymers outlined above, a natural polymer-chitosan, has emerged as a viable candidate for coating these nanoparticles. Aside from its properties of biodegradability, low toxicity, and immunogenicity, its amphiphilic nature provides the addition benefit of allowing hydrophobic PSs to be loaded onto the UCNPs while providing aqueous solubility (Cui et al. 2012a,b; Zhou et al. 2012).

Ceramic coatings such as silica are also popular with one of the first papers reporting UCNP-based PDT making use of this material (Qian et al. 2009; Zhang et al. 2007). Silica coating also provides stability in aqueous solutions and has low toxicity. Usually, the PS is deposited on the UCNP surface along with the silica such that it is encapsulated in the coating, with the oxygen and ROS molecules diffusing through its pores. In addition to silica coating, mesoporous silica can also be used. Here, the pore size can be controlled in the nanometer range, allowing molecules such as PSs to be loaded onto the surface of the particles through absorption alone (Idris et al. 2012). Moreover, multiple types of moieties can be co-loaded making this is an easy yet effective delivery system.

There are advantages and disadvantages to using either one of the materials described above. Polymer coatings though attractive in terms of safety and ease of modification, are known to quench upconversion fluorescence by 50% or more, with a greater quenching seen with increasing polymer concentration (Ungun et al. 2009). Silica coatings though less prone to do so are riddled with their own drawbacks. Even though physical absorption is an easy means of loading the cargo onto the particles, it reduces the degree of control one has in terms of the release characteristics of the cargo in a biological environment.

Before moving on to means of loading of PS onto UCNPs, it is prudent to first discuss the targeting moieties alluded to in the previous sections. As mentioned earlier, it is important to be able to deliver the PS to the right site for effective treatment. Although enhanced permeability and retention (EPR) effect, is effective in directing systemically delivered agents in a greater quantity to the tumor site, at times there is not sufficient accumulation for optimal therapeutic efficacy. When using nanoparticles as carriers, an effective targeting strategy is to coat them with certain groups that specifically interact with the cell type of interest. One of the most widely used targeting moieties is folic acid (FA). FA is relatively inexpensive, stable over a wide range of pH and temperature, and has high affinity for folate receptors often overexpressed in many human cancers. Cell lines such as human choriocarcinoma (JAR) cells, murine, S180 sarcoma tumors, HeLa human cervical cancer cells, B16F0 murine melanoma cells, Bel-7402 human hepatocellular carcinoma cells, and HT29 human colon cancer cells are reported to overexpress folate receptor (Idris et al. 2015). Thus, conjugating FA to UCNPs carrying the PSs, allows for an enhanced uptake by cancer cells (Cui et al. 2012b; Idris et al. 2012; Liu et al. 2012; X. Liu et al. 2013). Another receptor overexpressed in many cancers is cluster determinant 44. Targeting moieties such as hyaluronic acid have been used to target this receptor (Xu Wang et al. 2014). Besides these ubiquitous approaches, antibodies against proteins specific to a particular type of cancer can also be deployed. This is an effective targeting approach, albeit a more expensive one. Aptamers are relatively young players in this field but advantages such as ease of synthesis, flexibility in design, and chemical stability make them strong contenders (Tan et al. 2013). Furthermore, they have the natural ability to fold into G-quadruplex structures that can be stabilized by small-molecule ligands like porphyrin, providing a convenient means of loading porphyrin-based PSs in addition to their targeting ability (Shieh et al. 2010).

9.2.1.3 PS Loading Strategies

A key step in the UCNP-based PDT system is the loading of the PS onto the nanoparticle. Depending on the surface coating deployed, this can be done either through encapsulation, physical adsorption, or covalent linkage. Before going into details about these strategies, it is of importance to note some of the criteria for optimal PS loading. First, the loading strategy should allow for a high loading capacity so that a sufficient amount of PS is available for PDT. Second, it should permit good energy transfer between the UCNP core and PS. If the PS is encapsulated or trapped within a matrix, the matrix should allow oxygen and ROS to diffuse freely to the surrounding area. Lastly, there should be minimal leakage or premature release of PS, especially in case of systemic delivery (Idris et al. 2015).

The first strategy, encapsulation is one of the oldest. In this approach, the PS is not present on the surface of the particles or covalently linked to it, but is rather encapsulated in a matrix that coats the UCNPs or trapped between the UCNP core and surface via coatings such as lipids, polymers, etc. One of the first reports of UCNP-based PDT used encapsulation as a means of loading the PS on to the particles. The PS was trapped in a dense silica matrix achieved by mixing the PS in the reaction mixture while carrying out the silica coating process (Zhang et al. 2007). The amount of PS and thickness of the silica layer could be adjusted. Although a fast and relatively simple technique, efficient loading can only be achieved for cationic hydrophilic PSs owing to the negatively charged and hydrophilic nature of the silica matrix. This method also faces some challenges in terms of the ease of diffusion of ROS and oxygen through the matrix, which could impede the efficacy of this approach. Encapsulation-based loading has also been achieved using polymers and lipids. As mentioned earlier, chitosan is amphiphilic and provides dual advantages of loading hydrophobic PSs while presenting a hydrophilic surface for aqueous solubility. To this end, interactions between the hydrophobic part of this polymer and hydrophobic PS (such as zinc phthalocyanine (ZnPc)) have been used to trap the PS between the UCNP core and the hydrophilic part of chitosan present on the outer surface (Cui et al. 2012a,b) (Figure 9.2a). In another approach, both UCNPs and the PS

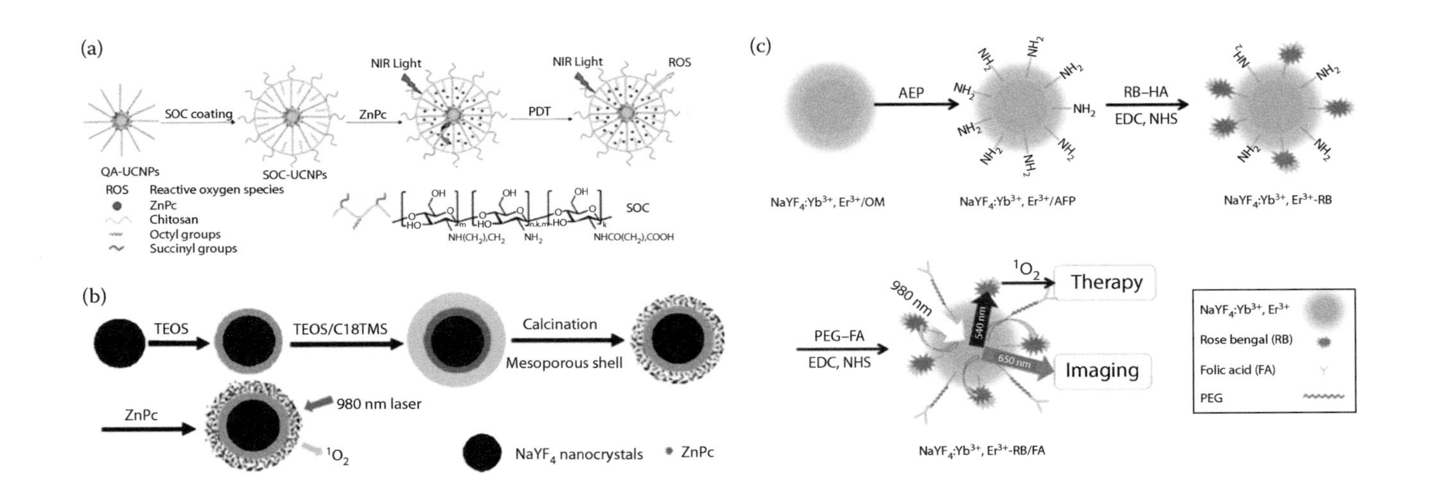

FIGURE 9.2
(a) Schematic showing synthesis of UCNPs encapsulating ZnPc. (b) Schematic showing loading of ZnPc via physical adsorption onto mesoporous silica-coated UCNPs, and (c) UCNPs carrying PS rose bengal attached to the surface covalently. (Reproduced with permission from Cui, S., H. Chen, H. Zhu, J. Tian, X. Chi, Z. Qian, S. Achilefu, and Y. Gu, *J. Mater. Chem.*, 22 (11), 4861–4873, 2012; Qian, H. S., H. C. Guo, P. Chi-Lui Ho, R. Mahendran, and Y. Zhang, *Small*, 5 (20), 2285–2290, 2009; Liu, K., X. Liu, Q. Zeng, Y. Zhang, L. Tu, T. Liu, X. Kong, Y. Wang, F. Cao, and S. A. G. Lambrechts, *ACS Nano*, 6 (5), 4054–4062, 2012, respectively.)

can be encapsulated within amphiphilic micelles. This method has a tunable loading capacity depending on concentration of PS (H. J. Wang et al. 2014), but with the caveat of PS leaking owing to micelle destabilization.

The most common strategy for loading the PS onto the UCNPs is through physical adsorption, primarily because of the ease and simplicity of this approach. Physical adsorption usually involves noncovalent attractive interactions such as hydrophobic interactions or electrostatic forces to attach the PS to the nanoparticle surface. For instance, PSs can be loaded into the pores of mesoporous silica-coated particles through physical adsorption (Figure 9.2b). The pore size of this layer can be adjusted allowing loading of differently sized molecules. In addition, the porous nature of this layer provides a large surface to volume ratio, enabling a greater loading capacity along with better permeability to oxygen and ROS molecules in comparison to silica coating alone (Idris et al. 2012; Qian et al. 2009). Physical adsorption through electrostatic interactions has also been used as a loading strategy. Using this method, a layer-by-layer self-assembly approach can be used, wherein oppositely charged layers can be alternately deposited on the nanoparticle surface, with one of the layers consisting of either the charged PS or a PS conjugated to a charged moiety. With this strategy multiple layers can be deposited, thereby affording a convenient means of controlling the amount of PS loaded. In one such example, a negatively charged Ce6-conjugated polymer was deposited along with a positively charged PAA–poly(allylamine hydrochloride) layer on the UCNP and the process repeated to yield up to three layers of the PS conjugate resulting in a Ce6 loading capacity of 3.4, 7.7, and 11.0 wt% for one-, two-, and three-layered structures, respectively (C. Wang et al. 2013).

The third strategy for loading is through covalent linkage of the PS to the nanoparticle (Figure 9.2c). Here, the PS is usually attached to the UCNP surface through cross-linking agents such as 1-ethyl-3-(3-dimethylaminopropyl)carbodiimide or N,N′-dicyclohexylcarbodiimide. Since the release of the PS from the UCNP surface is not strictly necessary for PDT, this approach provides a viable means of overcoming the problem of premature leakage or release of the PS faced by the other loading strategies mentioned above (Liu et al. 2012; Xu Wang et al. 2014). In fact, covalent linkage has been used for loading PSs onto silica-coated (Zhao et al. 2012) or mesoporous silica-coated UCNPs (Qiao et al. 2012) as well. Furthermore, covalent linkage in conjunction with physical adsorption has also been used to improve loading efficiency (Park et al. 2012). One of the major concerns of this approach is the fact that it requires chemical modification of the PS to enable conjugation, which can potentially reduce the efficacy of the PS. However, most studies have shown that the functional properties of the PS are retained following covalent linkage.

9.2.1.4 In Vitro and In Vivo PDT Using UCNPs

UCNP-based PDT has been demonstrated in cells and in animals, with the first successful proof of concept demonstration by Zhang et al. (2007). This group used the PS MC540 along with $NaYF_4$: Yb/Er UCNPs and used it to kill MCF-7/AZ breast cancer cells *in vitro*. These cells were incubated with the PS loaded UCNPs and irradiated with NIR light at 974 nm for 36–43 min. In subsequent years, several other groups have also reported UCNP-based PDT in cells with better optimization of PS loading, use of PSs with higher ROS yield, lower UCNP concentrations, and shorter NIR irradiation times at lower power densities. For instance, Chatterjee et al. used ZnPc as a PS and showed 50% cytotoxicity of HT29 human colon cancer cells using 0.733 mg/mL of UCNPs and NIR (980 nm) irradiation time of only 5 min (Chatterjee and Yong 2008). With most of these early reports, the method used for loading the PS onto the UCNPs was either through encapsulation or physical adsorption. In an effort to reduce the leakage of PS as seen with these approaches, covalent linkage to the nanoparticle surface was explored. By reducing PS leakage, a greater amount of the chemical could be delivered to the cells of interest, potentially allowing the use of a lower concentration of PS loaded UCNPs. For instance, Liu et al. (2012) covalently linked the PS rose bengal to the surface of PEG-coated UCNPs. With these particles, 50% of cell death of JAR cells could be achieved using a concentration of only 0.1 mg/mL and NIR (980 nm) irradiation time of 10 min (at 1.5 W/cm^2).

While the initial studies successfully demonstrated the ability of UCNPs as carriers and nanotransducers for PDT, later studies attempted to elucidate the mechanisms behind nanoparticle-mediated PDT. Guo et al. found that cells took up the nanoparticles in a concentration- and time-dependent fashion. When these cells were irradiated with NIR light, the resulting ROS (in this instance singlet oxygen, $^1O_2$) induced oxidative damage caused a change in the nuclear morphology as well as loss mitochondrial membrane integrity evident from the observed chromatin condensation, DNA fragmentation, and release of cytochrome c from the mitochondria. All these events are indicative of apoptosis mediated cell death (Guo et al. 2010).

The efficacy of UCNP-mediated PDT has also been demonstrated *in vivo* in mouse models. The earlier studies usually involved intratumoral injection of PS UCNP-loaded UCNPs. Later, systemic delivery of these particles was also reported with these particles usually modified with targeting moieties in addition to the PS. One of the first studies reporting *in vivo* PDT using UCNPs was by Wang et al. (2011). They used the PS Ce6 and $NaYF_4$:Yb, Er for killing $4T_1$ mouse breast cancer cells and *in vivo* for eradicating a $4T_1$ tumor grown in synergistic mice. The nanoparticles (43 mg/kg) were injected directly into the tumor and the site irradiated with a 980 nm laser for 30 min at 0.5 W/cm^2. This retarded tumor growth, a change not seen in untreated mice. Another report of intratumoral injection was by Cui et al., where they used ZnPc instead of Ce6 (Cui et al. 2012a). They were able to achieve significant retardation of tumor growth at a lower UCNP concentration (33 mg/kg), irradiation power density (0.4 W/cm^2), and NIR irradiation duration (15 min) though irradiation was carried out twice at different time points (30 min in total).

In most cases, the PS is loaded onto the UCNPs using one or more of the three strategies mentioned previously. These strategies have their own advantages and disadvantages with variability in terms of loading efficiency, leakage of PS, etc. In 2015, Zhang et al. reported a different approach to UCNP-based PDT. They uniformly surface coated a photocatalyst titanium dioxide (TiO_2) on a $NaYF_4$:Yb, Tm UCNP core (TiO_2–UCNP) (Lucky et al. 2015). Here, the titania coating itself was the PS. Using this nanoconstruct, cancer cells could effectively be destroyed both *in vitro* and *in vivo*. This approach allowed both a means of controlling the amount of the PS loaded and a way of stably loading it onto UCNPs.

Even though the intratumoral injection studies were beneficial in demonstrating the ability of UCNPs to act as transducers for PDT in solid tumors, their ability as carriers and for tumor specific targeting was brought forth by the subsequent studies reporting systemic injection. When it comes to systemic delivery, there are several other factors that need to be taken into consideration. This route involves the circulation of nanoparticles in the bloodstream before reaching the tumor site. Thus, it becomes necessary to modify the surface with stabilizers such as PEG to improve circulation time, avoid immune detection, etc. In addition, targeting moieties are also used to improve the accumulation of the nanoparticles at the tumor site. Idris et al. (2012) reported intravenous injection of FA–PEG-modified mesoporous silica-coated NaYF:Yb, Er UCNPs, co-loaded with PSs, MC540, and ZnPc (Figure 9.3). They used them for killing B16F0 melanoma in mice and found that that the UCNPs could activate both PSs (through their multiple visible emissions) with a single 980 nm laser excitation. Also, the simultaneous use of two PSs was more effective in reducing tumor volume than using a single one. A UCNP dose of 50 mg/kg and a long-exposure time of 60–120 min of the 980 nm light at 0.415 W/cm^2 was required for effective PDT treatment. Park et al. (2012) reported another systemic delivery system. They used a much lower UCNP dose of 5 mg/kg of the UCNP with only 5 min of exposure to 0.6 W/cm^2 of 980 nm light for effective inhibition of the U87-MG human glioblastoma tumor growth. Here, they used Ce6 as the PS, which was loaded onto the UCNPs through both physical adsorption and covalent linkage. This dual loading strategy resulted in high loading efficiency, thereby effectively lowering the dosage of PS loaded UCNPs needed for PDT.

9.2.2 Photochemical Internalization

A major challenge in the delivery of biomolecules is the poor efficiency of cytoplasmic delivery (due to poor endosomal escape) especially when using nanoparticles, since cells clear away the foreign nanoparticles, reducing the effective concentration of nanoparticles accumulated inside the cells. In addition, lysosomes are highly acidic (pH ~4.5) and contain nucleases that can degrade payload such as nucleic acids. Thus, poor endosomal release

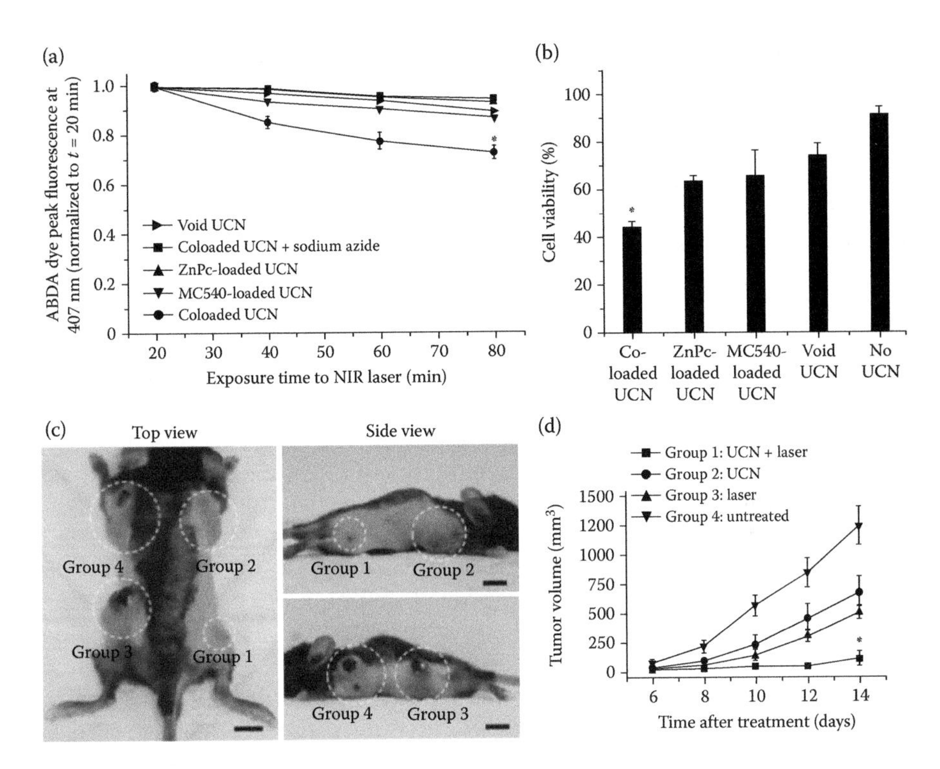

FIGURE 9.3
(a) Comparison of singlet oxygen production between coloaded, MC540-loaded, ZnPc-loaded, and void UCNPs under 980-nm NIR irradiation (2.5 W/cm^2) as determined by the decay of 9,10-anthracenediyl-bis(methylene)dimalonic acid (ABDA) fluorescence. (b) 3-(4,5-dimethylthiazol-2-yl)-5-(3-carboxymethoxyphenyl)-2-(4-sulfophenyl)-2H-tetrazolium assay for measuring cell viability to test efficacy of PDT treatment (by exposing cells that have taken up differentially loaded UCNPs to 2.5 W/cm^2 of 980-nm NIR laser for 40 min). (c) Representative gross photos of a mouse showing tumors (highlighted by dashed white circles) at 14 d after treatment with the conditions described for groups 1–4. Scale bars, 10 mm. (d) Tumor volumes in the four treatment groups at 6, 8, 10, 12, and 14 days after treatment to determine the effectiveness of the treatment in terms of tumor cell growth inhibition. Values are means ± scanning electron microscopy (SEM) ($n = 6$ mice per group). (Reproduced with permission from Idris, N. M., M. K. Gnanasammandhan, J. Zhang, P. C. Ho, R. Mahendran, and Y. Zhang, *Nat. Med.*, 18 (10), 1580–1585, 2012. http://www.nature.com/nm/journal/v18/n10/abs/nm.2933.html#supplementary-information.)

combined with harmful conditions in the late endosomes and lysosomes constitute a major hurdle in therapy (Berg et al. 2006). Various strategies have been used to overcome this issue and one such effective solution utilizing light is PCI.

This technology employs specific, preferably amphiphilic, photosensitizing compounds, which accumulate in the membranes of the endocytic vesicles. When the cells are illuminated with light of a specific wavelength, these photo sensitizers (PS) become excited and subsequently induce the formation of ROS, primarily singlet oxygen. The short range of action and short lifetime confines the damaging effect of these ROS to the production site (Oliveira et al. 2007). This localized effect induces the disruption of the endocytic vesicles, thereby releasing the entrapped therapeutic molecules into the cytosol before they are transferred to the lysosomes (Dominska and Dykxhoorn 2010). This technique thus provides an added dimension of spatial and temporal control over biomolecule delivery, making therapy more specific and effective.

9.2.2.1 Photosensitive Chemicals for PCI

Several different PSs have been reported for PCI with most of them based on porphyrin structures. The most popular of these is tetraphenylporphine disulfonate (TPPS2a) developed by Hogset et al. (PCI Biotech), and has been reported by several groups to improve therapeutic efficacy of drug and gene delivery systems incorporating this chemical (Boe et al. 2010; Fretz et al. 2007; Oliveira et al. 2007; Selbo et al. 2006). Recently, another PS TCPS2a (Berstad et al. 2012), developed by the same group is gaining prominence with several advantages over TPPS2a such as improved amphiphilicity and red shifted absorption (Lilletvedt et al. 2011). Some of the other PSs used in PCI include aluminum disulfonate phthalocyanine (AlPcS2a) (Jin et al. 2011), rhodamine (Gillmeister et al. 2011), etc. These PSs produce ROS similar to those used in PDT. However, these PSs accumulate preferentially in the endocytic vesicles, are used in minute concentrations producing ROS that is sufficient to rupture the endosomes but not kill the cell.

Initial demonstrations of PCI were done *in vitro* wherein cells were co-incubated with the PS and the molecule to be delivered. After about 18 h of incubation, the cells were irradiated with UV or visible light (depending on the PS used). These studies revealed that efficacy of biomolecule delivery into the cytoplasm and thus, the resulting therapeutic efficacy were higher with PCI. An *in vivo* demonstration of PCI in improving gene transfection efficiency was done by Nishiyama et al. They used a ternary complex composed of a core comprising of DNA packaged with cationic peptides which was enclosed in the anionic dendrimer phtalocyanine. Phtalocyanine acted as a PS in this study. With PCI, the gene transfection was enhanced 100 fold. The animal experiments were done in rats where the complexes were introduced via subconjuctival injection (in the eye) (Nishiyama et al. 2005). Although these studies demonstrated the utility of PCI in enhancing cytoplasmic delivery of biomolecules, the fact that the photosensitizers were sensitive to UV/visible light limited their *in vivo* potential.

In 2014, Jayakumar et al. synthesized novel NIR-to-UV/vis UCNPs made up of a ore–shell architecture, which upon NIR excitation emit across the UV and visible range. These emissions were used to excite multiple photoactive molecules simultaneously, namely PS (TPPS2a) for PCI (visible emission) and photoresponsive nucleic acid (anti-STAT3 photomorpholino) for gene knockdown

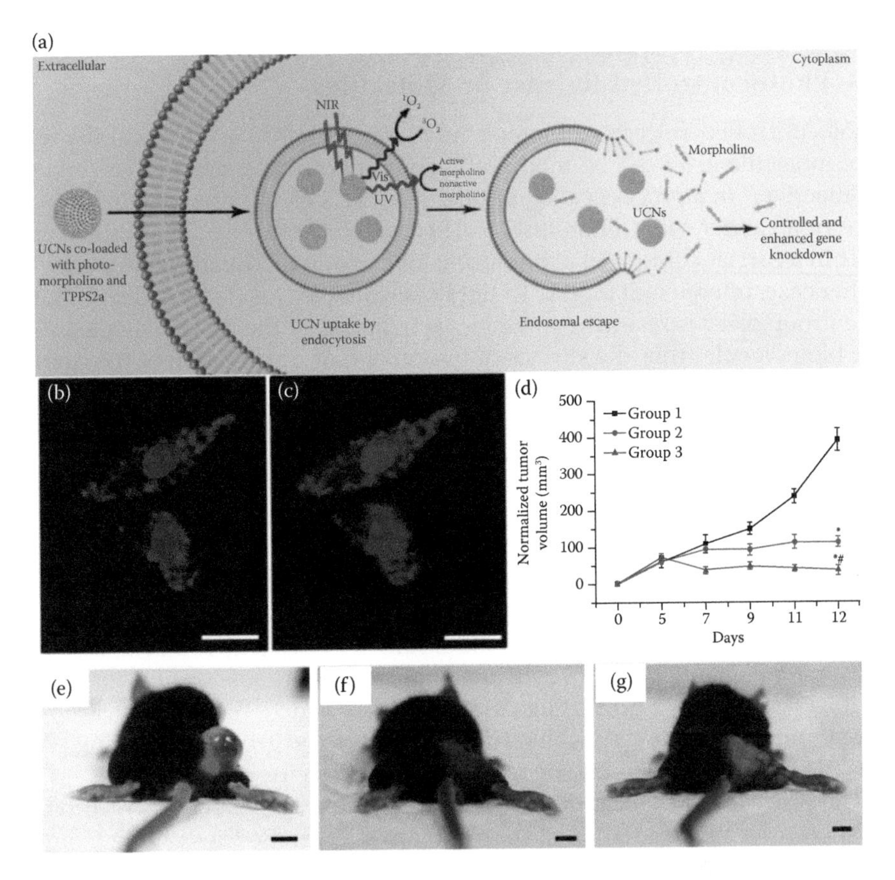

FIGURE 9.4
(a) Schematic showing simultaneous endosomal escape (PCI) and uncaging of nucleic acid using UCNPs. Distribution of UCNPs before (b) and after (c) NIR irradiation (scale bar: 5 μm): UCNPs, red; 2-(4-amidinophenyl)-1H-indole-6-carboxamidine (DAPI), blue. A more diffuse pattern shows cytosolic release after endosomal escape. (d) Change in tumor volume as a function of time to assess the effectiveness of simultaneous activation of TPPS2a (for PCI) and STAT3 photomorpholino for treating B16F0 melanoma tumor in C57 Bl/6 mice. Values are means ± SEM ($n = 6$ mice per group). Group 1, saline control; group 2, UCNPs loaded with photomorpholinos and irradiated with NIR laser; and group 3, UCNPs co-loaded with photomorpholinos and TPPS2a and irradiated with NIR laser. Representative gross photos of a mouse from each group 1–3 (e–g). Scale bar: 1 cm. (Reproduced with permission from Jayakumar, M. K. G., A. Bansal, K. Huang, R. Yao, B. N. Li, and Y. Zhang, *ACS Nano*, 2014.)

(UV emission) (Figure 9.4). These nanoparticles were coated with mesoporous silica and co-loaded with both the PS and nucleic acid. They were then added to B16F0 mouse melanoma cells *in vitro* or injected intratumorally into a B16F0 melanoma tumor grown in C57bl/6 mice. With PCI, the endosomal release of the anti-STAT3 photomorpholino was enhanced both *in vitro* and *in vivo*. This resulted in improved gene knockdown and thus reduction in tumor volume that was several folds higher (Jayakumar et al. 2014).

9.3 Photocontrolled Release of Molecules

Photocontrolled release of biomolecules can be achieved in several ways. The molecule itself can be modified using a photolabile group that renders it "inactive," in a process called photocaging or it can be attached to the surface of a carrier via a photolinker. Alternately, it can also be encapsulated within a carrier composed of photosensitive group modified components. In either case, upon irradiation with light of a suitable wavelength, the photolabile group cleaves resulting in either "activation" of the molecule, release of the biomolecule from the surface of the carrier, or dissociation of the carrier and subsequent release of cargo contained within, respectively. Photocaging is a popular technique for using light to achieve site-specific activation of biomolecules. It involves modifying the molecule to be delivered (plasmid/siRNA/drug, etc.) using a photolabile moiety. This renders the molecule nonfunctional (Sortino 2012). When the desired delivery site is irradiated with light of a suitable wavelength (usually UV light), this photolabile group gets released, thereby rendering the molecule functional or active again. This process of covalently linking the photolabile group to a biomolecule is termed "caging" and the biomolecule is said to be "caged" (Yu et al. 2010). In most biomolecules, certain key functional groups are responsible for bioactivity. Photocaging works by blocking these key functional groups (carboxyl groups, amino groups, phosphate moieties, hydroxyl groups, etc.) using photolabile molecules, thereby, hindering their bioactivity.

The process of photocaging allows us to achieve site-specific activation of biomolecules via irradiation of the region of interest. Besides caging biomolecules directly, photolabile groups that act as linkers have also been used for controlled release. This involves attaching the desired molecule to a carrier via a linker that can be cleaved upon irradiation with light of a suitable wavelength to which it is sensitive (Yang et al. 2013a). Although the manner in which the release of the desired molecule is achieved is different for this technique in comparison to photocaging, the basic premise of using a photolabile group remains the same with a similar end result. This section will describe some of the commonly used photolabile groups used in photocontrolled delivery systems, UCNP-based strategies for photocontrolled delivery, and application in cells and in animals.

9.3.1 Photolabile Groups in Photocontrolled Delivery

There are certain criteria that photolabile groups used in photocontrolled delivery systems intended for biological use need to meet. These include high quantum yield of the photoreaction, high absorption coefficient of the group, safe photochemical by-products, and solubility in aqueous solvents (Pelliccioli and Wirz 2002). Several different types of photolabile molecules have been developed for this purpose and can be divided into several categories depending on the mechanism of photolysis. These include: o-nitrobenzyl (NB) and related groups (e.g., nitrophenyl ethyl (NPE), o-NB, 1-(4,5-dimethoxy-2-nitrophenyl) diazoethane (DMNPE)), coumarin-4-yl-methyl and related groups (e.g., 7-methoxycoumarin-4-ylmethyl), p-hydroxy-phenacyl (pHP) group (a promising alternative to the NB-based groups to cage biomolecules), and other miscellaneous groups such as nitroindolinyl (NI), 4-methoxyl-7-nitroindolinyl, and benzoin (Pelliccioli and Wirz 2002; Yu et al. 2010) (Figure 9.5). These groups are modified through the addition of certain substitution groups in the basic framework of the photolabile

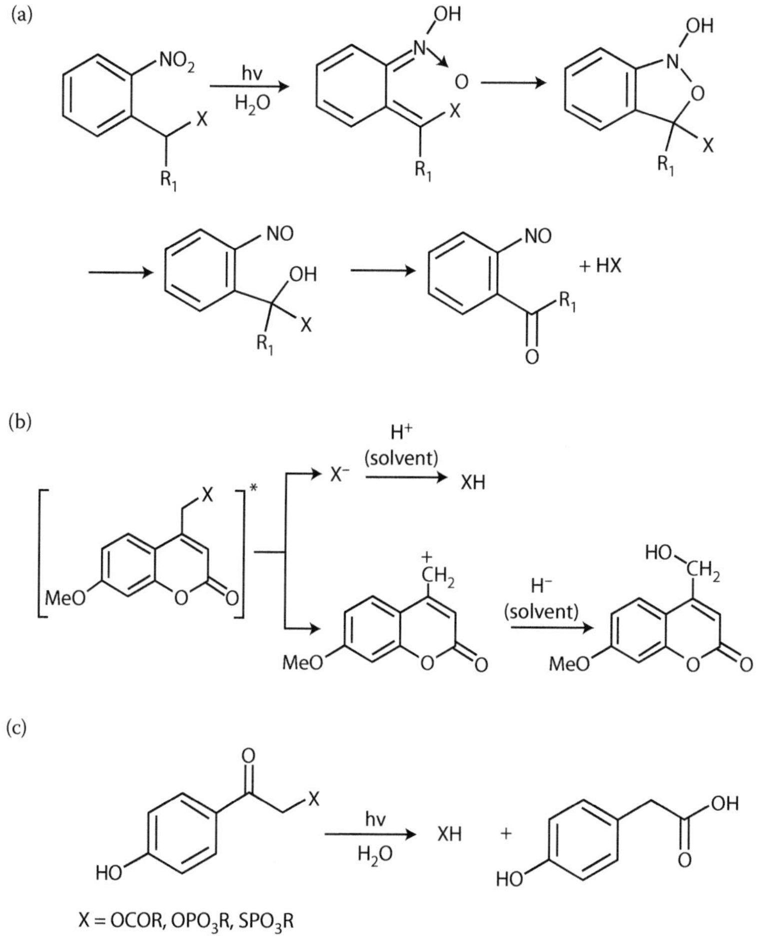

(a)

(b)

(c)

X = OCOR, OPO_3R, SPO_3R

FIGURE 9.5
Mechanism of photolysis of (a) NB (R_1 = H) or NPE (R_1 = Me) caged compounds, (b) coumarin-4-ylmethyl caged compounds, and (c) pHP caged carboxylates, phosphates, and thiolphosphates. (Reproduced with permission from Yu, H., J. Li, D. Wu, Z. Qiu, and Y. Zhang, *Chem. Soc. Rev.*, 39 (2), 464–473, 2010. doi: 10.1039/b901255a.)

molecule to attain the aforementioned properties desirable for biological use such as good aqueous solubility, long-wavelength absorption (redshift), and high absorption coefficient (Yu et al. 2010).

o-NB derivatives are the most common photolabile protecting groups. This group is synthetically incorporated into the molecule to be caged via linkage to a heteroatom (usually *O*, *S*, or *N*) as an ether, thioether, ester, amine, or similar functional group. Although very effective, NB cages have some distinct disadvantages. These include toxic by-products and slow release rates following excitation.

A promising alternative to o-NB derivatives is the pHP group. pHP is used to cage mostly carboxylates and phosphates and has a fast release rate following excitation. In addition, aqueous solubility, nontoxic by-products, high stability under physiological conditions, makes this group very attractive. However, a disadvantage is the relatively low absorption coefficient at wavelengths above 320 nm.

A relatively new player in this field is the Coumarin-4-ylmethyl group and its derivatives. It can be used to cage carboxylic acids, phosphates, amino group as well as the hydroxyl group. It has a high absorption coefficient and fast release rate making it suitable for a host of applications (Yu et al. 2010).

9.3.2 UCNP-Based Phototriggered Release *In Vitro* and *In Vivo*

Most of the photolabile groups described in the previous section respond to UV light. Not only is UV light toxic, it also has low tissue penetration, which makes it unsuitable for use in a clinical setting. Thus, the use of UCNPs as NIR to UV transducers is advantageous particularly for *in vivo* delivery and site-specific activation of bioactive molecules such as drugs, nucleic acids, etc. There are several strategies for employing UCNPs in photocontrolled delivery systems. They can be deployed as transducers for and carriers of photocaged biomolecules. When the region of interest is irradiated with NIR light, the upconverted light from the UCNPs uncages the biomolecules, rendering them functional. UCNPs can also be co-encapsulated with caged biomolecules within a larger particle (with a porous surface) in a "yolk–shell" approach, such that irradiation with NIR light uncages the biomolecule which then diffuses out of the pores of the larger particle in its active from. Lastly, UCNPs have also been co-encapsulated with the cargo to be delivered in hydrogels and micelles made of components that contain photolabile groups. Upon irradiation with NIR light, the photolabile group is cleaved due to the upconverted light produced by the UCNPs, resulting in disruption of the hydrogel or micelle and releasing the contents confined within.

One of the earliest reports of using UCNPs for NIR triggered uncaging was by Carling et al. (2010). They used a benzoin cage to modify a model molecule, which upon uncaging released carboxylic acid. Although not of therapeutic use, this study showed that UCNPs could be used as transducers for NIR-based uncaging of molecules. This generated a lot of interest in this field with subsequent reports of NIR mediated uncaging of small molecules and bioactive moieties for therapeutic use. For instance, Garcia and group reported photoactivation of caged nitric oxide with potential applications in cancer treatment (Garcia et al. 2012). Caged chemotherapeutic drugs can also be uncaged in a site-specific manner using UCNPs upon irradiation with NIR light. Dai et al. used $NaYF_4$: Yb/Tm @ $NaGdF_4$ UCNPs for photocontrolled delivery of the light-activated prodrug dipeptidyl-peptidase (DPP) (trans,trans,trans-$[Pt(N_3)_2\text{-}(NH_3)(py)(O_2CCH_2CH_2\text{–}COOH)_2]$) and multimodal imaging (Dai et al. 2013). The upconverted UV light from the UCNPs activated the prodrug to its active form. They used these prodrugs loaded particles *in vitro* in HeLa cells. Significant reduction in cell viability was seen when cells were incubated with drug-loaded UCNPs and irradiated with NIR. Irradiation with NIR light alone did not cause such a toxic effect. These prodrug loaded particles were also tested *in vivo* in a mouse model with H22 xenografts. NIR irradiation upon treatment with these prodrug-loaded particles caused excellent inhibition of tumor growth. In addition, the Gd doping allowed these particles to be used for MRI and the presence of Yb and Gd ions for computed tomography imaging. Thus, this platform allowed for both targeted delivery of a chemotherapeutic drug and multimodal imaging. Zhao et al. demonstrated an interesting approach for using UCNPs for photocontrolled drug delivery. They synthesized yolk–shell particles, each consisting of a single UCNP yolk and mesoporous silica shell. The coumarin-caged chemotherapeutic drug chlorambucil was adsorbed in the space between the UCNP yolk and mesoporous silica shell. Upon irradiation with NIR light, the upconverted UV light produced by the UCNPs uncaged the drug, which diffused out of the pores of the shell in its active form. This system was tested both *in vitro* and *in vivo* (Zhao et al. 2013) (Figure 9.6a–g). Instead of modifying the drug directly, photocontrolled delivery using mesoporous silica-coated UCNP can also be achieved by loading the drug of interest into the pores and then covering or "capping" the pore with a photolabile group (J. Liu et al. 2013; Yang et al. 2013b). Upon irradiation with NIR light, the upconverted light from the UCNPs cleaves the capping group, leaving the cargo free to diffuse out of the pores. Besides drugs and small molecules, UCNPs have also been used for photocontrolled delivery of nucleic acids such as plasmids, siRNA, photomorpholinos, etc. Jayakumar et al. demonstrated the use of NIR to UV UCNPs in achieving spatially and temporally controlled gene expression/knockdown with the use of caged siRNA and plasmid (Jayakumar et al. 2012). Green fluorescent protein (GFP) plasmid was caged using DMNPE [1-(4,5-DMNPE] and loaded onto mesoporous silica-coated UCNPs. When cells were incubated with these particles and irradiated with NIR light, strong GFP expression was observed which was absent in cells not irradiated with NIR light (Figure 9.6h–j). GFP expression in cells expressing this protein could also be knocked down using caged anti-GFP plasmid upon irradiation with NIR light. In addition, the toxicity of the mesoporous silica-coated UCNPs was minimal at the concentration

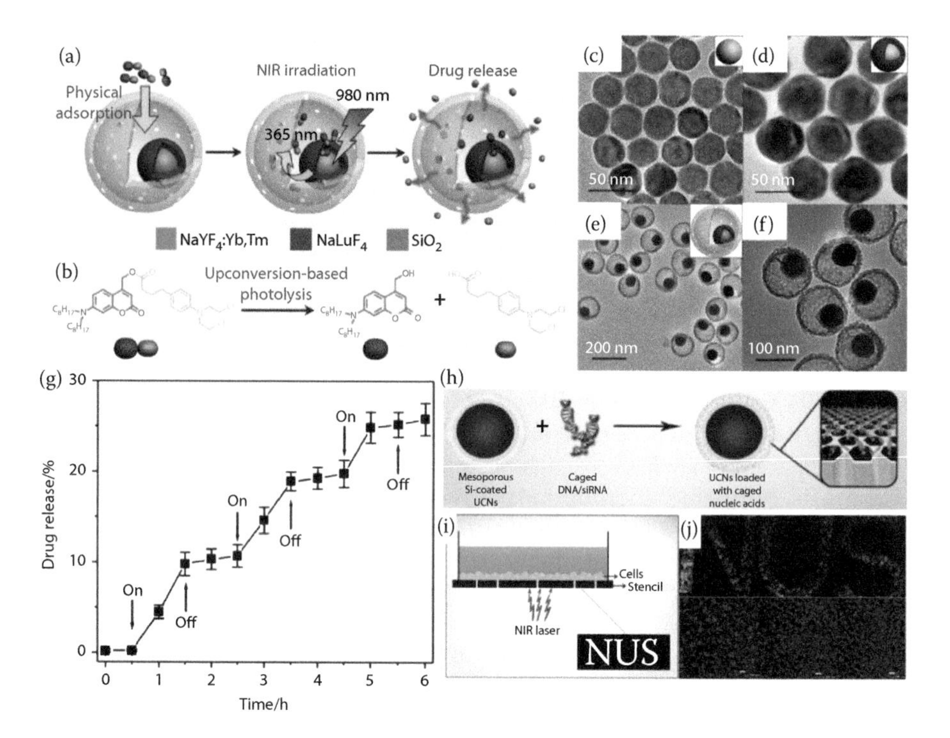

FIGURE 9.6
(a) Schematic illustration of the NIR-regulated upconversion-based drug delivery and (b) the photolysis of the prodrug under upconversion emission from the yolk–shell UCNPs (YSUCNPs). Transmission electron microscopy (TEM) images of the (c) NaYF:Yb,Tm, (d) NaYF:Yb,Tm@NaLuF, and (e, f) YSUCNP nanoparticles. (g) The photoregulated release of chlorambucil (drug) from YSUCNP-ACCh controlled by a 980 nm laser. "ON" and "OFF" indicate the initiation and termination of laser irradiation, respectively. The working power density of the 980 nm laser was 570 mW/cm^2. (h) Schematic showing loading of caged plasmid DNA/siRNA into the mesopores of UCNPs. (i) Schematic of setup showing the stencil and the position of the laser for patterning of cells transfected with caged pGFP. Inset shows the pattern (National University of Singapore [NUS]) on the stencil. (j) Composite image showing the GFP fluorescence from three different wells (one letter in each well of a 96-well plate) (scale bar, 200 μm) and live-cell DAPI staining of the same to show the cell confluence. (Reproduced with permission from Zhao, L., J. Peng, Q. Huang, C. Li, M. Chen, Y. Sun, Q. Lin, L. Zhu, and F. Li, *Adv. Funct. Mater.*, 2013. doi: 10.1002/adfm.201302133; Jayakumar, M. K., N. M. Idris, and Y. Zhang, *Proc. Natl. Acad. Sci. USA*, 109 (22), 8483–8488, 2012. doi: 10.1073/pnas.1114551109, respectively.)

used in the experiments (0.5 mg/mL). Thus, this method was able to achieve very efficient photo uncaging of nucleic acids using deep penetrating and nontoxic NIR light, thereby overcoming the limitations associated with traditional methods for photocontrolled gene delivery. Yang et al. also used NIR to UV UCNPs to demonstrate photocontrolled gene expression/knockdown. However, instead of loading photocaged nucleic acids, they functionalized the surface of their silica-coated UCNPs with a photoresponsive o-NB linker terminating with a positively charged alkyl amine group and used electrostatic interactions with negatively charged siRNA to achieve loading. Upon irradiation with NIR, the photolinker was cleaved resulting in siRNA release. This system was demonstrated *in vitro* by knocking down GFP using a Si-UCNP-photolinker-anti-Enhanced green fluorescent protein (EGFP) siRNA complex upon irradiation at 980 nm (Yang et al. 2013a).

In 2014, Jayakumar et al. also showed simultaneous photoactivation of two different molecules. They synthesized core–shell $NaYF_4$: Yb, Tm@$NaYF_4$: Yb, Er UCNPs that could emit across the UV and visible range upon irradiation with NIR light at 980 nm. These particles were coated with mesoporous silica and co-loaded with photomorpholino (anti-STAT3) and TPPS2a, a PS used for PCI. B16F0 murine melanoma cells (overexpressing STAT3) were incubated with these particles and irradiated with NIR light. The visible emission of the UCNPs activated TPPS2a resulting in improved endosomal escape (and release of cargo into the cytoplasm) and the UV emission activated the photomorphoino. This resulted a significant reduction in STAT3 levels and cell death as compared to delivery of photomorpholino alone. Similar results were seen *in vivo* in a mouse melanoma model.

UCNPs have also been used for disrupting photosensitive hydrogels and micelles containing the molecules to be delivered encapsulated within. Yan et al. (2011) synthesized micelles using block copolymers [poly(ethylene oxide)-block-poly(4,5-dimethoxy-2-NB methacrylate)] and encapsulated NaYF:Yb, Tm UCNP in them. Upon irradiation with an NIR laser, the UCNPs produced UV light which caused the photocleavage of o-NB groups in the micelle, micelle dissociation, and release of a dye molecule contained within. Although this study used a dye molecule to demonstrate the concept, this system can be applied for delivery of therapeutic cargo. The same group also used UCNPs ($NaYF_4$: Yb/Tm) to achieve temporally controlled release of biomolecules entrapped in a hydrogel (Yan et al. 2012). They prepared a hydrogel, with a cross-linked hybrid polyacrylamide–PEG structure held together by photoresponsive o-NB groups. The UCNPs and desired biomolecules were entrapped in this hydrogel. Upon irradiation with NIR at 980 nm, the UCNPs emitted UV light, which resulted in the cleavage of the UV sensitive o-NB groups, releasing the entrapped biomolecules in the process. The release of the biomolecules was dependent on the laser power, allowing for a stepwise (not continuous) release. It is important to note that these experiments made use of very high laser power (3–5 W) and for prolonged time periods (30 min and above). This is a potential limiting factor for the use of this method *in vivo*.

9.4 Photothermal Therapy

PDT though promising, has certain limitations in that it is dependent on the oxygen levels in the tumor. Consequently, it is potentially less effective in hypoxic conditions where the oxygen partial pressure can fall below 40 mmHg (common in large tumors) (Henderson and Fingar 1987; Mitchell et al. 1985; See et al. 1984). Thus, a therapy that is independent of oxygen levels is beneficial in this regard. PTT is one such modality. It is similar to PDT in that it uses a medium, which upon irradiation with light of a suitable wavelength results in cell death, with the major difference being the mechanism of cell kill. In case of PDT, the PS upon irradiation produces ROS but in PTT, the light absorbed is converted to heat. This raises the temperature of the local environment to more than 41°C (hyperthermia) and causes irreversible cell damage (Boulnois 1986; Nikfarjam et al. 2005)

The nonspecific damage to the surrounding healthy tissue had made it impractical to use hyperthermia as a mode of cancer therapy earlier but PTT has allowed researchers to use a noninvasive stimulus such as light to induce hyperthermia in a controlled and localized manner, thereby making it a viable and effective option that can be used to treat tumors even under hypoxic conditions (Boulnois 1986; C. Jin et al. 2013; Nikfarjam et al. 2005).

The key consideration for a PS or any light to heat transducer suitable for this technique is the efficacy of this conversion. Although naphthalocyanines and metal porphyrins have been used frequently, they are prone to photobleaching under laser irradiation, which greatly reduces their therapeutic potential (Camerin et al. 2005). Recently, gold nanoparticles have become increasingly popular as they offer excellent light to heat conversion due to the surface plasmon resonance (SPR) oscillation, high absorption cross section of NIR light, and good photostability. Thus, they can be used to achieve PTT in deeper tissues and for a prolonged duration of time (Hu et al. 2006; Huang et al. 2008; Nakamura et al. 2010; O'Connell et al. 2002). Other plasmonic metals like silver have also been used for the same.

9.4.1 UCNP-Based PTT

Besides the use of gold and silver nanoparticles, UCNPs coated with these materials have also been used for PTT. Use of UCNPs provides the added advantage of simultaneous imaging owing to their multiple visible and/or NIR emissions. Dong et al. demonstrated the use of silver-coated UCNPs for image guided PTT. They synthesized UCNPs with a $NaYF_4$ core doped with ytterbium, that is, Yb^{3+} (absorber/donor ion) and erbium, Er^{3+} (emitter/acceptor ion). These UCNPs are excited at NIR 980 nm and emit in the visible range (green and red regions). They were then coated with silver (Ag), resulting in a core–shell structure with a UCNP core and Ag shell. Ag, a plasmonic metal was chosen owing to its good photothermal properties and the thickness of the shell was tuned so that the SPR absorption was at 980 nm. These particles could be used for imaging (due to the visible emissions of the UCNPs) and PTT (due to the SPR of the Ag shell) simultaneously upon excitation with NIR at 980 nm. These particles were tested out *in vitro* but not *in vivo*. The toxicity of these particles, though lower than that of the UCNP core alone, was significant and increased with increasing particle concentration. In fact, the concentration used for *in vitro* therapy (1 mg/mL) caused about 50% cell death even without NIR irradiation, a serious drawback that might limit their *in vivo* potential (Dong et al. 2011).

UCNPs coated with gold have also been used for PTT. Qian et al. (2011) synthesized $NaYF_4$:Yb, Er/$NaYF_4$/silica (core/shell/shell) UCNPs (~70–80 nm) and deposited gold nanoparticles (~6 nm) on the surface of the silica shell of these nanoparticles. The UCNPs were excited by NIR and the upconverted green light was coupled with the surface plasmon of Au leading to rapid heat conversion, resulting in the destruction of BE (2)-C (neuroblastoma) cancer cells. Cheng et al. prepared multifunctional particles in which UCNPs were coated in a layer-by-layer manner first with iron oxide nanoparticles and then with a gold shell, followed by PEG coating to impart aqueous stability. The gold coating was utilized for NIR-mediated PTT while the UCNP luminescence and iron oxide allowed for dual-modal luminescence and MRI (Cheng et al. 2011). These particles were used for killing cancer cells *in vitro* and their potential for tumor imaging was analyzed *in vivo*. PTT *in vivo* was not demonstrated. The same group later demonstrated the use of these particles for *in vivo* PTT and multimodal imaging. In this study, the multifunctional UCNPs were intravenously injected and then magnetically targeted to the tumor site though the application of an external magnetic field, followed by irradiation of the tumor site with NIR light (Figure 9.7). This approach resulted in 8-fold higher tumor uptake of UCNPs and subsequently an enhanced tumor ablation effect (Cheng et al. 2012).

9.5 UCNPs in Combination Therapy

UCNPs can have multiple emission peaks and can be modified to carry multiple molecules simultaneously. This has led to an advent of UCNP-based combination therapies, wherein these particles are used as carriers and/or transducers for delivery and activation of multiple photoresponsive molecules for a synergistic therapeutic effect. For instance, there have been reports of UCNP-based therapies combining PDT and PTT, PDT and drug/gene delivery, as well as trimodal PDT, chemo- and radiotherapy. Y. Wang et al. (2013) used UCNPs covalently grafted with nanographene oxide (NGO) via bifunctional PEG for combined PDT and PTT along with tumor imaging . The NGO coating has a strong absorption coefficient in the NIR range and can be used for PTT. Next, the PS ZnPc was loaded onto the NGO coating. Upon irradiation with light at 808 nm, the NGO coating caused a localized heating effect resulting in cancer cell kill through PTT. Irradiation with 630 nm light-activated ZnPc and caused cell kill via PDT. Irradiating HeLa cells incubated with these nanocomposites with both 808 and 630 nm light

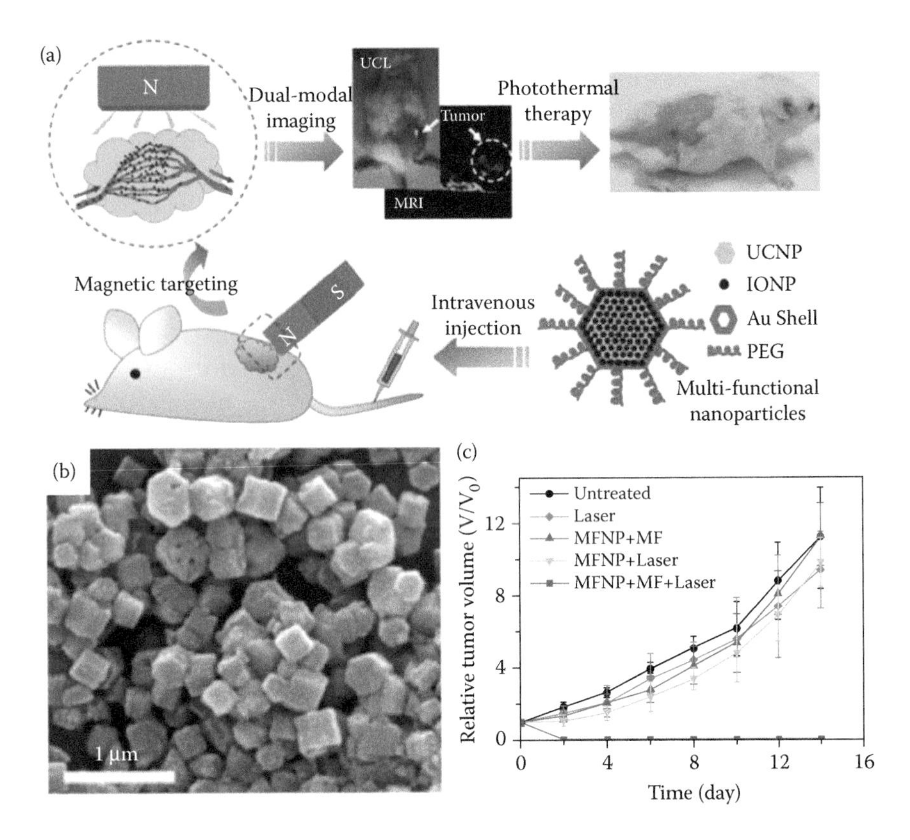

FIGURE 9.7
(a) Schematic illustration showing the composition of multifunctional nanoparticles (MFNP) and the concept of *in vivo* imaging-guided magnetically targeted PTT. The magnetic field around the tumor region induces local tumor accumulation of MFNPs. (b) An SEM image of MFNP-PEG. (c) Treatment of tumor *in vivo* using UCNP-mediated PTT. The growth of $4T_1$ tumors in different groups of mice after treatment. The tumor volumes were normalized to their initial sizes. For the treatment group, eight mice injected with MFNP-PEG were placed under the tumor-targeted magnetic field for 2 h and then exposed to an 808-nm laser at a power density of 1 W/cm^2 for 5 min. Other four groups of mice (seven mice per group) were used as controls: (1) no MFNP injection and no laser (untreated); (2) laser only without MFNP injection (laser); (3) injected with MFNP under the magnetic field but without no laser irradiation (MFNP + MF); and (4) injected with MFNP and exposed to the laser but without the magnetic targeting. Error bars were based on standard deviation (SD). (Reproduced with permission from Cheng, L., K. Yang, Y. Li, X. Zeng, M. Shao, S.-T. Lee, and Z. Liu, *Biomaterials*, 33 (7), 2215–2222, 2012.)

resulted in a synergistic cell kill effect, which was greater than that seen with either of the therapies individually. Interestingly, this study did not use the upconversion property of UCNPs for PTT or PDT; rather UCNPs were used as carriers and their upconverted light used for imaging alone.

Yuan and group reported combined chemotherapy and PDT triggered via upconverted light produced by UCNPs (Yuan et al. 2014). They covalently attached a chemotherapeutic drug doxorubicin (DOX) to a PEGylated conjugated polyelectrolyte (CPE) PS through a UV-cleavable o-NB linker. This CPE–DOX was used to encapsulate hydrophobic UCNPs in aqueous media. When irradiated with NIR light, the UV linker cleaved due to the upconverted UV light produced by the UCNPs. In addition, the upconverted visible light activated the photosensitzer, resulting in ROS production. Thus, by utilizing the multiple emission peaks of UCNPs, both photocontrolled delivery of a chemotherapeutic drug and PDT could be achieved. This system was demonstrated *in vitro* using U87-MG glioblastoma cells. A similar approach was used by Xin Wang et al (2014) for combined PDT and gene therapy.

A trimodal synergistic therapy using UCNPs was demonstrated by Fan et al. (2014). They synthesized Gd-doped UCNPs with a mesoporous silica core, with a cavity between the core and the shell. Hematoporphyrin (a PS that doubles up as a radiosensitizer) was encapsulated in the space between the UCNP core and mesoporous silica shell. A radiosensitizer/chemodrug docetaxel (Dtxl) was covalently grafted onto the shell. This nanocomposite was used for theranostics, that is, both diagnostics and therapy. Gd doping allowed for MRI, which in combination with upconversion luminescence was used for locating the tumor. NIR excitation and X-ray irradiation resulted in combined radio–chemo therapy and PDT. This nanotheranostic platform was tested both *in vitro* using HeLa cells and *in vivo* in $4T_1$ tumor bearing mice. Maximum reduction in tumor volume was seen in mice exposed to both NIR light and X-ray radiation (trimodal therapy) and was significantly higher than that observed for individual modalities (Figure 9.8).

9.6 Limitations of UCNPs

UCNPs have provided a simple yet effective means of using NIR light for photoactivation applications that had long relied on UV or visible light. Although this has brought these techniques closer to clinical use, UCNPs are not without their limitations. The quantum yield of these particles is very low, usually less than 1% (Cheng et al. 2013). As a result, a relatively high intensity of NIR is required to obtain emission levels sufficient for photoactivation and imaging. In addition, most UCNPs are excited by NIR at 980 nm, a wavelength close to the absorption peak of water molecules. This combined with the high laser power required for excitation can result in significant tissue heating. These problems pose a barrier to the widespread use of UCNPs in therapy. As a result, tremendous efforts are being made toward improving quantum yield and developing UCNPs with alternate excitation wavelengths.

Quantum yield of UCNPs has been improved through optimization of dopant ratios and dopant types use of different host materials and through coating with an undoped shell such as $NaYF_4$ or CaF_2. The addition of an undoped layer around the particles reduces nonradiative excitation losses

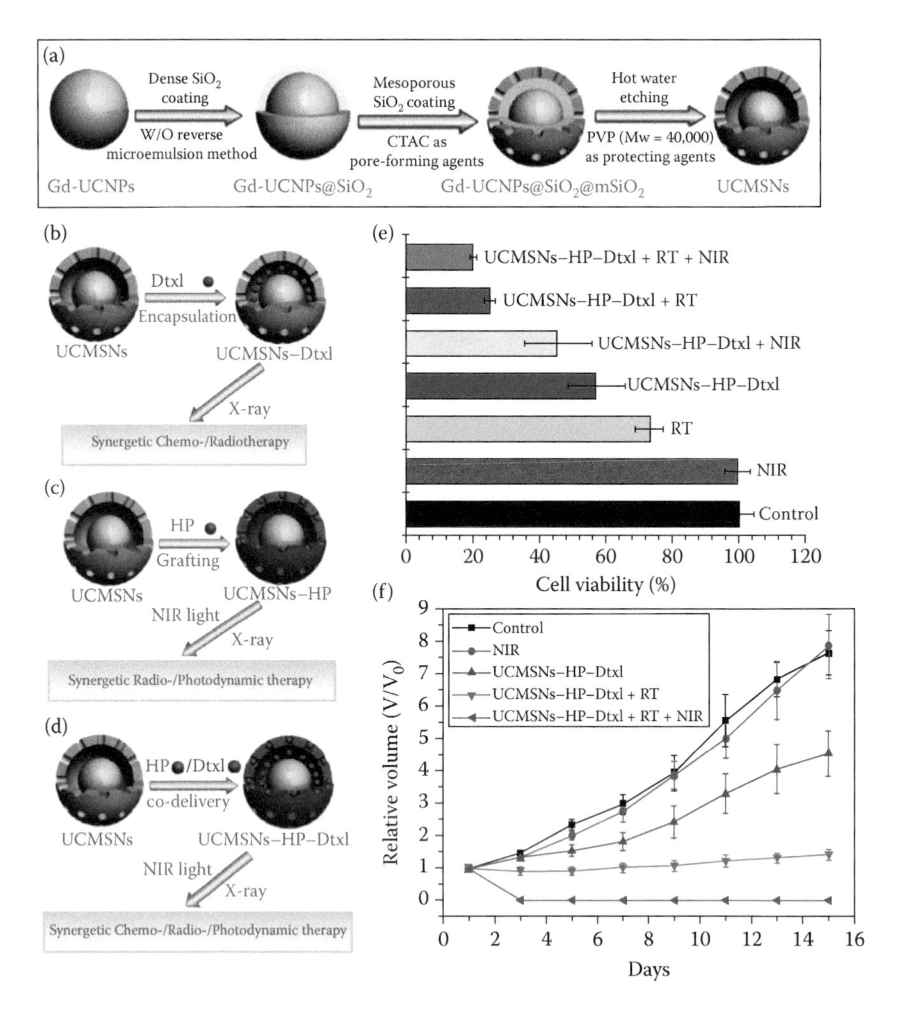

FIGURE 9.8
(a) Schematic illustration of the synthetic procedure of UCNPs core/mesoporous silica shell nanotheranostics (UCMSNs). Gd-UCNPs were prepared by epitaxial growth $NaGdF_4$ layer on $NaYF_4$:Yb/Er/Tm through a typical thermal decomposition process. Then, a dense silica layer was coated on Gd-UCNPs by a reverse microemulsion method producing Gd-UCNPs@SiO_2. Subsequently, a mesoporous silica shell was deposited on Gd-UCNPs@SiO_2 via the template of cetyltrimethylammonium chloride (CTAC) resulting in Gd-UCNPs@SiO_2@mSiO_2. Finally, UCMSNs were successfully fabricated based on a "surface-protected hot water etching" strategy. (b) Schematic diagram of synergetic chemo-/radiotherapy effects of free Dtxl/UCMSNs-Dtxl, (c) Synergetic radio-/PDT effects of UCMSNs–HP and (d) schematic diagram, and (e) *In vitro* evaluation of synergetic chemo-/radio-/PDT effects on HeLa cells after co-incubated with 5 mg/mL UCMSNs-HP-Dtxl. (f) *In vivo* evaluation of synergetic chemo-/radio-/PDT on $4T_1$ tumor bearing mice after intravenous injection of UCMSNs–HP–Dtxl. Tumor growth curves of $4T_1$ tumor bearing mice over a period of half a month after the corresponding treatments. Control groups received phosphate buffered saline (PBS). (Reproduced with permission from Fan, W., B. Shen, W. Bu, F. Chen, Q. He, K. Zhao, S. Zhang, L. Zhou, W. Peng, and Q. Xiao, *Biomaterials*, 35 (32), 8992–9002, 2014.)

at the particle surface thereby improving upconversion emission (Heer et al. 2004). Development of UCNPs with alternative excitation wavelengths different from the conventional 980 nm has been achieved through the use of novel dopants like neodymium (Nd^{3+}) and different lattice structures. UCNPs doped with neodymium ions (Nd^{3+}) can absorb at 800 nm, a wavelength at which water has a low absorption coefficient. In these particles, Nd^{3+} is used as a sensitizer. It absorbs maximally at 800 nm, transferring energy to Yb^{3+} which subsequently transfers it to the activator ions (Tm^{3+}, Er^{3+}, etc.), resulting in UV or visible upconversion emission. These particles usually have a core–shell, or core with multiple shells structure to physically separate the sensitizer and activator ions such that deleterious energy transfer between the dopants does not quench the upconversion emission efficiency (Shen et al. 2013; Y.-F. Wang et al. 2013; Wen et al. 2013; Xie et al. 2013). These Nd^{3+}-doped UCNPs are relatively new and have not gone through extensive *in vitro* and *in vivo* testing. It remains to be seen whether their photoactivation capabilities are comparable to the UCNPs that have been used thus far.

Using host matrices different from the conventionally used $NaYF_4$ also produces UCNPs with excitation wavelengths different from 980 nm. For instance, when $BaSr_2Y_6O_{12}$ is used as the host matrix, the resulting nanoparticles can be excited at 1540 nm. Using $LiYF_4$ as a host matrix results in UCNPs with excitation at 1490 nm. Furthermore, these $LiYF_4$ UCNPs have a relatively high quantum yield (1.2% for 85 nm particles at 10–150 W cm^2).

Another hurdle that curbs the clinical potential of UCNPs is the uncertainty in terms of systemic toxicity, their clearance, and long-term effects on the body. Though their safety *in vitro* and in short-term *in vivo* studies has been extensively reported, their long-term toxicity effects and potential bioaccumulation are largely unexplored. Such studies exploring long-term effects need to be undertaken for these particles to enter into clinical practice. Besides these roadblocks, the lack of standardization owing to the existence of numerous synthesis methods makes it difficult to draw comparisons between resulting UCNPs. Quality control in terms of uniformity in particle size, optical properties, etc. also serves as an impediment in the scaling up of UCNP synthesis. These problems can be addressed to a large extent through standardization of synthesis techniques and increased automation. This would allow for greater consistency in the quality of the particles synthesized besides also enabling easy scaling up of the synthesis procedure, all of which are important for bringing these particles to commercial use.

9.7 Conclusion and Outlook

Technologies like upconversion overcome a major limitation associated with traditional photoactivation techniques, namely the use of low penetrating or toxic UV/visible light for excitation by allowing the use of deeper penetrating and biologically friendly NIR light instead. This has made UCNPs nanotranducers of choice for phototherapeutic applications. These particles are excited by long-wavelength radiation (such as NIR) and can emit across a range of UV and visible light that can be tuned (depending on dopants and their ratios) in keeping with the absorption characteristics of the photoresponsive molecule in question. In the context of medicine, their unique properties like low autofluorescence, negligible photobleaching, tunability of emission peaks also enable simultaneous real time imaging, allowing one to track the delivery and distribution of therapeutic molecules that they carry.

Although promising, UCNPs do face certain limitations at present. Their quantum yield is low (<1%), thus a relatively high intensity of NIR is needed for activation, which may lead to potentially harmful heating effects owing to the high absorption of water at 970 nm, which is very close to the 980 nm light used to excite UCNPs. Efforts have been made to overcome some of these hurdles with innovations like the addition of an undoped shell such as $NaYF_4$ around the UCNPs to improve quantum yield, development of nanoparticles doped with neodymium ions that can be excited by 800 nm where the absorption coefficient of water is low, etc. Despite these efforts, UCNPs still have a long way to go before they can be put to clinical use. Batch-to-batch variability, uncertain systemic toxicity and clearance need to be addressed for their clinical and commercial potential to be realized.

References

Aebischer, A., M. Hostettler, J. Hauser, K. Krämer, T. Weber, H. U. Güdel, and H.-B. Bürgi. 2006. Structural and spectroscopic characterization of active sites in a family of light-emitting sodium lanthanide tetrafluorides. *Angew. Chem. Int. Ed.* 45 (17):2802–2806.

Allen, C. M., W. M. Sharman, and J. E. van Lier. 2001. Current status of phthalocyanines in the photodynamic therapy of cancer. *J. Porphyr. Phthalocya.* 5 (02):161–169.

Allison, R. R. and C. H. Sibata. 2008. Photofrin® photodynamic therapy: 2.0 mg/kg or not 2.0 mg/kg that is the question. *Photodiagnosis and Photodynamic Therapy* 5 (2):112–119.

Allison, R. R. and C. H. Sibata. 2010. Oncologic photodynamic therapy photosensitizers: A clinical review. *Photodiagnosis Photodyn. Ther.* 7 (2):61–75.

Ben-Hur, E. and I. Rosenthal. 1985. The phthalocyanines: A new class of mammalian cells photosensitizers with a potential for cancer phototherapy. *Int. J. Radiat. Biol. Relat. Stud. Phys. Chem. Med.* 47 (2):145–147.

Berg, K., A. Høgset, L. Prasmickaite et al. 2006. Photochemical internalization (PCI): A novel technology for activation of endocytosed therapeutic agents. *Med. Laser Appl.* 21 (4):239–250. doi: 10.1016/j.mla.2006.08.004.

Bernstein, Z. P., B. D. Wilson, A. R. Oseroff, C. M. Jones, S. E. Dozier, J. S. J. Brooks, R. Cheney, L. Foulke, T. S. Mang, and D. A. Bellnier. 1999. Photofrin photodynamic therapy for treatment of AIDS-related cutaneous Kaposi's sarcoma. *AIDS* 13 (13):1697–1704.

Berstad, M. B., A. Weyergang, and K. Berg. 2012. Photochemical internalization (PCI) of HER2-targeted toxins: Synergy is dependent on the treatment sequence. *Biochim. Biophys. Acta* 1820 (12):1849–1858. doi: 10.1016/j.bbagen.2012.08.027.

Boe, S., S. Saeboe-Larssen, and E. Hovig. 2010. Light-induced gene expression using messenger RNA molecules. *Oligonucleotides* 20 (1):1–6. doi: 10.1089/oli.2009.0209.

Bonnett, R. 1995. Photosensitizers of the porphyrin and phthalocyanine series for photodynamic therapy. *Chem. Soc. Rev.* 24 (1):19–33.

Boulnois, J.-L. 1986. Photophysical processes in recent medical laser developments: A review. *Laser. Med. Sci.* 1 (1):47–66.

Camerin, M., S. Rello, A. Villanueva, X. Ping, M. E. Kenney, M. A. J. Rodgers, and G. Jori. 2005. Photothermal sensitisation as a novel therapeutic approach for tumours: Studies at the cellular and animal level. *Eur. J. Cancer* 41 (8):1203–1212.

Carling, C.-J., F. Nourmohammadian, J.-C. Boyer, and N. R. Branda. 2010. Remote-control photorelease of caged compounds using near-infrared light and upconverting nanoparticles. *Angew. Chem. Int. Ed.* 49 (22):3782–3785.

Castano, A. P, P. Mroz, and M. R. Hamblin. 2006. Photodynamic therapy and anti-tumour immunity. *Nat. Rev. Cancer* 6 (7):535–545.

Chatterjee, D. K., M. K. Gnanasammandhan, and Y. Zhang. 2010. Small upconverting fluorescent nanoparticles for biomedical applications. *Small* 6 (24):2781–2795.

Chatterjee, D. K. and Z. Yong. 2008. Upconverting nanoparticles as nanotransducers for photodynamic therapy in cancer cells. *Nanomedicine (Future medicine)* 3(1): 73–82.

Cheng, L., C. Wang, and Z. Liu. 2013. Upconversion nanoparticles and their composite nanostructures for biomedical imaging and cancer therapy. *Nanoscale* 5 (1):23–37. doi: 10.1039/c2nr32311g.

Cheng, L., K. Yang, Y. Li, J. Chen, C. Wang, M. Shao, S.-T. Lee, and Z. Liu. 2011. Facile preparation of multifunctional upconversion nanoprobes for multimodal imaging and dual-targeted photothermal therapy. *Angew. Chem.* 123 (32):7523–7528.

Cheng, L., K. Yang, Y. Li, X. Zeng, M. Shao, S.-T. Lee, and Z. Liu. 2012. Multifunctional nanoparticles for upconversion luminescence/MR multimodal imaging and magnetically targeted photothermal therapy. *Biomaterials* 33 (7):2215–2222.

Cui, S., H. Chen, H. Zhu, J. Tian, X. Chi, Z. Qian, S. Achilefu, and Y. Gu. 2012a. Amphiphilic chitosan modified upconversion nanoparticles for in vivo photodynamic therapy induced by near-infrared light. *J. Mater. Chem.* 22 (11):4861–4873.

Cui, S., D. Yin, Y. Chen, Y. Di, H. Chen, Y. Ma, S. Achilefu, and Y. Gu. 2012b. *In vivo* targeted deep-tissue photodynamic therapy based on near-infrared light triggered upconversion nanoconstruct. *ACS Nano* 7 (1):676–688.

Dai, Y., H. Xiao, J. Liu, Q. Yuan, P. Ma, D. Yang, C. Li, Z. Cheng, Z. Hou, and P. Yang. 2013. *In vivo* multimodality imaging and cancer therapy by near-infrared light-triggered trans-platinum pro-drug-conjugated upconverison nanoparticles. *J. Am. Chem. Soc.* 135 (50):18920–18929.

Dominska, M. and D. M. Dykxhoorn. 2010. Breaking down the barriers: siRNA delivery and endosome escape. *J. Cell Sci.* 123 (8):1183–1189.

Dong, B., S. Xu, J. Sun, S. Bi, D. Li, X. Bai, Y. Wang, L. Wang, and H. Song. 2011. Multifunctional $NaYF_4$: Yb3+, Er3+@ Ag core/shell nanocomposites: Integration of upconversion imaging and photothermal therapy. *J. Mater. Chem.* 21 (17):6193–6200.

Dougherty, T. J., C. J. Gomer, B. W. Henderson, G. Jori, D. Kessel, M. Korbelik, J. Moan, and Q. Peng. 1998. Photodynamic therapy. *J. Natl. Cancer Inst.* 90 (12):889–905.

Fan, W., B. Shen, W. Bu, F. Chen, Q. He, K. Zhao, S. Zhang, L. Zhou, W. Peng, and Q. Xiao. 2014. A smart upconversion-based mesoporous silica nanotheranostic system for synergetic chemo-/radio-/photodynamic therapy and simultaneous MR/UCL imaging. *Biomaterials* 35 (32):8992–9002.

Fretz, M. M., A. Hogset, G. A. Koning, W. Jiskoot, and G. Storm. 2007. Cytosolic delivery of liposomally targeted proteins induced by photochemical internalization. *Pharm. Res.* 24 (11):2040–2047. doi: 10.1007/s11095-007-9338-9.

Garcia, J. V., J. Yang, D. Shen, C. Yao, X. Li, R. Wang, G. D. Stucky, D. Zhao, P. C. Ford, and F. Zhang. 2012. NIR-triggered release of caged nitric oxide using upconverting nanostructured materials. *Small* 8 (24):3800–3805.

Gillmeister, M. P., M. J. Betenbaugh, and P. S. Fishman. 2011. Cellular trafficking and photochemical internalization of cell penetrating peptide linked cargo proteins: A dual fluorescent labeling study. *Bioconjug. Chem.* 22 (4):556–566. doi: 10.1021/bc900445g.

Guo, H., H. Qian, N. M. Idris, and Y. Zhang. 2010. Singlet oxygen-induced apoptosis of cancer cells using upconversion fluorescent nanoparticles as a carrier of photosensitizer. *Nanomed. Nanotechnol. Biol. Med.* 6 (3):486–495.

Heer, S., K. Kömpe, H.-U. Güdel, and M. Haase. 2004. Highly efficient multicolour upconversion emission in transparent colloids of lanthanide-doped $NaYF_4$ nanocrystals. *Adv. Mater.* 16 (23–24):2102–2105.

Henderson, B. W. and T. J. Dougherty. 1992. How does photodynamic therapy work? *Photochem. Photobiol.* 55 (1):145–157.

Henderson, B. W. and V. H. Fingar. 1987. Relationship of tumor hypoxia and response to photodynamic treatment in an experimental mouse tumor. *Cancer Res.* 47 (12):3110–3114.

Hou, Z., Y. Zhang, K. Deng, Y. Chen, X. Li, X. Deng, Z. Cheng, H. Lian, C. Li, and J. Lin. 2015. UV-emitting upconversion-based TiO_2 photosensitizing nanoplatform: Near-infrared light mediated *in vivo* photodynamic therapy via mitochondria-involved apoptosis pathway. *ACS Nano* 9 (3):2584–2599.

Hu, M., J. Chen, Z. Y. Li, L. Au, G. V. Hartland, X. Li, M. Marquez, and Y. Xia. 2006. Gold nanostructures: Engineering their plasmonic properties for biomedical applications. *Chem. Soc. Rev.* 35 (11):1084–1094. doi: 10.1039/b517615h.

Huang, X., P. K. Jain, I. H. El-Sayed, and M. A. El-Sayed. 2008. Plasmonic photothermal therapy (PPTT) using gold nanoparticles. *Laser. Med. Sci.* 23 (3):217–228.

Idris, N. M., M. K. Gnanasammandhan, J. Zhang, P. C. Ho, R. Mahendran, and Y. Zhang. 2012. *In vivo* photodynamic therapy using upconversion nanoparticles as remote-controlled nanotransducers. *Nat. Med.* 18 (10):1580–1585. http://www.nature.com/nm/journal/v18/n10/abs/nm.2933.html#supplementary-information.

Idris, N. M., M. K. G. Jayakumar, A. Bansal, and Y. Zhang. 2015. Upconversion nanoparticles as versatile light nanotransducers for photoactivation applications. *Chem. Soc. Rev.* 44(6): 1449–1478.

Idris, N. M., S. S. Lucky, Z. Li, K. Huang, and Y. Zhang. 2014. Photoactivation of core–shell titania coated upconversion nanoparticles and their effect on cell death. *J. Mater. Chem. B* 2 (40):7017–7026.

Jayakumar, M. K. G., A. Bansal, K. Huang, R. Yao, B. N. Li, and Y. Zhang. 2014. Near-infrared-light-based nano-platform boosts endosomal escape and controls gene knockdown *in vivo*. *ACS Nano* 8(5):4848–4858.

Jayakumar, M. K., N. M. Idris, and Y. Zhang. 2012. Remote activation of biomolecules in deep tissues using near-infrared-to-UV upconversion nanotransducers. *Proc. Natl. Acad. Sci. USA* 109 (22):8483–8488. doi: 10.1073/pnas.1114551109.

Jin, C. S., J. F. Lovell, J. Chen, and G. Zheng. 2013. Ablation of hypoxic tumors with dose-equivalent photothermal, but not photodynamic, therapy using a nanostructured porphyrin assembly. *ACS Nano* 7 (3):2541–2550.

Jin, H., J. F. Lovell, J. Chen, K. Ng, W. Cao, L. Ding, Z. Zhang, and G. Zheng. 2011. Cytosolic delivery of LDL nanoparticle cargo using photochemical internalization. *Photochem. Photobiol. Sci.* 10 (5):810–816. doi: 10.1039/c0pp00350f.

Jin, S., L. Zhou, Z. Gu, G. Tian, L. Yan, W. Ren, W. Yin, X. Liu, X. Zhang, and Z. Hu. 2013. A new near infrared photosensitizing nanoplatform containing blue-emitting up-conversion nanoparticles and hypocrellin A for photodynamic therapy of cancer cells. *Nanoscale* 5 (23):11910–11918.

Juzeniene, A., Q. Peng, and J. Moan. 2007. Milestones in the development of photodynamic therapy and fluorescence diagnosis. *Photochem. Photobiol. Sci.* 6 (12):1234–1245.

Krämer, K. W., D. Biner, G. Frei, H. U. Güdel, M. P. Hehlen, and S. R. Lüthi. 2004. Hexagonal sodium yttrium fluoride based green and blue emitting upconversion phosphors. *Chem. Mater.* 16 (7):1244–1251.

Lilletvedt, M., H. H. Tonnesen, A. Hogset, S. A. Sande, and S. Kristensen. 2011. Evaluation of physicochemical properties and aggregation of the photosensitizers TPCS2a and TPPS2a in aqueous media. *Die Pharmazie Int. J. Pharm. Sci.* 66 (5):325–333.

Liu, J., W. Bu, L. Pan, and J. Shi. 2013. NIR-triggered anticancer drug delivery by upconverting nanoparticles with integrated azobenzene-modified mesoporous silica. *Angew. Chem. Int. Ed.* 52(16):4375–4379.

Liu, K., X. Liu, Q. Zeng, Y. Zhang, L. Tu, T. Liu, X. Kong, Y. Wang, F. Cao, and S. A. G. Lambrechts. 2012. Covalently assembled NIR nanoplatform for simultaneous fluorescence imaging and photodynamic therapy of cancer cells. *ACS Nano* 6 (5):4054–4062.

Liu, Q., Y. Sun, T. Yang, W. Feng, C. Li, and F. Li. 2011. Sub-10 nm hexagonal lanthanide-doped $NaLuF_4$ upconversion nanocrystals for sensitive bioimaging *in vivo*. *J. Am. Chem. Soc.* 133 (43):17122–17125.

Liu, X., C. N. Kim, J. Yang, R. Jemmerson, and X. Wang. 1996. Induction of apoptotic program in cell-free extracts: Requirement for dATP and cytochrome c. *Cell* 86 (1):147–157.

Liu, X., M. Zheng, X. Kong, Y. Zhang, Q. Zeng, Z. Sun, W. J. Buma, and H. Zhang. 2013. Separately doped upconversion-C 60 nanoplatform for NIR imaging-guided photodynamic therapy of cancer cells. *Chem. Commun.* 49 (31):3224–3226.

Lucky, S. S., N. M. Idris, Z. Li, K. Huang, K. C. Soo, and Y. Zhang. 2015. Titania coated upconversion nanoparticles for near-infrared light triggered photodynamic therapy. *ACS Nano* 9 (1):191–205.

Mitchell, J. B., S. McPherson, W. DeGraff, J. Gamson, A. Zabell, and A. Russo. 1985. Oxygen dependence of hematoporphyrin derivative-induced photoinactivation of Chinese hamster cells. *Cancer Res.* 45 (5):2008–2011.

Nakamura, T., A. Tamura, H. Murotani, M. Oishi, Y. Jinji, K. Matsuishi, and Y. Nagasaki. 2010. Large payloads of gold nanoparticles into the polyamine network core of stimuli-responsive PEGylated nanogels for selective and noninvasive cancer photothermal therapy. *Nanoscale* 2 (5):739–746.

Ni, D., J. Zhang, W. Bu, H. Xing, F. Han, Q. Xiao, Z. Yao, F. Chen, Q. He, and J. Liu. 2014. Dual-targeting upconversion nanoprobes across the blood–brain barrier for magnetic resonance/fluorescence imaging of intracranial glioblastoma. *ACS Nano* 8 (2):1231–1242.

Nikfarjam, M., V. Muralidharan, and C. Christophi. 2005. Mechanisms of focal heat destruction of liver tumors. *J. Surg. Res.* 127 (2):208–223.

Nishiyama, N., A. Iriyama, W. D. Jang et al. 2005. Light-induced gene transfer from packaged DNA enveloped in a dendrimeric photosensitizer. *Nat. Mater.* 4 (12):934–41. doi: 10.1038/nmat1524.

O'Connell, M. J., S. M. Bachilo, C. B. Huffman, V. C. Moore, M. S. Strano, E. H. Haroz, K. L. Rialon, P. J. Boul, W. H. Noon, and C. Kittrell. 2002. Band gap fluorescence from individual single-walled carbon nanotubes. *Science* 297 (5581):593–596.

Oliveira, S., M. M. Fretz, A. Hogset, G. Storm, and R. M. Schiffelers. 2007. Photochemical internalization enhances silencing of epidermal growth factor receptor through improved endosomal escape of siRNA. *Biochim. Biophys. Acta* 1768 (5):1211–1217. doi: 10.1016/j.bbamem.2007.01.013.

Ouyang, J., D. Yin, X. Cao, C. Wang, K. Song, B. Liu, L. Zhang, Y. Han, and M. Wu. 2014. Synthesis of $NaLuF_4$-based nanocrystals and large enhancement of upconversion luminescence of $NaLuF_4$: Gd, Yb, Er by coating an active shell for bioimaging. *Dalton Trans.* 43 (37):14001–14008.

Park, Y. I., H. M. Kim, J. H. Kim, K. C. Moon, B. Yoo, K. T. Lee, N. Lee, Y. Choi, W. Park, and D. Ling. 2012. Theranostic probe based on lanthanide-doped nanoparticles for simultaneous *in vivo* dual-modal imaging and photodynamic therapy. *Adv. Mater.* 24 (42):5755–5761.

Pelliccioli, A. P. and J. Wirz. 2002. Photoremovable protecting groups: Reaction mechanisms and applications. *Photochem.Photobiol. Sci.* 1 (7):441–458. doi: 10.1039/b200777k.

Peng, Q., K. Berg, J. Moan, M. Kongshaug, and J. M. Nesland. 1997. 5-Aminolevulinic acid-based photodynamic therapy: Principles and experimental research. *Photochem. Photobiol.* 65 (2):235–251.

Plaetzer, K., B. Krammer, J. Berlanda, F. Berr, and T. Kiesslich. 2009. Photophysics and photochemistry of photodynamic therapy: Fundamental aspects. *Laser. Med. Sci.* 24 (2):259–268.

Qian, H. S., H. C. Guo, P. C.-L. Ho, R. Mahendran, and Y. Zhang. 2009. Mesoporous-silica-coated up-conversion fluorescent nanoparticles for photodynamic therapy. *Small* 5 (20):2285–2290.

Qian, L. P., L. H. Zhou, H.-P. Too, and G.-M. Chow. 2011. Gold decorated $NaYF_4$: Yb, Er/$NaYF_4$/silica (core/shell/shell) upconversion nanoparticles for photothermal destruction of BE (2)-C neuroblastoma cells. *J. Nanopart. Res.* 13 (2):499–510.

Qiao, X.-F., J.-C. Zhou, J.-W. Xiao, Y.-F. Wang, L.-D. Sun, and C.-H. Yan. 2012. Triple-functional core–shell structured upconversion luminescent nanoparticles covalently grafted with photosensitizer for luminescent, magnetic resonance imaging and photodynamic therapy *in vitro*. *Nanoscale* 4 (15):4611–4623.

See, K. L., I. J. Forbes, and W. H. Betts. 1984. Oxygen dependency of photocytotoxicity with haematoporphyrin derivative. *Photochem. Photobiol.* 39 (5):631–634.

Selbo, P. K., A. Weyergang, A. Bonsted, S. G. Bown, and K. Berg. 2006. Photochemical internalization of therapeutic macromolecular agents: A novel strategy to kill multidrug-resistant cancer cells. *J. Pharmacol. Exp. Ther.* 319 (2):604–612. doi: 10.1124/jpet.106.109165.

Shen, J., G. Chen, A.-M. Vu, W. Fan, O. S. Bilsel, C.-C. Chang, and G. Han. 2013. Engineering the upconversion nanoparticle excitation wavelength: Cascade sensitization of tri-doped upconversion colloidal nanoparticles at 800 nm. *Adv. Opt. Mater.* 1 (9):644–650.

Shieh, Y.-A., S.-J. Yang, M.-F. Wei, and M.-J. Shieh. 2010. Aptamer-based tumor-targeted drug delivery for photodynamic therapy. *ACS Nano* 4 (3):1433–1442.

Sortino, S. 2012. Photoactivated nanomaterials for biomedical release applications. *J. Mater. Chem.* 22 (2):301. doi: 10.1039/c1jm13288a.

Tan, W., M. J Donovan, and J. Jiang. 2013. Aptamers from cell-based selection for bioanalytical applications. *Chem. Rev.* 113 (4):2842–2862.

Ungun, B., R. K. Prud'Homme, S. J. Budijon, J. Shan, S. F. Lim, Y. Ju, and R. Austin. 2009. Nanofabricated upconversion nanoparticles for photodynamic therapy. *Opt. Express* 17 (1):80–86.

Wang, C., L. Cheng, Y. Liu, X. Wang, X. Ma, Z. Deng, Y. Li, and Z. Liu. 2013. Imaging-guided pH-sensitive photodynamic therapy using charge reversible upconversion nanoparticles under near-infrared light. *Adv. Funct. Mater.* 23 (24):3077–3086.

Wang, C., H. Tao, L. Cheng, and Z. Liu. 2011. Near-infrared light induced *in vivo* photodynamic therapy of cancer based on upconversion nanoparticles. *Biomaterials* 32 (26):6145–6154.

Wang, C., D. Yin, K. Song, J. Ouyang, and B. Liu. 2014. Preparation of bi-functional $NaGdF_4$-based upconversion nanocrystals and fine-tuning of emission colors of the nanocrystals by doping with Mn^{2+}. *Vacuum* 107:311–315.

Wang, H. J., R. Shrestha, and Y. Zhang. 2014. Encapsulation of photosensitizers and upconversion nanocrystals in lipid micelles for photodynamic therapy. *Part. Part. Syst. Char.* 31 (2):228–235.

Wang, Xin, K. Liu, G. Yang, L. Cheng, L. He, Y. Liu, Y. Li, L. Guo, and Z. Liu. 2014. Near-infrared light triggered photodynamic therapy in combination with gene therapy using upconversion nanoparticles for effective cancer cell killing. *Nanoscale* 6 (15):9198–9205.

Wang, Xu, C.-X. Yang, J.-T. Chen, and X.-P. Yan. 2014. A dual-targeting upconversion nanoplatform for two-color fluorescence imaging-guided photodynamic therapy. *Anal. Chem.* 86 (7):3263–3267.

Wang, Y., H. Wang, D. Liu, S. Song, X. Wang, and H. Zhang. 2013. Graphene oxide covalently grafted upconversion nanoparticles for combined NIR mediated imaging and photothermal/photodynamic cancer therapy. *Biomaterials* 34 (31):7715–7724.

Wang, Y.-F., G.-Y. Liu, L.-D. Sun, J.-W. Xiao, J.-C. Zhou, and C.-H. Yan. 2013. Nd^{3+}-sensitized upconversion nanophosphors: Efficient *in vivo* bioimaging probes with minimized heating effect. *ACS Nano* 7 (8):7200–7206.

Wen, H., H. Zhu, X. Chen, T. F. Hung, B. Wang, G. Zhu, S. F. Yu, and F. Wang. 2013. Upconverting near-infrared light through energy management in core–shell–shell nanoparticles. *Angew. Chem.* 125 (50):13661–13665.

Xie, X., N. Gao, R. Deng, Q. Sun, Q.-H. Xu, and X. Liu. 2013. Mechanistic investigation of photon upconversion in Nd^{3+}-sensitized core–shell nanoparticles. *J. Am. Chem. Soc.* 135 (34):12608–12611.

Yan, B., J.-C. Boyer, N. R. Branda, and Y. Zhao. 2011. Near-infrared light-triggered dissociation of block copolymer micelles using upconverting nanoparticles. *J. Am. Chem. Soc.* 133 (49):19714–19717.

Yan, B., J.-C. Boyer, D. Habault, N. R. Branda, and Y. Zhao. 2012. Near infrared light triggered release of biomacromolecules from hydrogels loaded with upconversion nanoparticles. *J. Am. Chem. Soc.* 134 (40):16558–16561.

Yang, Y., F. Liu, X. Liu, and B. Xing. 2013a. NIR light controlled photorelease of siRNA and its targeted intracellular delivery based on upconversion nanoparticles. *Nanoscale* 5 (1):231–238. doi: 10.1039/c2nr32835f.

Yang, Y., B. Velmurugan, X. Liu, and B. Xing. 2013b. NIR photoresponsive cross-linked upconverting nanocarriers toward selective intracellular drug release. *Small* 9(17):2937–2944.

Yu, H., J. Li, D. Wu, Z. Qiu, and Y. Zhang. 2010. Chemistry and biological applications of photo-labile organic molecules. *Chem. Soc. Rev.* 39 (2):464–473. doi: 10.1039/b901255a.

Yuan, Y., Y. Min, Q. Hu, B. Xing, and B. Liu. 2014. NIR photoregulated chemo- and photodynamic cancer therapy based on conjugated polyelectrolyte–drug conjugate encapsulated upconversion nanoparticles. *Nanoscale* 6 (19):11259–11272.

Zhang, P., W. Steelant, M. Kumar, and M. Scholfield. 2007. Versatile photosensitizers for photodynamic therapy at infrared excitation. *J. Am. Chem. Soc.* 129 (15):4526–4527.

Zhao, L., J. Peng, Q. Huang, C. Li, M. Chen, Y. Sun, Q. Lin, L. Zhu, and F. Li. 2013. Near-infrared photoregulated drug release in living tumor tissue via yolk–shell upconversion nanocages. *Adv. Funct. Mat.* 24(3):363–371. doi: 10.1002/adfm.201302133.

Zhao, Z., Y. Han, C. Lin, D. Hu, F. Wang, X. Chen, Z. Chen, and N. Zheng. 2012. Multifunctional core–shell upconverting nanoparticles for imaging and photodynamic therapy of liver cancer cells. *Chem. Asian J.* 7 (4):830–837.

Zhou, A., Y. Wei, B. Wu, Q. Chen, and D. Xing. 2012. Pyropheophorbide A and c(RGDyK) comodified chitosan-wrapped upconversion nanoparticle for targeted near-infrared photodynamic therapy. *Mol. Pharmaceut.* 9 (6):1580–1589.

10

Upconverting Nanoparticles for Security Applications

A. Baride and J. Meruga

CONTENTS

10.1 Introduction 291
10.2 Upconverting Nanoparticles 292
10.3 Upconversion Security Printing Prospective 293
10.4 Synthesis, Functionalization, and Scale-up 298
    10.4.1 Hansen Solubility Parameters 299
10.5 Toxicity Assessment 299
10.6 Security Printing Applications 301
    10.6.1 Red, Green, and Blue Upconversion Printing 302
    10.6.2 Micro Upconverting QR Code Tags 304
    10.6.3 Color-Coded Multilayer Photopatterned Microstructures 305
    10.6.4 Micro Barcodes 305
    10.6.5 Tunable Lifetime Multiplexing of UCNPs 305
    10.6.6 NIR-to-NIR Upconversion Security Printing 308
    10.6.7 Scanning Laser Imaging of Upconversion Printing 312
    10.6.8 Multicolor Nano Barcoding in a Single Upconversion Crystal 313
    10.6.9 Upconversion Polymeric Nanofibers 314
10.7 Readability of Upconversion Prints 314
10.8 Conclusion 315
Acknowledgments 316
References 316

10.1 Introduction

Upconversion (UC) is a nonlinear process in which the excitation energy is absorbed by two or more photons that leads to the emission of light at shorter wavelength than the excitation wavelength. In this case, near-infrared (NIR) light, that is, 980 nm (lower energy, longer wavelength) is upconverted to visible light, 400–700 nm (higher energy, shorter wavelength) by UC material. In contrast to other emission processes, UC has high internal quantum efficiency (Haase and Schäfer 2011). In the 1950s, the UC effect was first discovered and studied intensively during the 1960s but up until 2003 it was observed only in bulk, transition metals or rare earth-doped phosphors which were suggested to have applications in the field of detectors, display, and laser materials (Bünzli and Eliseeva 2011). Since then UC materials have found their application in solar cells, waveguide amplifiers and later, they became applicable in rewritable optical storage nondestructive optical memory and many more potential scientific applications; and in the last decade, UC materials became well known in the biological field for bio sensing, drug delivery, etc. (Ho et al. 2003; Kim et al 2009; Shalav et al. 2005; Suyver et al. 2005; Vennerberg and Lin 2011; Wang and Bass 2004; Wang and Liu 2009; Wang et al. 2010; Zhou et al. 2015).

10.2 Upconverting Nanoparticles

Lanthanide-doped nanoparticles (NPs) became prominent in the late 1990s because of the development of nanotechnology research which has increased the interest in new synthesis routes that produced highly efficient NPs with narrow size distribution that are able to form a stable colloidal dispersion. Over the course of time, high quality upconverting nanoparticles (UCNPs) were routinely synthesized and properties like desired dispersibility, particle size, shape, crystallographic phase, and optical nature were tailored (Haase and Schäfer 2011). Upconverting NPs have the unique property of emitting visible light upon the excitation with NIR laser light. Upconversion systems have attracted biologists' attention because, unlike UV excitation, NIR excitation results in virtually zero background fluorescence (Wang et al. 2010). The advantage with upconversion NPs is that UCNPs have very high chemical stability, narrow emission bands, the lack of bleaching, the excitation wavelength falls in the biological window, high internal quantum efficiency, minimal effect of surface quenching, and the absence of blinking effects. However, one limiting aspect of UCNPs is difficulty in producing the color of interest. Because of their broad range of advantages, in recent years, UCNPs have been discussed as promising alternatives to organic fluorophores and quantum dots in the field of medical imaging and security printing. As the down-conversion security inks have become much easier to obtain and duplicate to produce counterfeits, UCNPs' security inks have been considered as the next generation of security. A variety of security printing applications with UCNPs have been introduced recently (Bao et al. 2012; Baride et al. 2015; Blumenthal et al. 2012; Kim and Kim 2015; Kumar et al. 2014; Lee et al. 2014; Lu et al. 2014; Meruga et al. 2015, 2012, 2014).

10.3 Upconversion Security Printing Prospective

The advantage with an upconversion system is the ability to tune the optical properties by altering the doping composition of lanthanide ions. A vast range of colors can be achieved by varying the doping composition of lanthanide ions. Moreover, in a few cases, the color emitted by an upconversion system depends on the excitation power.

The lanthanide ions with partially filled 4f electrons have discrete energy levels because of columbic repulsions and spin–orbital coupling of 4f electrons. The ladder-like discrete energy levels contribute to multiple emission bands. However, a unique combination of lanthanide ions is necessary to produce a distinct emission bands. For example, green emission produced by the 17%Yb^{3+}/3%Er^{3+} system has emission bands at 520 and 540 nm and also a 410 nm blue band and 655 nm red band in the visible region (Anderson et al. 2013). Blue emission produced by the Yb^{3+}/Tm^{3+} system has emission bands at 440 and 470 nm and also 650 and 800 nm in the UC region. Red emission produced by the Er^{3+}/Tm^{3+} system has only one emission band at the 655 nm region. The emission intensity of various bands depends on the excitation power density and the doping concentration. Figure 10.1a is the emission response in the visible region from Yb^{3+}/Er^{3+} (green), Yb^{3+}/Tm^{3+} (blue), and Er^{3+}/Tm^{3+} (red)-doped β-$NaYF_4$ crystals. In Figure 10.1b, the emission response of the Yb^{3+}/Tm^{3+} system at a different power density is

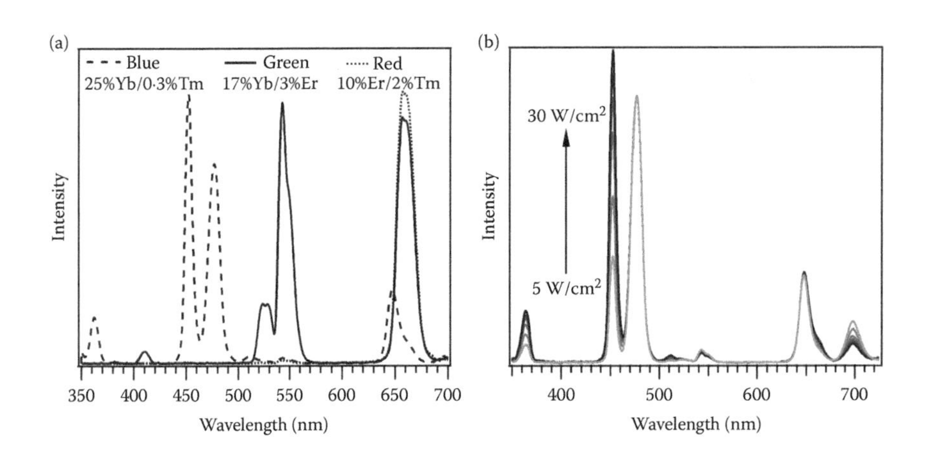

FIGURE 10.1
(a) UC emission response from β-$NaYF_4$ crystals doped with 17%Yb^{3+}/3%Er^{3+} (green), 25%Yb^{3+}/0.3%Tm^{3+} (blue), and 10%Er^{3+}/2%Tm^{3+} (red) ions excited with 980 nm laser. (b) UC emission response of nanocrystals doped with 25%Yb^{3+}/0.3%Tm^{3+} excided with 980 nm light (5–30 W/cm^2) power. The peak intensities are normalized to peak maxima of 470 nm corresponding to Tm: $^1G_4 \rightarrow {}^3H_6$ transition.

normalized to the 470 nm emission peak. Different emission bands of the Yb^{3+}/Tm^{3+} system, especially the 440 and 470 nm, respond differently to the excitation power density.

By now, we understand that the intensity of various bands is characterized by the doping composition and excitation power density. This feature adds an advantage to produce a finger-print type security feature, where it is very difficult to reproduce the emission profile at a given power density, even if one were able to identify the emitting species. A small variation in doping concentration also results in a drastic change in optical properties of the upconversion system, which can be used as a forensic feature in security printing (Baride et al. 2015; Lee et al. 2014; Lu et al. 2014).

Figure 10.2a is a comparison of emission responses from 10%Er^{3+} and 10%Er^{3+}/2%Tm^{3+}-doped systems. The 10%Er^{3+}-doped UCNPs emit green color and the spectral profile has a small 410 nm band, 520–540 nm band, and a 650 nm band in the visible region. The Tm^{3+} ion incorporated as a co-dopant, quenched the characteristic green emission and enhanced the red emission. Another example is in Figure 10.2b. The 25%Yb^{3+}/0.3%Tm^{3+}-doped system has two strong peaks in the blue region, a peak in the red region, and a very strong peak in the NIR (800 nm) region. Increasing doping concentration to 48%Yb^{3+}/2%Tm^{3+} resulted in the quenching of blue and red peaks and the appearance of a new peak at 700 nm. Moreover, the emission intensity of the NIR (800) nm peak is enhanced six fold facilitating better signal in upconversion imaging.

The luminescence lifetime of various emission bands is also greatly affected by alteration in the co-dopant concentration. Therefore, incorporation of

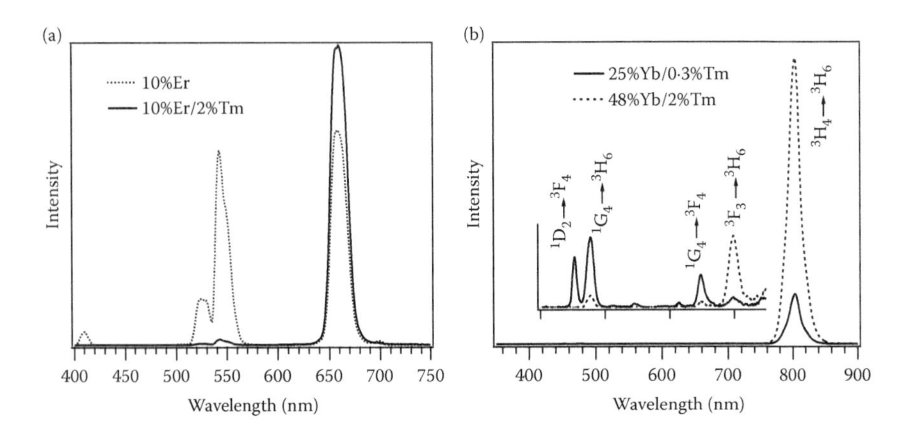

FIGURE 10.2
UC emission response from nanocrystals doped with (a): 10%Er^{3+} (green) and 10%Er^{3+}/2%Tm^{3+} (red) and (b): 25%Yb^{3+}/0.3%Tm^{3+} (blue) and 48%Yb^{3+}/2%Tm^{3+} (red) upon excited with 980 nm light. The variation in the doping concentration greatly affects the emission intensity of individual emission bands.

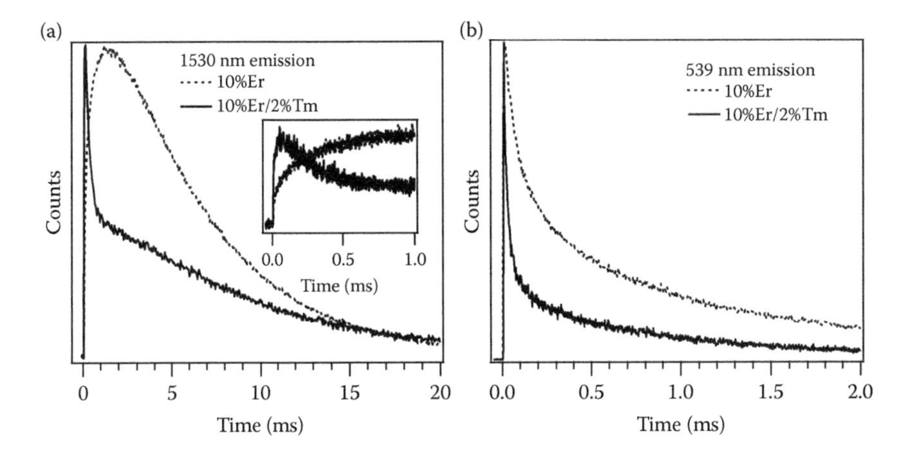

FIGURE 10.3
Luminescence decay of 10%Er and 10%Er/2%Tm-doped UCNPs in (a) NIR, 1530 nm and (b) green, 532 nm region. The luminescence lifetime of NIR and green emission of 10%Er UCNPs is greatly different from the 10%Er/2%Tm-doped UCNPs. The characteristic features of decay patterns can be used as a forensic feature in security printing.

temporal dimension to the upconversion emission makes the upconversion inks very difficult to reproduce. For example, Figure 10.3a and b is luminescence decay curves of NIR (1530 nm) and green (532 nm) from 10%Er and 10%Er/2%Tm-doped β-$NaYF_4$ UCNPs. The NIR and green luminescence decay of 10%Er/2%Tm UCNPs is significantly different from the 10%Er-doped UCNPs. A characteristic feature of rapid initial decay followed by a relatively slow secondary decay in the 10%Er/2%Tm-doped system can be used as a forensic feature in security printing.

Unlike organic dyes, upconversion emission is not significantly affected by the surrounding medium. In organic dyes, quenching of luminescence by the surround is always a concern, especially when dealing with solid-state molecules. Surface quenching is more in small nanocrystals but not practically observable in large nanocrystals. Nonradiative dipole–dipole coupling across the large crystals and the ligand is less predominant than the radiative (emission-and-reabsorption) process. The reason is that the distance (r) between the emitting lanthanide ion and the ligand can be as far as the radius of the particles (if the ion located at the center of the nanocrystal). From the energy transfer equation, the nonradiative energy transfer efficiency decreases $(R_0/r)^6$ as the distance increases (R_0 is the distance at which resonance energy transfer is 50% efficient).

The emitting ions that are present at the core barely participate in the nonradiative energy transfer (see Figure 10.4). In such case, the radiative energy transfer becomes more predominant than the nonradiative energy transfer process. Nevertheless, the necessary condition for both, the radiative energy transfer and the nonradiative energy transfer is spectral overlap

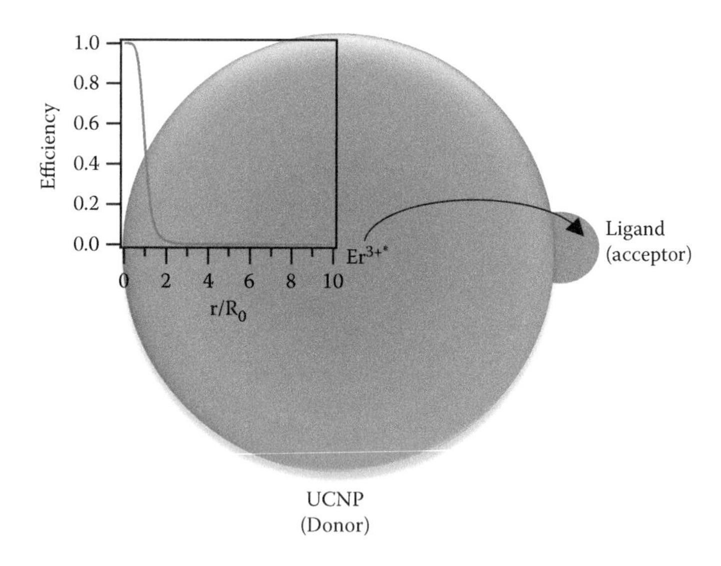

FIGURE 10.4
Schematic representation of the energy transfer efficiency as a function of distance between the surface ligand to the emitting ion that is present in the core.

integral, which is unaffected by the distance between the donor and acceptor. (Lakowicz 2006)

$$k_{ET} = \frac{k_R \kappa^2}{r^6}\left[\frac{9000(\ln 10)}{128\pi^5 \mathrm{N} n^4}\right]\int_0^{\infty} F_D(\lambda)\varepsilon_A(\lambda)\lambda^4 \partial\lambda \tag{10.1}$$

k_{ET}—energy transfer rate constant
k_R—radiative rate constant (k_R = Einstein coefficient $\boldsymbol{A}$)
r—distance between the donor and acceptor (in cm)
N—Avogadro number
n—refractive index of the medium
κ^2—dipole orientation factor
$\int_0^{\infty} F_D(\lambda)\varepsilon_A(\lambda)\lambda^4 \partial\lambda$—spectral overlap integral of donor emission and acceptor absorbance

Above equation is simplified to

$$k_{ET} = \frac{1}{\tau}\left[\frac{R_0}{r}\right]^6 \tag{10.2}$$

τ = lifetime of donor
R_0 = Förster distance, the distance at which resonance energy transfer is 50% efficient

Interestingly enough, the ratio of emission bands also varies with variation in the surrounding temperature. In Er^{3+} ion, two energy levels $^2H_{11/2}$ and $^4S_{3/2}$ are in quasi-thermal equilibrium. High temperature favors high-energy state, $^2H_{11/2}$, and vice versa. Because the excited state population distribution varies with the temperature, the emission intensities of $^2H_{11/2}$(520 nm) and $^4S_{3/2}$ (540 nm) energy states change in opposite direction (see Figure 10.5). Therefore, the study of change in the 520 nm/540 nm emission intensity ratio as a function of temperature has potential applications in monitoring temperature (Paez and Strojnik 2003).

$$\frac{N_i}{N} = \frac{e^{-E_i/kT}}{\sum_j e^{-E_j/kT}} \tag{10.3}$$

Ni = population distribution in the ith microstate
N = total population
Ei = energy of ith microstate
k = Boltzmann constant
T = temperature

Because of more variables such as doping composition, concentration of ions, excitation power, temperature of the surroundings, and the size of the nanocrystals defines the optical properties of the upconversion system, reproducing the optical properties without the prior knowledge of

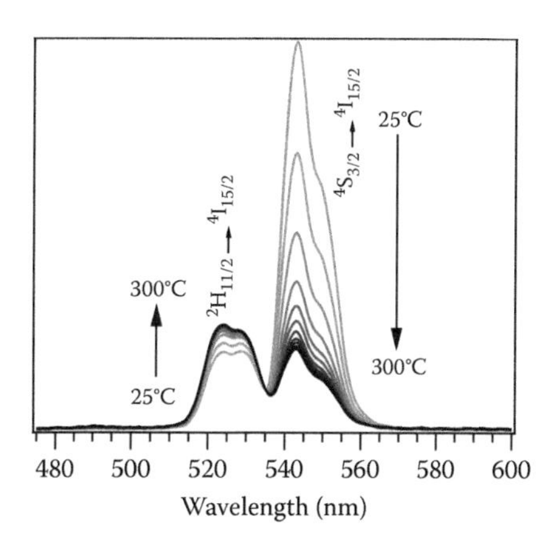

FIGURE 10.5
This figure represents the effect of temperature on the emission intensity ratio. The 520 nm emission band is from Er^{3+}: $^2H_{11/2} \rightarrow {}^4I_{15/2}$ transition and the 540 nm emission band is from Er^{3+}: $^4S_{3/2} \rightarrow {}^4I_{15/2}$ transition in Yb^{3+}/Er^{3+}-doped β-$NaYF_4$ nanocrystals. The Er ion $^2H_{11/2}$ and $^4S_{3/2}$ energy states are close enough for the Maxwell–Boltzmann distribution of excited state population to become more apparent.

composition and its effect is extremely difficult. Therefore, the complexities associated with the optical properties of the upconversion systems make the UCNPs very suitable for security printing applications.

10.4 Synthesis, Functionalization, and Scale-up

Lanthanides ions do not show UC property when they exist as single ion entities in a solution but inherit UC activity when embedded in host crystals. For the reason, it was not feasible to produce UC activity in solutions. UC applications were very much limited to solid-state physics and photonics until the breakthrough of nanotechnology. Thanks to nanotechnology, UC activity was brought into solutions by forming colloidal dispersions.

However, in the UCNPs because of their small size and high surface energy, aggregation occurs which reduces their stability in the solutions. During synthesis, the UCNPs are usually capped with nonpolar ligands such as oleic acid which prevents the aggregation by reducing the surface energy. Capping agents also allow them to disperse readily in nonpolar solvents. The surface polarity of the UCNPs can be changed by ligand exchange, which makes them compatible to a wide range of polar and non-polar solvents. And this increases the arena of UCNPs' applications. Post-synthetic surface modifications are used widely to functionalize the UCNPs with polar ligands, but there is no standard procedure to functionalize the UCNPs with the ligand of interest. The functionalization of UCNPs or any NPs depends on their dispersibility, ligand stability, and ligand-NP binding strength in a given solvent.

In situ functionalization of NPs is not suitable for UCNPs synthesis because high temperatures (>250°C) or high-pressure (autoclave) conditions were often used for NPs synthesis and the ligands tend to degrade at the reaction conditions. The post-synthetic modification approach is very popularly for UCNPs' functionalization. Post-synthetic functionalization of NPs can be achieved by the two following methods, (1) ligand modification and (2) ligand exchange. Ligand modification can be done either by a micelle-type coating or by chemical reaction. Polymer ligands and silica shell coatings are used to functionalize the UCNPs in a micelle-type coating. Ligand modification by chemical reaction can be performed by acid hydrolysis, amine–carboxylic acid bio-conjugation, Grignard reaction, free radical polymerization, cycloaddition, etc. These are a few examples of post-synthetic ligand modification by chemical reaction. Ligand exchange can be done either in homogeneous or in heterogeneous solvent systems. When both UCNPs and ligands are compatible with a common solvent, homogenous solvent systems are used for ligand exchange. And, when UCNPs and ligands are not compatible with a common solvent, ligand exchange in heterogeneous solvent systems (two or more solvents) is preferred for ligand exchange. The heterogeneous solvent system used can be either miscible or immiscible. In the immiscible solvent system, ligand exchange is achieved by the phase separation extraction method.

Nevertheless, no procedure specifically defines stability of the nanoparticulate dispersion in a given solvent system. The stability of the nanocrystals is a function of the NPs' ligand and solvent interaction. The zeta potential helps to determine the stability of the NPs' dispersion. The NPs' dispersion with zeta potential greater than –30 mV and less than +30 mV is considered to be unstable. However, it is not necessary that the NPs with high zeta potential can form a stable dispersion. The Hansen solubility parameters (HSP) of a solvent system help to theoretically predict the stability of the NPs' dispersion in a given solvent system.

10.4.1 Hansen Solubility Parameters

This approach describes the intermolecular interactions in terms of dispersive (δd), polar (δp), and hydrogen bonding (δh) forces. A fourth interaction parameter R_0 is used to describe the variance in δd, δp, and δh for materials that are larger than the molecule scale. By calculating Equations 10.1 and 10.2, it is possible to identify solvents that will disperse the UCNPs when the HSPs of a material are known

$$R_a^2 = 4(\delta d_2 - \delta d_1)^2 + (\delta p_2 - \delta p)^2 + (\delta h_2 - \delta h_1)^2 \quad (10.4)$$

$$RED = \frac{R_a}{R_0} \quad (10.5)$$

Using the forces δd, δp, and δh as axis, a three-dimensional solubility space can be constructed in which solvent molecules are plotted as individual points and solutes are plotted as spheres (radius R_0). If $RED < 1$, the UCNPs will be dispersible, and if $RED > 1$, the UCNPs will not be dispersible in the solvent (Petersen et al. 2014).

10.5 Toxicity Assessment

Researchers have proposed many potential applications using NPs in the past two decades. Now, these potential applications are close to becoming real applications. As we look forward to NP-based applications, the first question that arises is about health and environment safety. The unique ability of NP can only be limited to demonstration if the NPs pose more harm than benefit to health and the environment. For example, many biological applications were proposed using semiconductor quantum dots, however, because of cytotoxicity of the heavy metal chalcogenides, the applications are now pretty much limited to semiconductor devices. On other hand, UCNPs for biological applications were gaining popularity not only because of the ability to induce *in vivo* NIR excitation but also because of low toxicity. *In vivo* and *in vitro* toxicity assessment studies performed during the recent past indicate very little to no harm posed by UCNPs when incorporated into biological systems (Xing et al. 2012).

Xing et al. performed *in vitro* cell viability and *in vivo* bio-distribution and excretion of $NaYbF_4$:Tm^{3+} UCNPs. For *in vitro* studies, human liver cells (HL-7702) and murine macrophage cells (RAW264.7) were incubated with various concentration dispersions of UCNPs (0, 0.05, 0.1, 0.2, 0.4, 0.8, and 1.6 mgYb/mL) for 24 h and *in vitro* cytotoxicity was measured by 3-(4, 5-dimethylthiazol-2-yl)-2, 5-diphenyltetrazolium bromide (MTT) assay. More than 82.7% of the HL-7702 cells and 88.9% of the RAW264.7 cells survived even in the highly concentrated dispersion (1.6 mg Yb^{3+}/mL) which demonstrates low toxicity of UCNPs as shown in Figure 10.6.

For *in vivo* studies, 150 mg Yb^{3+}/kg of body mass equivalent UCNPs were injected intravenously. The author reported accumulation of UCNPs in major organs such as the liver and spleen, and a little concentration in heart, kidney, and lungs. UCNPs' concentration decreased to less than 20% in the first 7 days and disappeared within 30 days. Excretion of UCNPs was mainly observed in feces and urine (see Figure 10.7). Excretion through feces is more predominant than urine. Yb^{3+} fecal concentration fell below 20% μg/g within 7 days compared to the first day of uptake, 105% μg/g.

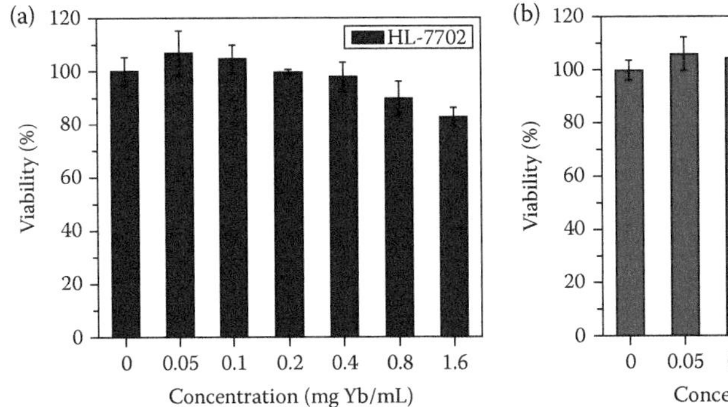

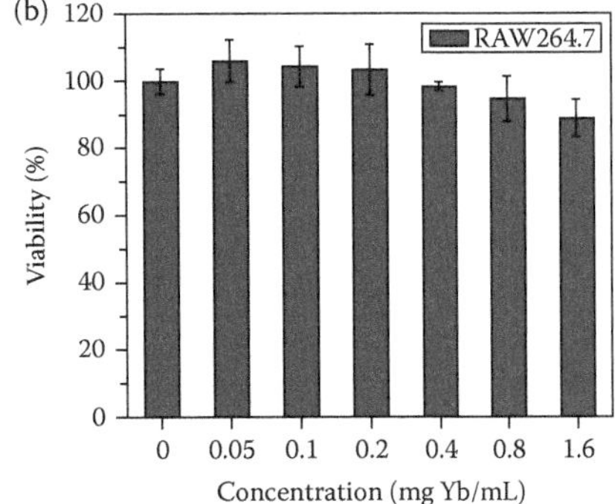

FIGURE 10.6
Effect of NPs dispersion concentration on *in vitro* cell viabilities of HL-7702 (a) and RAW264.7 cells (b). The UCNPs of different concentrations were incubated with the test cells for 24 h. (Reprinted from *Biomaterials*, 33 (21), Xing, H., W. Bu, Q. Ren et al. A $NaYbF_4$: Tm^{3+} nanoprobe for CT and NIR-to-NIR fluorescent bimodal imaging, 5384–5393, Copyright 2012, with permission from Elsevier.)

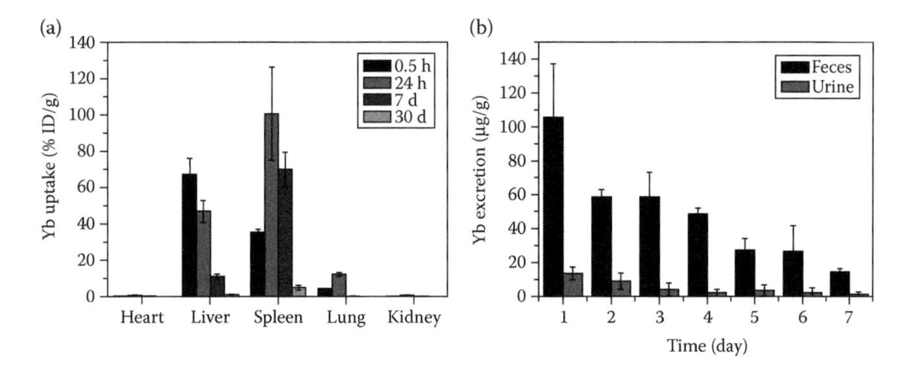

FIGURE 10.7
(a) Bio-distribution of NPs at different time points characterized by Yb^{3+} levels in various organs of mice after injection of UCNPs (150 mg Yb^{3+}/kg). (b) Yb^{3+} contents in mice excreta (feces and urine) during the first week after injection of UCNPs. (Reprinted from *Biomaterials*, 33 (21), Xing, H., W. Bu, Q. Ren et al. 2012. A $NaYbF_4$: Tm^{3+} nanoprobe for CT and NIR-to-NIR fluorescent bimodal imaging, 5384–5393, Copyright (2012), with permission from Elsevier.)

10.6 Security Printing Applications

UCNPs for security printing and anti-counterfeiting have attracted the attention of the scientific and nonscientific community alike because of scientific, security, and commercial interest. Security printing is the process of converting printed products using limited access materials along with unique technologies and a multiplicity of the printing process to produce end products that are secure in operating procedures, and manufacturing and distributing protocols. The process involved in the printing of security features serves as a protection against tampering, forgery, counterfeiting, and manipulation of data on items like banknotes, passports, tamper-evident labels, identity cards, etc. Some of the existing methods that are applied in the security printing industry are watermarks, digital watermarks, intaglio printing, micro printing, and printing with fluorescent inks (Blumenthal et al. 2012, Meruga et al. 2012).

According to the US Customs and Border Protection (CBP), counterfeiting costs governments and private industries billions of loss of value in currency, drugs, and documents (CBP 2015). The International Chamber of Commerce estimates that approximately 7% of the world trade is in counterfeit goods. It is estimated that annually, counterfeiting costs about $1.7 trillion globally and it has grown over 10,000% in the past two decades. In the United States alone, counterfeiting costs government and private industries about $200–$250 billion and 750,000 American jobs annually. The British think tank International Policy Network estimates approximately 700,000 deaths a year are caused by counterfeit malaria and tuberculosis drugs. The profits from counterfeiting have been linked to organized crime, drug trafficking, and terrorist activity.

There is a continuous need for new and innovative authentication applications in order to fight the growing problem of counterfeiting and security printing. UCNP inks are an effective tool for this.

10.6.1 Red, Green, and Blue Upconversion Printing

Oleic acid capped β-$NaYF_4$ NPs, doped with lanthanide combinations of Er^{3+}/Tm^{3+}, Yb^{3+}/Er^{3+}, and Yb^{3+}/Tm^{3+} that upconvert NIR into red, green, and blue, respectively, were synthesized. These red, green, and blue (RGB) emitting UCNPs were dispersed separately in a solvent mixture of toluene and methylbenzoate with poly(methyl methacrylate) (PMMA) as the binding agent and were used to formulate the inks. As shown in Figure 10.8, these RGB inks were printed individually as overlapping circles to demonstrate the covert RGB color system. This RGB color system theoretically demonstrates the ability of the primary RGB colors to produce any color when layered in controlled amounts (Meruga et al. 2014).

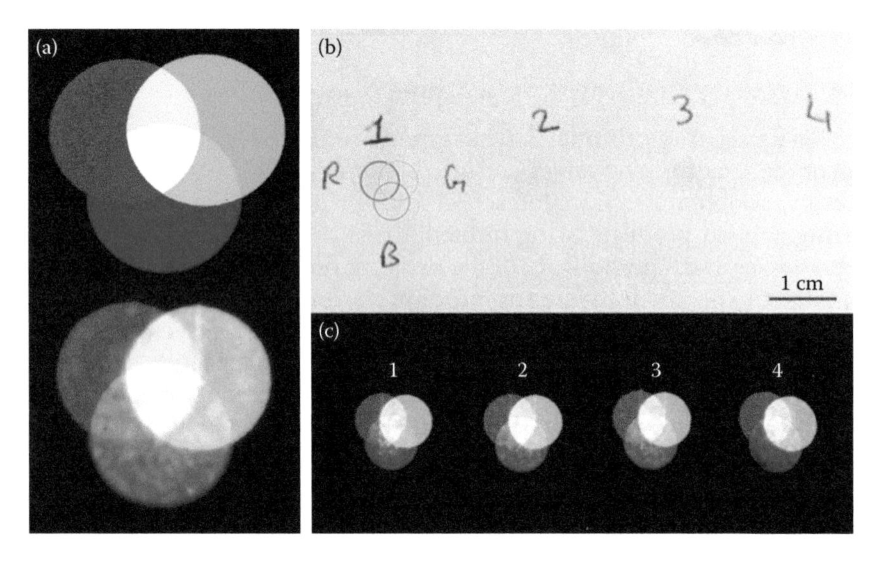

FIGURE 10.8
Demonstration of RGB color system using UCNPs inks. (a) (Top) Schematic representation of the RGB (primary) additive colors producing the secondary colors when overlapped. (Bottom) UC of RGB overlapping circles printed with the UCNPs inks producing secondary colors. (b) Paper (under ambient lighting) onto which the four sets of RGB patterns were printed. The location of the pattern 1 (overlapping circles) is indicated by colored rings which were added to the photograph using the "paint" program. This shows that the patterns are totally covert under ambient light. (c) Paper sample in (b) under 980 nm excitation in which the UC luminescence images of the four sets of RGB patterns are clearly visible. (Meruga, J. M., A. Baride, W. Cross et al. 2014. Red-green-blue printing using luminescence-upconversion inks. *J. Mater. Chem. C* 2 (12):2221–2227. doi: 10.1039/C3TC32233E. Reproduced by permission of The Royal Society of Chemistry.)

Using the RGB UCNPs inks, single, dual, and multicolor quick response (QR) codes with embedded additional security features and high payload capacity were printed (Meruga et al. 2015). The printing process involves generating the QR code with standard computer-aided design equipment which serves as the tool path for the aerosol jet printer, and the RGB UCNPs inks were sprayed depositing covert patterns onto substrates including paper, plastic film, office tape, glass, pills, feathers, etc. (Baride et al. 2015, Blumenthal et al. 2012, Meruga et al. 2012, 2014). These security features were deposited using direct-write technologies like the Optomec® aerosol jet printer.

By layering the RGB colors theoretically, one can generate any color and print almost any image covertly which is very hard to replicate (see Figure 10.9). This demonstration thus opens doors for printing complex

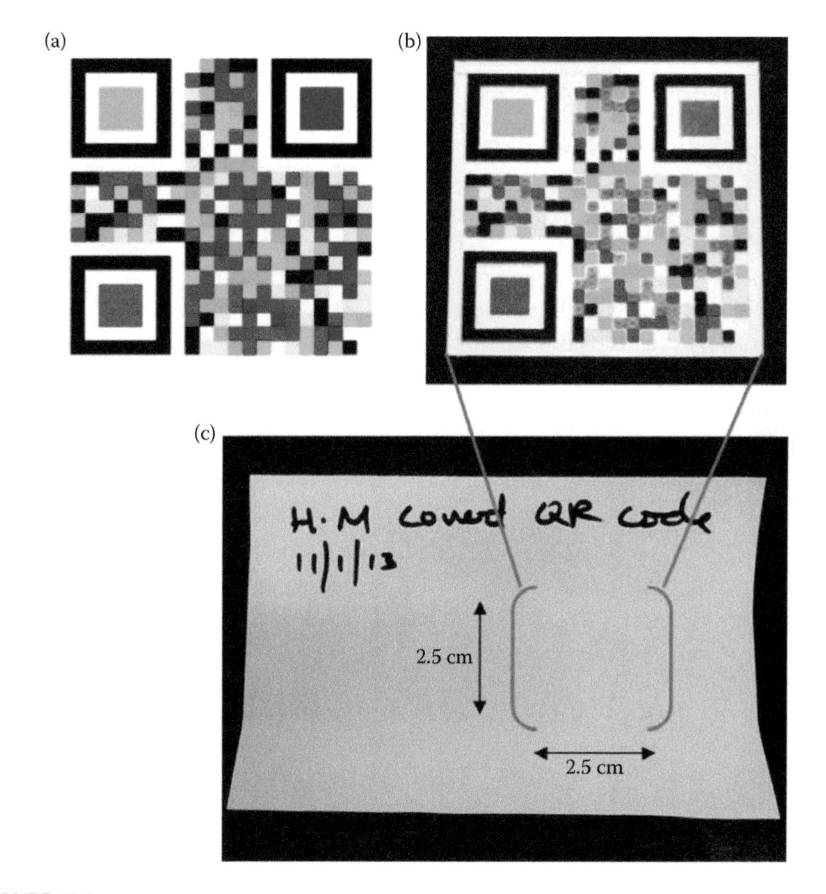

FIGURE 10.9
Multilayered covert QR code with high payload. (a) Simulated design of the layered multicolor QR code that has high payload capacity. (b) UC image of the multicolor covert QR code under NIR laser in a dark room. (c) Paper (under ambient lighting) on which the multicolor covert QR code was printed. (Reproduced from Meruga, J. M., C. Fountain, J. Kellar et al. *Int. J. Comput. Appl.* 37(1), 17–27, 2015. doi: 10.1080/1206212X.2015.1061254 with permission of the Taylor & Francis.)

covert images (using RGB UCNPs inks) in true colors which will be almost impossible to replicate (Meruga et al. 2012).

The security features using RGB UCNP inks were printed on a variety of surfaces including medical pills and cultural heritage artifacts (a painted turkey feather shown as a faux eagle feather) to verify their authenticity.

The multicolored QR code can provide three times as much storage space as a regular black and white QR code of the same size when read as a custom app designed for smart phones (Meruga et al. 2015).

10.6.2 Micro Upconverting QR Code Tags

Electrical nanoimprint lithography (e-NIL) is a parallel process that consists of simultaneous transferring of micro to nanopatterns from a conductive mold to a thermoplastic electret film and introducing negative or positive electrical charges into the bottom of those imprinted patterns. Using the e-NIL process, in a single run of approximately 15 min, more than 300,000 micro NP-based QR codes were generated on a substrate of 4 inch diameter (Ressier et al. 2012). UCNPs were directed into the micro QR code patterns by electrostatic assembly. Thus, micro upconverting QR codes were produced (see Figure 10.10). These QR codes when transferred to a transparent tape, which could be used as stickers to tag a variety of products. They are invisible to the naked eye but can be decoded and authenticated by UC and a smartphone application.

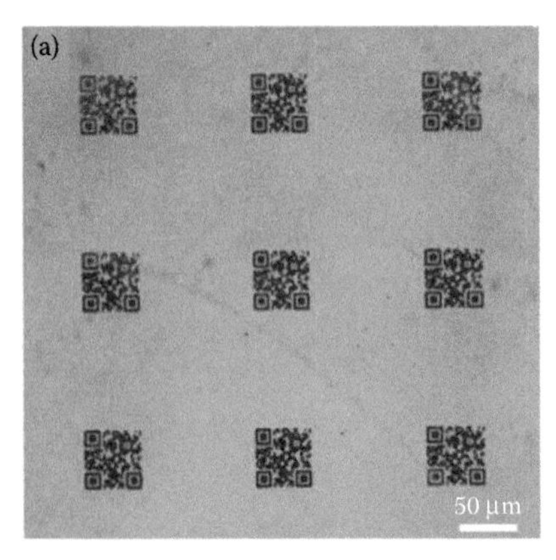

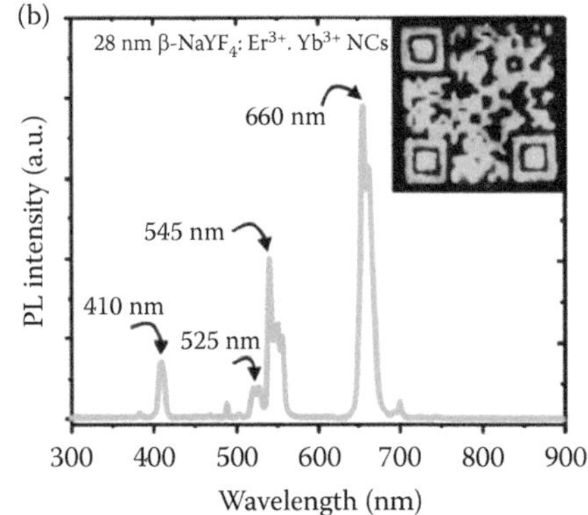

FIGURE 10.10
(a) Optical microscopy image of some 50 μm sized β-UCNPs-based QR codes present on the 4 inch wafer and (b) photoluminescence spectrum of the UC QR code (inset) UC image of a single 50 μm sized QR code under 980 nm laser. (Reproduced with permission of the IOP Publishing from High-throughput fabrication of anti-counterfeiting colloid-based photoluminescent microtags using electrical nanoimprint lithography. *Nanotechnology* 25 (34):345302.)

10.6.3 Color-Coded Multilayer Photopatterned Microstructures

A glass plate was spin coated (1200 rpm) with NP solution. The thickness of the NP solution is approximately 100 nm. A chrome/quartz photomask was placed on top of the spun film and the film was exposed to UV light (254 nm, 390 μW/cm, Spectroline Model ENF-260C Handheld UV Lamp) for 10 min at room temperature. Upon removal of the photomask, the exposed film was baked on a hot plate (at 100°C for 90s). This baked film was then immersion developed in anhydrous hexane for 30 s (at room temperature) and rinsed with fresh anhydrous hexane. With this treatment, the exposed areas remain on the glass surface and the nonexposed areas are removed leaving a pattern as shown in Figure 10.11 (Kim et al. 2009).

10.6.4 Micro Barcodes

Doyle et al. demonstrated the advantage of upconversion nanorods embedded in polyurethane acrylate (PUA) for security printing applications. The group synthesized nine spectrally distinct UC nanorods approximately 500 nm long by adjusting the doping composition of Yb/Er/Tm. The nanorods were loaded into PUA by a high-performance stop-flow lithography technique, where the nanorods suspended in the polymer precursor were passed through parallel channels in a laminar flow. A polymer bed formed from an open end of the channel was exposed to UV light and subjected to precise cutting to produce a multiband polymer strip as small as 200 μm. The author proposed that the multiband polymer strips that glow in various colors in different regions of the strip can be used as a color barcode as shown in Figure 10.12 (Lee et al. 2014).

10.6.5 Tunable Lifetime Multiplexing of UCNPs

Tunable luminescent lifetimes in the microsecond region were exploited to code individual UCNPs. Tm^{3+} ion concentration in Yb/Tm-doped $NaYF_4$ nanocrystals was varied from 0.2% to 8% to produce as many as eight different doping composition nanocrystals. The lifetime of blue emission (445 nm) from Tm: $^1D_2 \rightarrow {}^3F_4$ was evaluated at various time domains to distinguish the emission from different fluorophores. Figure 10.13 represents the lifetime distribution of Yb/Tm-doped $NaYF_4$ nanocrystals at different doping compositions (Lu et al. 2014).

The well separated blue emission lifetimes were decreased with the increase in Tm^{3+} ion concentration and the lifetimes are spread over the range of 25.6–662.4 μs. Discrete lifetimes are a result of varying the sensitizer–emitter distance by altering the doping composition of Yb/Tm ions. Because the lifetime is independent of both intensity and color, the lifetime-based imaging was used for security printing applications. Figure 10.14 demonstrates time domain luminescence security printing application.

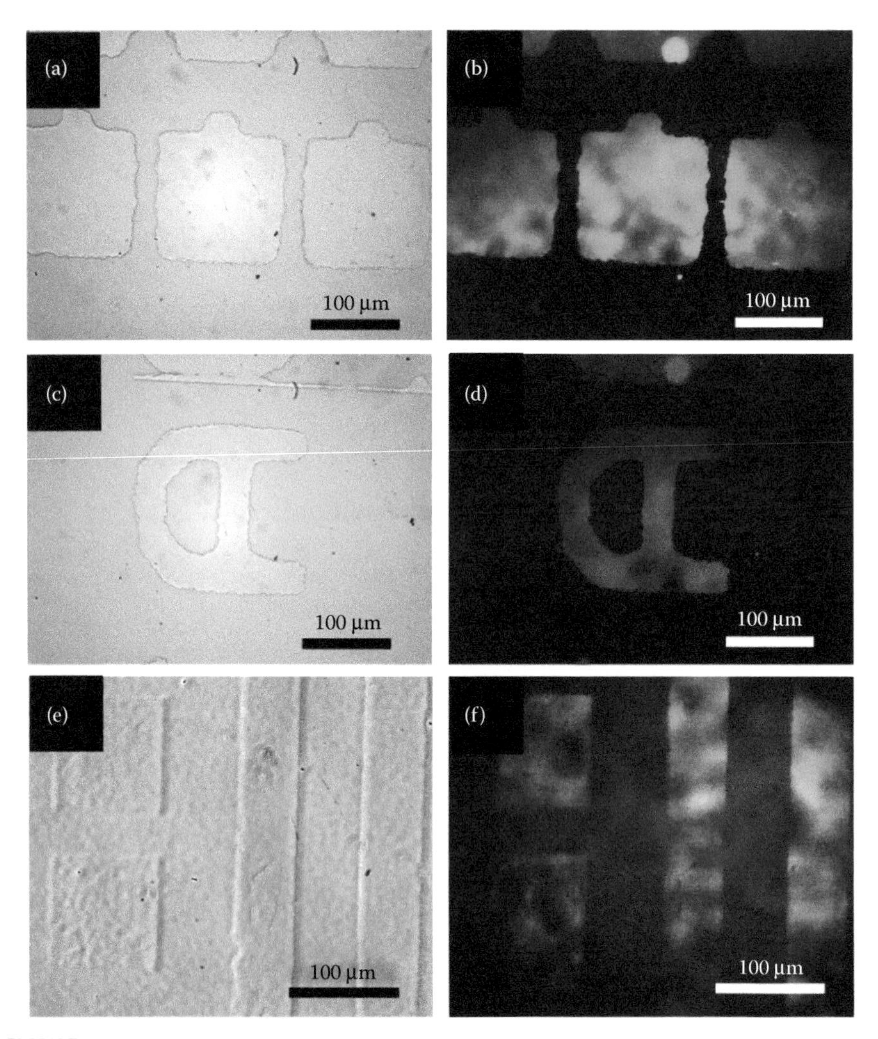

FIGURE 10.11
Optical microscopy images of UCNP patterns and their UC images. (a) Optical microscopy image of UCNPs ($NaYF_4$:Er^{3+}, Yb^{3+}). (b) UC image of pattern shown in (a). (c) Optical microscopy image of UCNPs ($NaYF_4$:Tm^{3+}, Yb^{3+}). (d) UC image of pattern shown in (b). (e) Optical microscopy image of bilayer of UCNPs: $NaYF_4$:Er^{3+}, Yb^{3+} (first layer, patterned) and NaYF4:Tm^{3+}, Yb^{3+} (second layer, no pattern). (f) UC image of pattern shown in (e). (Reproduced with permission of the IOP Publishing from Kim, W. J., M. Nyk, and P. N. Prasad. 2009. Color-coded multilayer photopatterned microstructures using lanthanide (III) ion co-doped $NaYF_4$ nanoparticles with upconversion luminescence for possible applications in security. *Nanotechnology* 20 (18):185301.)

Three overlapping pictures printed using UCNPs doped with Yb:Tm ratios of 20:4, 20:1, and 20:0.5 have lifetimes, 52, 159, and 455 μs, respectively. Normal luminescence color imaging does not show what is concealed in the layered picture (Figure 10.14a). However, the prints from each doping composition were decoded with time-resolved scanning cytometry, each layer having

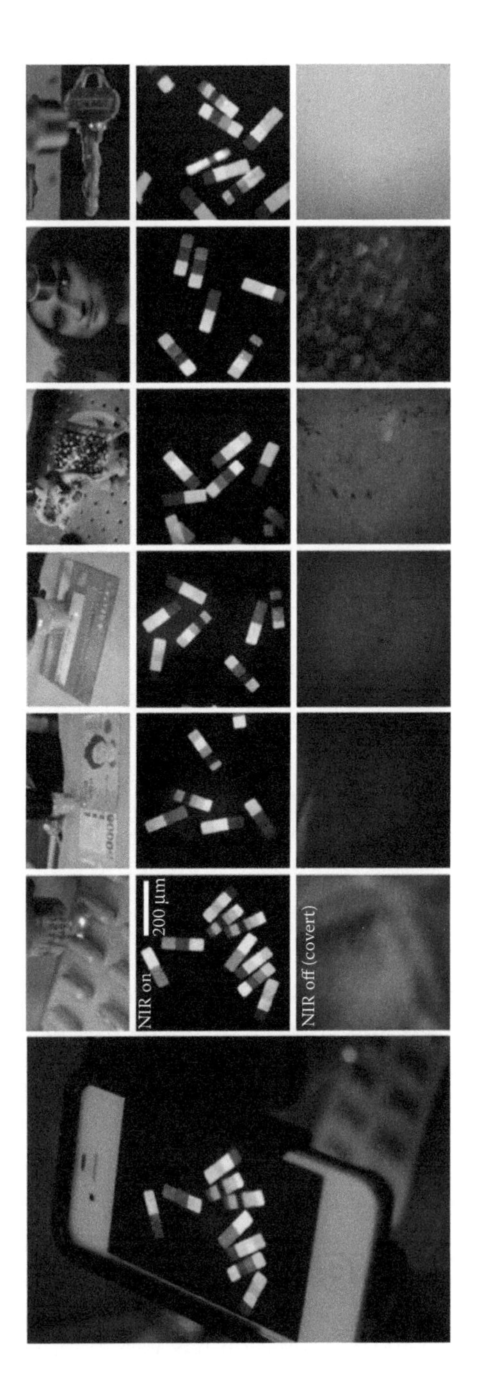

FIGURE 10.12
UC imaging of micro barcodes. Top: image was captured using a portable decoder (Apple iPhone 4S, 20 objective). Middle and bottom: images captured upon exposure to 1 W 980 nm laser excitation (middle); and under ambient lighting (bottom). (Reprinted by permission from Springer Nature, *Nat. Mater.*, Lee, J., P. W. Bisso, R. L. Srinivas et al. 2014. Universal process-inert encoding architecture for polymer microparticles. 13 (5):524–529. doi: 10.1038/nmat3938, Copyright 2014.)

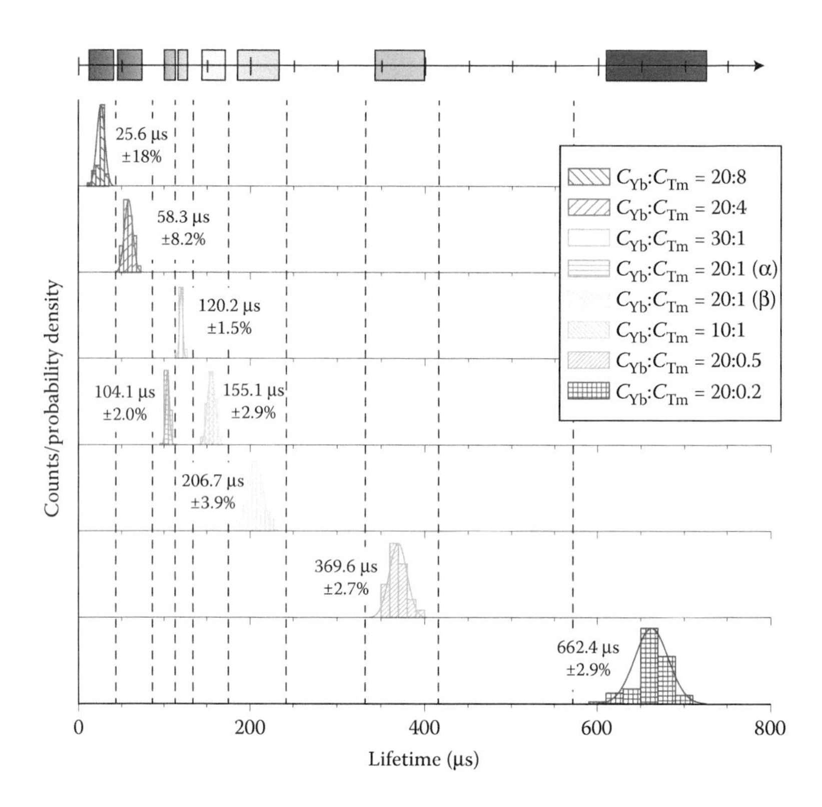

FIGURE 10.13
Blue luminescence lifetime (445 nm, Tm: $^1D_2 \rightarrow {}^3F_4$) from Yb/Tm-doped $NaYF_4$ nanocrystals at different doping combinations. The blue lifetime decreased with increase in Tm^{3+} ion concentration resulting in discrete lifetimes at different Yb/Tm combinations. (Reprinted by permission from Springer Nature, *Nat. Photonics*, Lu, Y., J. Zhao, R. Zhang et al. 2014. Tunable lifetime multiplexing using luminescent nanocrystals. 8 (1):32–36. doi: 10.1038/nphoton.2013.322, Copyright 2014.)

distinct lifetimes as shown in Figure 10.14b. This technique can be used as a security application and only authorized individuals who are aware of the luminescence lifetimes are able to access the secret information and are trained accordingly. Besides security printing, time-resolved scanning has potential applications in multichannel bio imaging, high-density data storage, etc.

10.6.6 NIR-to-NIR Upconversion Security Printing

The Yb/Tm-doped β-$NaYF_4$ nanocrystals are popular for their blue emission bands at 450 nm (Tm: $^1D_2 \rightarrow {}^3F_4$) and 475 nm (Tm: $^1G_4 \rightarrow {}^3H_6$). However, these nanocrystals also have a very strong emission band in the NIR region at 800 nm (Tm: $^3H_4 \rightarrow {}^3H_6$). The 800 nm emission has very interesting applications in security printing and biological applications. Because both 980 nm excitation light and 800 nm emission light are invisible to the naked eye and

(a)

(b)

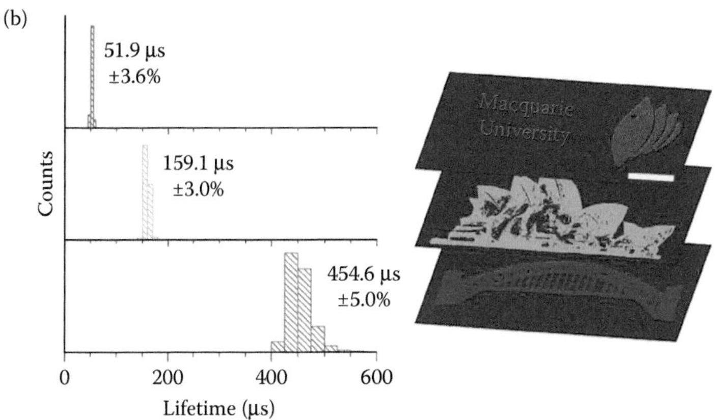

(c)

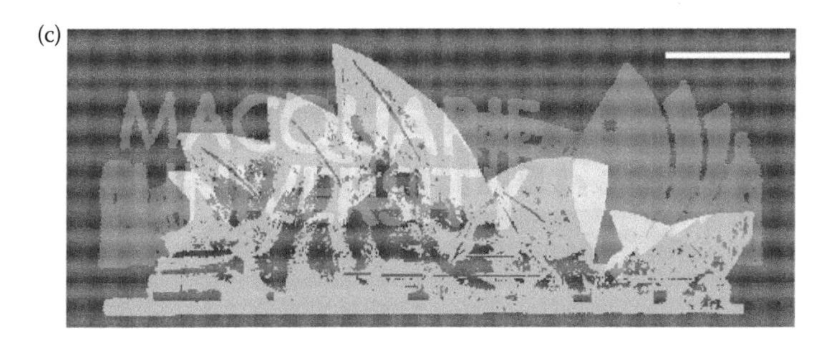

FIGURE 10.14
Demonstration of lifetime-encoded document security and photonic data storage. (a–c) Three overlapping patterns are printed with different Tm t-dots: (CYb:CTm) 20:4 for the "Macquarie University" logo, 20:1 for the Sydney Opera House image and 20:0.5 for the Sydney Harbour Bridge image. Intensity-based luminescence imaging only gives a complex picture (a), but time-resolved scanning separates the patterns based on the lifetime components of every pixel (b), so that genuine multiplexing information contained in the same overlapping space of the document can be decoded (c; pseudocolor is used to indicate the luminescence lifetime for each pixel). Scale bars (all images), 5 mm. (Reprinted by permission from Springer Nature, *Nat. Photonics*, Lu, Y., J. Zhao, R. Zhang et al. 2014. Tunable lifetime multiplexing using luminescent nanocrystals. 8 (1):32–36. doi: 10.1038/nphoton.2013.322, Copyright 2014.)

need a device such as a charge coupled device (CCD) camera to capture the 800 nm emission. Indeed, conventional imaging devices have a silicon CCD as a photo sensor, which has maximum photoresponse at that wavelength. Besides ease of imaging, the important merit of 800 nm emission is that the light can penetrate to a good extent through most organic molecules, including biological tissue and polymer films. However, the 800 nm emission usually observed in blue UCNPs (20–25%Yb/0.2–0.5%Tm) is not optimized for 800 nm emission. Baride et al. optimized the 800 nm emission by varying the Yb/Tm doping composition and reported NIR-to-NIR security printing applications (Baride et al. 2015). Figure 10.15 demonstrates potential applications of NIR-to-NIR upconversion in security printing.

The NIR-to-NIR-based security printing applications can be:

1. Direct printing of 800 nm emitting UCNPs on a substrate
2. Lamination of print between two polymer films that are transparent to NIR light (both 800 and 980 nm)
3. Protection of print with a NIR transparent polymer, such as epoxy
4. Protection of print with a polymer that is transparent to NIR light but opaque to visible light

Figure 10.16 is an example of the NIR-to-NIR security printing applications depicted in Figure 10.15.

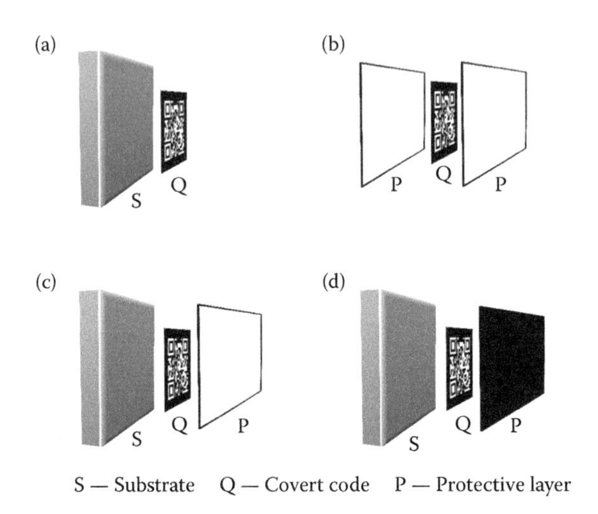

FIGURE 10.15
NIR-to-NIR upconversion applications in security printing. (a) Covert code printed on a substrate. (b) Covert code laminated between two transparent films. (c) Covert code on a substrate protected by a transparent laminate. (d) Covert code on a substrate protected by an opaque laminate. (Baride, A., J. M. Meruga, C. Douma et al. 2015. A NIR-to-NIR upconversion luminescence system for security printing applications. *RSC Adv.* 5 (123):101338–101346. doi: 10.1039/C5RA20785A. Reproduced by permission of The Royal Society of Chemistry.)

The 48%Yb/2%Tm-doped β-$NaYF_4$ UCNPs have as much as six times more 800 nm emission intensity than that of 25%Yb/0.03%Tm β-$NaYF_4$ UCNPs. Both doping combinations showed very similar powder dependence, which suggests that the 48%Yb/2%Tm is the optimum doping for the studied power range (5–35 W/cm^2). The NIR UCNPs printed on plain paper invisible to the naked eye are highlighted with a red-dotted box in Figure 10.16a. The printed feature produced a bright feature of 800 nm emission upon excitation with as low as 1.5 W/cm^2 power 980 nm continuous wave (CW) laser. The bright QR code feature captured with a NIR-sensitive CCD camera is shown in Figure 10.15a (left). Similarly, in Figure 10.16b, a QR code print of NIR-to-NIR upconversion ink embedded inside a transparent epoxy film is captured with a NIR-sensitive camera upon excitation with 980 nm light. Figure 10.16c is an image of the electronic circuit board that has NIR-to-NIR upconversion ink protected with epoxy polymer. The left side of the image is the QR code protected with visible light translucent film (black epoxy) and the right side of the image is the QR code protected with visible light transparent (clear epoxy). In either case, printed features are not visible to the naked eye but appear in the NIR-sensitive camera upon 980 nm excitation. Figure 10.16d left is the QR code captured with a NIR-sensitive

FIGURE 10.16
QR code (1 × 1 cm) printed with NIR-to-NIR upconversion ink on plain paper (a), and laminated between two films (b) is invisible to naked eye and is highlighted with red-dotted box. The QR codes appear bright because of 800 nm emission upon 980 nm excitation. (c) QR codes printed on an electronic circuit with NIR-to-NIR upconversion ink are coated with black (left) and transparent (right) epoxy polymer. (d) Upon 980 nm illumination, the NIR luminescent images are captured using a NIR-sensitive camera. The brightness of the image read through the black epoxy (left) is comparable to that read through the transparent epoxy (right). The QR codes show retention of integrity after epoxy coating. (Baride, A., J. M. Meruga, C. Douma, et al. 2015. A NIR-to-NIR upconversion luminescence system for security printing applications. *RSC Adv.* 5 (123):101338–101346. doi: 10.1039/C5RA20785A. Reproduced by permission of The Royal Society of Chemistry.)

camera upon excitation of the printed feature across the black epoxy protection film. Figure 10.16d right is the QR code from the printed feature excited across the transparent epoxy film.

The NIR-to-NIR upconversion in security printing provides covert security features that are either embedded inside a film or protected with a polymer coating or hidden under visible light translucent coating. Such security feature helps to enable strong security features, especially in the area of protecting identity of electronic circuit boards, integrated circuits, identity cards, etc.

10.6.7 Scanning Laser Imaging of Upconversion Printing

Upconversion print reading is an essential aspect of upconversion security printing. Because upconversion emission has low external quantum efficiency, upconversion imaging requires a high-energy excitation source, often lasers. Therefore, with scanning laser imaging, high resolution large area images can be collected. Khaydukov et al. demonstrated a few centimeters large scanning laser imaging of NIR (800) nm emission from β-$NaYF_4$:Yb/Tm core/$NaYF_4$ shell upconversion prints (Figure 10.17) exciting with 980 nm laser as low as 0.5 W/cm^2 power density (Khaydukov et al. 2015).

Also, the UCNPs used were functionalized with amphiphilic polymer and dispersed in aqueous ink base and the upconversion images were printed on an HP Photosmart C4283 standard inkjet printer.

FIGURE 10.17
Scanning laser image of a few centimeters large upconversion graphic images excited with 0.5 W/cm^2 980 nm laser and emission collected at 800 nm. (With kind permission from Springer Nature, *Nanotechnol. Russ.*, Khaydukov, E. V., V. V. Rocheva, K. E. Mironova et al. Biocompatible upconversion ink for hidden anticounterfeit labeling, 10 (11–12), 2015, 904–909, doi: 10.1134/S1995078015060051.)

10.6.8 Multicolor Nano Barcoding in a Single Upconversion Crystal

Multicolor nano barcoding where the upconversion crystals color were coded with certain colors in the middle of the crystal and a different color at the tips, is a good candidate for embedded security printing applications (Zhang et al. 2014). The dual-color system is achieved by altering the doping composition at the tips keeping the host lattice very much the same. To observe the difference in the luminescence between the tip and body of the crystal, 2 μm size rods were used. Different activators doping at the tips were achieved by a facile seed growth technique. Small α-$NaYF_4$ seed added to the reaction mixture of large β-$NaYF_4$ micro rod resulted in axial growth of β-$NaYF_4$ to produce dual-color micro rods.

In Figure 10.18, the dual-color systems, GRG, BRB, BGB, RBR, GBG, and RGR are produced by the primary color red (R), green (G), and blue (B) upon

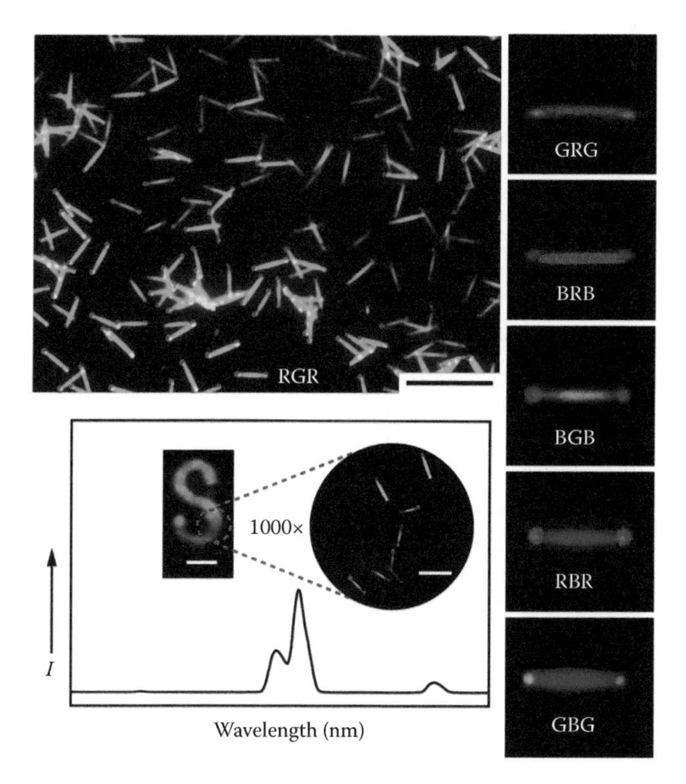

FIGURE 10.18
Dual-color upconversion micron rods achieved by varying the doping composition of lanthanide ions at the tips. A letter S printed with RGR micro rods printed by a stamping process shows unique red-tip and green-body rod features under microscopic observation. The R, G, and B represent red, green and blue, respectively, in RGR, GRG, BRB, BGB, RBR, and GBG. (Reprinted with permission from Zhang, Y., Zhang, L., Deng, R. et al. Multicolor barcoding in a single upconversion crystal. *J. Am. Chem. Soc.* 136 (13), 4893–4896. Copyright 2014, American Chemical Society.)

980 nm excitation. Green upconversion is achieved with 5%Yb/0.05%Er, red color with 50%Yb/0.05%Er, and blue color with 20%Yb/0.2Tm doping. Because of the unique color code system, the nano barcode incorporated micro rods have potential applications in upconversion security printing.

10.6.9 Upconversion Polymeric Nanofibers

Besides printing of upconversion inks, UCNPs are incorporated into nanofibers to produce an upconversion fabric. PMMA is a nontoxic biocompatible polymer and was used as a matrix material. A dispersion of UCNPs in 1,1,2,2-tetrachloroethane was mixed with a solution of 30 wt% PMMA in 1,1,2,2-tetrachlorethane:dimethylformamide (1:1) to produce a 2–6 wt% loading of UCNPs. The UCNPs/PMMA dispersion was subjected to electrospun fabrication to produce an upconversion fabric (Bao et al. 2012). The fabric has approximately 90 nm diameter particles embedded in approximately 316 nm dia. fiber. The fabric appeared green upon 980 nm excitation. UCNPs loaded fabric has very many applications in document security.

10.7 Readability of Upconversion Prints

Readability and imaging of upconversion prints plays a critical role in the commercialization of upconversion security printing applications. Usually, a high-energy NIR source is required to induce upconversion emission. Class IV lasers of more than a watt power are required to image the prints. Therefore, protective measurements are necessary while imaging the upconversion. The upconversion signal can be enhanced either by increasing the UCNPs loading or by increasing the excitation power density. In any case, to a minimum, 1 W/cm^2 power is required to perceivable emission. However, for NIR (800 nm) emission, where a silicon CCD camera is used to image the upconversion, readable images are produced with less than a W/cm^2 excitation power. The advantage with using the silicon CCD camera is that the silicon CCD has maximum sensitivity to light at around 800 nm.

Nevertheless, direct excitation and imaging of large images (>1 × 1 cm) become practically difficult due exponential decrease in the flux with increase in the beam width. A better approach for large area imaging is by scanning laser imaging or confocal imaging. Sophisticated devices with scanning laser imaging capabilities are required to capture large area images. However, for forensic-level investigation of optical properties, research-level spectroscopic instrumentation is necessary. For example, single photon counters are required to study the lifetime behavior of upconversion print.

The science aspect of upconversion systems and its potential in security printing is well established. The engineering aspect of developing the suitable devices to read the printed features is the next phase of upconversion security printing.

10.8 Conclusion

Upconversion nanomaterials are transparent to the visible light but emit strong visible and NIR light upon NIR excitation. The unique optical property of upconversion systems where the nanocrystals are not visible to the naked eye yet emit visible light upon excitation yield to security printing applications. The upconversion nanocrystals are different from dyes and pigments because the inorganic crystals have high thermal stability, photostability, and chemically inertness. Moreover, recent studies showed minimal to no toxicity effect of UCNPs. These attributes together make UCNPs suitable for security printing and anti-counterfeiting in wide range of substrates such as, currency, security documents, art , pills, integrated circuits, and so on.

The advantage with upconversion systems is the ability to tune the optical properties by altering the doping composition of lanthanide ions. In addition, a small variation in doping concentration also results in a drastic change in the optical properties of an upconversion system. Moreover, in a few cases, the color emitted by an upconversion system depends on the excitation power. Interestingly enough, the ratio of emission bands also varies with variations in the surrounding temperature. Because of too many variables that define optical properties, reproducing the optical properties without prior knowledge of composition and its effect is extremely difficult. The complexities that are associated with the optical properties of upconversion systems make the UCNPs very suitable for security printing applications.

A single-color upconversion QR code, a multicolor large volume capable upconversion QR code, a multicolor banded upconversion micro bar, multicolor upconversion micro rods, a NIR-to-NIR upconversion embedded in the material, NP-loaded fabrics, and a luminescence lifetime dependent security imaging are few of the many applications reported in the recent past.

The formulation of an upconversion system into a printable ink is an important aspect of security printing applications. NPs capped with nonpolar groups, such as oleic acid are directly formulated into an ink in a nonpolar solvent base or functionalized with polar groups and formulated into an ink in a polar solvent base. Appropriate binding agents were used to bind the NPs to the substrate. Nonprinting techniques include the embedment of UCNPs in polymer to produce nanofibers or polymer bars, and the electrostatic assembly of NPs.

Readability and imaging of upconversion prints is a very essential part of security printing. Because of low external quantum efficiency of upconversion systems, high-energy laser is required to image upconversion prints. Scanning laser imaging helps to achieve large area upconversion imaging. Nevertheless, NIR protective goggles are sufficient to visualize the visible features, if imaging is not required. However, research-level spectroscopic resources are necessary to perform detailed spectroscopic studies of upconversion prints, especially when dealing with forensic-level security.

The progress made in the past 5 years on upconversion NPs for security printing is promising. Many security printing models were proposed to protect the identity of wide range of substances, ranging from pill to electronic circuit boards, currency to art. Development of safe readers for the upconversion prints will bring upconversion systems for security printing much closer to commercialization.

Acknowledgments

Dr. Meruga would like to extend his appreciation to the following:

To the God for wisdom and perseverance in writing of this book chapter. In addition, he is also grateful to Dr. Aravind Baride, the co-author for this chapter; Dr. Jon Kellar, his supervisor and the SPACT group and SDSM&T MES Department; and the state and federal funding agencies for providing opportunity to conduct research that helped in writing this book chapter. His special thanks to the editor Dr. Claudia Altavilla for her help and support.

He owes his deep respect to his family: his wife Sony, son Ethan, his father Sundararao and mother Mani, for their endless encouragement and prayers.

References

Anderson, R. B., S. J. Smith, P. S. May et al. 2013. Revisiting the NIR-to-visible upconversion mechanism in β-$NaYF_4$:Yb^{3+},Er^{3+}. *J. Phys. Chem. Lett.* 5 (1):36–42. doi: 10.1021/jz402366r.

Bao, Y., Q. A. N. Luu, Y. Zhao et al. 2012. Upconversion polymeric nanofibers containing lanthanide-doped nanoparticles via electrospinning. *Nanoscale* 4 (23):7369–7375. doi: 10.1039/C2NR32204H.

Baride, A., J. M. Meruga, C. Douma et al. 2015. A NIR-to-NIR upconversion luminescence system for security printing applications. *RSC Adv.* 5 (123):101338–101346. doi: 10.1039/C5RA20785A.

Blumenthal, T., J. Meruga, P. S. May et al. 2012. Patterned direct-write and screen-printing of NIR-to-visible upconverting inks for security applications. *Nanotechnology* 23 (18):185305.

Bünzli, J.-C. G. and S. V. Eliseeva. 2011. Basics of lanthanide photophysics. In *Lanthanide Luminescence: Photophysical, Analytical and Biological Aspects*, edited by P. Hänninen and H. Härmä. Berlin Heidelberg: Springer.

CBP. 2015. *Intellectual Property Rights Seizure Statistics Fiscal Year 2014*. Washington, DC: US Customs and Border Protection.

Haase, M. and H. Schäfer. 2011. Upconverting nanoparticles. *Angew. Chem. Int. Ed.* 50 (26):5808–5829. doi: 10.1002/anie.201005159.

Ho, H., W. W. Wong, and S. Y. Wu. 2003. Multilayer optical storage disk based on the frequency up-conversion effect from rare-earth ions. *Opt. Eng.* 42 (8):2349–2353. doi: 10.1117/1.1588298.

Khaydukov, E. V., V. V. Rocheva, K. E. Mironova et al. 2015. Biocompatible upconversion ink for hidden anticounterfeit labeling. *Nanotechnol. Russ.* 10 (11–12):904–909. doi: 10.1134/S1995078015060051.

Kim, J.-H. and J.-H. Kim. 2015. Triple-emulsion microcapsules for highly efficient multispectral upconversion in the aqueous phase. *ACS Photonics* 2(5):633–638. doi: 10.1021/acsphotonics.5b00042.

Kim, W. J., M. Nyk, and P. N. Prasad. 2009. Color-coded multilayer photopatterned microstructures using lanthanide (III) ion co-doped $NaYF_4$ nanoparticles with upconversion luminescence for possible applications in security. *Nanotechnology* 20 (18):185301.

Kumar, P., J. Dwivedi, and B. K. Gupta. 2014. Highly luminescent dual mode rare-earth nanorod assisted multi-stage excitable security ink for anti-counterfeiting applications. *J. Mater. Chem. C* 2 (48):10468–10475. doi: 10.1039/C4TC02065K.

Lakowicz, J. R. 2006. *Principles of Fluorescence Spectroscopy*. 3rd ed. New York: Springer. http://dx.doi.org/10.1007/978-0-387-46312-4

Lee, J., P. W. Bisso, R. L. Srinivas et al. 2014. Universal process-inert encoding architecture for polymer microparticles. *Nat. Mater.* 13 (5):524–529. doi: 10.1038/nmat3938.

Liu, S., L. Zhang, T. Yang et al. 2014. Development of upconversion luminescent probe for ratiometric sensing and bioimaging of hydrogen sulfide. *ACS Appl. Mater. Inter.* 6:11013–11017. doi: 10.1021/am5035158.

Lu, Y., J. Zhao, R. Zhang et al. 2014. Tunable lifetime multiplexing using luminescent nanocrystals. *Nat. Photonics* 8 (1):32–36. doi: 10.1038/nphoton.2013.322.

Meruga, J. M., A. Baride, W. Cross et al. 2014. Red–green–blue printing using luminescence-upconversion inks. *J. Mater. Chem. C* 2 (12):2221–2227. doi: 10.1039/C3TC32233E.

Meruga, J. M., W. M. Cross, P. S. May et al. 2012. Security printing of covert quick response codes using upconverting nanoparticle inks. *Nanotechnology* 23 (39):395201.

Meruga, J. M., C. Fountain, J. Kellar et al. 2015. Multi-layered covert QR codes for increased capacity and security. *Int. J. Comput. Appl.* 37 (1):17–27. doi: 10.1080/1206212X.2015.1061254.

Paez, G. and M. Strojnik. 2003. Erbium-doped optical fiber fluorescence temperature sensor with enhanced sensitivity, a high signal-to-noise ratio, and a power ratio in the 520–530- and 550–560-nm bands. *Appl. Opt.* 42 (16):3251–3258. doi: 10.1364/AO.42.003251.

Petersen, J. B., J. Meruga, J. S. Randle et al. 2014. Hansen solubility parameters of surfactant-capped silver nanoparticles for ink and printing technologies. *Langmuir* 30 (51):15514–15519. doi: 10.1021/la502948b.

Ressier, L., E. Palleau, and S. Behar. 2012. Electrical nano-imprint lithography. *Nanotechnology* 23 (25):255302.

Shalav, A., B. S. Richards, T. Trupke et al. 2005. Application of $NaYF_4$: Er^{3+} up-converting phosphors for enhanced near-infrared silicon solar cell response. *Appl. Phys. Lett.* 86:16–19. doi: 10.1063/1.1844592.

Suyver, J. F., A. Aebischer, D. Biner et al. 2005. Novel materials doped with trivalent lanthanides and transition metal ions showing near-infrared to visible photon upconversion. *Opt. Mater.* 27 (6):1111–1130. doi: http://dx.doi.org/10.1016/j.optmat.2004.10.021.

Vennerberg, D. and Z. Lin. 2011. Upconversion nanocrystals: Synthesis, properties, assembly and applications. *Sci. Adv. Mater.* 3 (1):26–40. doi: 10.1166/sam.2011.1137.

Wang, F., D. Banerjee, Y. Liu et al. 2010. Upconversion nanoparticles in biological labeling, imaging, and therapy. *Analyst* 135 (8):1839–1854. doi: 10.1039/C0AN00144A.

Wang, F. and X. Liu. 2009. Recent advances in the chemistry of lanthanide-doped upconversion nanocrystals. *Chem. Soc. Rev.* 38 (4):976–989. doi: 10.1039/B809132N.

Wang, Q. H. and M. Bass. 2004. Photo-luminescent screens for optically written displays based on upconversion of near infrared light. *Electron. Lett.* 40 (16):987–988. doi: 10.1049/el:20045046.

Xing, H., W. Bu, Q. Ren et al. 2012. A $NaYbF_4$: Tm^{3+} nanoprobe for CT and NIR-to-NIR fluorescent bimodal imaging. *Biomaterials* 33 (21):5384–5393. doi: http://dx.doi.org/10.1016/j.biomaterials.2012.04.002.

Zhang, Y., Zhang, L., Deng, R., Tian, J., Zong, Y., Jin, D., and Liu, X. 2014. Multicolor barcoding in a single upconversion crystal. *J. Am. Chem. Soc.* 136 (13):4893–4896. doi:10.1021/ja5013646.

Zhou, B., B. Shi, D. Jin et al. 2015. Controlling upconversion nanocrystals for emerging applications. *Nat. Nanotechnol.* 10 (11):924–936. doi: 10.1038/nnano.2015.251.

11

Nanothermometry Using Upconverting Nanoparticles

Eva Hemmer and Fiorenzo Vetrone

CONTENTS

11.1 Introduction 319
11.1.1 Concept of Nanothermometry 319
11.1.2 Optical Nanothermometers 321
11.2 Lanthanide-Based UCNPs for Nanothermometry 323
11.2.1 Lanthanide-Based Upconversion 323
11.2.2 Theoretical Background: *LIR*, Boltzmann's Plot, Thermal Sensitivity, and Thermal Resolution 325
11.3 Recent Advances in UCNP-Based Nanothermometry 329
11.3.1 Er^{3+}-Based Nanothermometers 329
11.3.1.1 Er^{3+}-Based Upconverting Nanothermometers for Biomedical Applications 333
11.3.2 Tm^{3+}-Based Nanothermometers 334
11.3.3 Ho^{3+}-Based Nanothermometers 338
11.3.4 Nd^{3+}-Based Nanothermometers 340
11.3.5 Nanothermometers Based on the Combination of Different Ln^{3+} Ions 342
11.4 Nanothermometers Based on Multicomponent Nanoassemblies: Toward Sensitivity Enhancement and Multimodal Biomedical Applications 344
11.4.1 UCNP–Organic Hybrids for Improved Sensitivity 345
11.4.2 Multifunctional Nanoplatforms for Optical Heating and Thermal Sensing 346
11.5 Conclusions 350
References 351

11.1 Introduction

11.1.1 Concept of Nanothermometry

Bioimaging is a vital tool in the fields of biology and medicine. For instance, it enables the study of biological processes contributing to a better understanding of mechanisms triggering the development of diseases such as cancer. As a diagnostic tool, it is essential for the detection of diseases at a very early stage. Given the vast heterogeneity in diseases and therefore a patient's need for personalized diagnosis, the concept of multimodality is receiving increasing attention (Lee et al. 2012). In fact, multimodal imaging is a powerful tool combining the advantages of several imaging modalities while overcoming their intrinsic individual limitations. Among the most commonly applied techniques used for multimodal approaches are, for example, magnetic resonance imaging (MRI; pros: high spatial resolution, high penetration depth; cons: low sensitivity, long imaging time), computed tomography (CT; pros: high spatial resolution, high penetration depth; cons: radiation risk, not quantitative), ultrasound (US; pros: real time, low cost; cons: low resolution, operator-dependent analysis), positron emission tomography (PET; pros: high sensitivity, no penetration depth limit, quantitative, whole body scan; cons: radiation risk, high cost), single photon emission computed tomography (SPECT; pros: high sensitivity, no penetration depth limit; cons: radiation risk, low spatial resolution), and optical imaging (pros: high sensitivity, multicolor, high temporal resolution; cons: low spatial resolution, low penetration depth). This multimodal approach results in more powerful imaging tools providing increasingly reliable and accurate information, for instance increased sensitivity and higher spatial and temporal resolution when compared to single imaging modalities (Lee et al. 2012).

In recent years, it has been recognized that the temperature of a biological system is a crucial parameter that can act as a first indicator of disease or can be used to monitor biological processes. This has resulted in the emerging field of nanothermometry. Generally speaking, nanothermometry aims for the contact-less extraction of the local temperature of a given system with sub-micrometric spatial resolution (Jaque and Vetrone 2012). This is highly useful for the better understanding of processes whose dynamics and performance are dependent on the temperature of the local environment. Such thermal sensing at a well-defined local area of sub-micrometer dimensions finds specific interest in the biomedical field (Jaque et al. 2014b; Zhou et al. 2015) yet is by far not limited to biomedicine, and its rising attraction has also been reported in other fields such as micro/nanoelectronics, integrated photonics, and for the investigation of temperature and heat transfer in gaseous flows (Klimov et al. 2015; Mecklenburg et al. 2015; Perpiñà et al. 2016; Rothamer and Jordan 2012).

Many optical probes have been proven as suitable candidates for nanothermometry combining imaging capabilities with the advanced feature of temperature sensing, thus providing a powerful tool for novel multimodal thermal sensing, imaging, and potential diagnostic approaches. This chapter will first provide some theoretical background for optical nanothermometers and subsequently highlight the latest developments in the field of optical nanothermometry, with special emphasis on those probes that are based on lanthanide-doped upconverting nanoparticles (UCNPs).

11.1.2 Optical Nanothermometers

Optical nanothermometry is based on the temperature dependence of the spectral features of a phosphor. In other words, optical nanothermometry takes advantage of the fact that a phosphor's luminescence (the emission of light subsequent to the external excitation resulting in the population of electronically excited states) can be influenced by the surrounding temperature. More specifically, it is the temperature governed population probability of excited states and the subsequent generation of luminescence that is exploited in optical nanothermometry. Typical spectral features and how they are influenced by the local temperature are depicted in Figure 11.1. Based on these features, several types of nanothermometry can be described (Jaque and Vetrone 2012). Consequently, the analysis of the obtained spectral features allows us to draw conclusions about a specimen's temperature; and together with spatial optical imaging locally resolved thermal mapping of a sample becomes possible.

Luminescence intensity nanothermometry (Figure 11.1a) is based on the measurement of the intensity of the emitted light. Temperature variation changes the overall number of emitted photons, thus induces changes in the emission intensity. This is due to thermal activation of luminescence quenching mechanisms and/or increase in the non-radiative decay probabilities. While the emission intensity is rather easy to measure, a potential drawback of this approach is the sensitivity of the overall luminescence intensity to

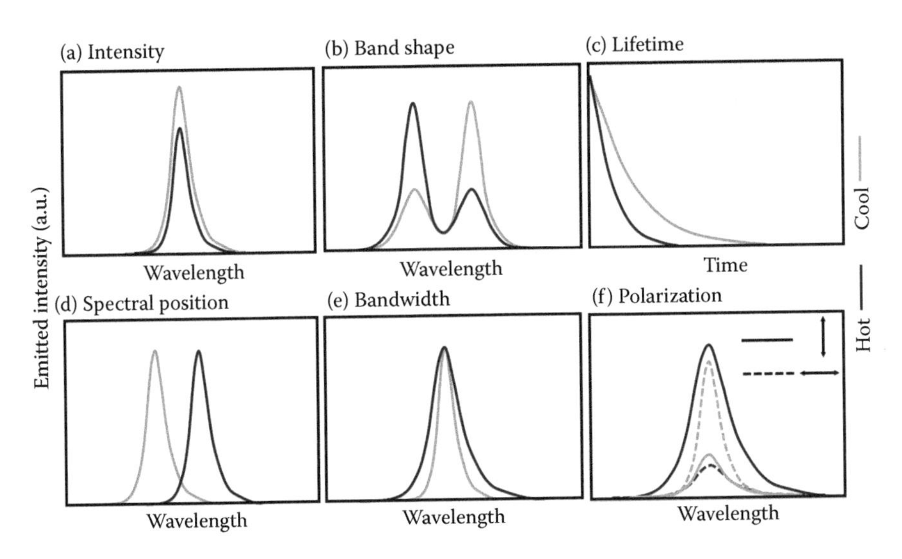

FIGURE 11.1
Schematic representation of the possible effects caused by a temperature increment on the luminescence. Dark gray lines correspond to higher temperatures. (Jaque, D. and F. Vetrone. 2012. Luminescence nanothermometry. *Nanoscale* 4:4301–26. Reproduced by permission of The Royal Society of Chemistry.)

other parameters such as probe concentration and/or the presence of other potential quenchers (e.g., in a biological sample). This makes a comparison of absolute intensities challenging. In order to overcome this, normalized spectra should be considered rather than absolute intensity values. An alternative approach is based on the analysis of intensity ratios between different spectral lines (Figure 11.1b; the herewith related luminescence intensity ratio (*LIR*) will be defined in Section 11.2.2). This type of nanothermometry, also referred to as *band-shape luminescence nanothermometry,* is most often applied with phosphors that exhibit electronic states from which emissions are generated that are very close in energy such that they are thermally coupled. Alternatively, the band shapes of two different emitters combined in one system can be used for such a ratiometric approach. The lifetime of the excited states is a further parameter that is widely employed for sensing and imaging, for instance in fluorescence lifetime microscopy (Cubeddu et al. 2002; Lakowicz et al. 1992) or in luminescence resonance energy transfer (LRET, also known as fluorescence or Förster resonance energy transfer, FRET) based assays (Mednitz and Hildebrandt 2014). *Luminescence lifetime* is defined as the time that the emitted luminescence intensity decays down to 1/e of its initial value after a pulsed excitation (assuming single exponential decay) (Wang et al. 2013). The time-dependent luminescence intensity I_t is related to the lifetime τ following Equation 11.1:

$$I_t = I_0 \cdot \exp\left(-\frac{t}{\tau}\right) \quad (11.1)$$

where I_0 is the luminescence intensity at time $t = 0$. The decay probabilities from electronic levels depend on a variety of factors, including temperature, which allows the use of lifetime data for thermal sensing (Figure 11.1c). Advantages of lifetime data over emission intensity data include their independence from phosphor concentrations and their potential for self-referencing. On the other hand, more complex read-out systems are required when compared to intensity measurements, and complications may occur when the decay profiles are not single-exponential. Further types of nanothermometry include *spectral luminescence nanothermometry* (Figure 11.1d) based on the temperature induced change of the spectral position of the emission lines; *bandwidth luminescence nanothermometry* (Figure 11.1e) exploiting line broadening upon temperature increase; and *polarization luminescence nanothermometry* (Figure 11.1f) based on the influence of the temperature on the polarization anisotropy (ratio between the luminescence intensities emitted at two orthogonal polarization states).

Thus, there are a variety of spectral features suitable for use in thermal sensing. Accordingly, remarkable phosphors and probe materials of large diversity were reported for optical thermal sensing. With regards to thermal sensing in, for example, biological systems or micro- and nanoelectronics, sub-micron size probes are required, leading to the development of nanothermometers based on organic dyes, quantum dots, and lanthanide ions (Ln^{3+}) (Brites et al. 2011, 2012; Jaque and Vetrone 2012; Jaque et al. 2014b; Quintanilla et al. 2016; Wang et al. 2013, X. Wang et al. 2015). Wang et al. (2002) reported for the first time the use of Eu^{3+} ions for lanthanide-based nanothermometry. Since then, worldwide research efforts have led to the development of nanothermometers exploiting various Ln^{3+} ions, among them Ln^{3+}-doped UCNPs that are recently emerging as truly exciting players in the field of thermal sensing and which will be described in more detail in the following sections.

11.2 Lanthanide-Based UCNPs for Nanothermometry

11.2.1 Lanthanide-Based Upconversion

The outstanding optical properties of the lanthanides are based on the involvement of the 4*f* orbital in the electronic configuration of the trivalent lanthanide ions (Ln^{3+}), which is characterized by an incompletely filled 4*f* shell, located inside the complete $5s^2$ and $5p^6$ shells. This results in a shielding of valence electrons, which are therefore only weakly affected by the environment. Consequently, when doped in appropriate host materials, the influence of the host lattice on the optical transitions within the 4*f* configuration is small, and narrow optical absorption and emission bands as well as long lifetimes of the excited electronic states of the Ln^{3+} are obtained. Due to the electromagnetic range of the energy levels of the Ln^{3+} ions (energy diagrams of selected Ln^{3+} relevant for upconversion-based nanothermometry are shown in Figure 11.2; for a complete overview, the reader is referred to the so-called *Dieke's Diagram* [Blasse and Grabmaier 1994; Dieke 1968; Wegh et al. 2000]), Ln^{3+}-doped inorganic host materials find application as phosphors covering the broad range from laser technologies and optical communication to the biomedical field and solar applications (Barnes 2004; Y. S. Liu et al. 2013; Wang et al. 2011; W. F. Yang et al. 2014; Zagumennyi et al. 2004).

One of the characteristic features of Ln^{3+}-doped host materials is the so-called upconversion process resulting in the emission of higher energy light upon excitation with light at lower energies (Auzel 2004). While several mechanisms (energy transfer upconversion, excited-state absorption [ESA], and photon avalanche) are possible to achieve upconversion, the one most often applied is based on energy transfer between Ln^{3+} ions. This typically involves doping of an inorganic host material with activator ions, such as erbium (Er^{3+}), thulium (Tm^{3+}), holmium (Ho^{3+}) or neodymium (Nd^{3+}), and sensitizer ions (e.g., Yb^{3+}) as for example depicted in Figure 11.2.

Upon excitation of Yb^{3+} ions with near-infrared light (NIR) (980 nm) from the $^2F_{7/2}$ ground state to the sole excited state ($^2F_{5/2}$), energy transfer to the

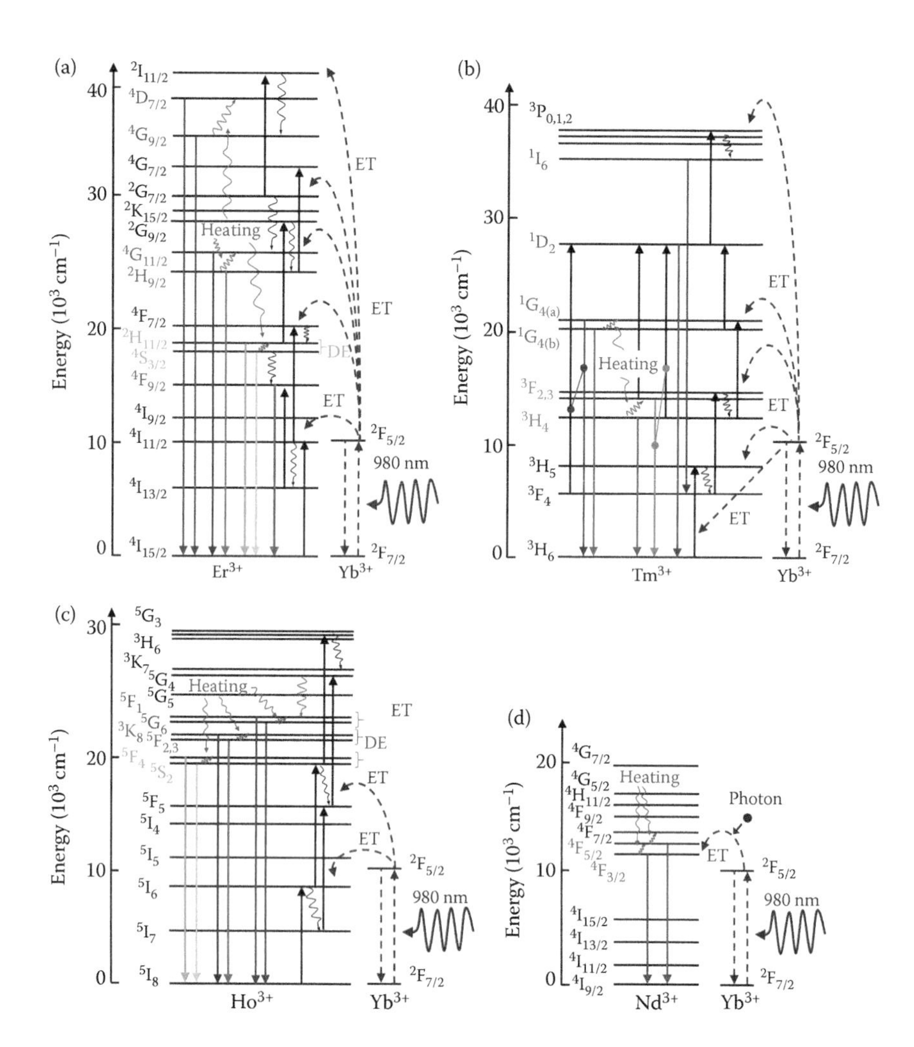

FIGURE 11.2
Mechanisms of optical thermal sensing through (a) Er^{3+}/Yb^{3+}, (b) Tm^{3+}/Yb^{3+}, (c) Ho^{3+}/Yb^{3+}, and (d) Nd^{3+}/Yb^{3+} energy transfer under 980 nm NIR excitation. "Heating" refers to the temperature increase necessary to induce the population of the higher energy levels. (Redrawn based on references Wang, R. et al. 2015. *Opt. Mater.* 43:18–24; Xu, W., Z. G. Zhang, and W. W. Cao. 2012b. *Opt. Lett.* 37:4865–67; Zheng, K. Z. et al. 2015. *Opt. Express* 23:7653–58; Quintanilla, M. et al. 2015. *J. Mater. Chem. C* 3:3108–13.)

activator ions in close proximity takes place; in case of the very well-studied Er^{3+}/Yb^{3+} system (Figure 11.2a) to the first intermediate excited state of Er^{3+}, $^4I_{11/2}$, which is resonant with the $^2F_{5/2}$ Yb^{3+} excited state. Subsequently, a second energy transfer from another Yb^{3+} ion is used for the excitation of the Ln^{3+} ions to the upper excited states (e.g., $^4F_{7/2}$ in case of Er^{3+}). This is followed by the non-radiative decay (multiphonon relaxation) to lower excited energy states (e.g., $^2H_{11/2}$ and $^4S_{3/2}$ green emitting states or $^4F_{9/2}$ red emitting state in case of Er^{3+}) from where radiative decay to the ground state occurs resulting in the emission of light of higher energy than used for excitation. It has to be taken into account that not all cases of energy transfer upconversion are characterized by a perfect match between the excited states of the sensitizer and the activator ions. For instance in case of Tm^{3+}, Ho^{3+} or Nd^{3+} (Figure 11.2b, d), energy transfer still takes place, however, small additional amounts of energy must be provided by or be released to the host lattice in the form of phonons.

Most interestingly, the population probability of some of the excited energy states is highly sensitive to temperature changes in the local environment of the Ln^{3+} ions (Figure 11.2). This allows for the exploitation of upconverting materials as optical nanothermometers, specifically as ratiometric nanothermometers. Further insight into the theoretical background of these temperature controlled processes will be provided in the following section.

11.2.2 Theoretical Background: *LIR*, Boltzmann's Plot, Thermal Sensitivity, and Thermal Resolution

As previously mentioned, the use of ratiometric nanothermometers exploiting the thermally induced variation of more than one emission intensity is a suitable tool to overcome drawbacks of techniques relying solely on one emission intensity. Seeking highly reliable intensity ratios, the *LIR* technique provides an attractive alternative tool (Quintanilla et al. 2011; Vetrone et al. 2010). This technique is based on the intensity ratio between two closely spaced energy levels that are thermally coupled. Generally, the energy gap between the two levels must be small, typically smaller than ~2000 cm^{-1}, in order to allow thermal coupling. The temperature governed population of the upper of the two levels involved is then described by Boltzmann's distribution. Under such conditions, *LIR* can be defined by Equation 11.2 (Quintanilla et al. 2011):

$$LIR = \frac{I_1}{I_2} = B \cdot \exp\left(-\frac{\Delta E}{k_B \cdot T}\right) \tag{11.2}$$

where ΔE is the energy gap between the emitting levels in close proximity I_1 (upper level) and I_2 (lower level). Temperature T is given in Kelvin. k_B is the Boltzmann factor. B is a constant that depends on the experimental system and the intrinsic spectroscopic parameters of the dopant/host pair (Equation 11.3):

$$B = \frac{c_1(v) \cdot A_1 \cdot g_1 \cdot h \cdot v_1}{c_2(v) \cdot A_2 \cdot g_2 \cdot h \cdot v_2} \tag{11.3}$$

where h is the Plank's constant, $A_{1,2}$ is the spontaneous emission rate of the level, $g_{1,2}$ its degeneracy, and $c_{1,2}(v)$ is the response of the detection system at

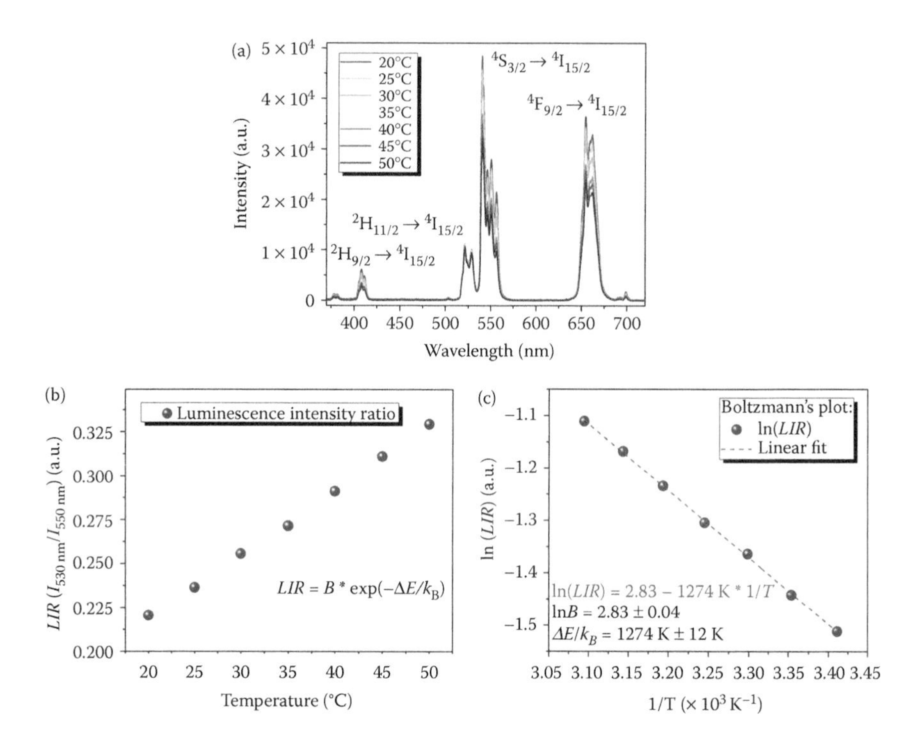

FIGURE 11.3
(a) Typical temperature dependent upconversion emission spectra of Er^{3+}/Yb^{3+} co-doped UCNPs (β-$NaGdF_4$ doped with 2 mol% Er^{3+} and 20 mol% Yb^{3+}; dispersed in H_2O; particle size: ~26 nm; excitation with a 980 nm laser diode; laser power: 330 mW). (b) *LIR* as a function of the temperature. (c) Boltzmann's plot. (Data taken from Hemmer, E. et al. 2015. *Chem. Mater.* 27:235–44.)

the emission frequency $v_{1,2}$. ΔE and B can be obtained from the least-squares fitting of the linear dependence of ln(*LIR*) versus the reciprocal absolute temperature, $1/T$, so-called Boltzmann's plot (Equation 11.4, Figure 11.3).

$$\ln(LIR) = \ln(B) - \frac{\Delta E}{k_B \cdot T} \tag{11.4}$$

The green emitting states $^2H_{11/2}$ and $^4S_{3/2}$ of Er^{3+} are among the most widely exploited energy levels for *LIR*-based nanothermometry with lanthanides (Figure 11.2a, Figure 11.3 and Section 11.3.1). By far less exploited is the use of Er^{3+}-based short-wavelength upconversion emission for thermal sensing. Xu et al. (2012b) verified the thermal coupling between the $^4G_{11/2}$ (384 nm) and $^2H_{9/2}$ (408 nm) excited states in Er^{3+}/Yb^{3+} co-doped $CaWO_4$, resulting in the shortest wavelength emissions for optical temperature measurement at the time. The well-separated emissions in the ultraviolet (UV) region at

384 and 408 nm (following a three-photon excitation process) are specifically favorable for high-temperature thermal sensing since the interference of background radiation can be avoided. Zheng et al. (2015) demonstrated that the $^4D_{7/2}$ (256 nm) and $^4G_{9/2}$ (276 nm) excited states of Er^{3+} ions doped into β-$NaLuF_4$ nanocrystals are thermally coupled and that their population ratio is fitted well by the Boltzmann's distribution. It was further shown that the thermal sensitivity of the five-photon upconversion emission in the UV region was superior to that of the green-based ratiometric nanothermometer ($^2H_{11/2} : {}^4S_{3/2}$) from room temperature to 330 K.

Moreover, the *LIR* technique is not limited to Er^{3+}-based UCNPs. Other (upconverting) Ln^{3+} ions possessing energy levels in close proximity whose thermal population is governed by Boltzmann's distribution include Tm^{3+}, Ho^{3+}, and Nd^{3+} (Figure 11.2b–d). Most recent research achievements involving these systems in nanothermometry will be highlighted in Section 11.3. Moreover, it must be noted that the *LIR* approach cannot only be applied between intensities originating from two emitting electronic levels separated by a small energy gap, but also to intensities that result from one emitting level if that specific level shows well-separated Stark sub-levels. For instance, the thermally coupled sub-Stark energy levels of the $^4F_{9/2}$ level in Er^{3+}, the $^3H_4$ excited state of the Tm^{3+} ions, or the high- and low-energy Stark sub-levels of the $^4F_{3/2}$ state of the Nd^{3+} ions (Dong et al. 2011; Liu et al. 2011; Rocha et al. 2014).

In order to describe the performance and suitability of a probe material as an optical temperature sensor, it is further necessary to introduce a measure to express the extent to which a thermally sensitive parameter changes upon temperature fluctuation, which is the *thermal sensitivity*, *S*. In this context, the *LIR* technique is a suitable tool to determine the thermal sensitivity of an Ln^{3+}-based nanothermometer, as it is defined by (Quintanilla et al. 2011)

$$S = \frac{\partial(LIR)}{\partial T} = LIR \cdot \left(\frac{\Delta E}{k_B \cdot T^2} \right) \quad (11.5)$$

where $LIR = I_1/I_2$ is determined by integration of the emission peaks and $\Delta E/k_B$ is given by the Boltzmann's plot (Figure 11.3c).

However, it must be taken into account that Equation 11.5 can only be applied if the population of the energy levels involved is Boltzmann governed. For instance in the case of surface functionalization or combination of UCNPs with additional energy acceptors or donors (such as organic dyes or quantum dots), partial quenching of the emission from the UCNPs can occur, and one cannot assume any longer Boltzmann's distribution between the energy levels. Under such conditions, it is still possible to estimate *S* by applying the basic definition of *S* as the thermal derivative of the *LIR*. However, a function different from the one given in Equation 11.5 for *LIR* has to be determined based on the experimental data for the ratio between the two intensities involved. This is possible by fitting the experimental *LIR* data, and the obtained fitted function, more precisely its thermal derivative, can then be applied in order to estimate the sensitivity of the system. (Hemmer et al. 2015). A further aspect that needs to be considered when estimating the sensitivity of an optical nanothermometer is optical heating. As shown by Rakov and Maciel (2012) continuous wave (cw) laser excitation of a probe at a given external temperature (i.e., the temperature at which the overall sample is kept by external heating/cooling) induces internal probe heating dependent on the applied laser power. This internal heating reduced the sensitivity of the thermal sensor. It was observed, however, that the use of pulsed excitation eliminated the optical heating effect and the maximum sensitivity of the investigated sensor increased from ~0.0056 K^{-1} for cw to ~0.0070 K^{-1} for pulsed excitation.

Finally, while the thermal sensitivity is a suitable parameter to describe the performance of a nanothermometer that enables the comparison of various systems (Figure 11.4 displays S of various upconverting materials [Quintanilla et al. 2016]), it has to be kept in mind that it does not take into account signal-to-noise-ratios or different emission quantum yields of different probes. However, these are essential experimental aspects and their consideration becomes necessary in terms of practical application. Therefore, an additional parameter has to be introduced to further describe the performance of an optical nanothermometer, namely the *thermal resolution*, ΔT_{min}.

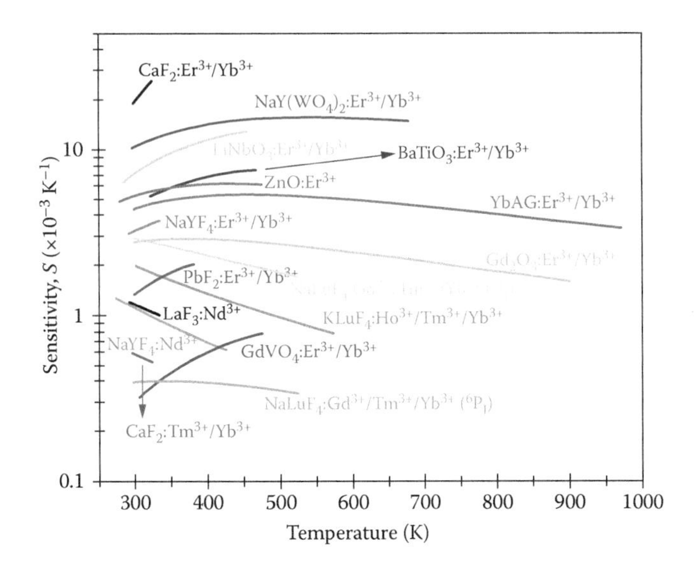

FIGURE 11.4
Thermal sensitivities of different upconverting materials proposed for *LIR*-based nanothermometry. (Quintanilla, M. et al. 2016. Luminescent nanothermometry with lanthanide-doped nanoparticles. In *Thermometry at the Nanoscale: Techniques and Selected Applications*, edited by L. D. Carlos, and F. Palacio, 124–66. Cambridge: The Royal Society of Chemistry. Reproduced by permission of Royal Society of Chemistry.)

The thermal resolution is defined as the minimum temperature that can be effectively resolved, and it can be statistically described for a given temperature by

$$\Delta T = \frac{\sigma}{S} \tag{11.6}$$

where σ is the standard derivation of several measurements and S is the sensitivity at this temperature (Quintanilla et al. 2016).

11.3 Recent Advances in UCNP-Based Nanothermometry

11.3.1 Er^{3+}-Based Nanothermometers

The dopant pair Er^{3+}/Yb^{3+} is one of the most intensively studied systems with regards to upconverting materials, and temperature dependent Er^{3+}/Yb^{3+}-based upconversion was reported for very diverse host materials, for instance, ceramics (F. Huang et al. 2015; Peng et al. 2012; Y. Yang et al. 2014; Y. M. Yang et al. 2014; Zou et al. 2015), glasses (dos Santos et al. 1999; Lai et al. 2010; Pandey et al. 2014), fluoride-based crystals (Chouahda et al. 2009; Jaque and Vetrone 2012; Sayoud et al. 2012; Vetrone et al. 2010), or composite materials such as UCNPs embedded in a glass matrix (Chen et al. 2015; Jiang et al. 2014).

Based on their studies of Er^{3+}/Yb^{3+}-doped Gd_2O_3, Singh et al. (2009) summarized the advantages of nanocrystalline host materials, namely their advanced thermal stability allowing for high-temperature sensing, and their chemical stability making this type of temperature sensor a permanent and robust device. Following this idea, low- and high-temperature sensing covering the broad region from 93 to 613 K was demonstrated for Y_2O_3:Er^{3+}/Yb^{3+} revealing its suitability as a thermal sensor for a very broad temperature range (Du et al. 2015b). Another example for recent achievements in the development of upconversion-based nanothermometers is the work by Gavrilovic et al. (2014) reporting a high relative sensitivity of 1.11% K^{-1} with an estimated thermal resolution of 1 K for Er^{3+}/Yb^{3+} co-doped $GdVO_4$UCNPs.

When developing optical thermal sensors, it is the obvious goal to achieve optimized thermal sensitivity. In this context, several parameters have been identified that play a key role for the design of high-performing thermal probes, including characteristics of (i) the host lattice (e.g., phonon energy, lattice distortions, and crystal field), (ii) the dopant ions (e.g., Ln^{3+} concentration, use of co-dopants), (iii) the particle morphology (size and shape), and (iv) the local probe environment (e.g., solvent, surface modification, and core–shell structures).

Regarding the host lattice, it was shown that annealing at higher temperatures affects the energy gap between the $^2H_{11/2}$ and $^4S_{3/2}$ levels of Er^{3+} ions doped into a ZnO matrix, resulting in a larger energy gap and a less efficient population of the $^2H_{11/2}$ level by thermalization when the probe was annealed (Wang et al. 2007). In general, the energy levels of Ln^{3+} ions in a solid host material experience Stark splitting, which occurs as the degenerate 4*f* levels are further split in the crystal field as a result of the local atomic configuration around the Ln^{3+} ions. Further, the degree of this Stark splitting influences the energy gap between two levels in close proximity (such as $^2H_{11/2}$ and $^4S_{3/2}$ in Er^{3+}). Thus, when annealing $ZnO:Er^{3+}$ at high temperature, the local atomic configuration around the optically active Er^{3+} ions was modified leading to a variation of the crystal field around the Er^{3+} ions and a change of the energy gap between the two energy levels. In a different study, L. Liu et al. (2013) suggested that the larger energy gap between the $^2H_{11/2}$ and $^4S_{3/2}$ levels of Er^{3+} in Y_2O_3 nanoparticles is due to the alteration of Er^{3+} clusters in the Y_2O_3 matrix after applying a re-crystallization procedure on the $Y_2O_3:Er^{3+}$ probe.

Further, seeking enhanced upconversion emission, the lattice phonon energy of the host material is an important factor. Since it strongly affects the upconversion emission, it is also expected to alter the sensitivity of a thermal sensor. In this context, Singh et al. (2015) investigated the optical and structural properties of three different probe materials, namely Er^{3+}/Yb^{3+} co-doped yttrium oxide (cubic $Y_{1.894}Yb_{0.1}Er_{0.006}O_3$), yttrium vanadate (tetragonal $Y_{0.947}Yb_{0.05}Er_{0.003}VO_4$), and yttrium phosphate ($Y_{0.947}Yb_{0.05}Er_{0.003}PO_4$). It was found that the thermal sensitivity and the *LIR* of the phosphate probe was significantly lower than that of the two other probes, most likely due to the higher phonon energy of the phosphate host lattice, enhancing non-radiative relaxation from the $^2H_{11/2}$ to the $^4S_{3/2}$ level, resulting in the increase of the $^4S_{3/2}$ originating emission when compared to the $^2H_{11/2}$ emission. Moreover, the authors report the vanadate-based phosphor $Y_{0.947}Yb_{0.05}Er_{0.003}VO_4$ as the *LIR*-based thermal sensor with the highest sensitivity (0.0105 K^{-1}) to date. Further work investigating the influence of the host lattice on the sensing performance was conducted by Liu et al. (2015) focusing on Yb^{3+}-sensitized Er^{3+}-doped Y_2O_3, YAG, and $LaAlO_3$ phosphors. It was found that the thermal sensitivity of the phosphors was proportional to the bond covalency of Y-O and La-O calculated following bond theory. These results highlight the importance of better understanding of the probes' structural and optical properties for the rational design of novel Ln-based nanothermometers.

Besides the host lattice, dopant concentration shows an effect on temperature sensing behavior. For instance, an increase in the maximal thermal sensitivity of Er^{3+}-doped Y_2O_3 from about 0.0028 K^{-1} to 0.0052 K^{-1} (at 741 K) was observed when the Er^{3+} doping concentration decreased from 12 mol% to lower doing concentration (0.5 and 1 mol%) (R. Wang et al. 2015). Since the energy gaps obtained from the Boltzmann's plots were substantially identical (1053 and 1030 cm^{-1}), the enhanced sensitivity must come from the parameter *B* in Equation 11.2. It was further shown that the larger *B* is attributed to a larger ratio of the spontaneous emission rates A_1/A_2 (Equation 11.3) in weakly doped samples. This is due to the fact that high doping concentrations can induce structural changes in the crystal lattice of the host material. In fact, these changes are expressed in the decrease of the cell parameters of the crystalline Y_2O_3 host lattice from 10.6035 Å to 10.5975 Å with the doping concentration increasing from 0.5 to 12 mol%. In general, the Ln^{3+} dopant ions replace stoichiometrically the metal ions in the host lattice, and changes of the host's crystal lattice are induced due to the differences in the ionic sizes of the dopant and the host ions. Eventually, these local structural changes affect the spontaneous emission rates, which are known to be very sensitive to the Ln^{3+} surrounding environment. In the same study, similar results were reported for Er^{3+}/Yb^{3+} co-doped $NaYF_4$.

Besides the concentration of the activator Ln^{3+} ion, co-doping of other ions into the host lattice has an effect on the Er^{3+}/Yb^{3+} upconversion emission and the performance of the resultant probe material as a thermal sensor. For instance, a set of studies discusses the effect of molybdenum (Mo) as co-dopant on the upconversion efficiency and the optical temperature sensing behavior, including Er/Mo:Yb_2TiO_7 (Cao et al. 2011), Er/Yb/Mo:$Na_{0.5}Bi_{0.5}TiO_3$, (Du and Yu 2015) Yb^{3+}/Er^{3+}:$AgLa(MoO_4)_2$, (Li et al. 2015) Er^{3+}/Yb^{3+}:$NaY(MoO_4)_2$ (Yang et al. 2015), or Er/Mo:$Yb_3Al_5O_{12}$ (Dong et al. 2012). The significant increase of the green upconversion emission has been assigned to an energy transfer process based on ground state absorption (GSA) and subsequent ESA in a Yb^{3+}–MnO_4^{2-} dimer complex, followed by a high excited-state energy transfer to the Er^{3+} ions. Moreover, it has to be noted that the temperature sensing properties of the Er^{3+}/Yb^{3+} ions in the described ceramic host materials were found to be enhanced by co-doping with Mo ions. As a further example showing the strong effect co-doping can have on the upconversion luminescence properties, Klier and Kumke (2015a) analyzed $NaYF_4$:Er^{3+}/Yb^{3+} co-doped with Gd^{3+} and revealed not only a strong dependence of the emission intensity distribution (e.g., green–red ratio) on the Gd^{3+} content, but also—most interesting with regard to nanothermometry—on the temperature profile of the luminescence intensity.

With regard to the influence of the probes' morphology on their thermal sensing performance, Dong et al. (2014) observed an increase of the thermal sensitivity when decreasing the size of Er^{3+}/Yb^{3+}:$NaYF_4$ upconverting microspheres from 1.6 to 0.7 µm ($S(1.6\ \mu m) = 24.7 \cdot 10^{-4}\ K^{-1}$; $S(0.7\ \mu m) = 36.8 \cdot 10^{-4}\ K^{-1}$). The size-dependent sensitivity to T was assigned to the increased surface to volume ratio for smaller spheres resulting in more Er^{3+} ions in close proximity to the sphere surface. Generally speaking, Ln^{3+} ions located at the surface of a particle undergo stronger electron–phonon interactions with changing temperature than the Ln^{3+} ions located inside. Consequently, in the case of the Er^{3+} doped into smaller spheres, the population process of $^4S_{3/2} \rightarrow {}^2H_{11/2}$ increases and the effect of the *LIR* ($I(^2H_{11/2})/I(^4S_{3/2})$) is enhanced. It must also be taken into account that undesired optical heating by the NIR excitation laser is more pronounced in smaller nano/microparticles than in larger ones, which results from the higher efficiency of a non-radiative relaxation process and enhanced electron–phonon interaction in nano/microparticles. Studies by Maciel et al. (2010) on Er^{3+}-doped $BaTiO_3$ UCNPs (26 and 58 nm in size) further confirmed the size dependence of the thermal sensing performance, while the sensitivity was found to be independent from the surrounding medium (comparing air, water, and glycerol).

Among upconverting nanomaterials, β-$NaYF_4$:Er^{3+}/Yb^{3+} UCNPs are commonly considered as one of the most efficient upconverting nanomaterials. However, their upconversion efficiency is still low (typically below 1 %) and efforts are undertaken for improvement. Upconversion efficiency enhancement is, for example, obtained by designing a core–shell structure where the Ln^{3+} ions are doped in the core and are well protected from the environment by an undoped shell. Vetrone et al. (2009) showed that the use of an active shell, doped with Yb^{3+}, allows for further upconversion enhancement. Since then, the concept of core–shell UCNPs has widely been applied. Nevertheless, the influence of a core–shell design on the thermal sensitivity of UCNPs has received much less attention. Li et al. (2014) compared the thermal behavior of core only, core-inert (undoped $NaYF_4$) shell, as well as core-active ($NaYF_4$:10% Yb^{3+}) shell β-$NaYF_4$:Er^{3+}/Yb^{3+} UCNPs. As expected, the inert and active shells significantly enhance the upconversion intensity of the UCNPs. Of special interest for application of these UCNPs as nanothermometers is the finding that the core–shell design does not alter their sensitivity to T. Importantly, it was found that the use of an active shell results in similar optical heating effect as in case of core only UCNPs. However, such undesired heating is almost negligible when an inert shell is applied, which can be attributed to the lack of NIR excitation light absorbing Yb^{3+} ions in the shell and the subsequent conversion of parts of absorbed energy into heat. The authors further showed that thermal-sensing properties of UCNPs are not affected by different dispersion media and that UCNPs exhibit high stability in broad pH range and salt conditions. These characteristics are highly suitable for biomedical applications.

In the context of biomedical applications, specifically with regard to tumor treatment and photothermal therapy approaches, Mahata et al. (2015) reported Er^{3+}/Yb^{3+} co-doped yttrium vanadate (YVO_4) UCNPs as highly sensitive nanothermometers (0.01169 K^{-1} at 380 K, to date claimed as the highest calculated sensitivity among reported results for inorganic nanosensors) and optical nanoheaters. While optical heating has previously been described as undesired, it also opens pathways for photothermal therapeutic approaches. In fact, Mahata et al. (2015) observed a temperature increase from 315 to 460 K under excitation of the UCNPs with 920 nm light. This excessive heat generation in the sample was assigned to the involved non-radiative processes, whereas the non-radiative decay rates increase exponentially with T. In addition, the previously mentioned stronger electron–phonon coupling in nanocrystalline particles when compared to larger particles or bulk materials enhances the non-radiative transition, providing additional contribution to the heat generation.

It becomes clear that, since the first mention of Ln^{3+}-based nanothermometry, much effort has been undertaken for a better understanding of the thermal behavior of (Er^{3+}-based) UCNPs and the design of highly sensitive thermal probes, resulting in promising candidates for applications in nanothermometry. In the following section, some of the most recent examples of Er^{3+}-doped UCNP-based nanothermometry for biomedical applications will be highlighted.

11.3.1.1 Er^{3+}-Based Upconverting Nanothermometers for Biomedical Applications

The first intracellular nanothermometry by use of UCNPs was realized by Vetrone et al. (2010). In this work, $NaYF_4$:Er^{3+}/Yb^{3+} UCNPs were incubated with human carcinoma (HeLa) cells in order to induce the cellular uptake of the nanothermometers. Subsequently, the optical transmission image of an individual HeLa cell was obtained at different inner temperatures set by use of a resistor-based heater. Upon changing of the applied voltage, and therewith of the internal temperature of the cells, the *LIR* (I_{525}/I_{545}) of the nanothermometer changed, where I_{525} and I_{545} are the relative upconverting emissions emanating from the thermally coupled $^2H_{11/2}$ and $^4S_{3/2}$ states, respectively. With the help of a previously recorded calibration curve (Figure 11.5a and b), the internal temperature of a single HeLa cell could be assigned (Figure 11.5c) and monitored until thermally induced cell death occurred.

Jaque et al. (2013) used a similar concept in order to monitor the thermal loading induced by tightly focused laser beams in both living cells and fluids. Very interesting from a biomedical application point of view, it was found that for the typical excitation intensities used in bioimaging experiments, pump-induced optical heating could be neglected. However, for larger excitation intensities (tens of MW cm^{-2}), the local thermal loading was found to be large enough to cause cell death.

Savchuk et al. (2014) developed fluorescence lifetime nanothermometers based on Er^{3+}/Yb^{3+}:NaY_2F_5O or Er^{3+}/Yb^{3+}: $NaYF_4$ UCNPs, whereas it was demonstrated that Er^{3+}/Yb^{3+}:NaY_2F_5O possesses a higher thermal sensitivity than Er^{3+}/Yb^{3+}:$NaYF_4$. Most importantly, the first experimental evidence on sub-tissue lifetime fluorescence thermal sensing by UCNPs in an *ex vivo* experiment was provided. Therefore, subsequent to sub-tissue injection of the Er^{3+}/Yb^{3+}:NaY_2F_5O UCNPs 1 mm below the surface of a chicken breast, local optical heating was induced by use of a 1090 nm heating laser (1.2 W), followed by recording of the fluorescence decay curves (excitation of Er^{3+}/Yb^{3+} by 980 nm). It was found that the lifetime measured inside the chicken breast was substantially reduced when the heating laser was on, providing evidence for the progressive heating of the tissue.

In a different approach, Schartner and Monro (2014) demonstrated localized temperature measurement using a probe (sodium zinc tellurite glass, doped with 1 mol% Er^{3+} and 9 mol% Yb^{3+}) at the tip of an optical fiber. Since

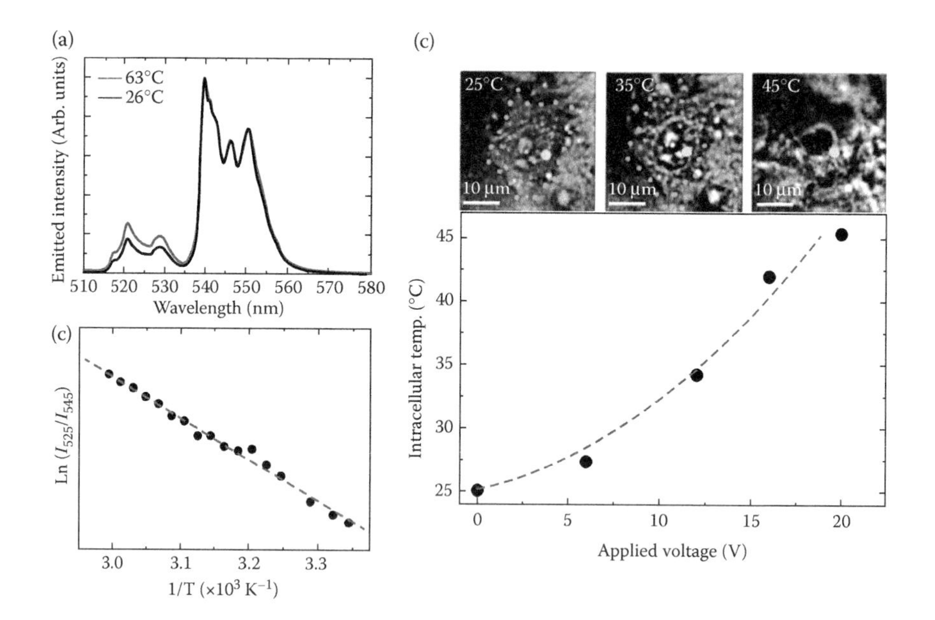

FIGURE 11.5
(a) Upconversion emission spectra obtained at two different cuvette temperatures (excitation at 920 nm) and (b) a plot of $\ln(I_{525}/I_{545})$ as a function of $1/T$ to calibrate the thermometric scale for the water-dispersible $NaYF_4:Er^{3+}/Yb^{3+}$ nanothermometers. (c) (Top) Optical transmission images of an individual HeLa cell at three inner temperatures. Cell death is observed at 45°C. (Bottom) Temperature of the HeLa cell determined by the Er^{3+} ion fluorescence in the $NaYF_4:Er^{3+}/Yb^{3+}$ UCNPs as a function of the applied voltage. (Reprinted with permission from Vetrone, F. et al. 2010. Temperature sensing using fluorescent nanothermometers. *ACS Nano* 4, 3254–58. Copyright 2010 American Chemical Society.)

the tip of a typical fiber is only 125 μm in size, this probe shows, for instance, potential for *in vivo* applications; namely thermal sensing of deeper tissue areas that are difficult to access and where the limited tissue penetration depth of visible light needs to be overcome. The point temperature sensing device exhibits a temperature resolution of approximately 0.1–0.3°C over the biological relevant range from 23°C to 39°C. Moreover, the application of a relatively thin probe layer with a high dopant concentration allowed to minimize the required excitation power to 300 μW (98 mW mm^{-2}), reducing the potential for heat induced damage to biological samples.

11.3.2 Tm^{3+}-Based Nanothermometers

As depicted in Figure 11.2b, Tm^{3+} ions show potential for upconversion-based thermal sensing based on the thermal coupling of their excited states such as $^3F_{2,3}$ and $^3H_4$. Taking advantage of this, Xing et al. (2014) proposed Tm^{3+}/Yb^{3+} co-doped $LiNbO_3$ single crystals as a highly sensitive optical temperature sensor based on the *LIR* technique. Due to the thermal population of the

$^3F_{2,3}$ level from the $^3H_4$ level, the emission intensity at 700 nm ($^3F_{2,3} \rightarrow {}^3H_6$) is greatly enhanced with increasing temperature, while the emission intensity at 800 nm ($^3H_4 \rightarrow {}^3H_6$) decreases.

Further, Tm^{3+} exhibits two distinct blue upconversion emission peaks at 477 and 488 nm that are assigned to the transition from the Stark sub-levels of the $^1G_4$ state to the $^3H_6$ ground state of Tm^{3+} ions. Variation of temperature causes a relevant change in the intensity ratio of these two luminescence lines. Exploiting the temperature induced population redistribution between these Stark levels of the $^1G_4$ state of Tm^{3+}, nanothermometry by use of, for example, Tm^{3+}/Yb^{3+}co-doped $NaNbO_3$ (Pereira et al. 2015) or $Na_2Y_2B_2O_7$ (Soni et al. 2015) nanoprobes has recently been reported.

Aside from the use of thermally coupled energy levels in close proximity to each other or of Stark sub-levels, an approach based on the thermally controlled population of energy levels in further distance was suggested by Zhou et al. (2014a) who designed a core–shell structure based on a Tm^{3+}/Yb^{3+} co-doped $NaYF_4$ core and a Pr^{3+}-doped $NaYF_4$ shell (Figure 11.6a).

The resultant core–shell UCNPs allowed for the exploitation of the temperature dependent population of different Tm^{3+} energy levels for nanothermometry, while the Pr^{3+}-doped shell was used to evaluate potential optical heating effects. Pr^{3+}-doped host materials are known for their thermally coupled energy levels, $^3P_2$ and $^3P_0$ suitable for *LIR*-based thermal sensing upon direct excitation with a blue laser (no upconversion process) (Zhou et al. 2014b). Importantly, there is no evidence for optical heating upon such direct excitation, which allows us to use the temperature dependent emission of Pr^{3+} ions to detect a potential heating of the core–shell structure caused by illumination with the 980 nm laser required for the excitation of Tm^{3+}/Yb^{3+} ions. Simultaneous excitation of the sample with 447 and 980 nm lasers revealed that optical heating can be neglected up to a laser power of 47 mW (spot size of the irradiated area: 2 mm × 3 mm). Subsequently, the different thermal behavior of the two upconversion emissions at 696 and 646 nm originating from the $^3F_{2,3}$ and $^1G_4$ excited states of Tm^{3+} ions were used for thermal sensing upon 980 nm excitation (Figure 11.6b). It was shown that the 696 nm emission ($^3F_{2,3} \rightarrow {}^3H_6$) increases monotonously with T due to the thermal population of the $^3F_{2,3}$ states from the lower $^3H_4$ state (Figure 11.6c). In contrast, the 646 nm emission ($^1G_4 \rightarrow {}^3F_4$) first increases with increasing temperature until 390 K, followed by a decrease. This behavior was explained as follows. Due to the energy mismatch between the Yb^{3+} and Tm^{3+} ions, the upconverted population of the $^1G_4$ level is sensitive to the participation of phonons, fostering the level's population when the temperature increases. However, phonon-assisted cross relaxation may be triggered by further temperature increase resulting in a decreasing $^1G_4$ population and consequently reduced emission intensity. Moreover, the $^1G_4$ state is populated from the $^3H_4$ state by an energy transfer process. A decreased $^3H_4$ population caused by thermal population of the $^3F_{2,3}$ levels will therefore further reduce the population of the $^1G_4$ level. Thus, it is a complex interplay between different phonon-assisted processes governing

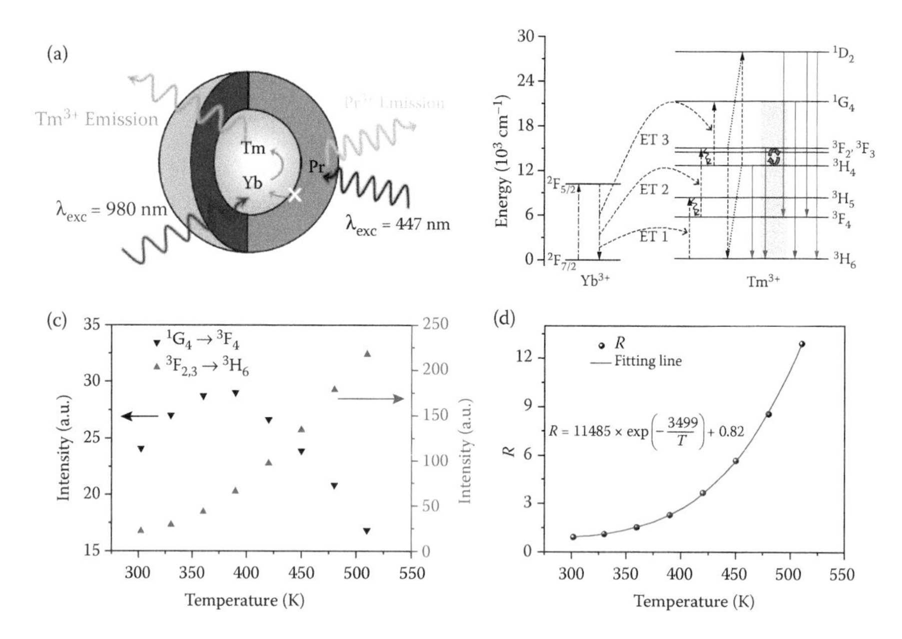

FIGURE 11.6
(a) Schematic diagram of the Tm^{3+}/Yb^{3+}:$NaYF_4$ core/Pr^{3+}:$NaYF_4$ shell UCNPs. (b) Schematic illustration of the upconversion processes in the UCNPs under 980 nm excitation. The dashed-dotted, dashed, curved, and full arrows indicate photon excitation, energy transfer, multi-phonon relaxation, and emission processes. (c) Integral emission intensities of $^3F_{2,3} \rightarrow {}^3H_6$ and $^1G_4 \rightarrow {}^3F_4$ transitions as a function of T. (d) T dependence of the ratio R (*LIR*) between these integral intensities. (Reproduced from Zhou, S.S. et al. 2014. *Opt. Lett.*, 39: 6687–90, with permission of The Optical Society.)

the population of the different Tm^{3+} energy levels, eventually resulting in the highly temperature sensitive emission from two not directly thermally coupled energy levels (Figure 11.6d: *LIR* as a function of T).

Aiming for photodynamic therapy (Bhaumik et al. 2015; Idris et al. 2015), for example for the treatment of skin cancer, UV-generating phosphors are highly attractive probes. However, UV light is strongly absorbed and scattered by biological tissue restricting the accessible regions when using external UV excitation sources. In contrast, absorption and scattering of NIR light by biological media is significantly reduced when compared to UV light (Anderson and Parrish 1981; Smith et al. 2009). Consequently, in order to broaden the field in which UV light can be applied, the generation of UV light under NIR light excitation was suggested. In this context, Tm^{3+}-doped UCNPs have been proven as excellent UV emitters following 980 nm laser excitation.

For example, Tm^{3+}/Yb^{3+} co-doped SrF_2 UCNPs were synthesized and their T-dependent emission properties were carefully analyzed by Quintanilla et al. (2015). Besides emission peaks in the visible region, Tm^{3+} shows two strong emission bands in the UV at around 370 and 350 nm, which can be assigned to the radiative transitions of the excited $^1D_2$ and $^1I_6$ levels.

The upconversion route that populates the $^1D_2$ state is often described as a combination of several energy transfer processes combining energy transfer from the $^2F_{5/2}$ Yb^{3+} excited state to Tm^{3+} ions and (due to the strong non-resonant character of this energy transfer) Tm^{3+}–Tm^{3+} cross-relaxation processes (Figure 11.2b). The population of $^1I_6$ takes place after a fifth energy transfer process from excited Yb^{3+} ions in the vicinity. Figure 11.7 shows the upconversion emission spectra obtained under 980 nm excitation and the peak intensities of the main emissions as a function of *T*.

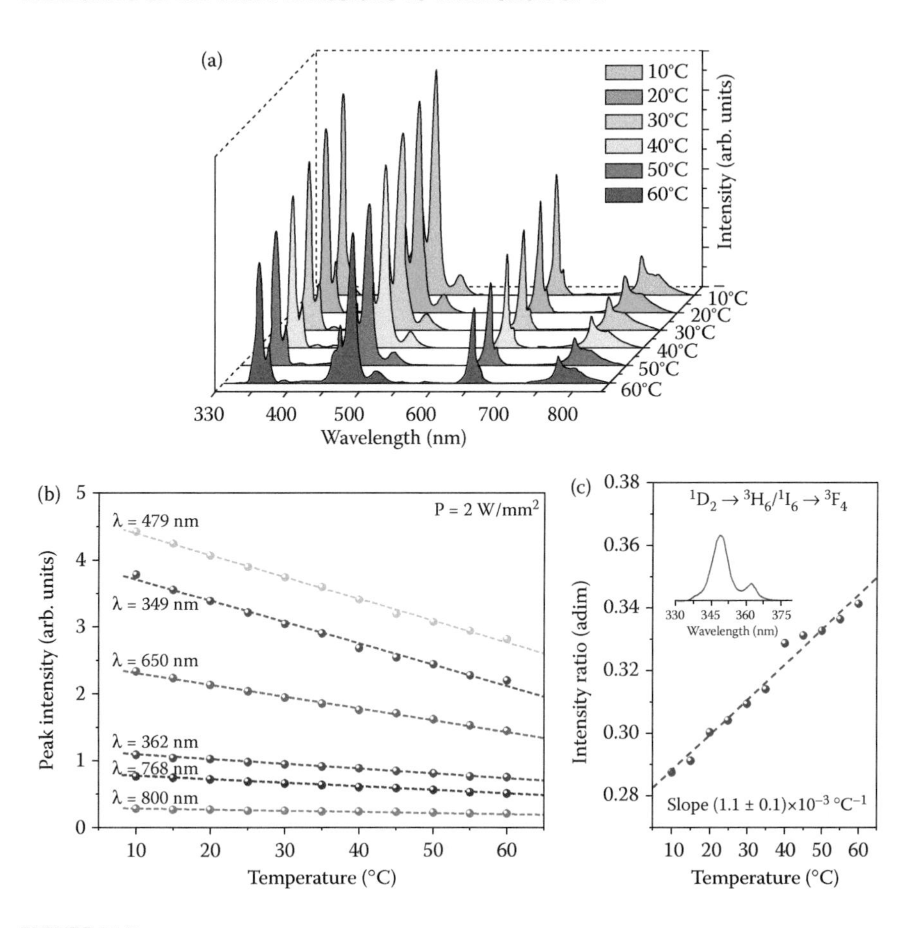

FIGURE 11.7
(a) Overall temperature dependence of the upconversion emission spectrum and (b) peak intensities of the main emission bands as a function of the temperature of SrF_2:Tm^{3+}/Yb^{3+} UCNPs in a colloidal dispersion in D_2O. The dashed lines in the lower graph have been added to guide the eye. (c) UV-based *LIR* ($^1D_2 \rightarrow {}^3H_6/{}^1I_6 \rightarrow {}^3F_4$). The inset shows the corresponding spectral range. The dashed lines are the linear fit of the experimental data. The corresponding slopes are also given in the graph. (Quintanilla, M. et al. 2015. Intense ultraviolet upconversion in water dispersible SrF_2: Tm^{3+}, Yb^{3+} nanoparticles: The effect of the environment on light emissions. *J. Mater. Chem.* C 3:3108–13. Reproduced by permission of The Royal Society of Chemistry.)

It becomes obvious that the overall emission intensity decreases with increasing temperature (Figure 11.7a). However, the extent to which this decrease happens is found to vary from peak to peak, especially in case of the NIR emission from the $^3H_4$ level and the UV emission attributed to the $^1D_2$ level (362 nm) (Figure 11.7b). These differences are related to different excitation paths in which the states are populated; increase of multiphonon relaxation processes on the one hand and favored phonon-assisted energy transfer processes on the other hand, particularly important in the population of the $^1D_2$ state by non-resonant Yb^{3+}–Tm^{3+} energy transfer and Tm^{3+}–Tm^{3+} cross-relaxation processes. Thus, in principle, thermal sensing would be possible by using graphs given in Figure 11.7b as temperature calibration curves. However, in order to overcome drawbacks related to intensity-based nanothermometers (see Introduction), the more reliable *LIR* technique has successfully been applied using the ratio between the intensities originating from the $^1D_2 \rightarrow {}^3H_6$ and $^1I_6 \rightarrow {}^3F_4$ transitions (Figure 11.7c).

11.3.3 Ho^{3+}-Based Nanothermometers

Optical temperature sensing based on upconversion emission has further been reported for Ho^{3+}-based phosphors. Most of the published work body is dedicated to Ho^{3+}-doped glasses or ceramic host materials (Du et al. 2015a, b; Singh 2007; Verma and Rai 2012; Xu et al. 2012a, 2013a), while studies on upconverting micro- or nanoparticles (e.g., Yb^{3+}/Ho^{3+} co-doped $NaLuF_4$ microcrystals [Zhou et al. 2014c] or Yb^{3+}/Ho^{3+}co-doped Y_2O_3 UCNPs [Lojpur et al. 2013]) are much less frequent. Generally, these nanothermometers are either based on the temperature dependent population of thermally coupled energy levels (green $^5F_4/^5S_2 \rightarrow {}^5I_8$ emissions, blue $^5F_{2,3}/^3K_8 \rightarrow {}^5I_8$ and $^5F_1/^5G_6 \rightarrow {}^5I_8$ emissions, Figure X.2c) or on the temperature dependence of the ratio between the overall green ($^5F_4/^5S_2 \rightarrow {}^5I_8$) and NIR ($^5F_4/^5S_2 \rightarrow {}^5I_7$) emission.

As an example of very recent work on Ho^{3+}-doped UCNPs for nanothermometry, a study by Savchuk et al. (2015) should be highlighted here. The authors developed a Ho^{3+}/Yb^{3+} co-doped KLuW nanothermometer that shows outstanding versatility since one single probe material offers different thermal sensing techniques, thus providing a way to corroborate temperature measurements. These include the *LIR* technique of two thermally coupled Stark sub-levels, the intensity ratio between the red and green luminescence bands, the observation of the change of the color of the emitted light arising from the UCNPs, and lifetime measurements (Figure 11.8). Figure 11.8a1 shows the temperature dependence of the red emission band that consists of two peaks assigned to the radiative transition from different Stark sub-levels of the $^5F_5$ energy level to the $^5I_8$ ground state of Ho^{3+}. Due to the small energy gap between the two thermally coupled Stark sub-levels, the upper sub-level can be thermally populated from the lower sub-level when the temperature increases. Consequently, the *LIR* (here denoted as

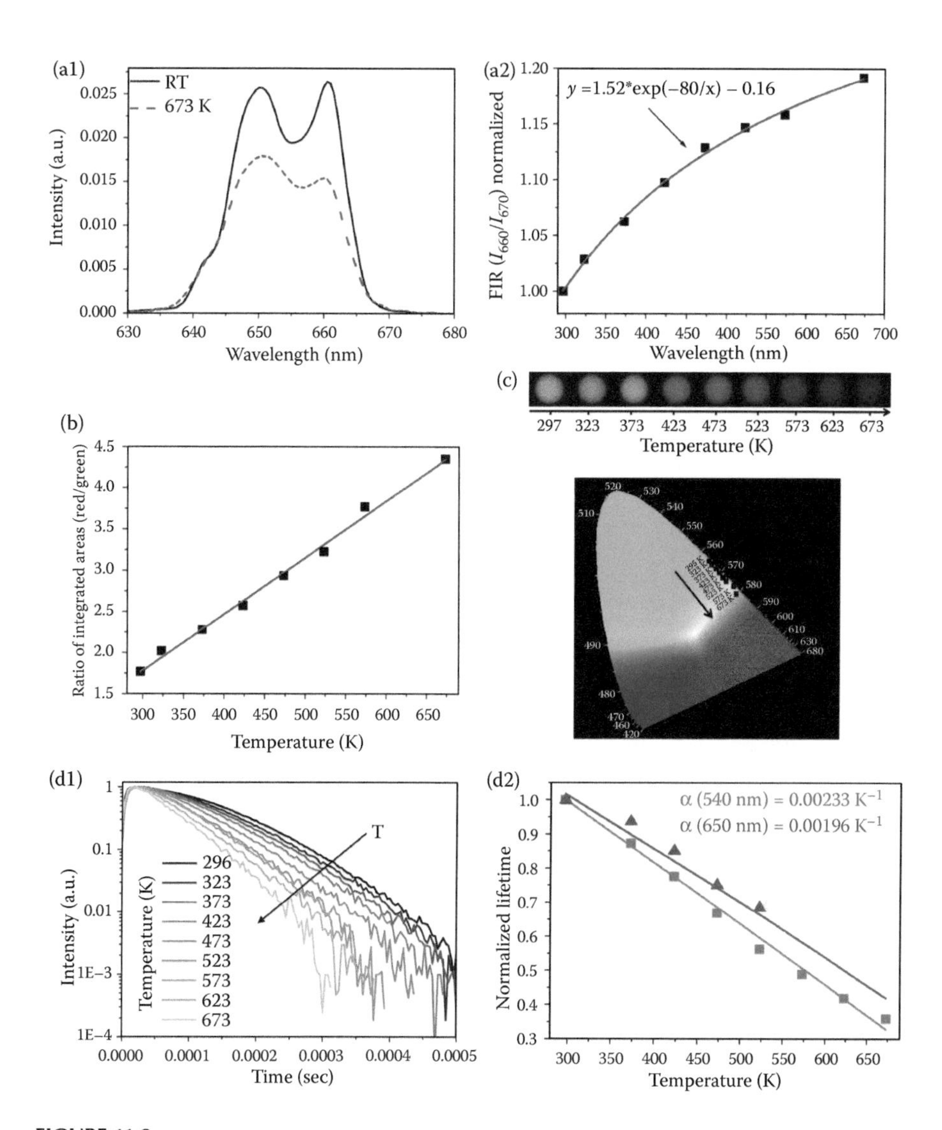

FIGURE 11.8

Ho^{3+}/Yb^{3+} co-doped KLuW-based nanothermometry. (a) Red Stark sub-level emission ($^5F_5 \rightarrow {}^5I_8$; a1: emission spectra at room temperature and 673 K; a2: *T*-dependent *LIR* of the I_{650}/I_{660}). (b) Ratio of integrated areas between the red ($^5F_5 \rightarrow {}^5I_8$) and green ($^5S_2/{}^5F_4 \rightarrow {}^5I_8$) emission bands as the function of *T*. (c) Color perception of the emission arising from the sample. (d) *T* dependence of the lifetime (d1: fluorescence decay curves corresponding to the 540 nm emission line associated with the $^5S_2/{}^5F_4$ radiative transition; d2: normalized lifetime values as a function of *T*). (Reprinted with permission from Savchuk, O. A. et al. 2015. Ho,Yb:KLu$(WO_4)_2$ nanoparticles: A versatile material for multiple thermal sensing purposes by luminescent thermometry. *J. Phys. Chem. C* 119, 18546–58. Copyright 2015 American Chemical Society.)

FIR, fluorescence intensity ratio) can be determined as a function of temperature, revealing that the experimental data can be well fitted by Boltzmann's distribution (Figure 11.8a2). In contrast to this intensity ratio of the emissions originating from the Stark sub-levels, the ratio of the integrated areas between the overall red ($^5F_5 \rightarrow {}^5I_8$) and green ($^5S_2/^5F_4 \rightarrow {}^5I_8$) emission bands (Figure 11.8b) follows a linear temperature dependence. A characteristic of such linear temperature dependence is the fact that it ensures a constant thermal sensitivity in the whole working temperature range of the thermal probe, which simplifies calibration of the nanothermometer. It must further be mentioned, that the two excited states considered here are located too far from each other for thermal coupling. Therefore, the populations of the green and red emitting excited states do not follow a Boltzmann's distribution, and the *LIR* technique cannot be applied. However, the population of the $^5F_5$ electronic level of Ho^{3+} mainly arises from the non-radiative relaxation from the $^5S_2$ and $^5F_4$ levels; a process that is temperature dependent. Thus, the $^5F_5$ and the $^5S_2$ as well as $^5F_4$ levels are considered as electronically coupled, which is expressed by the plot of the thermal evolution of I_{red}/I_{green} depicted in Figure 11.8b. It has further been observed that the color perception of the emission arising from the Ho^{3+}/Yb^{3+}:KLuW UCNPs changes from yellow to dark orange (Figure 11.8c), which was suggested as a new possibility for visual temperature determination, although the authors critically mention that the thermal resolution would be very wide, of the order of 50 K, if it is intended to be visualized by the naked eye. Finally, the prepared UCNPs were used for lifetime-based nanothermometry by measuring the decay curves corresponding to the 540 nm emission line associated with the $^5S_2/^5F_4$ radiative transition (Figure 11.8d1). Rising times in these curves indicate the occurrence of energy transfer processes, while the observation of a non-single exponential component indicates the existence of non-radiative processes that are responsible for the reduction of the lifetime with increasing temperature. Remarkably, the normalized lifetimes for the emission lines centered at 540 and 650 nm depicted in Figure 11.8d2 show a linear temperature dependence, thus a temperature independent thermal sensitivity. In fact, while the *LIR* technique revealed the highest sensitivity of the presented UCNPs at room temperature, the intensity ratio technique comparing the green and red emission intensities as well as the lifetime-based approach show a constant thermal sensitivity over the evaluated temperature range, which would allow for simplified calibration and for more precise measurements at higher temperatures.

11.3.4 Nd^{3+}-Based Nanothermometers

As a last representative of the group of Ln^{3+} ions applied in nanothermometry, Nd^{3+}-based phosphors should be briefly discussed. In case of Nd^{3+}/Yb^{3+} co-doped phosphors under 980 nm illumination, as depicted in Figure 11.2d, the population of the Nd^{3+} excited levels does not follow an upconversion mechanism but a phonon-assisted energy transfer process from the $^5F_{5/2}$ excited level of Yb^{3+} to the $^4F_{3/2}$ excited level of Nd^{3+} caused by the energy mismatch between the electronic levels involved. Subsequently, radiative relaxation can take place from the NIR emitting levels of Nd^{3+}. In fact, it is this emission of NIR light that trigged an increasing interest in Nd^{3+}-doped phosphors when seeking optical probes for biomedical application. The reasoning behind this is, again, the fact that NIR light is less scattered and absorbed by water and biological matter than visible or UV light resulting in advanced transparency of biological tissues toward NIR. This ultimately opens new avenues for the use of optical probes in the biomedical field. Moreover, as seen in Figure 11.2d, the energy gaps between the $^4F_{3/2}$, $^4F_{5/2}$, and $^4F_{7/2}$ levels are relatively small, so that the $^4F_{5/2}$ and $^4F_{7/2}$ states can be thermally populated from their lower energy levels. This makes them suitable for thermal sensing. For instance, the group of W. Cao studied the temperature dependence of the NIR emissions originated from the $^4F_{7/2}/^4S_{3/2} \rightarrow {}^4I_{9/2}$ (~750 nm), $^4F_{5/2}/^2H_{9/2} \rightarrow {}^4I_{9/2}$ (~803 nm), and $^4F_{3/2} \rightarrow {}^4I_{9/2}$ (~863 nm) transitions of Nd^{3+} ions in Nd^{3+}/Yb^{3+} co-doped oxyfluoride glass ceramic (Xu et al. 2013b) and in $CaWO_4$ powder (Xu et al. 2014) under 980 nm laser excitation.

In another study, Wawrzynczyk et al. (2012) used a 830 nm laser for the direct excitation of Nd^{3+} ions in heavily Nd^{3+}-doped $NaYF_4$ nanoparticles. Thermal sensing was reported by monitoring the absolute NIR luminescence intensity or by measuring the intensity ratio of the two Stark sub-levels of the $^4F_{3/2}$ multiplet in the Nd^{3+} ions. Based on their findings, the authors formulated two main advantages of the proposed nanothermometers. First, the possibility of using NIR light for the excitation and emission wavelength allows for a performance fully in the optical transmission tissue penetration window. This window, also known as the *biological window*, encompasses three distinct wavelength NIR regions, namely 700–950 nm (NIR-I), 1000–1350 nm (NIR-II), and 1550–1870 nm (NIR-III), in which the scattering and absorption of light by water, blood, and various biological tissues is significantly reduced when compared to UV or visible light (Anderson and Parrish 1981; Prodi et al. 2015; Sordillo et al. 2014; Smith et al. 2009). This allows for deeper penetration depth and enhanced propagation of NIR light through biological samples making NIR emitting probes highly attractive for optical bioimaging (Diao et al. 2015; Hemmer et al. 2013a, 2016; Yao et al. 2014). Second, the proposed nanothermometers provide the possibility for excitation intensity dependent heating (high excitation regime) on the one hand and thermal sensing (low excitation) on the other. Following this idea, recent research efforts focus on the development of Nd^{3+}-doped nanothermometers and nanoheaters for biomedical applications that operate in the NIR region taking advantage from reduced scattering and absorption by biological tissues when compared to UV or visible light (Carrasco et al. 2015; Ceron et al. 2015; Marciniak et al. 2015; Rocha et al. 2014).

11.3.5 Nanothermometers Based on the Combination of Different Ln^{3+} Ions

The use of co-dopants (e.g., Mn) for enhanced thermal sensing performance was mentioned in Section 11.3.1. In this section, the focus will be set on the combination of various Ln^{3+} ions and the advantage that can be taken from energy transfer processes between different Ln^{3+} ions for the development of novel nanothermometers.

Co-doping with two different Ln^{3+} ions with close laying energy levels in one host material and the use of the *LIR* between the two emissions originating from these levels for nanothermometry was reported by Pandey and Rai (2013). In $Ho^{3+}/Tm^{3+}/Yb^{3+}$ co-doped Y_2O_3, the $^5F_3$ excited level of Ho^{3+} and the $^1G_4$ excited level of Tm^{3+} are populated through energy transfer from the excited $^5F_{5/2}$ Yb^{3+} level to the Ho^{3+} and Tm^{3+} ions. Since these two excited states of Ho^{3+} and Tm^{3+} are very close to each other, exhibiting a narrow energy gap of 473 cm^{-1}, a thermal variation changes the population of the levels and consequently the emission intensities. It was shown that the population of both blue emitting levels follows Boltzmann's distribution, and the *LIR* technique was applied revealing a maximum sensitivity of $6.96 \cdot 10^{-3}$ K^{-1} at 303 K.

The combination of appropriate Ln^{3+} ions can further enable emissions that are not possible otherwise. For instance, Gd^{3+} ions in the ground state cannot absorb 980 nm photons directly because of the large energy gap between the ground state $^8S_{7/2}$ and the first excited, UV emitting states ^6P_J. In order to foster UV emission from the excited Gd^{3+} states, Zheng et al. (2013) prepared $NaLuF_4$ microcrystals that were co-doped with Yb^{3+} and Tm^{3+} in addition to Gd^{3+}. The use of Tm^{3+} ions as sensitizers resulted in the population of the excited states ^6I_J of Gd^{3+} through an energy transfer process from Tm^{3+} ($^3P_2 \rightarrow {}^3H_6$) to Gd^{3+} ($^8S_{7/2} \rightarrow {}^6I_J$) (Figure 11.9a).

Additionally, the ^6P_J and ^6D_J Gd^{3+} levels can be populated by non-radiative relaxation process of ^6I_J to ^6P_J and by energy transfer of Yb^{3+}/Gd^{3+} and Tm^{3+}/Gd^{3+}. With regards to nanothermometry, it is important that the narrow spacing between the Gd^{3+} excited levels favors the thermal population of the ^6P_J ($^6P_{5/2}$, $^6P_{7/2}$) and ^6I_J ($^6I_{13/2}$, $^6I_{15/2}$, $^6I_{11/2}$, $^6I_{9/2}$, $^6I_{7/2}$) sub-levels, allowing for the application of the *LIR* technique for UV-based thermal sensing. Maximum sensitivity values of about 0.0004 K^{-1} (at 333 K) where obtained when exploiting the *LIR* between $^6P_{5/2}$ and $^6P_{7/2}$ levels and 0.0029 K^{-1} (at 298 K) for the *LIR* between $^6I_{9/2}$ and $^6I_{7/2}$ levels. In addition to the suitability for thermal sensing, the potential for optical (based on Tm^{3+}) and MR imaging (based on Gd^{3+}) is identified, which makes the reported $Gd^{3+}/Tm^{3+}/Yb^{3+}$ co-doped $NaLuF_4$ phosphor a promising platform for multimodal biomedical applications.

Following a similar idea exploiting energy transfer between different Ln^{3+} ions, Zheng et al. (2014) designed a ratiometric optical nanothermometer that combines different Ln^{3+} ions in one core–shell UCNP. The combination of a Tm^{3+}/Yb^{3+} co-doped $NaGdF_4$ core with a Tb^{3+}/Eu^{3+} co-doped $NaGdF_4$

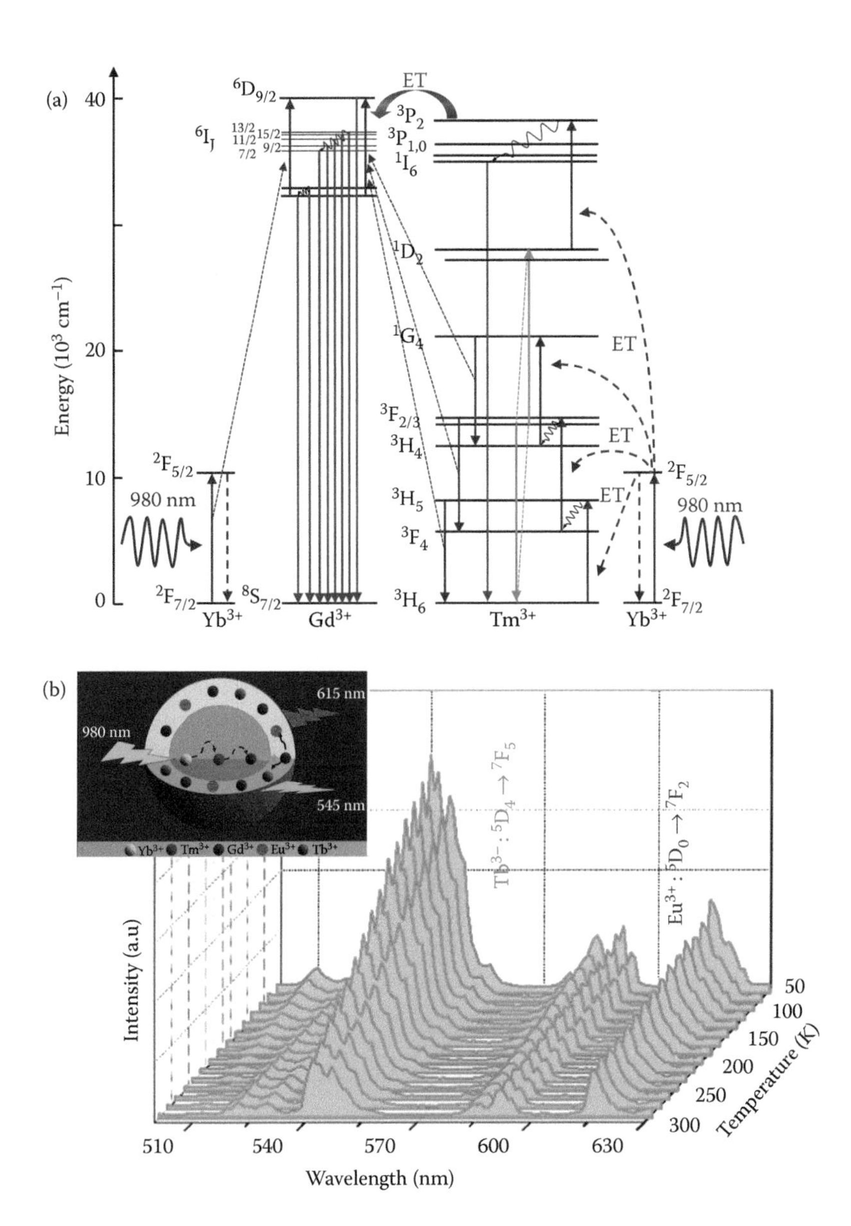

FIGURE 11.9
Multiple Ln^{3+} doping for UCNP-based nanothermometry. (a) Energy level diagrams of Yb^{3+}, Tm^{3+}, and Gd^{3+} ions, and possible upconversion processes. (Zheng, K. Z., Z. Y. Liu, C. J. Lv, and W. P. Qin. 2013. Temperature sensor based on the UV upconversion luminescence of Gd^{3+} in Yb^{3+}-Tm^{3+}-Gd^{3+} codoped $NaLuF_4$ microcrystals. *J. Mater. Chem. C* 1:5502–07. Reproduced by permission of The Royal Society of Chemistry.) (b) *T*-dependent green and red upconversion emission spectra of $NaGdF_4$:Tm^{3+}/Yb^{3+}@$NaGdF_4$:Tb^{3+}/Eu^{3+} core–shell nanoparticles under 980 nm excitation. (Zheng, S. H. et al. 2014. Lanthanide-doped $NaGdF_4$ core–shell nanoparticles for non-contact self-referencing temperature sensors. *Nanoscale* 6:5675–79. Reproduced by permission of The Royal Society of Chemistry.)

shell allows for the highly temperature sensitive generation of green and red emission upon NIR excitation (Figure 11.9b). The Tm^{3+} ions are excited to the $^1I_6$ level via a five-photon upconversion process from the excited Yb^{3+} ions in the $^5F_{5/2}$ level, followed by energy transfer to the $^6P_{7/2}$ Gd^{3+} level, from where energy is transferred in order to populate the ^5D_J levels of the Tb^{3+} and Eu^{3+} ions. The emission intensity originating from the $^5D_4 \rightarrow {}^7F_5$ transition (Tb^{3+} at 545 nm) shows a linear temperature dependence, whereas the $^5D_0 \rightarrow {}^7F_2$ transition (Eu^{3+} at 615 nm) is much less affected by temperature changes. Consequently, the *LIR* between these two emissions could be used for self-referencing nanothermometry.

Aiming for *in vivo* nanothermometry, the different optical properties of biological tissues, and scattering and absorption of the used light are a challenge. Addressing this, Klier and Kumke (2015b) exploited the increased penetration depth of ~800 nm light when compared to 980 nm light and investigated the influence of the excitation wavelength on the thermal behavior of Er^{3+}/Yb^{3+}:$NaYF_4$ UCNPs co-doped with Gd^{3+} and Nd^{3+}. In this system, Nd^{3+} ions are excited from their $^4I_{9/2}$ to the $^4F_{5/2}$ energy level, followed by a non-radiative relaxation step to the $^4F_{3/2}$ level of Nd^{3+}. Subsequent energy transfer takes place from this excited Nd^{3+} level to the $^2F_{5/2}$ level of the Yb^{3+} ions, followed by regular upconversion processes resulting in the characteristic green and red Er^{3+} emissions. Investigation of the heating and sensing capability of the Nd^{3+} co-doped probes revealed that the excitation with 795 nm induced no optical heating, while 980 nm illumination (170 mW excitation power for both wavelengths) resulted in a temperature increase of ΔT ~0.6 K, which may trigger cell death. The thermal sensitivity was found to be almost unaffected by the choice of excitation wavelength. Thus, Nd^{3+}-sensitized phosphors can offer an attractive alternative when optical heating must be avoided.

11.4 Nanothermometers Based on Multicomponent Nanoassemblies: Toward Sensitivity Enhancement and Multimodal Biomedical Applications

During the last few years, worldwide effort has been undertaken in the search for temperature sensitive phosphors that are suitable for thermal sensing down to the sub-micron scale. In this context, Ln^{3+}-based upconverting materials have been recognized as potential nanothermometers that are expected to open new pathways in the field of thermal sensing, with particular emphasis in the biomedical field, for example, sub-cellular thermal mapping or temperature monitoring in the frame of thermal therapeutic approaches. Seeking enhanced sensitivity in the biologically relevant temperature range and/or multimodality is a challenge that has been most recently addressed by researchers in the field. Here, the assembly of UCNP-based thermal sensors with a second moiety, such as an organic dye, a thermoresponsive polymer or metal nanostructures, provides attractive possibilities for the design of application-oriented multifunctional nanoplatforms.

11.4.1 UCNP–Organic Hybrids for Improved Sensitivity

Chen et al. (2013) used the concept of LRET in order to achieve a multicolor hybrid nanothermometer based on Er^{3+}/Yb^{3+} co-doped $NaYF_4$ UCNPs and the organic dye rhodamine 6G (R6G). Due to the spectral overlap between the green Er^{3+} emission and the absorption band of R6G, energy can be transferred resonantly from the Er^{3+} ions to the R6G dye. In the case where the energy donors (Er^{3+} ions) and the energy acceptors (R6G dye) are in close enough proximity, the energy transfer takes place non-radiatively. In addition, radiative energy transfer takes place where some of the emitted photons from the ${}^2H_{11/2} \rightarrow {}^4I_{15/2}$ and ${}^4S_{3/2} \rightarrow {}^4I_{15/2}$ transitions are reabsorbed by the R6G, and excite the electrons in R6G from the highest occupied molecular orbital to the lowest unoccupied molecular orbital. Subsequent radiative recombination gives rise to the R6G emission. Implementation of these UCNP–dye hybrids in self-assembled hemispherical microstructures led to an intensity-based nanothermometer demonstrating better thermal sensitivity than the pure UCNPs, which was assigned to the energy transfer between UCNPs and R6G.

Alternatively, J. C. Huang et al. (2015) opted for a hydrogel obtained by modification of Er^{3+}/Yb^{3+} co-doped $NaYF_4$ UCNPs with the thermoresponsive polymer poly(N-isopropylacrylamide) (pNIPAM) cross-linked with poly(acrylamide) (pAAm) that could be applied for intensity-based nanothermometry. At temperatures higher than the transition T of pNIPAM, the polymer chains undergo a reversible coil-globule phase transition resulting in increased opacity of the hydrogel. The observed thermal sensitivity was linked to the reduction of the detectable upconversion emission caused by this higher opacity of the hydrogel when T was higher than the transition T of pNIPAM.

The combination of thermoresponsive polymer pNIPAM and a LRET pair for thermal sensing was suggested by Hemmer et al. (2015). In their strategy, the surface of Er^{3+}/Yb^{3+} co-doped $NaGdF_4$ UCNPs (acting as energy donors) was modified with pNIPAM and subsequently conjugated with the organic dye FluoProbe532A (acting as an energy acceptor). Here, the use of pNIPAM as a temperature responsive linker between the UCNPs and the dye molecules adds additional thermal sensitivity to the system. Most importantly, it was shown that with the increase of the temperature, the collapse of the pNIPAM polymer chains led to closer proximity between the UCNP/dye pair (Figure 11.10a) that eventually results in the decrease of the green emission intensity due to more effective energy transfer processes (Figure 11.10b). In other words, the efficiency of the energy transfer between the donor UCNP and the acceptor dye is controlled by the spatial distance between both moieties, which is itself controlled by the temperature dependent collapsing

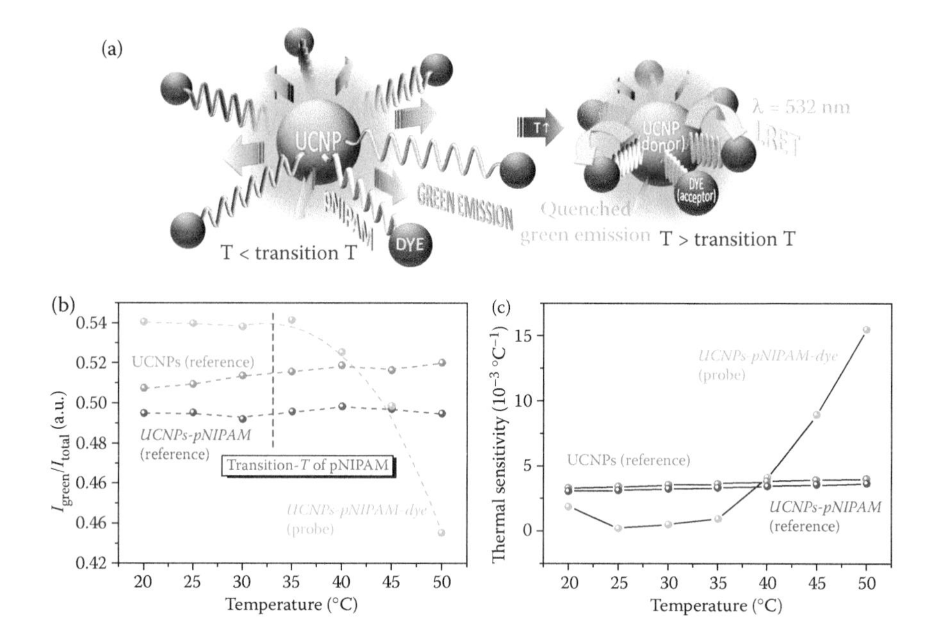

FIGURE 11.10
(a) Scheme of the LRET system based on Er^{3+}/Yb^{3+} co-doped $NaGdF_4$ UCNPs, thermoresponsive polymer pNIPAM and organic dye FluoProbe532A. (b) *T*-dependence of the relative green emission from Er^{3+}/Yb^{3+} co-doped $NaGdF_4$ UCNPs (energy donors) modified with pNIPAM and conjugated with an organic dye (energy acceptor) under 980 nm laser excitation. The *LIR* was defined as the ratio between the green emission intensity (wavelength region: 496–580 nm) versus the total upconversion emission (wavelength region: 300–840 nm). (c) Estimated thermal sensitivity of the UCNPs–pNIPAM–dye probe (reference probes: pNIPAM modified UCNPs without dye and unmodified UCNPs). (Reprinted with permission from Hemmer, E. et al. 2015. Temperature-induced energy transfer in dye-conjugated upconverting nanoparticles: A new candidate for nanothermometry. *Chem. Mater.* 27, 235–44. Copyright 2015 American Chemical Society.)

or stretching of the pNIPAM chains. Estimation of the thermal sensitivity revealed a significant increase when reaching temperatures above the polymer's transition temperature of 35°C (which is in close proximity to a *T* range of biological interest) compared to UCNPs without the dye acceptor unit (Figure 11.10c). The use of a thermoresponsive polymer with a tailored transition temperature below 35°C is expected to induce a shift of the observed improved thermal sensitivity toward the physiologically highly interesting range of ~37°C.

11.4.2 Multifunctional Nanoplatforms for Optical Heating and Thermal Sensing

The effect of optical heating in Ln^{3+}-doped UCNPs has already been mentioned in the previous sections of this chapter. We could see that, dependent on the conditions under which the nanothermometers are applied, this heating effect must be considered as good (e.g., when aiming for controlled heating at the sub-micron scale) or evil (e.g., when affecting the accuracy of the thermometer or inducing death of healthy cells). During the last few years, light-induced thermal heating has gained particular interest in the biomedical field since photothermal methods open new therapeutic approaches (Cheng et al. 2014; Jaque et al. 2014a; Melamed et al. 2015; Shanmugam et al. 2014). Furthermore, aiming for controlled heating at a (sub-)cellular level, plasmon-induced nanoheaters were recognized as attractive candidates. For the application of such nanoheaters in nanomedicine, high spatial (10^{-6} m) and temperature (10^{-1} degree) resolution is most crucial and still challenging (Debasu et al. 2013). This has a twofold advantage where (i) careful monitoring of the thermal profile guarantees strong enough heating in regions where desired, for instance in tumorous tissues, while (ii) simultaneously providing the spatial limitation to exactly the region requiring treatment in order to avoid damage of healthy tissues and cells in the vicinity. In this context, the combination of Ln^{3+}-doped nanoparticles as nanothermometers with gold nanostructures as nanoheaters has been suggested. For example, Rocha et al. (2013) provided the first demonstration of a single-beam sub-tissue-controlled heating process (sub-tissue hyperthermia process) based on the combination of Nd^{3+}:LaF_3 nanoparticles and gold nanorods (GNRs) that act as nanothermometers, and nanoheaters respectively. Shortly after, the decoration of Er^{3+}/Yb^{3+} co-doped Gd_2O_3 nanorods with gold nanoparticles (AuNPs) as an all-in-one nanoplatform was suggested by Debasu et al. (2013). This approach combined for the first time nanoheaters and nanothermometers in one composite nanoassembly (Figure 11.11a). When the heaters and the sensors are separate in space, for instance when the individual components are dispersed together in solution, an average temperature of the sample volume under irradiation is measured since the recorded emission spectra include the contribution of those emitting particles that are away from the heaters, and thus undergo less heating than those particles in close proximity to the heaters. However, when the heaters and the sensors constitute a single integrating nanoplatform, such as in the presented nanoassembly, the absolute local temperature can be measured as the heaters and sensors are not spatially separated from each other. In these nanoassemblies, the surface temperature detected by the upconverting nanorods was adjusted by controlling the amount of AuNPs on their surface, where a higher concentration of AuNPs induced a more pronounced heating effect. Concerning the heating mechanism, evidence was found for direct (980 nm, used for excitation of the Er^{3+}) and multiphoton absorption, probably involving interband transitions of Au. Aside, heating was assigned to the Au absorption of the green Er^{3+} emission (510 to 565 nm), which is in resonance with the localized surface plasmon resonance of Au. The local temperature was then monitored with the help of the *LIR* technique using the Boltzmann's distribution of the intensity ratio between the ${}^2H_{11/2} \rightarrow {}^4I_{15/2}$ and ${}^4S_{3/2} \rightarrow {}^4I_{15/2}$ Er^{3+} transitions.

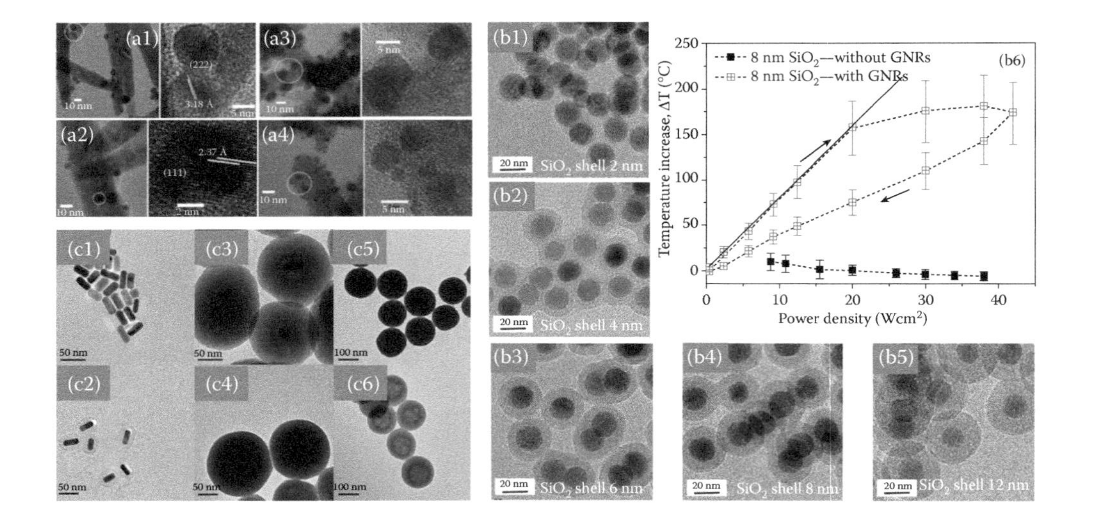

FIGURE 11.11

Selected nanoplatforms for optical heating and thermal sensing based on Ln^{3+} upconverting and gold nanostructures. (a) Transmission electron microscopy (TEM) images of an all-in-one Er^{3+}/Yb^{3+}:Gd_2O_3-AuNPs nanocomposites with different nominal gold amount: (a1) 1.25, (a2) 2.5, (a3) 12.5, and (a4) 25 μmol Au. The images on the right-hand side magnify the regions depicted by the white circles on the left. (Debasu, M. L. et al.: All-in-one optical heater-thermometer nanoplatform operative from 300 to 2000 K based on Er^{3+} emission and blackbody radiation. *Adv. Mater.* 2013. 25. 4868–74. Copyright Wiley-VCH Verlag GmbH & Co. KGaA. Reproduced with permission.) (b) Plasmonic heating and thermal sensing based on Er^{3+}/Yb^{3+} co-doped $NaGdF_4$and GNRs: (b1–b5) TEM image of Er^{3+}/Yb^{3+}:$NaGdF_4$ with different SiO_2 shell thicknesses of 2–12 nm. (b6) Calculated temperatures at different excitation powers for 8 nm SiO_2-coated samples with and without GNRs. The dashed line is added to guide the eye. The straight line is the theoretical fit of the data before deformation of the GNRs (Rohani, S. et al.: Enhanced luminescence, collective heating, and nanothermometry in an ensemble system composed of lanthanide-doped upconverting nanoparticles and gold nanorods. *Adv. Opt. Mater.* 2015. 3. 1606–13. Copyright Wiley-VCH Verlag GmbH & Co. KGaA. Reproduced with permission.) (c) Nanoheating and -sensing GNR-Er^{3+}/Yb^{3+}:$NaYF_4$ nanocomposite: TEM images of (c1) GNRs, (c2) SiO_2-coated GNRs, (c3) GNRs with a SiO_2 and Er^{3+}/Yb^{3+}:$Y(OH)CO_3$ shell, (c4) GNRs with a SiO_2 and Er^{3+}/Yb^{3+}:$NaYF_4$ shell, (c5) Er^{3+}/Yb^{3+}:$Y(OH)CO_3$, and (c6) Er^{3+}/Yb^{3+}:$NaYF_4$ hollow nanoshells. (Huang, Y., Federico R., and Fiorenzo V. 2015. A single multifunctional nanoplatform based on upconversion luminescence and gold nanorods. *Nanoscale* 7:5178–85. Reproduced by permission of The Royal Society of Chemistry.)

Obtained thermal resolutions reached from 0.3 to 2.0 K in the range 300–1050 K including the biologically relevant temperature range.

Further, in the search for methods to improve the photoluminescence efficiency of UCNPs, plasmon enhancement by use of Au nanostructures has been reported. In fact, GNRs create a localized electromagnetic field that can enhance the emission intensity from UCNPs. Here, the distance between the upconverter and the Au nanostructure is an essential parameter. Very recently, Rohani et al. (2015) developed a nanothermometry–nanoheating platform based on SiO_2-coated Er^{3+}/Yb^{3+} co-doped $NaGdF_4$ UCNPs and GNRs. The longitudinal surface plasmon resonance of the GNRs was tuned to 980 nm, in resonance with the Yb^{3+} absorption wavelength, so the GNRs and the UCNPs can be simultaneously excited. Moreover, the variation of the SiO_2 shell thicknesses allowed tuning of the minimum distance between the UCNPs and the GNR (Figure 11.11b1–5). Investigating the effect of the SiO_2 shell thickness on the plasmon-induced upconversion enhancement, an optimized shell thickness of 8 nm was found. Besides, excitation with 980 nm triggers heat generation from the GNRs that is controlled by the chosen laser power and monitored by the *LIR* technique using the $^2H_{11/2} \rightarrow {}^4I_{15/2}$ and $^4S_{3/2} \rightarrow {}^4I_{15/2}$ emission bands of Er^{3+} ions. While no heat increase was observed in SiO_2-coated UCNPs without GNRs, regardless of the applied excitation power, a temperature increase by 4°C to 170°C was observed in the sample consisting of SiO_2-coated UCNPs and GNRs upon increase of the laser power density from 2 W cm^{-2} to 42 W cm^{-2} (Figure 11.11b6). Up to a power density of 20 W cm^{-2}, linear behavior of the temperature increase was observed, while reduced heat generation was found for higher power densities. This observation was assigned to the deformation of GNRs at elevated temperatures, which is an important finding in terms of practical applications of nanoheaters for the higher temperature range. Overall, upconversion enhancement and effective heating performance of GNRs as well as *in situ* thermal sensing with UCNPs was demonstrated, providing a new perspective to nanoscale sensors that can be potentially used in photothermal therapy.

A different strategy to combine plasmonic heating and optical nanothermometry presented by Y. Huang et al. (2015) consists of a multifunctional nanoplatform composed of a GNR core and an outer hollow shell of upconverting Er^{3+}/Yb^{3+} co-doped $NaYF_4$ (Figure 11.11c). In this approach, the aspect ratio of the GNRs corresponds to a longitudinal surface plasmon resonance at 654 nm, which is in good overlap with the 980 nm induced red $^4F_{9/2} \rightarrow {}^4I_{15/2}$ emission from Er^{3+} ions doped into the outer shell. Thus, upon optical illumination with 980 nm, the upconverting shell emits light of ~660 nm wavelength that subsequently is absorbed by the GNRs due to surface plasmon resonance and eventually triggers the rapid plasmonic heat formation in the GNRs. The energy transfer from the excited Er^{3+} to the GNRs was further obvious from a quenching of the red emission band in the upconversion spectrum by 61.8% when compared to nanoshell Er^{3+}/Yb^{3+}:$NaYF_4$ structures without GNRs. Simultaneously, the thermal change was monitored by use of the green emission bands ($^2H_{11/2} \rightarrow {}^4I_{15/2}$ and $^4S_{3/2} \rightarrow {}^4I_{15/2}$) from Er^{3+} ions following the *LIR* technique revealing a temperature increase by 9°C upon excitation of the GNRs-Er^{3+}/Yb^{3+}:$NaYF_4$ core–shell system with 980 nm. Finally, the authors went one step further evaluating the potential of their nanostructure for drug delivery. Therefore, the anticancer drug doxorubicin was loaded into the GNRs-Er^{3+}/Yb^{3+}:$NaYF_4$ nanocomposites and the drug release profile evaluated. It was found that the photothermal effect (thus increased temperature) as described before, could trigger fast drug release, especially at low pH values, conditions that are found in tumor extracellular environments, which makes the suggested strategy promising for applications in cancer therapy.

11.5 Conclusions

In conclusion, since the first mention of Ln^{3+}-based nanothermometry in 2002 by Wang et al. (2002) the field of optical nanothermometry has clearly seen significant impact through the development of novel dopant/host combinations for improved sensitivity in application-oriented temperature ranges, assembly of multifunctional nanoplatforms allowing for simultaneous heating, thermal sensing and imaging, and first proofs of concept by *in vivo* and *ex vivo* applications.

Challenges that will have to be addressed by future studies include simultaneous control over the thermal and spatial resolution, especially when aiming for nanothermometry at a sub-cellular level in the biomedical field. Previous achievements raise high expectations that lanthanide-doped UCNPs provide the necessary performance to fulfill this task. Yet, for biomedical applications, an additional vital aspect is the nanothermometers' biocompatibility. The various strategies applying surface modification, for instance with poly(ethylene glycol), amine groups, or SiO_2, (Hemmer et al. 2013a, b, 2014; Ju et al. 2012; Shan et al. 2008; Zhang et al. 2015) have been followed to address potential toxicity issues of UCNPs. However, surface modification of UCNPs raises new questions such as its influence on thermal performance of the sensors or its stability when optical heating applies. As shown, uncontrolled optical heating leading to thermally induced damage of healthy biological tissue can be overcome by the choice of the appropriate excitation wavelength. Moreover, NIR light in the biological window has been recognized as advantageous over visible light for bioimaging applications (Diao et al. 2015; Hemmer et al. 2013a, 2016; Yao et al. 2014) since it poses advanced penetration capabilities through biological tissues due to reduced scattering and absorption by water and biological tissues. Consequently, NIR-based nanothermometers will be of particular interest for *in vivo* applications where exciting and emitted light have to pass through thicker layers of various biological tissues.

This led to the first application of the concept of nanothermometry in the NIR region. Based on the very promising results from these early studies, a future shift from upconversion-based to NIR light-based nanothermometry may be likely. Finally, the combination of obtained sensing and heating nanostructures with other modalities such as the potential for magnetic imaging or drug delivery will result in truly multimodal platforms.

Overall, the striking recent challenges of today's nanothermometers, by continuously expanding on the important results previously obtained in the field of lanthanide-based nanothermometry, is expected to significantly contribute to the development of innovative non-toxic multimodal theranostic (combining diagnostic and therapeutic approaches) materials and devices.

References

Anderson, R. R., and J. A. Parrish. 1981. The optics of human-skin. *J. Invest. Dermatol.* 77:13–19.

Auzel, F. 2004. Upconversion and anti-Stokes processes with f and d ions in solids. *Chem. Rev.* 104:139–73.

Barnes, N. P. 2004. Lanthanide series lasers—Near infrared. In *Handbook of Laser Technology and Applications: Laser Design and Laser Systems*, edited by C. E. Webb, and J. D. C. Jones, 383–410. Boca Raton: CRC Press.

Bhaumik, J., A. K. Mittal, A. Banerjee, Y. Chisti, and U. C. Banerjee. 2015. Applications of phototheranostic nanoagents in photodynamic therapy. *Nano Res.* 8:1373–94.

Blasse, G, and B. C. Grabmaier. 1994. *Luminescent Materials.* Berlin, Heidelberg: Springer-Verlag.

Brites, C. D., P. P. Lima, N. J. Silva et al. 2012. Thermometry at the nanoscale. *Nanoscale* 4:4799–829.

Brites, C. D. S., P. P. Lima, N. J. O. Silva et al. 2011. Lanthanide-based luminescent molecular thermometers. *New J. Chem.* 35:1177–83.

Cao, B. S., Y. Y. He, Z. Q. Feng, Y. S. Li, and B. Dong. 2011. Optical temperature sensing behavior of enhanced green upconversion emissions from Er-Mo:$Yb_2Ti_2O_7$ nanophosphor. *Sens. Actuators B* 159:8–11.

Carrasco, E., B. del Rosal, F. Sanz-Rodriguez et al. 2015. Intratumoral thermal reading during photo-thermal therapy by multifunctional fluorescent nanoparticles. *Adv. Funct. Mater.* 25:615–26.

Ceron, E. N., D. H. Ortgies, B. del Rosal et al. 2015. Hybrid nanostructures for high-sensitivity luminescence nanothermometry in the second biological window. *Adv. Mater.* 27:4781–87.

Chen, D. Q., Z. Y. Wan, Y. Zhou et al. 2015. Bulk glass ceramics containing Yb^{3+}/Er^{3+}:beta-$NaGdF_4$ nanocrystals: Phase-separation-controlled crystallization, optical spectroscopy and upconverted temperature sensing behavior. *J. Alloy. Compd.* 638:21–28.

Chen, R., V. D. Ta, F. Xiao, Q. Zhang, and H. Sun. 2013. Multicolor hybrid upconversion nanoparticles and their improved performance as luminescence temperature sensors due to energy transfer. *Small* 9:1052–57.

Cheng, L., C. Wang, L. Z. Feng, K. Yang, and Z. Liu. 2014. Functional nanomaterials for phototherapies of cancer. *Chem. Rev.* 114:10869–939.

Chouahda, Z., J. P. Jouart, T. Duvaut, and M. Diaf. 2009. The use of the green emission in Er^{3+}-doped CaF_2 crystals for thermometry application. *J. Phys.: Condens. Mat.* 21:5.

Cubeddu, R., D. Comelli, C. D'Andrea, P. Taroni, and G. Valentini. 2002. Time-resolved fluorescence imaging in biology and medicine. *J. Phys. D* 35:R61–R76.

Debasu, M. L., D. Ananias, I. Pastoriza-Santos et al. 2013. All-in-one optical heater-thermometer nanoplatform operative from 300 to 2000 K based on Er^{3+} emission and blackbody radiation. *Adv. Mater.* 25:4868–74.

Diao, S., G. Hong, A. L. Antaris et al. 2015. Biological imaging without autofluorescence in the second near-infrared window. *Nano Res.* 8:3027–34.

Dieke, G. H. 1968. *Spectra and Energy Levels of Rare Earth Ions in Crystals.* New York: Interscience Publishers.

Dong, B., B. Cao, Y. He et al. 2012. Temperature sensing and *in vivo* imaging by molybdenum sensitized visible upconversion luminescence of rare-earth oxides. *Adv. Mater.* 24:1987–93.

Dong, B., R. N. Hua, B. S. Cao et al. 2014. Size dependence of the upconverted luminescence of $NaYF_4$:Er,Yb microspheres for use in ratiometric thermometry. *Phys. Chem. Chem. Phys.* 16:20009–12.

Dong, N. N., M. Pedroni, F. Piccinelli et al. 2011. NIR-to-NIR two-photon excited CaF_2: Tm^{3+},Yb^{3+} nanoparticles: Multifunctional nanoprobes for highly penetrating fluorescence bio-imaging. *ACS Nano* 5:8665–71.

dos Santos, P. V., M. T. de Araujo, A. S. Gouveia-Neto, J. A. M. Neto, and A. S. B. Sombra. 1999. Optical thermometry through infrared excited upconversion fluorescence emission in Er^{3+}- and Er^{3+}-Yb^{3+}-doped chalcogenide glasses. *IEEE J. Quantum Electron.* 35:395–99.

Du, P., L. H. Luo, and J. S. Yu. 2015a. Low-temperature thermometry based on upconversion emission of Ho/Yb-codoped $Ba_{0.77}Ca_{0.23}TiO_3$ ceramics. *J. Alloy. Compd.* 632:73–77.

Du, P., L. H. Luo, Q. Y. Yue, and W. P. Li. 2015b. The simultaneous realization of high- and low-temperature thermometry in Er^{3+}/Yb^{3+}-codoped Y_2O_3 nanoparticles. *Mater. Lett.* 143:209–11.

Du, P. and J. S. Yu. 2015. Effect of molybdenum on upconversion emission and temperature-sensing properties in $Na_{0.5}Bi_{0.5}TiO_3$:Er/Yb ceramics. *Ceram. Int.* 41:6710–14.

Gavrilovic, T. V., D. J. Jovanovic, V. Lojpur, and M. D. Dramicanin. 2014. Multifunctional Eu^{3+}- and Er^{3+}/Yb^{3+}-doped $GdVO_4$ nanoparticles synthesized by reverse micelle method. *Sci. Rep.* 4:9.

Hemmer, E., A. Benayas, F. Légaré, and F. Vetrone. 2016. Exploiting the biological windows: Current perspectives on fluorescent bioprobes emitting above 1000 nm. *Nanoscale Horiz.* 1:168–84.

Hemmer, E., M. Quintanilla, F. Légaré, and F. Vetrone. 2015. Temperature-induced energy transfer in dye-conjugated upconverting nanoparticles: A new candidate for nanothermometry. *Chem. Mater.* 27:235–44.

Hemmer, E., N. Venkatachalam, H. Hyodo et al. 2013a. Upconverting and NIR emitting rare earth based nanostructures for NIR-bioimaging. *Nanoscale* 5:11339–61.

Hemmer, E., F. Vetrone, and K. Soga. 2014. Lanthanide-based nanostructures for optical bioimaging: Small particles with large promise. *MRS Bull.* 39:960–64.

Hemmer, E., T. Yamano, H. Kishimoto et al. 2013b. Cytotoxic aspects of gadolinium oxide nanostructures for up-conversion and NIR bioimaging. *Acta Biomater.* 9:4734–43.

Huang, F., Y. Gao, J. C. Zhou, J. Xua, and Y. S. Wang. 2015. Yb^{3+}/Er^{3+} co-doped $CaMoO_4$: A promising green upconversion phosphor for optical temperature sensing. *J. Alloy. Compd.* 639:325–29.

Huang, J. C., B. Z. He, Z. H. Cheng, and L. Zhou. 2015. Upconverting PAAm/PNIPAM/$NaYF_4$:Yb:Er hydrogel with enhanced luminescence temperature sensitivity. *J. Lumin.* 160:254–57.

Huang, Y., F. Rosei, and F. Vetrone. 2015. A single multifunctional nanoplatform based on upconversion luminescence and gold nanorods. *Nanoscale* 7:5178–85.

Idris, N. M., M. K. G. Jayakumar, A. Bansal, and Y. Zhang. 2015. Upconversion nanoparticles as versatile light nanotransducers for photoactivation applications. *Chem. Soc. Rev.* 44:1449–78.

Jaque, D., L. M. Maestro, E. Escudero et al. 2013. Fluorescent nano-particles for multiphoton thermal sensing. *J. Lumin.* 133:249–53.

Jaque, D., L. Martinez Maestro, B. del Rosal et al. 2014a. Nanoparticles for photothermal therapies. *Nanoscale* 6:9494–530.

Jaque, D., B. del Rosal, E. M. Rodriguez et al. 2014b. Fluorescent nanothermometers for intracellular thermal sensing. *Nanomedicine* 9:1047–62.

Jaque, D. and F. Vetrone. 2012. Luminescence nanothermometry. *Nanoscale* 4:4301–26.

Jiang, S., P. Zeng, L. Q. Liao et al. 2014. Optical thermometry based on upconverted luminescence in transparent glass ceramics containing $NaYF_4:Yb^{3+}/Er^{3+}$ nanocrystals. *J. Alloy. Compd.* 617:538–41.

Ju, Q., D. Tu, Y. Liu et al. 2012. Amine-functionalized lanthanide-doped $KGdF_4$ nanocrystals as potential optical/magnetic multimodal bioprobes. *J. Am. Chem. Soc.* 134:1323–30.

Klier, D. T. and M. U. Kumke. 2015a. Upconversion luminescence properties of $NaYF_4$:Yb:Er nanoparticles codoped with Gd^{3+}. *J. Physi. Chem. C* 119:3363–73.

Klier, D. T. and M. U. Kumke. 2015b. Upconversion $NaYF_4$:Yb:Er nanoparticles co-doped with Gd^{3+} and Nd^{3+} for thermometry on the nanoscale. *RSC Adv.* 5:67149–56.

Klimov, N. N., S. Mittal, M. Berger, and Z. Ahmed. 2015. On-chip silicon waveguide Bragg grating photonic temperature sensor. *Opt. Lett.* 40:3934–36.

Lai, B. Y., L. Feng, J. Wang, and Q. A. Su. 2010. Optical transition and upconversion luminescence in Er^{3+} doped and Er^{3+}-Yb^{3+} co-doped fluorophosphate glasses. *Opt. Mater.* 32:1154–60.

Lakowicz, J. R., H. Szmacinski, K. Nowaczyk, K. W. Berndt, and M. Johnson. 1992. Fluorescence lifetime imaging. *Anal. Biochem.* 202:316–30.

Lee, D. E., H. Koo, I. C. Sun et al. 2012. Multifunctional nanoparticles for multimodal imaging and theragnosis. *Chem. Soc. Rev.* 41:2656–72.

Li, D. D., Q. Y. Shao, Y. Dong, and J. Q. Jiang. 2014. Temperature sensitivity and stability of $NaYF_4:Yb^{3+}$, Er^{3+} core-only and core–shell upconversion nanoparticles. *J. Alloy. Compd.* 617:1–6.

Li, T., C. F. Guo, S. S. Zhou, C. K. Duan, and M. Yin. 2015. Highly sensitive optical thermometry of Yb^{3+}-Er^{3+} codoped $AgLa(MoO_4)_2$ green upconversion phosphor. *J. Am. Ceram. Soc.* 98:2812–16.

Liu, G. F., L. L. Fu, Z. Y. Gao et al. 2015. Investigation into the temperature sensing behavior of Yb^{3+} sensitized Er^{3+} doped Y_2O_3, YAG and $LaAlO_3$ phosphors. *RSC Adv.* 5:51820–27.

Liu, L., Y. J. Chen, X. L. Zhang, Z. G. Zhang, and Y. X. Wang. 2013. Improved optical thermometry in Er^{3+}:Y_2O_3 nanocrystals by re-calcination. *Opt. Commun.* 309:90–94.

Liu, L., Y. X. Wang, X. R. Zhang et al. 2011. Optical thermometry through green and red upconversion emissions in $Er^{3+}/Yb^{3+}/Li^{+}$:ZrO_2 nanocrystals. *Opt. Commun.* 284:1876–79.

Liu, Y. S., D. T. Tu, H. M. Zhu, E. Ma, and X. Y. Chen. 2013. Lanthanide-doped luminescent nano-bioprobes: From fundamentals to biodetection. *Nanoscale* 5:1369–84.

Lojpur, V., M. Nikolic, L. Mancic, O. Milosevic, and M. D. Dramicanin. 2013. Y_2O_3:Yb,Tm and Y_2O_3:Yb,Ho powders for low-temperature thermometry based on up-conversion fluorescence. *Ceram. Int.* 39:1129–34.

Maciel, G. S., M. Alencar, C. B. de Araujo, and A. Patra. 2010. Upconversion emission of $BaTiO_3$:Er^{3+} nanocrystals: Influence of temperature and surrounding medium. *J. Nanosci. Nanotechnol.* 10:2143–48.

Mahata, M. K., K. Kumar, and V. K. Rai. 2015. Er^{3+}-Yb^{3+} doped vanadate nanocrystals: A highly sensitive thermographic phosphor and its optical nanoheater behavior. *Sens. Actuators B* 209:775–80.

Marciniak, L., A. Bednarkiewicz, M. Stefanski et al. 2015. Near infrared absorbing near infrared emitting highly-sensitive luminescent nanothermometer based on Nd^{3+} to Yb^{3+} energy transfer. *Phys. Chem. Chem. Phys.* 17:24315–21.

Mecklenburg, M., W. A. Hubbard, E. R. White et al. 2015. Nanoscale temperature mapping in operating microelectronic devices. *Science* 347:629–32.

Mednitz, I. and N. Hildebrandt. 2014. *FRET—Förster Resonance Energy Transfer, from Theory to Applications*. Weinheim, Germany: Wiley-VCH.

Melamed, J. R., R. S. Edelstein, and E. S. Day. 2015. Elucidating the fundamental mechanisms of cell death triggered by photothermal therapy. *ACS Nano* 9:6–11.

Pandey, A. and V. K. Rai. 2013. Optical thermometry using FIR of two close lying levels of different ions in Y_2O_3:Ho^{3+}-Tm^{3+}-Yb^{3+} phosphor. *Appl. Phys. B* 113:221–25.

Pandey, A., V. K. Rai, V. Kumar, V. Kumar, and H. C. Swart. 2015. Upconversion based temperature sensing ability of Er^{3+}-Yb^{3+} codoped $SrWO_4$: An optical heating phosphor. *Sens. Actuators B* 209:352–58.

Pandey, A., S. Som, V. Kumar et al. 2014. Enhanced upconversion and temperature sensing study of Er^{3+}-Yb^{3+} codoped tungsten-tellurite glass. *Sens. Actuators B* 202:1305–12.

Peng, D. F., X. S. Wang, C. N. Xu et al. 2012. Bright upconversion luminescence and increased Tc in $CaBi_2Ta_2O_9$:Er high temperature piezoelectric ceramics. *J. Appl. Phys.* 111:5.

Pereira, A. F., K. U. Kumar, W. F. Silva et al. 2015. Yb^{3+}/Tm^{3+} co-doped $NaNbO_3$ nanocrystals as three-photon-excited luminescent nanothermometers. *Sens. Actuators B* 213:65–71.

Perpiñà, X., M. Vellvehi, and X. Jordà 2016. Thermal issues in microelectronics. In *Thermometry at the Nanoscale: Techniques and Selected Applications*, edited by. L. D. Carlos, and F. Palacio, 383–436. Cambridge: The Royal Society of Chemistry.

Prodi, L., E. Rampazzo, F. Rastrelli, A. Speghini, and N. Zaccheroni. 2015. Imaging agents based on lanthanide doped nanoparticles. *Chem. Soc. Rev.* 44:4922–52.

Quintanilla, M., A. Benayas, R. Naccache, and F. Vetrone. 2016. Luminescent nanothermometry with lanthanide-doped nanoparticles. In *Thermometry at the Nanoscale: Techniques and Selected Applications*, edited by L. D. Carlos, and F. Palacio, 124–66. Cambridge: The Royal Society of Chemistry.

Quintanilla, M., I. X. Cantarelli, M. Pedroni, A. Speghini, and F. Vetrone. 2015. Intense ultraviolet upconversion in water dispersible SrF_2: Tm^{3+}, Yb^{3+} nanoparticles: The effect of the environment on light emissions. *J. Mater. Chem. C* 3:3108–13.

Quintanilla, M., E. Cantelar, F. Cussó, M. Villegas, and A. C. Caballero. 2011. Temperature sensing with up-converting submicron-sized $LiNbO_3$:Er^{3+}/Yb^{3+} particles. *Appl. Phys. Express* 4:022601.

Rakov, N., and G. S. Maciel. 2012. Three-photon upconversion and optical thermometry characterization of Er^{3+}:Yb^{3+} co-doped yttrium silicate powders. *Sens. Actuators B* 164:96–100.

Rocha, U., C. Jacinto, W. F. Silva et al. 2013. Subtissue thermal sensing based on neodymium-doped LaF_3 nanoparticles. *ACS Nano* 7:1188–99.

Rocha, U., K. U. Kumar, C. Jacinto et al. 2014. Nd^{3+} doped LaF_3 nanoparticles as self-monitored photo-thermal agents. *Appl. Phys. Lett.* 104:053703.

Rohani, S., M. Quintanilla, S. Tuccio et al. 2015. Enhanced luminescence, collective heating, and nanothermometry in an ensemble system composed of lanthanide-doped upconverting nanoparticles and gold nanorods. *Adv. Opt. Mater.* 3:1606–13.

Rothamer, D. A., and J. Jordan. 2012. Planar imaging thermometry in gaseous flows using upconversion excitation of thermographic phosphors. *Appl. Phys. B* 106:435–44.

Savchuk, O. A., J. J. Carvajal, M. C. Pujol et al. 2015. Ho,Yb:$KLu(WO_4)_2$ nanoparticles: A versatile material for multiple thermal sensing purposes by luminescent thermometry. *J. Phys. Chem. C* 119:18546–58.

Savchuk, O. A., P. Haro-Gonzalez, J. J. Carvajal et al. 2014. Er:Yb:NaY_2F_5O up-converting nanoparticles for sub-tissue fluorescence lifetime thermal sensing. *Nanoscale* 6:9727–33.

Sayoud, A., J. P. Jouart, N. Trannoy, M. Diaf, and T. Duvaut. 2012. Temperature measurements inside an Er^{3+}-Yb^{3+} co-doped fluoride crystal heated by a NIR laser diode and probed by red-to-green upconversion. *J. Lumin.* 132:566–69.

Schartner, E. P., and T. M. Monro. 2014. Fibre tip sensors for localised temperature sensing based on rare earth-doped glass coatings. *Sensors* 14:21693–701.

Shan, J., J. Chen, J. Meng et al. 2008. Biofunctionalization, cytotoxicity, and cell uptake of lanthanide doped hydrophobically ligated $NaYF_4$ upconversion nanophosphors. *J. Appl. Phys.* 104:094308.

Shanmugam, V., S. Selvakumar, and C. S. Yeh. 2014. Near-infrared light-responsive nanomaterials in cancer therapeutics. *Chem. Soc. Rev.* 43:6254–87.

Singh, A. K. 2007. Ho^{3+}:TeO_2 glass, a probe for temperature measurements. *Sens. Actuators A*136:173–77.

Singh, A. K., P. K. Shahi, S. B. Rai, and B. Ullrich. 2015. Host matrix impact on Er^{3+} upconversion emission and its temperature dependence. *RSC Adv.* 5:16067–73.

Singh, S. K., K. Kumar, and S. B. Rai. 2009. Er^{3+}/Yb^{3+} codoped Gd_2O_3 nano-phosphor for optical thermometry. *Sens. Actuators A* 149:16–20.

Smith, A. M., M. C. Mancini, and S. Nie. 2009. Bioimaging: Second window for *in vivo* imaging. *Nat. Nanotechnol.* 4:710–11.

Soni, A. K., R. Dey, and V. K. Rai. 2015. Stark sublevels in Tm^{3+}-Yb^{3+} codoped $Na_2Y_2B_2O_7$ nanophosphor for multifunctional applications. *RSC Adv.* 5:34999–5009.

Sordillo, L. A., Y. Pu, S. Pratavieira, Y. Budansky, and R. R. Alfano. 2014. Deep optical imaging of tissue using the second and third near-infrared spectral windows. *J. Biomed. Opt.* 19:056004.

Verma, R. K., and S. B. Rai. 2012. Laser induced optical heating from Yb^{3+}/Ho^{3+}:$Ca_{12}Al_{14}O_{33}$ and its applicability as a thermal probe. *J. Quant. Spectrosc. Radiat. Transfer* 113:1594–600.

Vetrone, F., R. Naccache, V. Mahalingam, C. G. Morgan, and J. A. Capobianco. 2009. The active-core/active-shell approach: A strategy to enhance the upconversion luminescence in lanthanide-doped nanoparticles. *Adv. Funct. Mater.* 19:2924–29.

Vetrone, F., R. Naccache, A. Zamarron et al. 2010. Temperature sensing using fluorescent nanothermometers. *ACS Nano* 4:3254–58.

Wang, G. F., Q. Peng, and Y. D. Li. 2011. Lanthanide-doped nanocrystals: Synthesis, optical-magnetic properties, and applications. *Accounts Chem. Res.* 44:322–32.

Wang, R., X. L. Zhang, F. Liu, Y. J. Chen, and L. Liu. 2015. Concentration effects on the FIR technique for temperature sensing. *Opt. Mater.* 43:18–24.

Wang, S. P., S. Westcott, and W. Chen. 2002. Nanoparticle luminescence thermometry. *J. Phys. Chem. B* 106:11203–09.

Wang, X. D., O. S. Wolfbeis, and R. J. Meier. 2013. Luminescent probes and sensors for temperature. *Chem. Soc. Rev.* 42:7834–69.

Wang, X., X. G. Kong, Y. Yu, Y. J. Sun, and H. Zhang. 2007. Effect of annealing on upconversion luminescence of ZnO:Er^{3+} nanocrystals and high thermal sensitivity. *J. Phys. Chem. C* 111:15119–24.

Wang, X., Q. Liu, Y. Bu et al. 2015. Optical temperature sensing of rare-earth ion doped phosphors. *RSC Adv.* 5:86219–36.

Wawrzynczyk, D., A. Bednarkiewicz, M. Nyk, W. Strek, and M. Samoc. 2012. Neodymium(III) doped fluoride nanoparticles as non-contact optical temperature sensors. *Nanoscale* 4:6959–61.

Wegh, R. T., A. Meijerink, R. J. Lamminmaki, and J. Holsa. 2000. Extending Dieke's diagram. *J. Lumin.* 87–89:1002–04.

Xing, L. L., Y. L. Xu, R. Wang, W. Xu, and Z. G. Zhang. 2014. Highly sensitive optical thermometry based on upconversion emissions in Tm^{3+}/Yb^{3+} codoped $LiNbO_3$ single crystal. *Opt. Lett.* 39:454–57.

Xu, W., X. Y. Gao, L. J. Zheng, Z. G. Zhang, and W. W. Cao. 2012a. Short-wavelength upconversion emissions in Ho^{3+}/Yb^{3+} codoped glass ceramic and the optical thermometry behavior. *Opt. Express* 20:18127–37.

Xu, W., Q. T. Song, L. J. Zheng, Z. G. Zhang, and W. W. Cao. 2014. Optical temperature sensing based on the near-infrared emissions from Nd^{3+}/Yb^{3+} codoped $CaWO_4$. *Opt. Lett.* 39:4635–38.

Xu, W., Z. G. Zhang, and W. W. Cao. 2012b. Excellent optical thermometry based on short-wavelength upconversion emissions in Er^{3+}/Yb^{3+} codoped $CaWO_4$. *Opt. Lett.* 37:4865–67.

Xu, W., H. Zhao, Y. X. Li et al. 2013a. Optical temperature sensing through the upconversion luminescence from Ho^{3+}/Yb^{3+} codoped $CaWO_4$. *Sens. Actuators B* 188:1096–100.

Xu, W., H. Zhao, Z. G. Zhang, and W. W. Cao. 2013b. Highly sensitive optical thermometry through thermally enhanced near infrared emissions from Nd^{3+}/Yb^{3+} codoped oxyfluoride glass ceramic. *Sens. Actuators B* 178:520–24.

Yang, W. F., X. Y. Li, D. Z. Chi, H. J. Zhang, and X. G. Liu. 2014. Lanthanide-doped upconversion materials: Emerging applications for photovoltaics and photocatalysis. *Nanotechnology* 25:16.

Yang, X. X., Z. L. Fu, Y. M. Yang et al. 2015. Optical temperature sensing behavior of high-efficiency upconversion: Er^{3+}-Yb^{3+} co-doped $NaY(MoO_4)_2$ phosphor. *J. Am. Ceram. Soc.* 98:2595–600.

Yang, Y. M., C. Mi, F. Y. Jiao et al. 2014. A novel multifunctional upconversion phosphor: Yb^{3+}/Er^{3+} codoped La_2S_3. *J. Am. Ceram. Soc.* 97:1769–75.

Yang, Y., C. Mi, F. Yu et al. 2014. Optical thermometry based on the upconversion fluorescence from Yb^{3+}/Er^{3+} codoped La_2O_2S phosphor. *Ceram. Int.* 40:9875–80.

Yao, J., M. Yang, and Y. X. Duan. 2014. Chemistry, biology, and medicine of fluorescent nanomaterials and related systems: New insights into biosensing, bioimaging, genomics, diagnostics, and therapy. *Chem. Rev.* 114:6130–78.

Zagumennyi, A. I., V. A. Mikhailov, and I. A. Shcherbakov. 2004. Rare earth ion lasers—Nd^{3+}. In *Handbook of Laser Technology and Applications: Laser Design and Laser Systems*, edited by C. E. Webb, and J. D. C. Jones, 353–82. Boca Raton: CRC Press.

Zhang, J. P., F. Y. Liu, T. Li, X. X. He, and Z. X. Wang. 2015. Surface charge effect on the cellular interaction and cytotoxicity of $NaYF_4$:Yb^{3+}, Er^{3+}@SiO_2 nanoparticles. *RSC Adv.* 5:7773–80.

Zheng, K. Z., Z. Y. Liu, C. J. Lv, and W. P. Qin. 2013. Temperature sensor based on the UV upconversion luminescence of Gd^{3+} in Yb^{3+}-Tm^{3+}-Gd^{3+} codoped $NaLuF_4$ microcrystals. *J. Mater. Chem. C* 1:5502–07.

Zheng, K. Z., W. Y. Song, G. H. He, Z. Yuan, and W. P. Qin. 2015. Five-photon UV upconversion emissions of Er^{3+} for temperature sensing. *Opt. Express* 23:7653–58.

Zheng, S. H., W. B. Chen, D. Z. Tan et al. 2014. Lanthanide-doped $NaGdF_4$ core–shell nanoparticles for non-contact self-referencing temperature sensors. *Nanoscale* 6:5675–79.

Zhou, H., M. Sharma, O. Berezin, D. Zuckerman, and M. Y. Berezin. 2016. Nanothermometry: From microscopy to thermal treatments. *Chem. Phys. Chem.* 17:27–36.

Zhou, S. S., G. C. Jiang, X. Y. Li et al. 2014a. Strategy for thermometry via Tm^{3+}-doped $NaYF_4$ core–shell nanoparticles. *Opt. Lett.* 39:6687–90.

Zhou, S. S., G. C. Jiang, X. T. Wei et al. 2014b. Pr^{3+}-doped beta-$NaYF_4$ for temperature sensing with fluorescence intensity ratio technique. *J. Nanosci. Nanotechnol.* 14:3739–42.

Zhou, S. S., S. Jiang, X. T. Wei et al. 2014c. Optical thermometry based on upconversion luminescence in Yb^{3+}/Ho^{3+} co-doped $NaLuF_4$. *J. Alloy. Compd.* 588:654–57.

Zou, H., X. S. Wang, Y. F. Hu et al. 2015. Optical thermometry based on the upconversion luminescence from Er doped $Bi_7Ti_4TaO_{21}$ ferroelectric ceramics. *J. Mater. Sci.: Mater. Electr.* 26:6502–05.

Index

A

Absorption, optimization of, 27–29
Activator distribution, shells allow optimization of, 215–218
Activator ions, 20, 69, 262, 283, 323, 324
Active core–active shell structure, enhancing UCL with, 178–183
  active shell containing Nd^{3+} sensitizers, 181–183
  active shell containing Yb^{3+} sensitizers, 178–181
Active-core@active-shell (ACAS), 198, 202, 212, 214, 239
Active-core–active-shell upconverting nanoparticles, 195–236
  background and driving forces, 196–198
  biomedical applications of UCNPs, 203–205
  conclusion, 236–239
  core–shell UCNPs, 212–221
    shell improves excitation schemes, 218–221
    shell increases functionality and biocompatibility, 212–214
    shell protects NPs from surface quenching, 214–215
    shells allow optimization of activator distribution, 215–218
    synthesis, properties, and issues, 221–236
  lanthanide-doped NPs, 198
    in biomedical sciences, 198–202
    issues and solutions, 206–212
Active core–inert shell structure, enhancing UCL with, 172–178
  heterogeneous core–shell UCNPs with enhanced UCL, 174–178
  homogeneous core–shell UCNPs with enhanced UCL, 172–174
Active ions concentration, increasing, 211–212
Active shell
  containing Nd^{3+} sensitizers, 181–183
  containing Yb^{3+} sensitizers, 178–181
Adamantaneacetic acid (ADAA), 114
Alkoxides, superconductors and, 10–11
Aminoethanephosphonic acid, 81
Aminophenylboronic acid (APBA), 125
Amphiphilic coatings, 72–74
Amphiphilic polymers, 74, 89, 112, 231, 312
"Antenna effect", 7
Anti-Stokes mechanisms, 196, 197
Anti-Stokes (upconversion) emission under NIR photoexcitation, 200
Apolar ligands, 111
Aptamers, 121, 264
Arrhenius, 3
Atomic number, 4
Auzel, Francois, 11, 196

B

Band-shape luminescence nanothermometry, 322
Bandwidth luminescence nanothermometry, 322
β-cyclodextrin (CD), 114
β-diketonate complexes, 7–9
Bidentate nitrates, 6
Bioimaging, 121–122, 197, 319–320
Biological window, 341
Biomedical applications, 37, 239, 332
  Er^{3+}-based upconverting nanothermometers for, 333–334
  lanthanide-doped NPs, 198–202
  multimodal, 344–350
  of UCNPs, 203–205
Boltzmann's plot, 326, 327, 330, 340, 347
Bragg's law, 4–5
Brauner, Bohuslav, 4
B16F0 murine melanoma cells, 263, 277

C

$CaGd_3F_{11}$ nanoparticles, 47
Calcium fluoride, 225
Carboxytetramethylrhodamine (TAMRA), 88
Carcinoembryonic antigen (CEA), 76–77
Cetyltrimethylammonium bromide (CTAB), 113, 230
Chemotherapy, drug delivery and, 126–129
Chitosan, 264
Color-coded multilayer photopatterned microstructures, 305
Combination therapy, UCNPs in, 279–281
Combustion synthesis, 51–52
Conjugated polyelectrolyte (CPE), 281
Cooperative sensitization upconversion (CSU), 20, 22–23
Cooperative upconversion (CUC), *see* Cooperative sensitization upconversion (CSU)
Coprecipitation method, 40–42
- fluorides, 41
- oxides, 41–42
- oxyfluorides, 42

Core@shell layer-by-layer structures, design of, 202
Core@shell nanocrystals, 55, 209
Core@shell NPs, 200, 214, 215, 228, 234, 236
Core–multishell nanostructure for enhanced UCL, 183–186
Core–multishell UCNPs, seed-mediated epitaxial growth of, 171
Core NPs, methods to confirm shell formation upon, 165, 169, 233–235
Core–shell architectures, 55–60
Core–shell UCNPs, 212–221
- crystalline homo-and hetero-structural shells, 232–233
- host materials for cores, 224–227
  - calcium fluoride, 225
  - metal oxide, 226–227
  - RE fluoride, 225–226
  - sodium fluoride, 224–225
- methods to confirm shell formation upon core NPs, 233–235
- nonhomogenous shell composition, 236
- properties, and issues, 221–236
- seed-mediated epitaxial growth of, 169–171
- shell improves excitation schemes, 218–221
- shell increases functionality and biocompatibility, 212–214
- shell protects NPs from surface quenching, 214–215
- shells allow optimization of activator distribution, 215–218
- shell synthesis methods and compositions, 227–229
- silica amorphous layer, 229–232
- synthesis, 221–223

Cores, host materials for, 224–227
- calcium fluoride, 225
- metal oxide, 226–227
- RE fluoride, 225–226
- sodium fluoride, 224–225

Cross-relaxation (CR) processes, 21–22, 204, 217–218
Crystalline homo-and hetero-structural shells, 232–233
Crystallographic structure tuning, 207–208
Cubic Er_2O_3 nanostructures, 48
CyTE-777-triethoxysilane, 90–91

D

Dickinson, R. G., 5
Dieke's diagram, 323, 324
Dimercaptosuccinic acid, 81
Direct ligand exchange, 112, 231
Direct luminescence quenching, 172
Down conversion (DC), 19, 196, 292
Doxorubicin (DOX), 209, 280, 350
Drug delivery, chemotherapy and, 126–129
Dyes, conjugation to, 87–94
Dye-sensitized nanoparticles, 29

E

EDTA, 5–6, 41
Electrical nanoimprint lithography (e-NIL), 304
Electron energy loss spectroscopy (EELS), 234, 235
Emission, optimization of, 32–34
Energy migration-induced quenching, 172
Energy migration-mediated upconversion (EMU), 20, 23–24, 219–221
Energy transfer, optimization of, 29–32
Energy transfer probability, 21
Energy transfer upconversion (ETU), 20, 21, 29, 104, 105, 196
Epitaxial core–shell nanostructures, upconversion enhancement, 163–187
  active core–active shell structure, enhancing UCL with, 178–183
    active shell containing Nd^{3+} sensitizers, 181–183
    active shell containing Yb^{3+} sensitizers, 178–181
  conclusion, 186–187
  core–multishell nanostructure for enhanced UCL, 183–186
  enhancing UCL with active core–inert shell structure, 172–178
    heterogeneous core–shell UCNPs with enhanced UCL, 174–178
    homogeneous core–shell UCNPs with enhanced UCL, 172–174
  introduction, 163–164
  nanochemistry for well-defined core–shell UCNPs, 165
    seed-mediated epitaxial growth of core–multishell UCNPs, 171
    seed-mediated epitaxial growth of core–shell UCNPs, 169–171
    synthesis of core UCNPs, nanochemistry for, 165–168
Er^{3+}-based nanothermometers, 329–333
Er^{3+}-based upconverting nanothermometers, 333–334
Ethylene diamine-tetra methylene phosphonic acid (EDTMP), 122
Europium complexes, 8
Excitation density, 28, 31
Excitation rate of state, 28
Excitation schemes, shell improves, 218–221
Excited state absorption (ESA), 20–21, 104–105, 196

F

Fluorescence imaging, 121
Fluorescence lifetime nanothermometers, 333
Fluorescent dyes, 70, 87, 89
Fluorides, 168
  coprecipitation method, 41
  solvo(hydro)thermal, 46–47
  thermolysis, 43–44
Folic acid (FA), 48, 78, 117, 263
Förster resonance energy transfer (FRET), 87, 200, 322
4f–4f intra-configurational transitions, 196

G

Gadolin, Johan, 3
Gadolinite, *see* Ytterbite
Gadolinium-based magnetic resonance imaging (MRI), 10
Gd^{3+} ions, 56, 170, 203, 207–208, 228, 235, 342
Gold nanoparticles (AuNPs), 80, 81, 120, 279, 347
Gold nanorods (GNRs), 347–350
Green fluorescent protein (GFP), 77, 275, 277

H

Hansen solubility parameters (HSPs), 299
Hart, Alan, 7
hCy3 dye, 89
Heterogeneous core–shell UCNPs, 170
  with enhanced UCL, 174–178
  seed-mediated epitaxial growth of, 170–171
High-angle annular dark-field (HAADF), 174, 234, 235

High-coordination numbers, 4–5
Hinckley, C. C., 9
Ho^{3+}-based nanothermometers, 338–340
Hoard, J. L., 6
Homogeneous core–shell UCNPs with enhanced UCL, 172–174
Homogenous epitaxial core–shell UCNPs, 170
Host lattice, properties of, 224
Host matrices, 108, 149, 261–262, 283
Human carcinoma (HeLa) cells, 333
Hyaluronated fullerene (HAC_{60}), 125
Hydro(solvo)thermal method, 109, 167–168, 221
Hydrophobic ligands, 103, 111, 223
Hydrophobic UCNPs, surface modifications of, 72–76, 94–95
  amphiphilic coatings, 72–74
  encapsulation with silica, 74–75
  ligand exchange, 75–76
Hydrothermal strategy, 110
  fluorides, 46–47
  oxides, 47, 48
  in synthesis of UCNPs, 71
1-hydroxyethane-1,1-diphosphonic acid (HEDP), 111–112

I

Igepal CO-520, 74, 78, 230
Inductively coupled plasma mass spectrometry (ICP-MS), 174
International Chamber of Commerce, 301
Ionic liquids (ILs), 52–54
IR laser beams, through pulse width of, 152–153

J

James, Charles, 4
Judd–Ofelt (J–O) theory, 25, 33

L

Lanthanide-based upconversion, 323–325
Lanthanide bis(trimethylsilyl)amides, 9–10
Lanthanide chelates, 8
Lanthanide-doped NPs, 37, 201, 292
  in biomedical sciences, 198–202
  crystallographic structure tuning, 207–208
  drawbacks, 206
  increasing active ions concentration, 211–212
  novel possibilities of, 196
  plasmonic and photonic effects, 210–211
  "rescue" methods, 206–207
  surface passivation with (Un) doped crystalline, 208–209
Lanthanide emitters, combination of, 150–152
Lanthanides, 3, 20, 38, 196, 298
  β-diketonate complexes, 7–9
  complexing agents for, 5–7
  EDTA complexes, 5–6
  high-coordination numbers, 4–5
  low-coordination numbers, 9–10
  MRI agents, 10
  positioning of, 4
  superconductors and alkoxides, 10–11
  upconversion and nanomaterials, 11–12
Lanthanide upconversion nanoparticles, tuning optical properties of, 139–141
  conclusion, 158
  introduction, 139–141
  tuning emission properties, 149–157
    multicolor arrays, 150–155
    single-band upconversion emission, 155–157
    tuning emission lifetime, 156–157
  tuning excitation wavelengths, 141–142
    Nd^{3+}-sensitized 745 nm LED excitation, 147
    Nd^{3+}-sensitized 800 nm laser excitation, 142–147
    organic dye-sensitized NIR excitation, 147–149
Lanthanum fluoride, 53, 225–226
Laser power, 153–155

Ligand, 102
exchange, 56, 58, 75–76, 79, 82, 87, 94, 111–112, 298–299
organic, 103, 108, 109, 115, 118, 120, 126
in preparation of nanomaterial, 109–111
role in colloidal stability, 111–114
role in optical properties of nanomaterial, 114–115
Ligand anchoring group, 108–109
Ligand field theory, 6
Liquid–solid–solution (LSS) approach, 167
Ln^{3+} ions, 20, 32, 196, 201, 202, 323–325, 330, 331, 342–344
Low-coordination numbers, 9–10
Luminescence, β-diketonate complexes and, 7–9
Luminescence intensity nanothermometry, 321–322
Luminescence intensity ratio (LIR), 322, 325–326, 327
Luminescence lifetime, 70, 71, 294–295, 322
Luminescence resonance energy transfer (LRET), 202, 322, 345, 346

M

Magnetic nanoparticles (MNPs), 82
Malonic acid (MA), 110, 111
Manhattan Project, 5
Mendeleev, 4
Merocyanine 540, 89
Mesoporous silica ($mSiO_2$), 75, 91–92, 93, 204, 209, 230, 271, 275, 281
Metal oxide, 226–227
Micro barcodes, 305
Micro upconverting QR codes, 304
Multichelating polymers, 109
Multicolor arrays, 150–155
by combination of lanthanide emitters, 150–152
through pulse width of IR laser beams, 152–153
by tuning laser power, 153–155
Multicolor nano barcoding in single upconversion crystal, 313–314
Multimodal imaging, 320
Multi-step synthesis strategy, 232–233

N

Nano-architectures, synthesis of, 231–232
Nanoassemblies, nanothermometers based on multicomponent, 344–350
Nanochemistry for well-defined core–shell UCNPs, 165–168
seed-mediated epitaxial growth of core–multishell UCNPs, 171
seed-mediated epitaxial growth of core–shell UCNPs, 169–171
synthesis of core UCNPs, 165–168
hydro(solvo)thermal method, 167–168
Ostwald ripening, 166–167
thermal decomposition, 165–166
Nanographene oxide (NGO), 279
Nanohybrid functionality, synergism in, 115–129
bioimaging, 121–122
drug delivery and chemotherapy, 126–129
encapsulation versus ionic and covalent binding, 115–117
functionality, 115
sensing, 118–121
targeting, 117–118
therapy, 122–123
photodynamic therapy, 123–126
photothermal therapy, 126
Nanomaterials
ligand in preparation of, 109–111
role of ligand in optical properties of, 114–115
UC luminescence of, 26–34
emission optimization, 32–34
optimization of absorption, 27–29
optimization of energy transfer, 29–32
upconversion and, 11–12
Nano-sized UC materials, 26–27
Nanostructures, 27, 37, 163
Nanotechnology, 23, 140, 298

Nanothermometry using upconverting nanoparticles, 319–351
conclusion, 350–351
lanthanide-based UCNPs for nanothermometry, 323–329
lanthanide-based upconversion, 323–325
LIR, Boltzmann's Plot, thermal sensitivity, and thermal resolution, 325–329
nanothermometers based on nanoassemblies, 344–350
nanoplatforms for optical heating and thermal sensing, 346–350
UCNP–organic hybrids for improved sensitivity, 345–346
nanothermometry concept, 319–320
optical nanothermometers, 321–323
UCNP-based nanothermometry, advances in, 329–344
Er^{3+}-based nanothermometers, 329–333
Er^{3+}-based upconverting nanothermometers, 333–334
Ho^{3+}-based nanothermometers, 338–340
nanothermometers based on combination of Ln^{3+} ions, 342–344
Nd^{3+}-based nanothermometers, 340–341
TM^{3+}-based nanothermometers, 334–338
$NaYF_4$ nanocrystals, 46–47, 69–70, 156, 165, 168, 221
Nd^{3+}-based nanothermometers, 340–341
Nd^{3+}-sensitized 745 nm LED excitation, 147
Nd^{3+}-sensitized 800 nm laser excitation, 142–147
Nd^{3+} sensitizers, active shell containing, 181–183
Near-infrared (NIR) light, 19–20, 69, 70, 102, 163, 204–205
Neodymium, 3, 4, 219, 283
Neodymium bromate, $[Nd(H_2O)_9](BrO_3)_3$, 5, 6
NIR-based phototherapy, need for, 255–257
NIR-to-NIR emitting system, 55, 56, 58
NIR-to-NIR upconversion security printing, 308–312
Nonhomogenous shell composition, 236
Nonradiative multiphonon relaxation (NMPR) rate, 25, 55
Nucleic acids, conjugation to, 79–86
N-vinylpyrrolidone (NVP), 128

O

Octadecene (ODE), 165, 166
Oleic acid (OA), 42, 165, 166–167, 298, 302, 315
Oleylamine (OM), 42, 165, 222
o-NB derivatives, 274
o-nitrobenzyl bromide, 78, 273
Optical heating, multifunctional nanoplatforms for, 346–350
Optical nanothermometers, 321–323
"Optical window", 19–20, 70, 142, 156, 201
Optimization
of absorption, 27–29
of emission, 32–34
energy transfer, 29–32
Optimized thermal sensitivity, 329
Organic capping, 108, 109, 121
Organic dye-sensitized NIR excitation, 147–149
Organic ligand, 108–109
in colloidal stability of UCNPs, 111–114
functionality, role of nanoparticles in, 115
in nanomaterial preparation, 109–111
Ostwald ripening, 43, 56, 71–72, 166–167, 174
nanochemistry for synthesis of UCNPs, 166–167
in synthesis of UCNPs, 71–72
Oxalic acid (OXA), 110, 111
Oxides
coprecipitation method, 41–42
solvo(hydro)thermal, 47–49
thermolysis, 43–44
Oxyfluorides
coprecipitation method, 42
thermolysis, 43–44

P

Phonon-induced energy loss, 32
Phosphate nanoparticles, 41–42
Phospholipids, 73–74
Photoactivation, 255–256
Photochemical internalization (PCI), 256, 257, 268–271
  definition, 256
  photosensitive chemicals for, 270–271
Photocontrolled delivery, photolabile groups in, 272–274
Photodynamic therapy (PDT), 123–126, 256, 257–268, 336
  as antimicrobial tool, 258
  NIR initiated, 204–205
  PS loading strategies, 264–266
  PSs for, 259–261
  tumoricidal effect of, 257–258
  UCNPs for, 261–264
  in vitro and *in vivo* PDT using UCNPs, 266–268
Photofrin®, 259, 260
Photoinduced ROS production, 257–271
  photochemical internalization, 268–271
  photodynamic therapy, 257–268
Photolabile groups in photocontrolled delivery, 272–274
Photoluminescence (PL) dynamics, 24–25
Photon avalanche (PA), 20, 22, 104, 105, 196
Photosensitive chemicals, for photochemical internalization, 270–271
Photo sensitizers (PSs), 256, 257, 269
  excitation wavelength for, 260
  loading strategies, 264–266
    covalent linkage, 266
    encapsulation, 264–265
    physical adsorption, 266
  for PDT, 259–261
Phototherapy, upconversion nanoparticles for, 255–284
  conclusion, 283–284
  limitations of UCNPs, 281–283
  NIR-based phototherapy, need for, 255–257
  photocontrolled release of molecules, 272–277
    photolabile groups in photocontrolled delivery, 272–274
    UCNP-based phototriggered release *in vitro* and *in vivo*, 274–277
  photoinduced ROS production, 257–271
    photochemical internalization, 268–271
    photodynamic therapy, 257–268
  photothermal therapy, 277–279
    UCNP-based PTT, 278–279
  UCNPs in combination therapy, 279–281
Photothermal therapy (PTT), 126, 256, 277–279
  UCNP-based, 278–279
Phthalocyanine dye, 260
Physical adsorption, 264, 266
Plasmonic and photonic effects, 210–211
Polarization luminescence nanothermometry, 322
Poly(acrylamide) (pAAm), 345
Poly(acrylic acid) (PAA), 75, 82, 88, 109, 112, 117, 128, 231
Poly(allylamine) (PAAm), 87, 111
Poly-(allylamine hydrochloride) (PAH), 113
Polyethylene glycol (PEG), 47, 202, 262
Poly(ethylene-glycol)-diacrylate (PEGDA), 128
Polyethyleneimine (PEI), 41, 262
Poly(maleic anhydride-alt-1-octadecene) (PMAO), 74, 112
Polymer encapsulation, 231
Poly(methyl methacrylate) (PMMA), 302, 314
Poly(N-isopropylacrylamide) (pNIPAM), 345, 346
Poly(sodium 4-styrenesulfonate) (PSS), 113
Porphyrins, 90, 259, 260, 278
Promethium, 4
Protein conjugation, 76–79
Protoporphyrin IX (PPIX), 260
$Pt(NH_3)2Cl_2$, 5

Q

QR code tags, micro upconverting, 304
Quantum dots (QDs), 84–85, 140

R

Ratiometric detection, 202
Ratiometric nanothermometers, 325, 327
Reactive oxygen species (ROS), 123, 213, 256, 257
Red, green, and blue (RGB) emitting UCNPs, 302–304
RE fluoride, 225–226
R6G dye, 345

S

Samarium, 4
Scanning laser imaging of upconversion printing, 312
Scanning transmission electron microscopy (STEM), 174
Security applications, upconverting nanoparticles for, 291–316
  conclusion, 315
  introduction, 291–292
  security printing applications, 301–314
  synthesis, functionalization and scale-up, 298–299
    Hansen solubility parameters, 299
  toxicity assessment, 299–301
  upconversion prints, readability of, 314–315
  upconversion security printing prospective, 293–298
Security printing applications, 301–314
  color-coded multilayer photopatterned microstructures, 305
  definition, 301
  micro barcodes, 305
  micro upconverting QR code tags, 304
  multicolor nano barcoding in single upconversion crystal, 313–314
  NIR-to-NIR upconversion security printing, 308–312
  red, green, and blue upconversion printing, 302–304
  scanning laser imaging of upconversion printing, 312
  tunable lifetime multiplexing of UCNPs, 305–308
  upconversion polymeric nanofibers, 314
Seed-mediated epitaxial growth
  of core–multishell UCNPs, 171
  of core–shell UCNPs, 169–171
    heterogeneous core–shell UCNPs, 170–171
    homogenous epitaxial core–shell UCNPs, 170
"Self-focusing by Ostwald ripening", 56
Sesquioxides, 41, 45, 50
Shell precursors, 169, 170, 171, 232–233
Shell synthesis methods and compositions, 227–229
Silica amorphous layer, 229–232
Silica, encapsulation with, 74–75
Silica-shell formation, 57–60
Single-band upconversion emission, 155–156
Single upconversion crystal, multicolor nano barcoding in, 313–314
Sodium RE fluoride systems ($NaREF_4$), 224–225
Sol–gel technique, 49–51
Solid phase-based assay formats, 82–83
Solvo(hydro)thermal, 45–49
  fluorides, 46–47
  oxides, 47–49
Spectral luminescence nanothermometry, 322
Spectroscope, 4
Staphylococcal enterotoxin B (SEB), 82
Stark splitting, 50, 330
Stöber method, 50, 230
Stokes emission in NIR, 201
Stokes VIS emission under UV excitation, 198–200
Succinic acid (SA), 110, 111,
Superconductors, alkoxides and, 10–11
Surface deactivations, 172
Surface passivation with (Un) doped crystalline, 208–209

Surface quenching, shell protects NPs from, 214–215
Synthetic methods, for luminescent nanomaterials, 40
combustion, 51–52
coprecipitation method, 40–42
ionic liquids, 52–54
sol–gel, 49–51
solvo(hydro)thermal, 45–49
thermolysis, 42–45

T

Targeting, 117–118
Tartaric acid (TA), 110, 111
Tetraethyl orthosilicate (TEOS), 74
Tetraphenylporphine disulfonate (TPPS2a), 270
Tetraprotonic acid, *see* 1-hydroxyethane-1,1-diphosphonic acid (HEDP)
Therapy, 122
photodynamic therapy, 123–126
photothermal therapy, 126
Thermal decomposition, 12, 169, 171
nanochemistry for synthesis of UCNPs, 165–166
reaction, 221–223
Thermal resolution, 325, 328–329, 349
Thermal sensing, 320, 322–323, 332, 335, 341, 342
multifunctional nanoplatforms for, 346–350
Thermal sensitivity, 325, 327–328, 330, 344–346
Thermolysis, 42–43
fluorides, 43–44
oxides, 45
oxyfluorides, 45
in synthesis of UCNPs, 71
Time-dependent luminescence intensity, 322–323
TM^{3+}-based nanothermometers, 334–338
Toxicity assessment, 299–301
Transmission electron microscopy (TEM), 77, 170, 174, 222
Trioctylphosphine (TOP), 110
Trioctylphosphine oxide (TOPO), 110, 222, 223
Tunable luminescent lifetimes, 305–308
Tuning emission lifetime, 156–157
Tuning emission properties, 149–157
multicolor arrays, 150–155
by combination of lanthanide emitters, 150–152
through pulse width of IR laser beams, 152–153
by tuning laser power, 153–155
single-band upconversion emission, 155–157
tuning emission lifetime, 156–157
Tuning excitation wavelengths, 141–142
Nd^{3+}-sensitized 745 nm LED excitation, 147
Nd^{3+}-sensitized 800 nm laser excitation, 142–147
organic dye-sensitized NIR excitation, 147–149
TWEEN 80, 74, 113

U

UC, *see* Upconversion (UC)
UC core–shell structure NPs, strategies for constructing, 169
UC efficiency, 26, 28, 32
UC emission, 19, 24, 33, 39, 47, 49, 52
as nonlinear process, 31
plasmonic resonance with, 33–34
surface characteristics of, 31–32
UC luminescence quantum yield (QY), 26
UC material multicolor emission and tuning, 107–108
UCNP@CB@dye, 116–117
UCNP-based nanothermometry, advances in, 329
Er^{3+}-based nanothermometers, 329–333
Er^{3+}-based upconverting nanothermometers, 333–334
Ho^{3+}-based nanothermometers, 338–340
nanothermometers based on combination of Ln^{3+} ions, 342–344
Nd^{3+}-based nanothermometers, 340–341
TM^{3+}-based nanothermometers, 334–338

UCNP–DNA conjugate, 80
UCNPs, *see* Upconversion nanoparticles (UCNPs)
Ultraviolet (UV), 19, 69, 104, 326
Upconversion (UC), 291–292
Upconversion emission quantum yield, 198
Upconversion luminescence (UCL), 19, 20, 26–34, 105, 122, 164
  approaches in enhancing, 26–34
    emission optimization, 32–34
    optimization of absorption, 27–29
    optimization of energy transfer, 29–32
  challenges and recent progress in, 26–34
  dynamics, 24–25
  emission, 20
  future perspective, 34
  origin of, 20
  photoluminescence dynamics of, 24–25
  physics of, 20
  principles of, 19–20
    EMU mechanism, 23–24
    traditional UC mechanisms, 20–23
  UC efficiency and measurement techniques, 26
Upconversion materials, 106–107
Upconversion nanomaterials, 11–12, 37–38, 332
  core–shell architectures, 55–60
    silica-shell formation, 57–60
  importance of composition, 39
  synthetic strategies, 40–54
    combustion, 51–52
    coprecipitation method, 40–42
    ionic liquids, 52–54
    sol–gel, 49–51
    solvo(hydro)thermal, 45–49
    thermolysis, 42–45
Upconversion nanoparticles (UCNPs), 69, 80, 102, 139–140, 164, 256
  based phototriggered release *in vitro* and *in vivo*, 274–277
  based PTT, 278–279
  biomedical applications of, 203–205
    active application, 204
    modulation application, 203–204
    passive applications, 203
  in combination therapy, 279–281
  conjugation to dyes, 87–94
    detection of thrombin, 83–84
  for determination of HIV antibodies, 85
  emission with NIR light, 102
  enhancing emission of, 107
  functionality to, via organic capping, 116
  immobilization of nucleotides to, 85
  ligand anchoring group, 108–109
  ligand role in colloidal stability of, 111–114
  limitations of, 281–283
  luminescence efficiency of, 107
  nanocarrier, 88
  nanochemistry for synthesis of core, 165–168
    hydro(solvo)thermal method, 167–168
    Ostwald ripening, 166–167
    thermal decomposition, 165–166
  nanothermometry using, 319–351
    conclusion, 350–351
    lanthanide-based UCNPs for nanothermometry, 323–329
    nanothermometers based on nanoassemblies, 344–350
    nanothermometry concept, 319–320
    optical nanothermometers, 321–323
    UCNP-based nanothermometry, advances in, 329–344
  nucleic acids, conjugation to, 79–86
  organic capping of, 109
  organic hybrids for improved sensitivity, 345–346
  organic ligand and, 103, 109
    functionality, role of NPs in, 115
    ligand in preparation of nanomaterial, 109–111
    optical properties of nanomaterial, role of ligand in, 114–115
    role of ligand in colloidal stability of, 111–114

synergism in nanohybrid functionality, 115–129
for PDT, 261
for phototherapy, 255–284
conclusion, 283–284
limitations of UCNPs, 281–283
NIR-based phototherapy, need for, 255–257
photocontrolled release of molecules, 272–277
photoinduced ROS production, 257–271
photothermal therapy, 277–279
UCNPs in combination therapy, 279–281
protein conjugation, 76–79
quantum yield of, 281, 283
for security applications, 291–316
conclusion, 315
introduction, 291–292
security printing applications, 301–314
synthesis, functionalization and scale-up, 298–299
toxicity assessment, 299–301
upconversion prints, readability of, 314–315
upconversion security printing prospective, 293–298
sensing, 118–121
(bio)chemical sensing, 118, 120
chemical sensing, 118
nanohybrids, 118–119
physical sensing, 118
surface modifications of hydrophobic, 72–76, 94–95
amphiphilic coatings, 72–74
encapsulation with silica, 74–75
ligand exchange, 75–76
synthesis of, 71–72
hydrothermal strategy, 71
Ostwald ripening, 71–72
thermolysis, 71
targeting, 117–118
tunable lifetime multiplexing of, 305–308
UC material multicolor emission and tuning, 107–108
upconversion materials, 106–107
upconversion phenomenon, 104–106
in vitro and *in vivo* PDT using, 266–268
water-dispersible, 119–120
Upconversion phenomenon, 104–106
Upconversion polymeric nanofibers, 314
Upconversion printing, scanning laser imaging of, 312
Upconversion prints, readability of, 314–315
Upconversion quantum yield (UCQY), 164, 177
Upconversion security printing prospective, 293–298
Urbain, Georges, 4, 8
US Customs and Border Protection (CBP), 301

V

van-der-Waals (vdW) interactions, 73
Veggel, Van, 41
von Welsbach, Carl Auer, 4

W

Weissman, S. I., 7
Werner, Alfred, 5
Wyckoff, R. W. G., 5

X

X-ray spectroscopy, 4, 234

Y

Yb^{3+} sensitizers, active shell containing, 178–181
Yokota–Tanimoto model, 25
Ytterbite, 3
Yttrium isopropoxide, 11
Yttrium oxide (Y_2O_3), 3, 226, 330

Z

Zusman–Burshteĭn model, 25